Lecture Notes in Computer Science   9237

Commenced Publication in 1973
Founding and Former Series Editors:
Gerhard Goos, Juris Hartmanis, and Jan van Leeuwen

More information about this series at http://www.springer.com/series/7407

Emmanuel Vincent · Arie Yeredor
Zbyněk Koldovský · Petr Tichavský (Eds.)

Latent Variable Analysis and Signal Separation

12th International Conference, LVA/ICA 2015
Liberec, Czech Republic, August 25–28, 2015
Proceedings

 Springer

Editors
Emmanuel Vincent
Inria
Villers-les-Nancy
France

Arie Yeredor
Tel Aviv University
Tel-Aviv
Israel

Zbyněk Koldovský
Technical University of Liberec
Liberec
Czech Republic

Petr Tichavský
The Czech Academy of Sciences
Prague
Czech Republic

ISSN 0302-9743   ISSN 1611-3349 (electronic)
Lecture Notes in Computer Science
ISBN 978-3-319-22481-7   ISBN 978-3-319-22482-4 (eBook)
DOI 10.1007/978-3-319-22482-4

Library of Congress Control Number: 2015945320

LNCS Sublibrary: SL1 – Theoretical Computer Science and General Issues

Springer Cham Heidelberg New York Dordrecht London

Printed on acid-free paper

Springer International Publishing AG Switzerland is part of Springer Science+Business Media
(www.springer.com)

Preface

This volume collects the papers presented at the 12th International Conference on Latent Variable Analysis and Signal Separation, LVA/ICA 2015. The conference was held during August 25–28, 2015, and was hosted by the Technical University of Liberec, Czech Republic, on the occasion of the 20th anniversary of the Faculty of Mechatronics, Informatics, and Interdisciplinary Studies.

Since its debut in 1999 under the banner of Independent Component Analysis and Blind Source Separation (ICA), the LVA conference series has attracted hundreds of researchers and practitioners and it has continuously broadened its horizons. Today it encompasses a host of additional forms and models of general mixtures of latent variables. Theories and tools borrowing from the fields of signal processing, applied statistics, machine learning, linear and multilinear algebra, numerical analysis and optimization, and numerous application fields offer exciting interdisciplinary interactions.

From 81 submitted papers, 61 were accepted as oral (29 papers) and poster (32 papers) presentations. The conference program put forward five special topics: tensor-based methods for blind signal separation; deep neural networks for supervised speech separation/enhancement; joint analysis of multiple datasets, data fusion, and related topics; advances in nonlinear blind source separation; and sparse and low-rank modeling for acoustic signal processing. Regular topics included theory (dictionary and manifold learning, optimization algorithms, performance analsysis, etc.); audio applications; and biomedical and other applications. A prize (sponsored by Conexant Systems) was awarded to the best student paper in the field of audio signal processing, and the contributions of the student nominees reflected the liveliness of this research topic.

The Organizing Committee was pleased to invite leading experts in these fields for keynote lectures: Tülay Adali (University of Maryland, Baltimore County, USA), Rémi Gribonval (Inria, France), and DeLiang Wang (Ohio State University, USA). Aware of the growing interest in emerging, as well as in classic LVA-related topics among novice and veteran researchers alike, the Organizing Committee decided to precede the conference by a two-day Summer School on Latent Variable Analysis and Signal Separation, with lectures given by Vicente Zarzoso (University of Nice Sophia Antipolis, France), Arie Yeredor (Tel Aviv University, Israel), Martin Haardt (Ilmenau University of Technology, Germany), Ali Taylan Cemgil (Boğaziçi University, Turkey), Emmanuel Vincent (Inria, France), Emanuël Habets (International Audio Laboratories, Germany), Mikkel N. Schmidt (Technical University of Denmark), and Antoine Deleforge (University of Erlangen, Germany).

This year's conference also provided a forum for the 5th Signal Separation Evaluation Campaign (SiSEC 2015). SiSEC 2015 successfully continued the series of evaluation campaigns initiated at ICA 2007, in London. Compared with previous campaigns, it featured two new datasets: full-length professionally produced music recordings and asynchronous recordings of speech mixtures.

The success of LVA/ICA 2015 was the result of the hard work of many people, whom we warmly thank here. First, we wish to thank the authors and the members of the Program Committee, without whom this high-quality volume would not exist. We also express our gratitude to the members of the International LVA Steering Committee for their continued support to the conference, as well as to the SiSEC 2015 organizers and to the local organization committee.

June 2015

Emmanuel Vincent
Arie Yeredor
Zbyněk Koldovský
Petr Tichavský

Organization

General Chairs

Zbyněk Koldovský   Technical University of Liberec, Czech Republic
Petr Tichavský   The Czech Academy of Sciences, Czech Republic

Program Chairs

Emmanuel Vincent   Inria, France
Arie Yeredor   Tel Aviv University, Israel

Special Sessions Chair

Shoji Makino   University of Tsukuba, Japan

Special Sessions Organizers

Afsaneh Asaei   Idiap Research Institute, Switzerland
Lieven De Lathauwer   KU Leuven, Belgium
Yannick Deville   University of Toulouse, France
Sharon Gannot   Bar-Ilan University, Israel
Christian Jutten   Gipsa-Lab, France
Dana Lahat   Gipsa-Lab, France
DeLiang Wang   Ohio State University, USA

Overseas Liaison

Andrzej Cichocki   RIKEN, Japan

SiSEC Chair

Nobutaka Ono   National Institute of Informatics/SOKENDAI, Japan

Program Committee

Tülay Adali   University of Maryland, Baltimore County, USA
André Almeida   Federal University of Ceará, Brazil
Shoko Araki   NTT Communication Science Laboratories, Japan
Afsaneh Asaei   Idiap Research Institute, Switzerland
Nancy Bertin   CNRS, France
Suchita Bhinge   University of Maryland, Baltimore County, USA

Sofiane Boudaoud	TU Compiègne, France
Zois Boukouvalas	University of Maryland, Baltimore County, USA
Charles Casimiro Cavalcante	Federal University of Ceará, Brazil
Ali Taylan Cemgil	Boğaziçi University, Turkey
Gilles Chabriel	University of Toulon, France
Jonathon Chambers	Loughborough University, UK
Pierre Comon	CNRS, France
Fengyu Cong	Dalian University of Technology, China
Sergio Cruces	University of Seville, Spain
Otto Debals	KU Leuven, Belgium
Lieven De Lathauwer	KU Leuven, Belgium
Yannick Deville	University of Toulouse, France
Scott Douglas	Southern Methodist University, USA
Iván Durán Díaz	University of Seville, Spain
Darren Emge	US Army Edgewood Chemical Biological Center, USA
Sharon Gannot	Bar-Ilan University, Israel
Jürgen Geiger	Huawei European Research Center, Germany
Xiaofeng Gong	KU Leuven, Belgium
Rémi Gribonval	Inria, France
Martin Haardt	Ilmenau University of Technology, Germany
Emanuël Habets	International Audio Laboratories, Germany
Christian Jutten	Joseph Fourier University, France
Amar Kachenoura	University Rennes 1, France
Juha Karhunen	Aalto University, Finland
Yosi Keller	Bar-Ilan University, Israel
Vojislav Kecman	Virginia Commonwealth University, USA
Srdan Kitic	Inria, France
Martin Kleinsteuber	TU Munich, Germany
Zbyněk Koldovský	Technical University of Liberec, Czech Republic
Dorothea Kolossa	Ruhr-University of Bochum, Germany
Ivica Kopriva	Rudjer Boskovich Institute, Croatia
Dana Lahat	Gipsa-Lab, France
Jonathan Laney	University of Maryland, Baltimore County, USA
Jonathan Le Roux	Mitsubishi Electric Research Laboratories, USA
Yuri Levin-Schwarz	University of Maryland, Baltimore County, USA
Qiu-Hua Lin	Dalian University of Technology, China
Antoine Liutkus	Inria, France
Ali Mansour	ENSTA Bretagne, France
Anke Meyer-Baese	Florida State University, USA
Éric Moreau	University of Toulon, France
Rami Mowakeaa	University of Maryland, Baltimore County, USA
Pejman Mowlaee	Graz University of Technology, Austria
Shinichi Nakajima	TU Berlin, Germany
Mohsen Naqvi	Loughborough University, UK
Francesco Nesta	Conexant Systems, USA

Guido Nolte	Universitätsklinikum Hamburg-Eppendorf, Germany
Nobutaka Ono	National Institute of Informatics/SOKENDAI, Japan
Alexey Ozerov	Technicolor Research and Innovation Laboratories, France
Bryan Pardo	Northwestern University, USA
Anh-Huy Phan	RIKEN, Japan
Mark Plumbley	University of Surrey, UK
Pavel Rajmic	Brno University of Technology, Czech Republic
Nima Reyhani	Aalto University, Finland
François Rigaud	Audionamix, France
Tapani Ristaniemi	University of Jyväskylä, Finland
Saeid Sanei	University of Surrey, UK
Hao Shen	TU Munich, Germany
Xizhi Shi	Shanghai Jiao Tong University, China
Paris Smaragdis	University of Illinois at Urbana-Champaign, USA
Václav Šmídl	The Czech Academy of Sciences, Czech Republic
Mikael Sorensen	KU Leuven, Belgium
Pablo Sprechmann	Courant Institute, USA
Maja Taseska	International Audio Laboratories, Germany
Petr Tichavský	The Czech Academy of Sciences, Czech Republic
Frederik Van Eeghem	KU Leuven, Belgium
Ricardo Vigário	Aalto University, Finland
Emmanuel Vincent	Inria, France
DeLiang Wang	Ohio State University, USA
Wai Look Woo	Newcastle University, UK
Arie Yeredor	Tel Aviv University, Israel
Tatsuya Yokota	RIKEN, Japan
Vicente Zarzoso	University of Nice Sophia Antipolis, France
Andreas Ziehe	TU Berlin, Germany
Zeinab Zohny	Loughborough University, UK

Sponsors

Technicolor
Faculty of Mechatronics, Informatics and Interdisciplinary Studies,
    Technical University of Liberec
Jablotron
Conexant Systems
Sony

Contents

Joint Analysis of Multiple Datasets, Data Fusion, and Related Topics

Advances in Nonlinear Blind Source Separation

Sparse and Low-Rank Modeling for Acoustic Signal Processing

Theory

Audio Applications

Biomedical and Other Applications

Tensor-Based Methods
for Blind Signal Separation

Stochastic and Deterministic Tensorization for Blind Signal Separation

Otto Debals[1,2,3]($\boxtimes$) and Lieven De Lathauwer[1,2,3]

[1] Department of Electrical Engineering (ESAT) – STADIUS Center
for Dynamical Systems, Signal Processing and Data Analytics, KU Leuven,
Kasteelpark Arenberg 10, 3001 Leuven, Belgium
Otto.Debals@esat.kuleuven.be, Lieven.DeLathauwer@kuleuven-kulak.be
[2] Group Science, Engineering and Technology, KU Leuven Kulak,
E. Sabbelaan 53, 8500 Kortrijk, Belgium
[3] iMinds Medical IT, KU Leuven, Kasteelpark Arenberg 10, 3001 Leuven, Belgium

Abstract. Given an instantaneous mixture of some source signals, the blind signal separation (BSS) problem consists of the identification of both the mixing matrix and the original sources. By itself, it is a non-unique matrix factorization problem, while unique solutions can be obtained by imposing additional assumptions such as statistical independence. By mapping the matrix data to a tensor and by using tensor decompositions afterwards, uniqueness is ensured under certain conditions. Tensor decompositions have been studied thoroughly in literature. We discuss the matrix to tensor step and present tensorization as an important concept on itself, illustrated by a number of stochastic and deterministic tensorization techniques.

Keywords: Blind source separation · Independent component analysis · Tensorization · Canonical polyadic decomposition · Block term decomposition · Higher-order tensor · Multilinear algebra

1 Blind Signal Separation and Matrix Data

The separation of sources from observed data is a well-known problem in signal processing, known as blind signal separation (BSS). The linear BSS problem consists of the decomposition of an observed data matrix $\mathbf{X} \in \mathbb{K}^{K \times N}$ as

$$\mathbf{X} = \mathbf{M}\,\mathbf{S} = \sum_{r=1}^{R} \mathbf{m}_r \cdot \mathbf{s}_r^{\mathrm{T}}, \tag{1}$$

in which $\mathbf{M} \in \mathbb{K}^{K \times R}$ is the mixing matrix and $\mathbf{S} \in \mathbb{K}^{R \times N}$ is the observed source matrix. The vector $\mathbf{m}_r$ is the rth column of $\mathbf{M}$ and $\mathbf{s}_r^{\mathrm{T}}$ is the rth row of $\mathbf{S}$. For each signal N samples are available. The set $\mathbb{K}$ stands for either $\mathbb{R}$ or $\mathbb{C}$. Furthermore, additive noise can be represented by a matrix $\mathbf{N} \in \mathbb{K}^{K \times N}$.

Equation (1) is a decomposition of the data matrix $\mathbf{X}$ in rank-1 terms, where each term corresponds to the contribution of one particular source. Except in

E. Vincent et al. (Eds.): LVA/ICA 2015, LNCS 9237, pp. 3–13, 2015.
DOI: 10.1007/978-3-319-22482-4_1

the case of a single source with $R = 1$, it is well-known that such a decomposition is not unique. Uniqueness appears by imposing additional constraints on the matrices. Acclaimed matrix decompositions with well-understood uniqueness conditions are the singular value decomposition (imposing column-wise orthogonality) and the QR and RQ factorizations (imposing triangularity and column-wise orthonormality). However, in the light of BSS, the constraints from these well-known decompositions are both too restrictive and unnatural. For instance, it is uncommon that the mixing matrix is known to be triangular, as it is uncommon that both mixing vectors and source vectors are mutually orthogonal. We are facing here what is called the factor indeterminacy problem in Factor Analysis (FA) [31]. One needs to resort to other assumptions and matrix decompositions, specifically tailored to the BSS problem.

One of the more realistic constraints for BSS is nonnegativity: nonnegative matrix factorization (NMF) is a decomposition in which the entries of the factor matrices are nonnegative [9,29,38,41]. Nonnegativity is natural for concentrations, number of occurrences, pixel intensities, frequencies, etc. Sparse component analysis (SCA) is also gaining in popularity [6,47]. In SCA, the source matrix $\mathbf{S}$ is assumed to be sparse. Note that nonnegativity in itself does not ensure uniqueness; often, one uses additional sparsity [23,25,28,32]. For dense data sets, SCA is mostly applied after a sparsifying transformation such as the wavelet transformation [17].

2   Blind Signal Separation and Tensor Data

A tensor is a higher-order generalization of vectors (boldface lowercase letters) and matrices (boldface uppercase letters). It is denoted by a calligraphic letter, e.g., $\mathcal{X}$, and is a multiway array of numerical values $x_{i_1 i_2 \cdots i_N} = \mathcal{X}(i_1, i_2, \ldots, i_N) = (\mathcal{X})_{i_1, i_2, \ldots, i_N}$ where $\mathcal{X} \in \mathbb{K}^{I_1 \times I_2 \times \cdots \times I_N}$. By fixing all but a single index, one obtains a mode-n vector, e.g., $\mathbf{a} = \mathcal{X}(i_1, \ldots, i_{n-1}, :, i_{n+1}, \ldots, i_N) \in \mathbb{K}^{I_n}$. A diagonal tensor only has nonzeros on the entries of which all the indices are equal.

The third-order counterpart of Eq. (1) is a decomposition of a tensor $\mathcal{X} \in \mathbb{K}^{I \times J \times K}$ in R rank-1 terms with $\mathbf{A} \in \mathbb{K}^{I \times R}$, $\mathbf{B} \in \mathbb{K}^{J \times R}$ and $\mathbf{C} \in \mathbb{K}^{K \times R}$:

$$\mathcal{X} = \sum_{r=1}^{R} \mathbf{a}_r \otimes \mathbf{b}_r \otimes \mathbf{c}_r = \mathcal{I} \cdot_1 \mathbf{A} \cdot_2 \mathbf{B} \cdot_3 \mathbf{C}, \tag{2}$$

in which $\otimes$ denotes the tensor (outer) product, $\cdot_i$ denotes the tensor-matrix product in the ith mode and $\mathcal{I} \in \mathbb{K}^{R \times R \times R}$ denotes a diagonal tensor with ones on the diagonal and zeros elsewhere. For all index values, we have that $x_{ijk} = \sum_{r=1}^{R} a_{ir} b_{jr} c_{kr}$. Equation (2) gives a polyadic decomposition (PD) of $\mathcal{X}$. If R is minimal, it is defined as the rank of $\mathcal{X}$ and the decomposition is called a canonical polyadic decomposition (CPD). It has been proven that the CPD is unique under relatively mild conditions, typically expressing that the rank-1 terms are "sufficiently different" while not necessitating additional constraints such as nonnegativity [21,22,37].

Recently, the block term decomposition (BTD) has been introduced [13,16]. Instead of decomposing a tensor in rank-1 terms, it is written as a linear combination of tensors with low multilinear rank. The multilinear rank of a tensor $\mathcal{X}$ is an N-tuple $(R_1, R_2, \ldots, R_N)$ with R_n the mode-n rank, defined as the dimension of the subspace spanned by the mode-n vectors of $\mathcal{X}$. A special instance of the BTD is the decomposition of a tensor $\mathcal{X} \in \mathbb{K}^{I \times J \times K}$ in rank-$(L_r, L_r, 1)$ terms for which uniqueness under mild conditions has been proven [13,14]. We then have

$$\mathcal{X} = \sum_{r=1}^{R} \mathbf{E}_r \otimes \mathbf{c}_r = \sum_{r=1}^{R} (\mathbf{A}_r \mathbf{B}_r^{\mathrm{T}}) \otimes \mathbf{c}_r, \tag{3}$$

with matrices $\mathbf{E}_r = \mathbf{A}_r \mathbf{B}_r^{\mathrm{T}} \in \mathbb{K}^{I \times J}$ of rank L_r. The matrices $\mathbf{A}_r \in \mathbb{K}^{I \times L_r}$ and $\mathbf{B}_r \in \mathbb{K}^{J \times L_r}$ have full column rank, and we have nonzero $\mathbf{c}_r \in \mathbb{K}^K$ for all r.

Tensor methods for BSS receive their success from the uniqueness of tensor decompositions such as the CPD and the BTD. These are becoming standard tools for BSS and have been applied in many domains such as telecommunication, array processing and chemometrics [8,15,35,36,45].

3 Tensorization of Matrix Data

Tensor techniques require the availability of tensor data. Matrix data obviously remain more common than tensor data. Nevertheless, the techniques may still be used for BSS after the data matrix is mapped to a tensor. The mapping to the tensor domain translates the assumptions made for BSS, with the subsequent tensor decompositions having the possibility of ensuring uniqueness. While the uniqueness and algorithms of tensor decompositions have received a lot of attention lately, we discuss different tensorization techniques. A clear overview is necessary to benefit from the advantages of tensor techniques for matrix data.

What is essential about the mappings, is that linear transformations are used that map the sources to matrices or tensors that (approximately) have low (multilinear) rank under a certain working hypothesis. The (multi)linearity of the transformation is necessary to retain a linear mixture of the sources and avoid the introduction of inseparable terms, while the low-rank structure enables us to apply the tensor decompositions of the previous section.

In a first subsection, we discuss a stochastic tensorization technique using higher-order statistics. The second subsection describes the use of parameter variation for tensorization, illustrated with second-order statistics. Three different deterministic techniques relying on Hankelization, Löwnerization and segmentation are discussed in Sects. 3.3, 3.4 and 3.5, respectively. Note that the uses of higher-order statistics and second-order statistics for BSS are well known, both applying tensorization in a different way. Other tensorization techniques are known too but not described here, see e.g. [1,33]. For each tensorization technique described, the multilinearity, working hypothesis, applied tensor decomposition and higher-order representation of each source are reported. Uniqueness results, which we omit because of brevity, can be found in more detailed literature.

3.1 Higher-Order Statistics

Higher-order statistics (HOS) are fundamental for independent component analysis (ICA), in which one separates the observations in mutually statistically independent sources. This technique for BSS is highly renowned and has been applied in a diversity of domains [7,10,11,39,40]. Within the different types of higher-order statistics, especially cumulants are compelling. They are able to separate non-Gaussian, mutually independent sources. For simplicity, we assume stationary, identically distributed signals. Consider a zero-mean stochastic signal vector $\mathbf{u}(t) \in \mathbb{K}^K$. We give the explicit definition of the fourth-order cumulant:

$$\left(\mathcal{C}_{\mathbf{u}}^{(4)}\right)_{i_1 i_2 i_3 i_4} \triangleq \mathrm{E}\left\{u_{i_1} u_{i_2}^* u_{i_3}^* u_{i_4}\right\} - \mathrm{E}\left\{u_{i_1} u_{i_2}^*\right\} \mathrm{E}\left\{u_{i_3}^* u_{i_4}\right\}$$
$$- \mathrm{E}\left\{u_{i_1} u_{i_3}^*\right\} \mathrm{E}\left\{u_{i_2}^* u_{i_4}\right\} - \mathrm{E}\left\{u_{i_1} u_{i_4}\right\} \mathrm{E}\left\{u_{i_2}^* u_{i_3}^*\right\}, \qquad (4)$$

with $\mathcal{C}_{\mathbf{u}}^{(4)} \in \mathbb{K}^{K \times K \times K \times K}$. Cumulants have very interesting properties, enabling the use of tensor decompositions for BSS [40]. First of all, the expression in Eq. (4) satisfies multilinearity (it gives a quadrilinear mapping) as requested from the introduction of the section: if $\mathbf{x}(t) = \mathbf{M}\mathbf{s}(t) + \mathbf{n}(t)$ then in the fourth-order case we have:

$$\mathcal{C}_{\mathbf{x}}^{(4)} = \mathcal{C}_{\mathbf{s}}^{(4)} \cdot_1 \mathbf{M} \cdot_2 \mathbf{M}^* \cdot_3 \mathbf{M}^* \cdot_4 \mathbf{M} + \mathcal{C}_{\mathbf{n}}^{(4)}. \qquad (5)$$

Second, higher-order cumulants of a Gaussian variable are zero. Under the assumption of Gaussian noise, $\mathcal{C}_{\mathbf{n}}^{(4)}$ from Eq. (5) becomes a zero tensor.

The working hypothesis in ICA with HOS is that the sources are non-Gaussian and mutually statistically independent. Then, the higher-order source cumulant $\mathcal{C}_{\mathbf{s}}^{(4)}$ from Eq. (5) is a diagonal tensor, with kurtoses κ_{s_r} as diagonal entries for $1 \leq r \leq R$. Hence, under the working hypothesis, Eq. (5) admits a CPD with a rank R:

$$\mathcal{C}_{\mathbf{x}}^{(4)} = \sum_{r=1}^{R} \kappa_{s_r} \, \mathbf{m}_r \otimes \mathbf{m}_r^* \otimes \mathbf{m}_r^* \otimes \mathbf{m}_r + \mathcal{C}_{\mathbf{n}}^{(4)}, \qquad (6)$$

with $\mathbf{M}$ satisfying the uniqueness conditions. The separation of the source vectors and mixing vectors in Eq. (1) has been translated to the identification of rank-1 terms in Eq. (6) as each source contributes a rank-1 term to the CPD.

A variant of applying a CPD in (6) is to use a maximal diagonalization technique [10] or the joint approximate diagonalization of eigenmatrices method (JADE) [7]. They are used in conjunction with a prewhitening step using the second-order covariance matrix.

3.2 Parameter Variation

Given some matrix data, one can perform a (multilinear) transformation depending upon a parameter to generate a set of matrices. After stacking them, a third-order tensor is obtained which can be decomposed to identify the underlying unknown components. It is used in the decoupling of multivariate polynomials [24] but also in BSS with the second-order blind identification (SOBI) algorithm [3] and variants. In SOBI, the set of matrices consists of lagged covariance matrices. Let us define $\mathbf{C}_{\mathbf{u}}(\tau) = \mathrm{E}\left\{\mathbf{u}(t)\mathbf{u}(t+\tau)^{\mathrm{H}}\right\} \in \mathbb{K}^{K \times K}$ as the covariance matrix with a lag τ of a stochastic signal vector $\mathbf{u}(t) \in \mathbb{K}^K$. Observe that this gives a bilinear transformation: if $\mathbf{x}(t) = \mathbf{M}s(t) + \mathbf{n}(t)$, then $\mathbf{C_x}(\tau) = \mathbf{MC_s}(\tau)\mathbf{M}^H + \mathbf{C_n}(\tau)$. For multiple lags $\tau_1, \ldots, \tau_L$ we then have:

$$
\begin{cases}
\mathbf{C_x}(\tau_1) = \mathbf{MC_s}(\tau_1)\mathbf{M}^H + \mathbf{C_n}(\tau_1), \\
\quad \vdots \\
\mathbf{C_x}(\tau_L) = \mathbf{MC_s}(\tau_L)\mathbf{M}^H + \mathbf{C_n}(\tau_L).
\end{cases}
\tag{7}
$$

The working hypothesis made by SOBI is that the source signals are mutually uncorrelated but individually correlated for the different lags $\tau_1, \ldots, \tau_L$.[1] Then, the corresponding lagged covariance matrices of the sources are diagonal matrices. Hence, the matrices $\mathbf{M}$ and $\mathbf{M}^*$ simultaneously diagonalize the lagged covariance matrices of $\mathbf{x}(t)$ in (7) [12]. Let us define $\sigma_{s_r}^2(\tau_l)$ as the autocovariance of source $s_r(t)$ for the given lag τ_l. We collect them for each source in a vector $\sigma_{s_r}^2 \in \mathbb{K}^L$ for all τ_l, $1 \leq l \leq L$. By stacking $\mathbf{C_x}(\tau_l)$ in the third dimension of a tensor $\mathcal{C_x}$ and assuming the noise level is low, a CPD emerges:

$$
\mathcal{C_x} = \sum_{r=1}^{R} \mathbf{m}_r \otimes \mathbf{m}_r^* \otimes \sigma_{s_r}^2 + \mathcal{C_n} = \mathcal{I} \cdot_1 \mathbf{M} \cdot_2 \mathbf{M}^* \cdot_3 \mathbf{\Sigma} + \mathcal{C_n},
\tag{8}
$$

in which $\mathbf{\Sigma} \in \mathbb{K}^{L \times R}$ contains the columns $\sigma_{s_r}^2$ for $1 \leq r \leq R$. Note that each source contributes a rank-1 term to $\mathcal{C_x}$. In [12], the connection between simultaneous matrix diagonalization and CPD is discussed.

A variant for nonstationary sources of the SOBI tensorization method is the stacking of a set of covariance matrices computed for different time frames [42].

3.3 Hankelization

Consider an exponential signal $f(k) = az^k$ arranged in a Hankel matrix $\mathbf{H}$. The matrix appears to have rank one:

$$
\mathbf{H} = \begin{bmatrix} f(0) & f(1) & f(2) & \cdots \\ f(1) & f(2) & f(3) & \cdots \\ f(2) & f(3) & f(4) & \cdots \\ \vdots & \vdots & \vdots \end{bmatrix} = a \begin{bmatrix} 1 \\ z \\ z^2 \\ \vdots \end{bmatrix} \begin{bmatrix} 1 & z & z^2 & \cdots \end{bmatrix}.
\tag{9}
$$

These simple exponential functions can be generalized to exponential polynomials, which are functions that can be written as sums and/or products of exponentials, sinusoids and/or polynomials. They have a broad relevance: for (multidimensional) harmonic retrieval, direction-of-arrival estimation, sinusoidal carriers in telecommunication, etc. [26,33,34,43,44]. Furthermore, they can be used to model various signal shapes. The idea is analogous to the approximation of functions with the well-known Taylor series expansion. Figures 1 and 2 show approximations of a sigmoid and Gaussian function through Hankelization.

---

[1] Note that the autocorrelation is not required for each source for each of the lags.

It has been shown that for an exponential polynomial signal of degree δ, the corresponding Hankel matrix will have rank δ [46]. The Hankel tensorization technique exists in mapping each row of the observed data matrix $\mathbf{X}$ from (1) to a Hankel matrix which is being stacked in a third-order tensor $\mathcal{H}_{\mathbf{X}}$. With $\mathbf{H}_{\mathbf{s}_r}$ the Hankel matrix of the rth source $\mathbf{s}_r$, we have because of linearity that

$$\mathcal{H}_{\mathbf{X}} = \sum_{r=1}^{R} \mathbf{H}_{\mathbf{s}_r} \otimes \mathbf{m}_r = \sum_{r=1}^{R} (\mathbf{A}_r \mathbf{B}_r^{\mathsf{T}}) \otimes \mathbf{m}_r. \tag{10}$$

The latter transition is based on the working hypothesis that the rth source can be approximated by an exponential polynomial of (low) degree L_r. Each matrix $\mathbf{H}_{\mathbf{s}_r}$ has (low) rank L_r then, and we have full column rank matrices $\mathbf{A}_r \in \mathbb{K}^{I \times L_r}$ and $\mathbf{B}_r \in \mathbb{K}^{J \times L_r}$. Hence, after the Hankel-tensorization (or Hankelization), a decomposition in rank-$(L_r, L_r, 1)$ terms like in Eq. (10) can be applied. Each source contributes a tensor with low multilinear rank $(L_r, L_r, 1)$.

3.4   Löwnerization

Another class of functions suitable for BSS is the set of rational functions, able to take on a very wide range of shapes. An illustration is given in Figs. 1 and 2 by approximating a sigmoid and Gaussian function. Rational functions have the same connection with Löwner matrices as exponential polynomials have with Hankel matrices [2, 27]. Given a function $f(t)$ sampled on $N = I+J$ points which are divided in two distinct point sets $X = \{x_1, \ldots, x_I\}$ and $Y = \{y_1, \ldots, y_J\}$, we define the entries of the Löwner matrix $\mathbf{L} \in \mathbb{K}^{I \times J}$ as follows:

$$\forall i, j: \quad l_{i,j} = \frac{f(x_i) - f(y_j)}{x_i - y_j}. \tag{11}$$

It has been shown in [18, 19] that an equivalent formulation as in Eq. (10) can be made: because of the linearity of the Löwner transformation, the tensor $\mathcal{L}_{\mathbf{X}}$, obtained by mapping every row of the observed data matrix $\mathbf{X}$ to a Löwner matrix and stacking these matrices, can be written as a linear combination of the Löwner matrices of the sources. Under the working hypothesis that the rth source can be modeled as a rational function of (low) degree L_r, the corresponding Löwner matrix will have (low) rank L_r. Like in the Hankel case, a BTD is obtained where the rth source contributes a rank-$(L_r, L_r, 1)$ term to $\mathcal{L}_{\mathbf{X}}$.

3.5   Segmentation

Segmentation is a general term used to denote the reshaping of a vector into a matrix, i.e., extracting small segments and stacking them after each other. Consider the following exponential vector: $\begin{bmatrix} 1 & z & z^2 & z^3 & z^4 & z^5 \end{bmatrix}$. If it is reshaped to a matrix, the latter has rank one:

$$\begin{bmatrix} 1 & z & z^2 & z^3 & z^4 & z^5 \end{bmatrix} \rightarrow \begin{bmatrix} 1 & z & z^2 \\ z^3 & z^4 & z^5 \end{bmatrix} = \begin{bmatrix} 1 \\ z^3 \end{bmatrix} \begin{bmatrix} 1 & z & z^2 \end{bmatrix}. \tag{12}$$

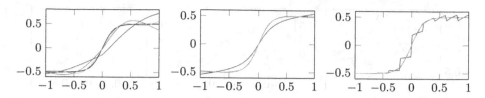

Fig. 1. Approximation of a sigmoid function $f(t) = \frac{1}{1+e^{-10t}}$. It is sampled uniformly 100 times in $[-1, 1]$ (——). To the left, an approximation with exponential polynomials is used by Hankelizing the samples. In the middle, Löwnerization is applied. To the right, segmentation with $I = J = 10$ is used. The tensorized matrix is approximated by a low-rank matrix through truncation of the singular value decomposition, after which the underlying signal is calculated from this low-rank matrix. Approximations for ranks $R = 1$ (——), $R = 2$ (——) and $R = 3$ (——) are shown.

Focusing on BSS, let us now reshape the kth row of the observed data matrix $\mathbf{X} \in \mathbb{K}^{K \times N}$ to a matrix $\mathbf{E}_{x_k} \in \mathbb{K}^{I \times J}$ with $N = I \times J$ for $k = 1, \ldots, K$, and stack these matrices in a tensor $\mathcal{X} \in \mathbb{K}^{I \times J \times K}$. The transformation is clearly linear. Let us start from the assumption that the segmented matrix of each source has rank one, as in Eq. (12). One obtains the following CPD:

$$\mathcal{X} = \sum_{r=1}^{R} \mathbf{E}_{s_r} \otimes \mathbf{m}_r = \sum_{r=1}^{R} \mathbf{a}_r \otimes \mathbf{b}_r \otimes \mathbf{m}_r. \tag{13}$$

with rank-1 matrices $\mathbf{E}_{s_r} = \mathbf{a}_r \otimes \mathbf{b}_r$ and vectors $\mathbf{a}_r \in \mathbb{K}^I$ and $\mathbf{b}_r \in \mathbb{K}^J$. This is equivalent to stating that the rth source signal can be written as a Kronecker product $\mathbf{a}_r^\mathsf{T} \otimes \mathbf{b}_r^\mathsf{T}$ for $r = 1, \ldots, R$, with the Kronecker product for row vectors $\mathbf{u} \in \mathbb{K}^{1 \times I}$, $\mathbf{v} \in \mathbb{K}^{1 \times J}$ defined as $\mathbf{u} \otimes \mathbf{v} = \begin{bmatrix} u_1 \mathbf{v} & u_2 \mathbf{v} & \cdots & u_I \mathbf{v} \end{bmatrix}$.

Although the hypothesis is fulfilled when the sources are, for instance, exponential functions, it is quite restrictive. By increasing the assumed rank $L_r \geq 1$ of the reshaped matrices $\mathbf{E}_{s_r}$, we obtain a BTD in rank-$(L_r, L_r, 1)$ terms:

$$\mathcal{X} = \sum_{r=1}^{R} \mathbf{E}_{s_r} \otimes \mathbf{m}_r = \sum_{r=1}^{R} (\mathbf{A}_r \mathbf{B}_r^\mathsf{T}) \otimes \mathbf{m}_r, \tag{14}$$

with matrices $\mathbf{A}_r \in \mathbb{K}^{I \times L_r}$ and $\mathbf{B}_r \in \mathbb{K}^{J \times L_r}$. Adding a subscript l to denote the lth column of the matrices $\mathbf{A}_r$ and $\mathbf{B}_r$, the working hypothesis now becomes that the source signals can be modeled as, or approximated by, sums of Kronecker products: $\mathbf{s}_r = \sum_{l=1}^{L_r} \mathbf{a}_{r,l}^\mathsf{T} \otimes \mathbf{b}_{r,l}^\mathsf{T}$. An example of a source exactly displaying this structure is a sine wave, which can be written as a sum of two Kronecker products. Other functions can be approximated too, e.g. sigmoid and Gaussian functions, illustrated in Figs. 1 and 2. While each source contributed a rank-1 term to $\mathcal{X}$ for the first hypothesis, it now contributes a term with low multilinear rank $(L_r, L_r, 1)$.

Note that because of the segmentation and the structure of the low-rank decompositions, a nonnegligible compression is obtained in the number of

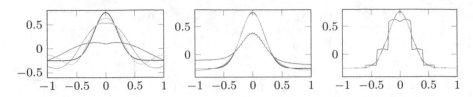

Fig. 2. Approximation of a Gaussian function $f(t) = e^{-\frac{1}{2}(5t)^2}$, sampled uniformly 100 times in $[-1, 1]$ (——). An equal procedure as in Figure 1 is used, with Hankelization (left), Löwnerization (middle) and segmentation (right) for ranks $R = 1$ (——), $R = 2$ (——) and $R = 3$ (——). As in Fig. 1, the exponential method is not very suitable for signals with horizontal asymptotes.

underlying variables. This is especially useful for big data systems, with many observed samples or signals. The technique has been described in [4,5] for large-scale BSS problems, including a generalization for higher-order segmentation. Segmentation of signal vectors to matrices or tensors has been successfully applied in various domains before, such as biomedical signal processing [20] and scientific computing for large-scale models with high dimensions and a very high number of numerical values [30].

4   Discussion and Conclusion

In many techniques for blind signal separation (BSS), multilinear algebra is used to recover the mixing vectors and the original source signals. Given only an observed data matrix, a transformation is made to higher-order structures called tensors. This paper introduces the tensorization step as an important concept by itself, as many results concerning tensorization have appeared in the literature in a disparate manner and have not been discussed as such. Higher-order statistics and second-order statistics, for example, are well-known to solve BSS, but apply tensorization in a significantly different way. Many links to multilinear algebra from other existing BSS techniques have not yet been established. Because of space limitations, the presentation of the idea has been restricted to instantaneous mixtures of one-dimensional sources. A following paper will discuss generalizations such as multidimensional sources or convolutive mixtures.

Acknowledgements. The research is funded by (1) a Ph.D. grant of the Agency for Innovation by Science and Technology (IWT), (2) Research Council KU Leuven: CoE PFV/10/002 (OPTEC), (3) F.W.O.: projects G.0830.14N and G.0881.14N, (4) the Belgian Federal Science Policy Office: IUAP P7/19 (DYSCO II, Dynamical systems, control and optimization, 2012–2017), (5) EU: The research leading to these results has received funding from the European Research Council under the European Union's Seventh Framework Programme (FP7/2007–2013) / ERC Advanced Grant: BIOTENSORS (no. 339804). This paper reflects only the authors' views and the Union is not liable for any use that may be made of the contained information.

References

1. Acar, E., Aykut-Bingol, C., Bingol, H., Bro, R., Yener, B.: Multiway analysis of epilepsy tensors. Bioinformatics **23**(13), i10–i18 (2007)
2. Antoulas, A.C., Anderson, B.D.O.: On the scalar rational interpolation problem. IMA J. Math. Control Inf. **3**(2–3), 61–88 (1986)
3. Belouchrani, A., Abed-Meraim, K., Cardoso, J., Moulines, E.: A blind source separation technique using second-order statistics. IEEE Trans. Sig. Process. **45**(2), 434–444 (1997)
4. Boussé, M., Debals, O., De Lathauwer, L.: Deterministic blind source separation using low-rank tensor approximations. Internal report 15–59, ESAT-STADIUS, KU Leuven, Belgium, April 2015
5. Boussé, M., Debals, O., De Lathauwer, L.: A novel deterministic method for large-scale blind source separation. In: Proceedings of the 23rd European Signal Processing Conference (EUSIPCO 2015, Nice, France), August 2015. Accepted for publication
6. Bruckstein, A.M., Donoho, D.L., Elad, M.: From sparse solutions of systems of equations to sparse modeling of signals and images. SIAM Rev. **51**(1), 34–81 (2009)
7. Cardoso, J.F., Souloumiac, A.: Blind beamforming for non-gaussian signals. IEE Proceedings F Radar Sig. Process. **140**(6), 362–370 (1993)
8. Cichocki, A., Mandic, D., Phan, A.H., Caiafa, C., Zhou, G., Zhao, Q., De Lathauwer, L.: Tensor decompositions for signal processing applications: from two-way to multiway component analysis. IEEE Sig. Process. Mag. **32**(2), 145–163 (2015)
9. Cichocki, A., Zdunek, R., Phan, A., Amari, S.: Nonnegative matrix and tensor factorizations: applications to exploratory multi-way data analysis and blind source separation. Wiley, Hoboken (2009)
10. Comon, P.: Independent component analysis, a new concept? Sig. Process. **36**(3), 287–314 (1994)
11. Comon, P., Jutten, C.: Handbook of Blind Source Separation: Independent Component Analysis and Applications. Academic Press, New York (2010)
12. De Lathauwer, L.: A link between the canonical decomposition in multilinear algebra and simultaneous matrix diagonalization. SIAM J. Matrix Anal. Appl. **28**(3), 642–666 (2006)
13. De Lathauwer, L.: Decompositions of a higher-order tensor in block terms – Part II: definitions and uniqueness. SIAM J. Matrix Anal. Appl. **30**(3), 1033–1066 (2008)
14. De Lathauwer, L.: Blind separation of exponential polynomials and the decomposition of a tensor in rank-$(L_r, L_r, 1)$ terms. SIAM J. Matrix Anal. Appl. **32**(4), 1451–1474 (2011)
15. De Lathauwer, L., Castaing, J.: Tensor-based techniques for the blind separation of DS-CDMA signals. Sig. Process. **87**(2), 322–336 (2007)
16. De Lathauwer, L., Nion, D.: Decompositions of a higher-order tensor in block terms – Part III: alternating least squares algorithms. SIAM J. Matrix Anal. Appl. **30**(3), 1067–1083 (2008)
17. De Vos, M.: Decomposition methods with applications in neuroscience. Ph.D. thesis, KU Leuven (2009)
18. Debals, O., Van Barel, M., De Lathauwer, L.: Blind signal separation of rational functions using Löwner-based tensorization. In: IEEE Proceedings on International Conference on Acoustics, Speech and Signal Processing, pp. 4145–4149. April 2015. Accepted for publication

19. Debals, O., Van Barel, M., De Lathauwer, L.: Löwner-based blind signal separation of rational functions with applications. Internal report 15–44, ESAT-STADIUS, KU Leuven, Belgium, March 2015

20. Deburchgraeve, W., Cherian, P., De Vos, M., Swarte, R., Blok, J., Visser, G.H., Govaert, P., Van Huffel, S.: Neonatal seizure localization using PARAFAC decomposition. Clin. Neurophysiol. **120**(10), 1787–1796 (2009)

21. Domanov, I., De Lathauwer, L.: On the uniqueness of the canonical polyadic decomposition of third-order tensors – Part I: basic results and uniqueness of one factor matrix. SIAM J. Matrix Anal. Appl. **34**(3), 855–875 (2013)

22. Domanov, I., De Lathauwer, L.: On the uniqueness of the canonical polyadic decomposition of third-order tensors – Part II: uniqueness of the overall decomposition. SIAM J. Matrix Anal. Appl. **34**(3), 876–903 (2013)

23. Donoho, D., Stodden, V.: When does non-negative matrix factorization give a correct decomposition into parts? In: Advances in Neural Information Processing Systems (2003)

24. Dreesen, P., Ishteva, M., Schoukens, J.: Decoupling multivariate polynomials using first-order information and tensor decompositions. SIAM J. Matrix Anal. Appl. **36**(2), 864–879 (2015)

25. Eggert, J., Korner, E.: Sparse coding and NMF. IEEE Proc. Int. Joint Conf. Neural Netw. **4**, 2529–2533 (2004)

26. Elad, M., Milanfar, P., Golub, G.H.: Shape from moments – an estimation theory perspective. IEEE Trans. Sig. Process. **52**(7), 1814–1829 (2004)

27. Fiedler, M.: Hankel and Löwner matrices. Linear Algebra Appl. **58**, 75–95 (1984)

28. Georgiev, P., Theis, F., Cichocki, A.: Sparse component analysis and blind source separation of underdetermined mixtures. IEEE Trans. Neural Netw. **16**(4), 992–996 (2005)

29. Gillis, N.: Nonnegative matrix factorization: complexity, algorithms and applications. Ph.D. thesis, UCL (2011)

30. Grasedyck, L.: Polynomial approximation in hierarchical Tucker format by vector tensorization, April 2010

31. Harman, H.H.: Modern Factor Analysis, 3rd edn. University of Chicago Press, Chicago (1976)

32. Hoyer, P.O.: Non-negative matrix factorization with sparseness constraints. J. Mach. Learn. Res. **5**, 1457–1469 (2004)

33. Hunyadi, B., Camps, D., Sorber, L., Van Paesschen, W., De Vos, M., Van Huffel, S., De Lathauwer, L.: Block term decomposition for modelling epileptic seizures. EURASIP J. Adv. Sig. Process. **2014**(1), 1–19 (2014)

34. Jiang, T., Sidiropoulos, N.D., ten Berge, J.M.: Almost-sure identifiability of multidimensional harmonic retrieval. IEEE Trans. Sig. Process. **49**(9), 1849–1859 (2001)

35. Kolda, T.G., Bader, B.W.: Tensor decompositions and applications. SIAM Rev. **51**(3), 455–500 (2009)

36. Kroonenberg, P.: Applied Multiway Data Analysis, vol. 702. Wiley-Interscience, Hoboken (2008)

37. Kruskal, J.B.: Three-way arrays: rank and uniqueness of trilinear decompositions, with application to arithmetic complexity and statistics. Linear Algebra Appl. **18**(2), 95–138 (1977)

38. Lee, D., Seung, H., et al.: Learning the parts of objects by non-negative matrix factorization. Nature **401**(6755), 788–791 (1999)

39. McCullagh, P.: Tensor Methods in Statistics, vol. 161. Chapman and Hall, London (1987)

40. Nikias, C.L., Petropulu, A.P.: Higher-Order Spectra Analysis: A Nonlinear Signal Processing Framework. PTR Prentice Hall, Englewood Cliffs (1993)
41. Paatero, P., Tapper, U.: Positive matrix factorization: a non-negative factor model with optimal utilization of error estimates of data values. Environmetrics **5**(2), 111–126 (1994)
42. Pham, D.T., Cardoso, J.F.: Blind separation of instantaneous mixtures of nonstationary sources. IEEE Trans. Sig. Process. **49**(9), 1837–1848 (2001)
43. Roemer, F., Haardt, M., Del Galdo, G.: Higher order SVD based subspace estimation to improve multi-dimensional parameter estimation algorithms. In: Fortieth Asilomar Conference on Signals, Systems and Computers, pp. 961–965. IEEE (2006)
44. Sidiropoulos, N.D.: Generalizing Caratheodory's uniqueness of harmonic parameterization to N dimensions. IEEE Trans. Inf. Theory **47**(4), 1687–1690 (2001)
45. Smilde, A.K., Bro, R., Geladi, P., Wiley, J.: Multi-way Analysis with Applications in the Chemical Sciences. Wiley Chichester, UK (2004)
46. Vandevoorde, D.: A fast exponential decomposition algorithm and its applications to structured matrices. Ph.D. thesis, Rensselaer Polytechnic Institute, Troy, NY (1998)
47. Zibulevsky, M., Pearlmutter, B.: Blind source separation by sparse decomposition in a signal dictionary. Neural Comput. **13**(4), 863–882 (2001)

Block-Decoupling Multivariate Polynomials Using the Tensor Block-Term Decomposition

Philippe Dreesen[1]([✉]), Thomas Goossens[1,2], Mariya Ishteva[1],
Lieven De Lathauwer[2,3], and Johan Schoukens[1]

[1] Vrije Universiteit Brussel (VUB), Brussels, Belgium
philippe.dreesen@gmail.com
[2] Department of Electrical Engineering (ESAT/STADIUS) and iMinds Medical IT,
KU Leuven, Leuven, Belgium
[3] KU Leuven Campus Kortrijk, Kortrijk, Belgium

Abstract. We present a tensor-based method to decompose a given set of multivariate functions into linear combinations of a set of multivariate functions of linear forms of the input variables. The method proceeds by forming a three-way array (tensor) by stacking Jacobian matrix evaluations of the function behind each other. It is shown that a block-term decomposition of this tensor provides the necessary information to block-decouple the given function into a set of functions with small input-output dimensionality. The method is validated on a numerical example.

Keywords: Multivariate polynomials · Multilinear algebra · Tensor decomposition · Block-term decomposition · Waring decomposition

1 Introduction

1.1 Problem Statement

The problem we study in the current paper is how to decompose a given multivariate vector-valued function $\mathbf{f}(\mathbf{u})$ into a (parametric) representation of the form

$$\mathbf{f}(\mathbf{u}) = \begin{bmatrix} \mathbf{W}_1 \cdots \mathbf{W}_R \end{bmatrix} \begin{bmatrix} \mathbf{g}_1(\mathbf{V}_1^T \mathbf{u}) \\ \vdots \\ \mathbf{g}_R(\mathbf{V}_R^T \mathbf{u}) \end{bmatrix}, \tag{1}$$

where $\mathbf{g}_i(\mathbf{x}_i) : \mathbb{R}^{m_i} \to \mathbb{R}^{n_i}$ map from m_i inputs to n_i outputs, $\mathbf{W}_i \in \mathbb{R}^{N \times n_i}$ and $\mathbf{V}_i \in \mathbb{R}^{M \times m_i}$, with $i = 1, \ldots, R$. Figure 1 is a schematical representation of the proposed structure. The case in which all $\mathbf{g}_i(\mathbf{x}_i)$ are univariate functions is related to the Waring decomposition [1,9] and is discussed in [5]. The current paper considers the case of *block-decoupling* with the internal functions $\mathbf{g}_i(\mathbf{x}_i)$ being multivariate vector-valued functions. It is assumed that the decomposition (1) exists (in the exact sense).

© Springer International Publishing Switzerland 2015
E. Vincent et al. (Eds.): LVA/ICA 2015, LNCS 9237, pp. 14–21, 2015.
DOI: 10.1007/978-3-319-22482-4_2

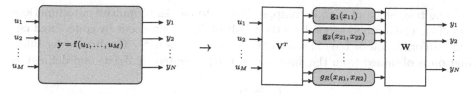

Fig. 1. The block-decoupling problem statement. From the polynomial mapping $\mathbf{y} = \mathbf{f}(\mathbf{u})$ we wish to find $\mathbf{V} = [\mathbf{V}_i]$ and $\mathbf{W} = [\mathbf{W}_i]$ and the mappings $\mathbf{g}_i(\mathbf{x}_i)$ such that $\mathbf{f}(\mathbf{x}) = \sum_{i=1}^{R} \mathbf{W}_i \mathbf{g}_i(\mathbf{V}_i^T \mathbf{u})$.

1.2   When and Why is a Block-Decoupling Favorable?

A block-decoupling (1) is a natural representation of a nonlinear mapping when inherent coupling among some internal variables exists, for instance due to underlying physics. Rather than unraveling the function into univariate branches, solely to be able to decouple the variables, it may be desirable to keep sets of variables together (see Example 1). Moreover, the introduction of (possibly many) internal branches may *increase* the parametric complexity of the function representation, which is undesirable. Therefore, block-decoupling (1) may also contribute to reducing parametric complexity.

Let us look at a simple case where we derive 'manually' from a coupled function its fully decoupled representation. We will see that full decoupling requires the introduction of several branches $g_i(x_i)$. This example serves as a justification to prefer a block-decoupling over full decoupling.

Example 1. To fully decouple the function $f(u_1, u_2) = u_1^2 u_2$, one needs to introduce three univariate branches. Indeed, it is easy to see that we have

$$u_1^2 u_2 = \frac{1}{6}\left((u_1 + u_2)^3 - (u_1 - u_2)^3\right) - \frac{1}{3}u_2^3,$$

from which we conclude that $f(u_1, u_2) = u_1^2 u_2$ can be fully decoupled as the sum of three univariate functions $g_1(x_1) = 1/6x_1^3$, with $x_1 = u_1 + u_2$, $g_2(x_2) = -1/6x_2^3$ with $x_2 = u_1 - u_2$ and $g_3(x_3) = -1/3x_3^3$ with $x_3 = u_2$. In more complicated cases, full decoupling may require the introduction of more univariate functions $g_i(x_i)$ than block-decoupled vector-valued functions $\mathbf{g}_i(\mathbf{x}_i)$.   ◇

2   Method

2.1   Block-Diagonalization of Jacobian Matrices

We assume that $\mathbf{f}(\mathbf{u})$ can be written as in (1). Although we will describe the method for the case that $\mathbf{f}(\mathbf{u})$ is polynomial, the method can easily be generalized to the non-polynomial case, and is applicable as long as the derivatives of $\mathbf{f}(\mathbf{u})$ can be obtained.

The task at hand is to decompose $\mathbf{f}(\mathbf{u})$ into blocks of multivariate functions as in (1). The method generalizes the result of [5] and proceeds by collecting first-order information of $\mathbf{f}(\mathbf{u})$ in a set of sampling points $\mathbf{u}^{(k)}$. The first-order information is obtained from the Jacobian of $\mathbf{f}(\mathbf{u})$, denoted by $\mathbf{J_f}(\mathbf{u})$ and defined as

$$\mathbf{J_f}(\mathbf{u}) = \begin{bmatrix} \frac{\partial f_1}{\partial u_1}(\mathbf{u}) & \cdots & \frac{\partial f_1}{\partial u_M}(\mathbf{u}) \\ \vdots & \ddots & \vdots \\ \frac{\partial f_N}{\partial u_1}(\mathbf{u}) & \cdots & \frac{\partial f_N}{\partial u_M}(\mathbf{u}) \end{bmatrix}. \tag{2}$$

Applying the chain rule of differentiation to $\mathbf{f}(\mathbf{u}) = \sum_{i=1}^{R} \mathbf{W}_i \mathbf{g}_i(\mathbf{V}_i^T \mathbf{u})$ leads to

$$\mathbf{J_f}(\mathbf{u}) = \begin{bmatrix} \mathbf{W}_1 \ldots \mathbf{W}_R \end{bmatrix} \begin{bmatrix} \mathbf{J_{g_1}}(\mathbf{V}_1^T \mathbf{u}) & & \\ & \ddots & \\ & & \mathbf{J_{g_R}}(\mathbf{V}_R^T \mathbf{u}) \end{bmatrix} \begin{bmatrix} \mathbf{V}_1^T \\ \vdots \\ \mathbf{V}_R^T \end{bmatrix}, \tag{3}$$

where the $\mathbf{J_{g_i}}(\mathbf{x}_i)$ are defined similar to (2).

2.2 Computing $\mathbf{W}_i$, $\mathbf{V}_i$ and $\mathcal{H}_i$

From (3) it follows that finding from the Jacobian evaluations $\mathbf{J_f}(\mathbf{u}^{(k)})$ the matrices $\mathbf{W}_i$, $\mathbf{V}_i$ and the functions $\mathbf{g}_i(\mathbf{x}_i)$, amounts to solving a simultaneous matrix block-diagonalization problem. By evaluating the Jacobian of $\mathbf{f}(\mathbf{u})$ in a set of K sampling points $\mathbf{u}^{(k)}$ we obtain a collection of Jacobian matrices $\mathbf{J_f}(\mathbf{u}^{(k)})$, $k = 1, \ldots, K$, which are stacked behind each other into the $N \times M \times K$ tensor $\mathcal{J} = \{\mathbf{J_f}(\mathbf{u}^{(1)}), \ldots, \mathbf{J_f}(\mathbf{u}^{(K)})\}$. The recent years have seen an increased research interest in tensor decompositions [2,8], which can be seen as higher-order extensions of well-known matrix decompositions such as the singular value decomposition [6]. The tensor decomposition that will be of interest for the current task is the block-term decomposition (BTD) in rank$(n_i, m_i, \cdot)$-terms [3,4,10,12], as it can be used to compute the simultaneous block-diagonalization of the Jacobian tensor $\mathcal{J}$. The BTD of $\mathcal{J}$ in rank$(n_i, m_i, \cdot)$-terms is the decomposition of $\mathcal{J}$ into

$$\mathcal{J} = \sum_{i=1}^{R} \mathcal{H}_i \bullet_1 \mathbf{W}_i \bullet_2 \mathbf{V}_i, \tag{4}$$

where $\bullet_i$ denotes the mode-i tensor product, and $\mathbf{W}_i$ and $\mathbf{V}_i$ are defined as above. The $n_i \times m_i \times K$ core tensors $\mathcal{H}_i$ contain in the slices the Jacobians $\mathbf{J_{g_i}}(\mathbf{x}^{(k)})$, with $\mathbf{x}_i^{(k)} = \mathbf{V}_i^T \mathbf{u}^{(k)}$. Figure 2 gives a graphical overview of the method.

2.3 Uniqueness

A lack of *global uniqueness* of the BTD can be expected because one can introduce nonsingular transformations $\mathbf{S}_i$ and $\mathbf{T}_i$ in the R terms of (4) to obtain the (equivalent) decomposition $\mathcal{J} = \sum_{i=1}^{R} \left(\mathcal{H}_i \bullet_1 \mathbf{T}_i^{-1} \bullet_2 \mathbf{S}_i^{-1} \right) \bullet_1 (\mathbf{W}_i \mathbf{T}_i) \bullet_2 (\mathbf{V}_i \mathbf{S}_i)$.

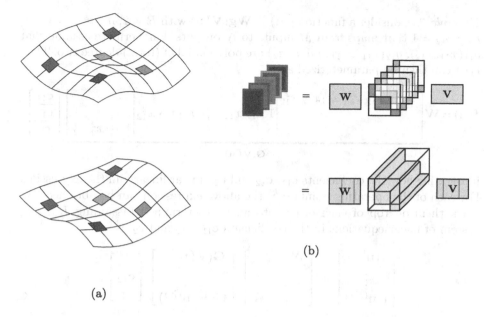

(b)

(a)

Fig. 2. Visual representation of the decomposition method. From the first-order information of $\mathbf{f}(\mathbf{u})$ a tensor consisting of Jacobian matrices is constructed. The block-term decomposition of this tensor results in the factors $\mathbf{V}_i$, $\mathbf{W}_i$ and the core tensors $\mathcal{H}_i$ from which the decoupling of $\mathbf{f}(\mathbf{u})$ can be found.

The uniqueness properties of BTD are discussed in [3,4], however, the case $\mathrm{rank}(n_i, m_i, \cdot)$ is not included. It is expected that uniqueness conditions along the lines of [4] can be obtained for the $\mathrm{rank}(n_i, m_i, \cdot)$ case, but this is beyond the scope of the current paper. During numerical experiments (using Tensorlab [11]) we have not encountered uniqueness issues—it seems safe to claim that cases with relatively small R are not problematic. In terms of decomposition (1), the effects of rotational ambiguities due to $\mathbf{S}_i$ and $\mathbf{T}_i$ are easy to understand as well. Let us consider the $R = 1$ case $\mathbf{f}(\mathbf{u}) = \mathbf{W}\mathbf{g}(\mathbf{V}^T\mathbf{u})$, in which we insert $\mathbf{S}^T$ and $\mathbf{T}$ and their inverses as $\mathbf{f}(\mathbf{u}) = \mathbf{W}\mathbf{T}\mathbf{T}^{-1}\mathbf{g}(\mathbf{S}^{-T}\mathbf{S}^T\mathbf{V}^T\mathbf{u}) = \widetilde{\mathbf{W}}\widetilde{\mathbf{g}}(\widetilde{\mathbf{V}}^T\mathbf{u})$, where $\widetilde{\mathbf{W}} = \mathbf{W}\mathbf{T}$, $\widetilde{\mathbf{V}}^T = \mathbf{S}^T\mathbf{V}^T$ and $\widetilde{\mathbf{g}}(\mathbf{x}) = \mathbf{T}^{-1}\mathbf{g}(\mathbf{S}^{-T}\mathbf{x})$. Both representations are equivalent, and the factors $\mathbf{V}$ and $\mathbf{W}$ can only be obtained up to linear transformations. The internal function $\mathbf{g}(\mathbf{x})$ has undergone both a change of input variables due to $\mathbf{S}^{-T}$ as well as a linear transformation at the output due to $\mathbf{T}^{-1}$, but the identified $\widetilde{\mathbf{g}}$ is still polynomial of the same degree as the true $\mathbf{g}$.

2.4   Recovering the Coefficients of $\mathbf{g}_i(\mathbf{x}_i)$

A parameterization of the internal functions $\mathbf{g}_i(\mathbf{x})$ can be obtained using interpolation. Since the internal functions $\mathbf{g}_i(\mathbf{x})$ are polynomial, the coefficients of $\mathbf{g}_i(\mathbf{x})$ can be obtained from solving a system of linear equations. We will illustrate the main idea by means of a simple example, from which a general method can easily be derived.

Example 2. Consider a function $\mathbf{f}(\mathbf{u}) = \mathbf{W}\mathbf{g}(\mathbf{V}^T\mathbf{u})$ with $R = 2$, $m_1 = 2$, $m_2 = 1$, $n_1 = n_2 = 1$ that maps from M inputs to N outputs. Furthermore assume that $g_{11}(x_{11}, x_{12})$, $g_{12}(x_{11}, x_{12})$ and $g_2(x_2)$ are polynomial of (total) degree two. Then $\mathbf{f}(\mathbf{u})$ can then be parameterized as

$$\mathbf{f}(\mathbf{u}) = \mathbf{W}\underbrace{\left[\begin{array}{c}1\ x_{11}\ x_{12}\ x_{11}^2\ x_{11}x_{12}\ x_{12}^2 \\ \hline 1\ x_{11}\ x_{12}\ x_{11}^2\ x_{11}x_{12}\ x_{12}^2 \\ \hline 1\ x_2\ x_2^2\ x_2^3\end{array}\right]}_{\mathbf{G}(\mathbf{V}^T\mathbf{u})}\begin{bmatrix}\mathbf{c}_{11} \\ \mathbf{c}_{12} \\ \mathbf{c}_2\end{bmatrix},$$

illustrating how the coefficients $\mathbf{c}_{11}$, $\mathbf{c}_{12}$ and $\mathbf{c}_2$ appear linearly in the expression. For each of the operating points $\mathbf{u}^{(k)}$ the above expression can be obtained. We stack them on top of each other into an overdetermined (assuming $K \gg 1$) system of linear equations in the coefficients $\mathbf{c}_{11}$, $\mathbf{c}_{12}$ and $\mathbf{c}_2$ as

$$\begin{bmatrix}\mathbf{f}(\mathbf{u}^{(1)}) \\ \vdots \\ \mathbf{f}(\mathbf{u}^{(K)})\end{bmatrix} = \begin{bmatrix}\mathbf{W} & & \\ & \ddots & \\ & & \mathbf{W}\end{bmatrix}\begin{bmatrix}\mathbf{G}(\mathbf{V}^T\mathbf{u}^{(1)}) \\ \vdots \\ \mathbf{G}(\mathbf{V}^T\mathbf{u}^{(K)})\end{bmatrix}\begin{bmatrix}\mathbf{c}_{11} \\ \mathbf{c}_{12} \\ \mathbf{c}_2\end{bmatrix}. \qquad \diamond$$

2.5 Algorithm Summary

The complete block-decoupling procedure is summarized as follows.

1. Evaluate the Jacobian matrix $\mathbf{J_f}(\mathbf{u})$ in a set of K sampling points $\mathbf{u}^{(k)}$, $k = 1, \ldots, K$ (Sects. 2.1 and 2.2).
2. Stack the Jacobian matrices into an $N \times M \times K$ tensor $\mathcal{J}$ (Sect. 2.2).
3. Compute the rank$(n_i, m_i, \cdot)$ block-term decomposition of $\mathcal{J}$, resulting in the factors $\mathbf{W}_i$, $\mathbf{V}_i$ and the core tensors $\mathcal{H}_i$ (Sect. 2.2).
4. Recover the coefficients of the internal functions $\mathbf{g}_i(\mathbf{x}_i)$ by solving a linear system (Sect. 2.4).

3   Numerical Example

We will now illustrate the method by means of a numerical example.

Example 3. We assume that a multivariate vector-valued function $\overline{\mathbf{f}}(\mathbf{s})$ is given that has an underlying representation of the form (1)

$$\overline{\mathbf{f}}(\mathbf{u}) = \overline{\mathbf{W}}\overline{\mathbf{g}}(\overline{\mathbf{V}}^T\mathbf{u}) = \overline{\mathbf{w}}_1\overline{g}_1(\overline{\mathbf{V}}_1^T\mathbf{u}) + \overline{\mathbf{w}}_2\overline{g}_2(\overline{\mathbf{v}}_2^T\mathbf{u}), \tag{5}$$

with $\overline{\mathbf{V}} = \left[\overline{\mathbf{V}}_1 | \overline{\mathbf{v}}_2\right]$ and $\overline{\mathbf{W}} = \left[\overline{\mathbf{w}}_1 | \overline{\mathbf{w}}_2\right]$ as

$$\overline{\mathbf{V}} = \begin{bmatrix}1 & 0 & 1 \\ -2 & 1 & -2 \\ 3 & -1 & 0 \\ -1 & 1 & 3\end{bmatrix}, \quad \overline{\mathbf{W}} = \begin{bmatrix}0 & 1 \\ 1 & 3 \\ -1 & 2 \\ 3 & 0\end{bmatrix}, \tag{6}$$

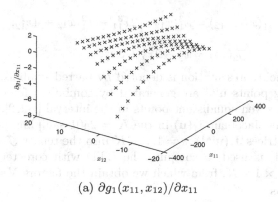

(a) $\partial g_1(x_{11}, x_{12})/\partial x_{11}$

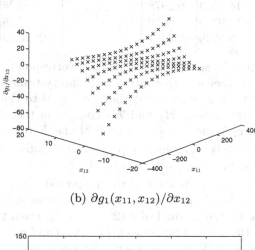

(b) $\partial g_1(x_{11}, x_{12})/\partial x_{12}$

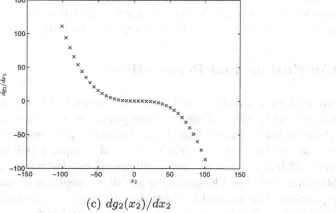

(c) $dg_2(x_2)/dx_2$

Fig. 3. Jacobians of $g_1(x_{11}, x_{12})$ and $g_2(x_2)$, obtained from the core tensors $\mathcal{H}_1$ and $\mathcal{H}_2$, which were computed using the BTD (with $\mathbf{x}_i = \mathbf{V}_i^T \mathbf{u}$).

and

$$\bar{g}_1(x_{11}, x_{12}) = x_{11}^3 x_{12} - 2x_{11}^3 - x_{11}^2 x_{12} + 4x_{12}^2,$$

$$\bar{g}_2(x_2) = x_2^4 - 2x_2^3 + 3x_2^2,$$

(7)

in which the 'true' representation is denoted by barred symbols.

The sampling points $\mathbf{u}^{(k)}$ are generated by combining for each of the four inputs $u_1, \ldots, u_4$ seven equidistant points in the interval $[-2, 2]$, such that $K = 7^4$. We sample the Jacobian $\mathbf{J_f(u)}$ in the $K = 2401$ sampling points and stack the Jacobian matrices $\mathbf{J_f(u}^{(k)})$, $k = 1, \ldots, K$ into the tensor $\mathcal{J}$.

Tensorlab [11] is used to compute the BTD with core tensor dimensions $1 \times 2 \times K$ and $1 \times 1 \times K$, from which we obtain the factors $\mathbf{V} = [\mathbf{V}_1|\mathbf{v}_2]$ and $\mathbf{W} = [\mathbf{w}_1|\mathbf{w}_2]$ as

$$\mathbf{V} = \begin{bmatrix} 7.5051 & -5.2297 & -3.1489 \\ -8.2850 & 14.7523 & 6.2978 \\ 15.7901 & -19.9820 & 0.0000 \\ -0.7799 & 9.5226 & -9.4467 \end{bmatrix}, \quad \mathbf{W} = \begin{bmatrix} 0.0000 & 1.4249 \\ -9.8728 & 4.2748 \\ 9.8728 & 2.8499 \\ -29.6183 & 0.0000 \end{bmatrix}. \quad (8)$$

Notice that the factors $\mathbf{V}$ and $\mathbf{W}$ do not exactly correspond to the underlying factors $\overline{\mathbf{V}}$ and $\overline{\mathbf{W}}$, but they are equal up to a similarity transformation. For the vectors $\mathbf{v}_2$, $\mathbf{w}_1$ and $\mathbf{w}_2$ this means that they are equal to the underlying ones up to scaling. The core tensors $\mathcal{H}_1$ and $\mathcal{H}_2$ contain in their frontal slices the Jacobians of $g_1(x_{11}, x_{12})$ and $g_2(x_2)$, for each of the K operating points, i.e., $\mathbf{x}_i^{(k)} = \mathbf{V}_i^T \mathbf{u}^{(k)}$. Figure 3 is a graphical representation obtained by plotting the entries in the fibers of $\mathcal{H}_i$ versus $\mathbf{x}_i^{(k)} = \mathbf{V}_i^T \mathbf{u}^{(k)}$.

We then compute the coefficients of the recovered $g_1(x_{11}, x_{12})$ and $g_2(x_2)$ from the solution of a Vandermonde-like linear system as in Sect. 2.4 (resulting in a norm-wise error on the residual of 2.1207×10^{-7}). From the recovered $\mathbf{V}_1$, $\mathbf{v}_2$, $\mathbf{w}_1$ and $\mathbf{w}_2$, and the internal functions $g_1(x_{11}, x_{12})$ and $g_2(x_2)$ we reconstruct the function $\mathbf{f(u)} = \mathbf{w}_1 g_1(\mathbf{V}_1^T \mathbf{u}) + \mathbf{w}_2 g_2(\mathbf{v}_2^T \mathbf{u})$ with a relative norm-wise error on the coefficients of 2.7562×10^{-10} comparing to $\overline{\mathbf{f}}(\mathbf{u})$. $\Diamond$

4   Conclusions and Perspectives

We have presented a method to decouple a given set of multivariate polynomials into linear combinations of multivariate polynomials with smaller dimensionality, acting on linear forms of the input variables. By considering the first-order information of the given function in a set of sampling points, we have shown that the problem reduces to the simultaneous block-diagonalization of a set of Jacobian matrices. The block-term tensor decomposition is used to compute the decomposition. The method is illustrated on a numerical example.

Ongoing work is concerned with applying the block-decoupling method to nonlinear block-oriented system identification, where we investigate how to unravel from a black-box nonlinear state-space model the nature of the static nonlinearities [7]. Other open questions include how the decoupling method can be used to simplify or approximate a given multivariate vector-valued function, and how uncertainty on the function $\mathbf{f}(\mathbf{u})$ can be taken into account.

Acknowledgments. This work was supported in part by the Fund for Scientific Research (FWO-Vlaanderen), by the Flemish Government (Methusalem), by the Belgian Government through the Inter-university Poles of Attraction (IAP VII) Program, by the ERC Advanced Grant SNLSID under contract 320378, by the ERC Advanced Grant BIOTENSORS under contract 339804, by the ERC Starting Grant SLRA under contract 258581, by the Research Council KU Leuven: CoE PFV/10/002 (OPTEC), by FWO projects G.0830.14N, G.0881.14N, and G.0280.15N and by the Belgian Federal Science Policy Office: IUAP P7/19 (DYSCO II, Dynamical systems, control and optimization, 2012–2017). Mariya Ishteva is an FWO Pegasus Marie Curie Fellow.

References

1. Carlini, E., Chipalkatti, J.: On Waring's problem for several algebraic forms. Comment. Math. Helv. **78**, 494–517 (2003)
2. Cichocki, A., Mandic, D.P., Phan, A.H., Caiafa, C.F., Zhou, G., Zhao, Q., De Lathauwer, L.: Tensor decompositions for signal processing applications: From two-way to multiway component analysis. IEEE Sig. Process. Mag. **32**(2), 145–163 (2015)
3. De Lathauwer, L.: Decompositions of a higher-order tensor in block terms–Part I: Lemmas for partitioned matrices. SIAM J. Matrix Anal. Appl. **30**, 1022–1032 (2008)
4. De Lathauwer, L.: Decompositions of a higher-order tensor in block terms–Part II: Definitions and uniqueness. SIAM J. Matrix Anal. Appl. **30**, 1033–1066 (2008)
5. Dreesen, P., Ishteva, M., Schoukens, J.: Decoupling multivariate polynomials using first-order information and tensor decompositions. SIAM J. Matrix Anal. Appl. **36**(2), 864–879 (2015)
6. Golub, G.H., Van Loan, C.F.: Matrix Computations, 3rd edn. Johns Hopkins University Press, Baltimore (1996)
7. Goossens, T.: Partial decoupling of multivariate polynomials in nonlinear system identification. Master's thesis, KU Leuven (2015)
8. Kolda, T.G., Bader, B.W.: Tensor decompositions and applications. SIAM Rev. **51**(3), 455–500 (2009)
9. Oeding, L., Ottaviani, G.: Eigenvectors of tensors and algorithms for Waring decomposition. J. Symb. Comp. **54**, 9–35 (2013)
10. Sorber, L., Van Barel, M., De Lathauwer, L.: Optimization-based algorithms for tensor decompositions: Canonical polyadic decomposition, decomposition in rank-$(L_r, L_r, 1)$ terms, and a new generalization. SIAM J. Optim. **23**(2), 695–720 (2013)
11. Sorber, L., Van Barel, M., De Lathauwer, L.: Tensorlab v2.0, January 2014. URL: http://www.tensorlab.net/
12. Tichavský, P., Phan, A.H., Cichocki, A.: Non-orthogonal tensor diagonalization, a tool for block tensor decompositions (2014). (preprint arXiv:1402.1673v2)

A Polynomial Formulation for Joint Decomposition of Symmetric Tensors of Different Orders

Pierre Comon[1,2], Yang Qi[1,2], and Konstantin Usevich[1,2(✉)]

[1] University Grenoble Alpes, GIPSA-Lab, 38000 Grenoble, France
[2] CNRS, GIPSA-Lab, 38000 Grenoble, France
{pierre.comon,yang.qi,konstantin.usevich}@gipsa-lab.grenoble-inp.fr

Abstract. We consider two models: simultaneous CP decomposition of several symmetric tensors of different orders and decoupled representations of multivariate polynomial maps. We show that the two problems are related and propose a unified framework to study the rank properties of these models.

Keywords: Coupled CP decomposition · Polynomial decoupling · Generic rank · X-rank

1 Introduction

Tensor decompositions became an important tool in engineering sciences and data analysis. Several models require tensor decompositions with additional constraints (coupled decompositions or structured tensors), but the properties of these constrained decompositions are not so well understood.

In this paper, we consider two models of this kind: (i) simultaneous CP decomposition of symmetric tensors of different orders (motivated by blind source separ ation) and (ii) decoupling of multivariate polynomials (motivated by problems of identification of nonlinear dynamical systems). We show that these two models are strongly related, and that the notion of rank in these models enjoys many properties similar to tensor rank.

First we define a source separation model in Sect. 1.1, and next the polynomial decomposition model in Sect. 1.2. Finally, the organization and contributions of the paper are described in Sect. 1.3.

1.1 Blind Source Separation and Independent Component Analysis

Consider a linear mixing model [6] in source separation

$$\mathbf{x} = A\mathbf{s},$$

This work is supported by the ERC project "DECODA" no.320594, in the frame of the European program FP7/2007–2013.

E. Vincent et al. (Eds.): LVA/ICA 2015, LNCS 9237, pp. 22–30, 2015.
DOI: 10.1007/978-3-319-22482-4_3

where A is an (unknown) mixing matrix

$$A = (\mathbf{a}_1 \cdots \mathbf{a}_r) \in \mathbb{K}^{n \times r},$$

$\mathbb{K} = \mathbb{R}$ or $\mathbb{C}$, and $\mathbf{s} = (s_1 \cdots s_r)^\top$ is the vector of independent (real or complex) random variables. Then the cumulants of $\mathbf{x}$ up to order d can be expanded as

$$
\begin{aligned}
\mathcal{C}_\mathbf{x}^{(1)} &= c_{1,1}\mathbf{a}_1 + \cdots + c_{1,r}\mathbf{a}_r, \\
\mathcal{C}_\mathbf{x}^{(2)} &= c_{2,1}\mathbf{a}_1 \otimes \mathbf{a}_1 + \cdots + c_{2,r}\mathbf{a}_r \otimes \mathbf{a}_r, \\
&\vdots \\
\mathcal{C}_\mathbf{x}^{(d)} &= c_{d,1}\mathbf{a}_1 \otimes \cdots \otimes \mathbf{a}_1 + \cdots + c_{d,r}\mathbf{a}_r \otimes \cdots \otimes \mathbf{a}_r,
\end{aligned}
\tag{1}
$$

where $c_{j,k}$ is the j-th cumulant of the random variable s_k [9].

In algebraic algorithms for blind source separation, typically a relaxed version of the decomposition problem (1) is considered. For example, in some approaches, a single cumulant (e.g., fourth order) is considered; in others the problem is reduced to decomposition of a partially symmetric tensor, see [6,9] for an overview. In most methods the structure of the joint decomposition (1) is lost, which we aim to avoid in this paper.

We should note that there exist few algorithms for blind source separation which use simultaneous diagonalization of symmetric tensors. In [8] a special case of $d = 4$, $n = 2$ is considered, and fourth- and third-order cumulants are simultaneously diagonalized by finding a common kernel of two matrices. In [7], a similar idea is used for combining cumulants of higher orders. (In [7] the case of $n > 2$ sensors is also considered, but is treated suboptimally.) A theoretical framework for joint decomposition of cumulant tensors is also addressed in [4], but without proposing numerical algorithms.

1.2 Block-Structured Models of Nonlinear Systems

A common problem in nonlinear system identification is to decompose a multivariate nonlinear mapping $F : \mathbb{R}^n \to \mathbb{R}^m$ in a block-structured form as a linear map followed by univariate nonlinear transformations, the outputs of which are linearly mixed again, see Fig. 1. This problem appears in identification of nonlinear state-space models [18] and parallel Wiener-Hammerstein systems [16].

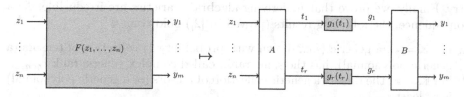

Fig. 1. Decomposition of a multivariate function in a block-structured form.

If the multivariate function is represented as a polynomial, and the scalar nonlinear functions are also polynomials, then the decomposition in Fig. 1 becomes a polynomial decomposition problem, which we describe formally below.

Let $\mathbb{K}$ be $\mathbb{R}$ or $\mathbb{C}$. By $\mathbb{K}_d[\mathbf{z}]$ we denote the space of homogeneous polynomials of degree d, and by $\mathbb{K}_{\leq d}[\mathbf{z}]$ the space of polynomials of degree $\leq d$. Consider a multivariate polynomial map $F : \mathbb{K}^n \to \mathbb{K}^m$, i.e., a vector $F(\mathbf{z}) = \left(f_1(\mathbf{z}) \cdots f_m(\mathbf{z})\right)^\top$ of multivariate polynomials ($f_i \in \mathbb{K}_{\leq d}[\mathbf{z}]$) in variables $\mathbf{z} = \left(z_1 \cdots z_n\right)^\top$. We say that F has a *decoupled representation*, if it can be expressed as

$$F(\mathbf{z}) = B \cdot \mathbf{g}(A^\top \mathbf{z}), \tag{2}$$

where

$$A = \left(\mathbf{a}_1 \cdots \mathbf{a}_r\right) \in \mathbb{K}^{n \times r}, \quad B = \left(\mathbf{b}_1 \cdots \mathbf{b}_r\right) \in \mathbb{K}^{m \times r},$$

are transformation matrices, and $\mathbf{g} : \mathbb{K}^r \to \mathbb{K}^r$ is defined as

$$\mathbf{g}(t_1, \ldots, t_r) = \left(g_1(t_1) \cdots g_r(t_r)\right)^\top$$

where g_k are nonhomogeneous univariate polynomials of degree $\leq d$.

The decomposition (2) is exactly the one depicted in Fig. 1, and can be also equivalently represented as

$$F(\mathbf{z}) = \mathbf{b}_1 g_1(\mathbf{a}_1^\top \mathbf{z}) + \cdots + \mathbf{b}_r g_r(\mathbf{a}_r^\top \mathbf{z}). \tag{3}$$

Recently, two different, but related methods were proposed for solving the decoupling problem in the case $m > 1$, see [11,18]. Both methods are based on CP decomposition of a non-symmetric tensor constructed from the coefficients of the polynomial mapping. We also should note that there exist other tensor-based methods for identifying block-structured systems [13], which operate with structured tensors.

1.3   Contributions of this Paper

The first aim of this paper is to show that the joint CP decomposition described in Sect. 1.1 is a special case of the polynomial decomposition from Sect. 1.2. Next, we show that both models can be viewed as a special case of X-rank decomposition: a powerful concept proposed recently in [2]. This concept provides a unified framework for studying properties of rank of the models (minimal r in (1) or (3)), and reformulate these questions in the language of algebraic geometry. Finally, we prove that underlying algebraic varieties are irreducible. As a consequence, the following results (proved in [2]) hold true.

1. For $\mathbb{K} = \mathbb{C}$, a generic (i.e., drawn with probability 1) collection of tensors (a generic polynomial), has the same rank, called complex generic rank $r_{gen,\mathbb{C}}$.
2. For $\mathbb{K} = \mathbb{R}$, the rank of a generic collection of tensors (or a generic polynomial) is at least $r_{gen,\mathbb{C}}$.
3. For the both real and complex fields, the maximal rank is at most twice the generic complex rank, i.e., $r_{max,\mathbb{R}}, r_{max,\mathbb{C}} \leq 2r_{gen,\mathbb{C}}$.

2 Polynomial Decompositions

2.1 Symmetric Tensors and Polynomials

There is a one-to-one correspondence between symmetric $\overbrace{n \times \cdots \times n}^{s}$ tensors and homogeneous polynomials of degree s [5]:

$$T(\mathbf{z}) = \mathcal{C} \times_1 \mathbf{z} \cdots \times_s \mathbf{z} \in \mathbb{K}_s[\mathbf{z}]. \tag{4}$$

Now assume that the tensor $\mathcal{C}$ admits a CP decomposition

$$\mathcal{C} = c_1 \mathbf{a}_1 \otimes \cdots \otimes \mathbf{a}_1 + \cdots + c_r \mathbf{a}_r \otimes \cdots \otimes \mathbf{a}_r. \tag{5}$$

Then, by (4), decomposition (5) is equivalent to the decomposition

$$T(\mathbf{z}) = c_1 \ell_1^d(\mathbf{z}) + \cdots + c_r \ell_r^d(\mathbf{z}), \tag{6}$$

where $\ell_k(\mathbf{z}) := \mathbf{a}_k^\top \mathbf{z}$ is a linear form. The decomposition (6) is called Waring decomposition [5].

2.2 Decomposition of Polynomials

By equivalence between (5) and (6), the system (1) can be rewritten as

$$\begin{aligned}
T^{(1)}(\mathbf{z}) &= c_{1,1} \ell_1(\mathbf{z}) + \cdots + c_{1,r} \ell_r(\mathbf{z}), \\
T^{(2)}(\mathbf{z}) &= c_{2,1} \ell_1^2(\mathbf{z}) + \cdots + c_{2,r} \ell_r^2(\mathbf{z}), \\
&\vdots \\
T^{(d)}(\mathbf{z}) &= c_{d,1} \ell_1^d(\mathbf{z}) + \cdots + c_{d,r} \ell_r^d(\mathbf{z}).
\end{aligned} \tag{7}$$

Now define the non-homogeneous polynomial $F \in \mathbb{K}_{\leq d}[\mathbf{z}]$ as

$$F(\mathbf{z}) = T^{(1)}(\mathbf{z}) + \cdots + T^{(d)}(\mathbf{z}), \tag{8}$$

Then from (7) it is easy to see that simultaneous Waring decomposition (7) (hence, the simultaneous symmetric CP decomposition (1)) is equivalent to the following problem: Given a multivariate polynomial $F \in \mathbb{K}_{\leq d}[\mathbf{z}]$, find minimal r, $g_k \in \mathbb{K}_{\leq d}[t]$ (univariate polynomials) and $\mathbf{a}_k \in \mathbb{K}^n$ such that

$$F(\mathbf{z}) = \sum_{k=1}^{r} g_k(\ell_k(\mathbf{z})), \tag{9}$$

where $\ell_k = \mathbf{a}_k^\top \mathbf{z}$ and $g_k(t) = c_{0,k} + c_{1,k} t + \cdots + c_{d,k} t^d$.

Note 1. Evidently, decomposition (9) is a special case of (3) with $m = 1$. Vice versa, any decomposition of the form (3) with $m = 1$ can be reduced to (9). Indeed, we can always assume that the linear transformation B is equal to $B = (1 \cdots 1)$, without loss of generality.

The authors are aware of only one work [1] which studies the theoretical properties of (9), and more precisely the maximal rank. Also, a practical algorithm for computation of the decomposition (9) was proposed recently in [17].

3 X-rank Decompositions

Here we recall a general definition of X-rank [2]. We will try to show how the decompositions in Sects. 1.1 and 1.2 may fit in the X-rank framework.

Let W be a vector space over $\mathbb{K}$, and $\mathbb{P}W$ be the corresponding projective space. Let $X \subset \mathbb{P}W$ be a nondegenerate projective variety and $\widehat{X}$ be an affine cone over X. Then for any v in $W \setminus \{0\}$ we can define the X-rank

$$\operatorname{rank}_X(v) = \min r : v = \widehat{x}_1 + \cdots + \widehat{x}_r, \quad \widehat{x}_k \in \widehat{X}. \tag{10}$$

The variety X (and its affine cone $\widehat{X}$) represents the set of rank-one terms.

Let us fix the variety X. The maximal X-rank is defined as

$$r_{max} := \max_{v \in W} \operatorname{rank}_X(v).$$

The typical ranks $r_{typ,k}$,

$$r_{typ,0} < \cdots < r_{typ,n_{typ}} \leq r_{max},$$

are all the numbers such that the sets $\{v \in W \mid \operatorname{rank}_X(v) = r_{typ,k}\}$ have non-empty interior in Euclidean topology (see also [2]). Informally speaking, the typical ranks are the X-ranks that appear with non-zero probability if we draw randomly the vector v from a continuous probability distribution on W.

For X-ranks, the following basic results are known [2].

Theorem 1 ([2], **Theorems 1, 3**). *If $r_{typ,0}$ is the smallest typical (real or complex) rank, then $r_{max} \leq 2r_{typ,0}$.*

Theorem 2 ([2], **page 1**). *If $\mathbb{K} = \mathbb{C}$ and X is an irreducible variety, then there exists a unique typical rank (called generic rank, and denoted by $r_{gen}^{\mathbb{C}}$).*

Theorem 3 ([2], **Theorem 2**). *If $\mathbb{K} = \mathbb{R}$ and X is an irreducible variety, and $X_{\mathbb{C}}$ is its complexification, then the smallest typical real rank equals the generic rank, i.e., $r_{typ,0}^{\mathbb{R}} = r_{gen}^{\mathbb{C}}$.*

It is easy to show that decompositions (3) and (9) can be viewed as special cases of (10), as pointed out below.

1. **Rank-One Polynomials** (9): take $W = \mathbb{K}_{\leq d}[\mathbf{z}]$ and

$$\widehat{X} := \{f(\mathbf{z}) \in W \mid f(\mathbf{z}) = g(\mathbf{a}^\top \mathbf{z}), g(t) = \sum_{j=0}^{d} c_j t^j, \quad \mathbf{a} \in \mathbb{K}^n\}. \tag{11}$$

2. **Rank-One Polynomial Maps** (3): take $W = (\mathbb{K}_{\leq d}[\mathbf{z}])^m$ and

$$\widehat{X} := \{F(\mathbf{z}) \in W \mid F(\mathbf{z}) = \mathbf{b}g(\mathbf{a}^\top \mathbf{z}), g(t) = \sum_{j=0}^{d} c_j t^j, \mathbf{a} \in \mathbb{K}^n, \mathbf{b} \in \mathbb{K}^m\}. \tag{12}$$

Although we expressed the rank-one sets in (11) and (12), it is not immediate that we can use Theorems 1–3. We still need to prove that these sets are algebraic varieties and are irreducible. This is exactly the goal of the following section.

4   Irreducibility and Generic Rank

4.1   Algebraic Description

Here we provide an alternative (algebraic) description of the sets (11) and (12). For a finite-dimensional vector space V over $\mathbb{K}$ we denote by S^dV the space of symmetric multilinear forms. (In particular, if V is isomorphic to $\mathbb{K}^n$, then S^dV is isomorphic to $\mathbb{K}_d[z_1, \ldots, z_n]$).

Rank-One Polynomials. Consider the following map.

$$
\begin{aligned}
\psi_1 : V \times \mathbb{K}^d &\to \overbrace{V \oplus S^2V \oplus \cdots \oplus S^dV}^{W_1 :=} \\
(\mathbf{a}, (c_1, \cdots, c_d)) &\mapsto (c_1\mathbf{a}, c_2\mathbf{a}^2, \ldots, c_d\mathbf{a}^d).
\end{aligned}
\tag{13}
$$

Next, we define $\widehat{X}_1$ as the image of ψ_1:

$$
\widehat{X}_1 := \psi_1(V \times \mathbb{K}^d).
\tag{14}
$$

It is easy to see that (14) corresponds to (11) (with the constant part of the polynomials removed).

Rank-One Polynomial Maps. Now consider the following map.

$$
\begin{aligned}
\psi_m : \mathbb{K}^m \times V \times \mathbb{K}^d &\to \overbrace{(V \oplus S^2V \oplus \cdots \oplus S^dV)^m}^{W_m :=} \\
(\mathbf{b}, \mathbf{a}, (c_1, \cdots, c_d)) &\mapsto
\begin{matrix}
(b_1c_1\mathbf{a}, & b_1c_2\mathbf{a}^2, & \cdots & b_1c_d\mathbf{a}^d, \\
\vdots & & & \vdots \\
b_mc_1\mathbf{a}, & b_mc_2\mathbf{a}^2, & \cdots & b_mc_d\mathbf{a}^d).
\end{matrix}
\end{aligned}
$$

Next, we define $\widehat{X}_m$ as the image of ψ_m:

$$
\widehat{X}_m := \psi_m(\mathbb{K}^m \times V \times \mathbb{K}^d) = \mathbb{K}^m \otimes \widehat{X}_1,
\tag{15}
$$

where $\widehat{X}_1$ is defined in (14). It is easy to see that (15) corresponds to (12) (with the constant parts of the polynomial maps removed).

Note 2. Since $\widehat{X}_1 = \psi_1(V \times \mathbb{K}^d) \subset W_1$ is the affine cone of a projective variety $X_1 \subset \mathbb{P}W_1$, then $\widehat{X}_m$ is the affine cone of the image of the Segre embedding $f : \mathbb{P}^{m-1} \times X_1 \to \mathbb{P}(\mathbb{K}^m \otimes W_1)$ defined by $f([b], [x]) = [b \otimes x]$, where $x \in \widehat{X}_m \subset W_m$.

4.2   Bundle Description and Irreducibility of X

Proposition 1. *The set $\widehat{X}_1$ defined in (14) is an irreducible affine algebraic variety. Consequently, it is an affine cone over a projective variety.*

Proof. Let $\mathbb{P}V$ be a complex projective space of dimension $n - 1$. Define ϕ : $\mathbb{P}V \to \mathbb{P}V \times \cdots \times \mathbb{P}S^d V$ by $\phi([v]) = ([v], \ldots, [v^d])$, then ϕ is an embedding (*i.e.*, $\mathbb{P}V$ is isomorphic to $\phi(\mathbb{P}V)$). Denote the image of ϕ by M.

Let T_W be the tautological line bundle (called canonical line bundle in [15, Sect. 2]) on a projective space $\mathbb{P}W$, *i.e.*, $T_W = \{([w], w) \in \mathbb{P}W \times W \,|\, w \in W\}$, where $\mathbb{P}W \times W$ is a trivial vector bundle over $\mathbb{P}W$.

Next, let $p_i : \mathbb{P}V \times \cdots \times \mathbb{P}S^d V \to \mathbb{P}S^i V$ be the i-th natural projection, then $D = \bigoplus_{i=1}^{d} p_i^* T_{S^i V}$ is a vector bundle of rank d over $\mathbb{P}V \times \cdots \times \mathbb{P}S^d V$, where p_i^* is the pull-back map induced by p_i [15]. Let E denote the restriction of D on M, thus E is a closed sub-variety of D.

Finally, define $\widetilde{\psi}_1 : E \to V \oplus \cdots \oplus S^d V$ by

$$\widetilde{\psi}_1([v], \ldots, [v^d], c_1 v, \ldots, c_d v^d) = (c_1 v, \ldots, c_d v^d).$$

It is easy to see that $\widehat{X}_1 = \widetilde{\psi}_1(E)$.

Since each $\mathbb{P}S^i V$ is complete ([14, Definition 7.1]) by [14, Theorem 7.22], $\mathbb{P}V \times \cdots \times \mathbb{P}S^d V$ is complete by [14, Sect. 7.5], then

$$\bigoplus_{i=1}^{d} p_i^*(\mathbb{P}S^i V \times S^i V) \to V \oplus \cdots \oplus S^d V \tag{16}$$
$$(([\alpha_1], \ldots, [\alpha_d]), \beta_1, \ldots, \beta_d) \mapsto (\beta_1, \ldots, \beta_d)$$

is proper by [14, Sect. 7.16a], where $\alpha_i, \beta_i \in S^i V$. Thus by [14, Sect. 7.17], the restriction to $D \to V \oplus \cdots \oplus S^d V$ is proper, and then $\widetilde{\psi}_1 : E \to V \oplus \cdots \oplus S^d V$ is proper. By definition of properness, $\widetilde{\psi}_1$ is universally closed, so $\widehat{X}_1 = \widetilde{\psi}_1(E)$ is closed, *i.e.*, $\widehat{X}_1$ is an affine variety. Because E is irreducible, $\widehat{X}_1$ is also irreducible.   □

By Note 2 from Sect. 4.1 we have the following corollary.

Corollary 1. $\widehat{X}_m$ *defined in* (15) *is also irreducible, and is an affine cone over a projective variety.*

Finally, from Proposition 1 and Corollary 1, we have that Theorems 1–3 can be applied for decompositions (9) and (3). In particular, for these decompositions there exists a complex generic rank (equal to the minimal real typical rank).

4.3   Generic Rank for Bivariate Polynomials

Finally, in the case $n = 2$ and $m = 1$, the variety X_1 (corresponding to the affine cone $\widehat{X}_1$ defined in (14)) is a special case of the rational normal scroll [10,12]. Using the results [3] on dimension of the r-th secant variety $\sigma_r(X)$ of a rational normal scroll X, we can explicitly find the complex generic rank for this case.

Proposition 2. *The generic rank for bivariate polynomials is equal to*

$$r_{gen} := \left\lceil \frac{2d + 7}{2} - \frac{\sqrt{8d + 17}}{2} \right\rceil - 1.$$

Proof. By [3, p. 359], the dimension of $\sigma_r(X_1) \subset \mathbb{P}^N$ is $\min\{N, N - \frac{(d-r+1)(d-r+2)}{2} + r\}$ (note that there is an incorrect sign in the original paper [3]). Thus the generic rank r_{gen} is the maximal $r \in \{1, \ldots, d\}$ such that

$$N - \frac{(d-r+2)(d-r+3)}{2} + r - 1 < N$$

which is equivalent to $r < \frac{2d+7}{2} - \frac{\sqrt{8d+17}}{2}$, *i.e.*, $r_{gen} = \left\lceil \frac{2d+7}{2} - \frac{\sqrt{8d+17}}{2} \right\rceil - 1$. □

References

1. Białynicki-Birula, A., Schinzel, A.: Representations of multivariate polynomials as sums of polynomials in linear forms. Colloq. Math. **112**(2), 201–233 (2008)
2. Blekherman, G., Teitler, Z.: On maximum, typical and generic ranks. Math. Ann. **362**(3–4), 1021–1031 (2015)
3. Catalano-Johnson, M.L.: The possible dimensions of the higher secant varieties. Am. J. Math. **118**, 355–361 (1996)
4. Comon, P.: Contrasts, independent component analysis, and blind deconvolution. Int. J. Adapt. Control Sig. Proc. **18**(3), 225–243 (2004)
5. Comon, P., Golub, G., Lim, L.H., Mourrain, B.: Symmetric tensors and symmetric tensor rank. SIAM. J. Matrix Anal. Appl. **30**(3), 1254–1279 (2008)
6. Comon, P., Jutten, C.: Handbook of Blind Source Separation, Independent Component Analysis and Applications. Academic Press (Elsevier), Oxford (2010)
7. Comon, P., Rajih, M.: Blind identification of under-determined mixtures based on the characteristic function. Sig. Process. **86**(9), 3334–3338 (2006)
8. De Lathauwer, L., Comon, P., De Moor, B., Vandewalle, J.: ICA algorithms for 3 sources and 2 sensors. In: Proceedings of the IEEE Signal Processing Workshop on Higher-Order Statistics, 1999, pp. 111–120 (1999)
9. De Lathauwer, L.: Algebraic methods after prewhitening. In: Comon, P., Jutten, C. (eds.) Handbook of Blind Source Separation, Independent Component Analysis and Applications, p. 831. Academic Press (Elsevier), Oxford (2010)
10. De Poi, P.: On higher secant varieties of rational normal scrolls. Le Matematiche **51**(1), 3–21 (1997)
11. Dreesen, P., Ishteva, M., Schoukens, J.: Decoupling multivariate polynomials using first-order information. SIAM. J. Matrix Anal. Appl. **36**(2), 864–879 (2015)
12. Harris, J.: Algebraic Geometry: A First Course. Springer, Berlin (1992)
13. Kibangou, A.Y., Favier, G.: Tensor analysis-based model structure determination and parameter estimation for block-oriented nonlinear systems. IEEE J. Sel. Topics Sig. Proc. **4**(3), 514–525 (2010)
14. Milne, J.: Algebraic Geometry version 6.00, 24 August 2014. http://www.jmilne.org/math/CourseNotes/AG.pdf
15. Milnor, J., Stasheff, J.: Characteristic Classes. Annals of Mathematics Studies. Princeton University Press, Princeton (1974)
16. Tiels, K., Schoukens, J.: From coupled to decoupled polynomial representations in parallel Wiener-Hammerstein models. In: 52nd IEEE Conference on Decision and Control, 10–13 December 2013, Florence, Italy, pp. 4937–4942 (2013)

17. Usevich, K.: Decomposing multivariate polynomials with structured low-rank matrix completion. In: 21st International Symposium on Mathematical Theory of Networks and Systems, 7–11 July 2014, Groningen, The Netherlands, pp. 1826–1833 (2014)
18. Van Mulders, A., Vanbeylen, L., Usevich, K.: Identification of a block-structured model with several sources of nonlinearity. In: Proceedings of the 14th European Control Conference (ECC 2014), pp. 1717–1722 (2014)

Rank Splitting for CANDECOMP/PARAFAC

Anh-Huy Phan[1]([⊠]), Petr Tichavský[2], and Andrzej Cichocki[1]

[1] Lab for Advanced Brain Signal Processing, Brain Science Institute - RIKEN,
Saitama, Japan
phan@brain.riken.jp
[2] Institute of Information Theory and Automation, Prague, Czech Republic
tichavsk@utia.cas.cz

Abstract. CANDECOMP/PARAFAC (CP) approximates multiway data by a sum of rank-1 tensors. Our recent study has presented a method to rank-1 tensor deflation, i.e. sequential extraction of rank-1 tensor components. In this paper, we extend the method to block deflation problem. When at least two factor matrices have full column rank, one can extract two rank-1 tensors simultaneously, and rank of the data tensor is reduced by 2. For decomposition of order-3 tensors of size $R \times R \times R$ and rank-R, the block deflation has a complexity of $\mathcal{O}(R^3)$ per iteration which is lower than the cost $\mathcal{O}(R^4)$ of the ALS algorithm for the overall CP decomposition.

Keywords: Canonical polyadic decomposition · PARAFAC · Tensor deflation

1 Introduction

An important property in matrix factorisations like eigenvalue decomposition, is that rank-1 matrix components can be sequentially estimated via a deflation method, such as the power iteration method. The matrix deflation procedure is possible because subtracting the best rank-1 term from a matrix reduces the matrix rank. Unfortunately, this sequential extraction procedure in general is not applicable to decompose a rank-R tensor [1].

In our recent study [2,3], we have introduced a tensor decomposition which is able to extract a rank-1 tensor from a high rank tensor. The method is based on the rank-1 plus multilinear-$(R-1, R-1, R-1)$ block tensor decomposition, but with a smaller number of parameters, basically two vectors per modes. This paper extends the rank-1 tensor extraction to block tensor deflation or rank splitting which splits a high rank-R tensor into two tensors of smaller ranks. In particular, we develop an alternating subspace update (ASU) algorithm to

P. Tichavský—The work of P. Tichavský was supported by The Czech Science Foundation through Project No. 14-13713S.
A. Cichocki—Also affiliated with the EE Dept., Warsaw University of Technology and with Systems Research Institute, Polish Academy of Science, Poland.

© Springer International Publishing Switzerland 2015
E. Vincent et al. (Eds.): LVA/ICA 2015, LNCS 9237, pp. 31–40, 2015.
DOI: 10.1007/978-3-319-22482-4_4

extract a multilinear rank-(2,2,2) tensor from a rank-R tensor. Since decomposition of a $2 \times 2 \times 2$ tensor can be found in closed-form, we can straightforwardly obtain the desired rank-1 components. The proposed algorithm estimates only 4 vectors and two scalars per dimension with a computational complexity of $\mathcal{O}(R^3)$. Moreover, it also requires a lower space cost than algorithms for the ordinary CANDECOMP/PARAFAC (CPD).

The paper is organised as follows. A tensor decomposition for the block tensor deflation or rank splitting is presented in Sect. 2. The proposed algorithm is presented in Sect. 3. Simulations in Sect. 4 will verify validity and performance of the proposed algorithm. Section 5 concludes the paper.

2 Preliminaries

Throughout the paper, we shall denote tensors by bold calligraphic letters, e.g., $\mathcal{A} \in \mathbb{R}^{I_1 \times I_2 \times \cdots \times I_N}$, matrices by bold capital letters, e.g., $\mathbf{A} = [a_1, a_2, \ldots, a_R] \in \mathbb{R}^{I \times R}$, and vectors by bold italic letters, e.g., a_j. The Kronecker product is denoted by $\otimes$. Inner product of two tensors is denoted by $\langle \mathcal{X}, \mathcal{Y} \rangle = \mathrm{vec}(\mathcal{X})^T \mathrm{vec}(\mathcal{Y})$. Contraction between two tensors along modes-m, where $m = [m_1, \ldots, m_K]$, is denoted by $\langle \mathcal{X}, \mathcal{Y} \rangle_m$, whereas $\langle \mathcal{X}, \mathcal{Y} \rangle_{-n}$ represents contraction along all modes but mode-n [4].

The mode-n matricization of tensor $\mathcal{Y}$ is denoted by $\mathbf{Y}_{(n)}$. The mode-n multiplication of a tensor $\mathcal{Y} \in \mathbb{R}^{I_1 \times I_2 \times \cdots \times I_N}$ by a matrix $\mathbf{U} \in \mathbb{R}^{I_n \times R}$ is denoted by $\mathcal{Z} = \mathcal{Y} \times_n \mathbf{U} \in \mathbb{R}^{I_1 \times \cdots \times I_{n-1} \times R \times I_{n+1} \times \cdots \times I_N}$. Products of a tensor $\mathcal{Y}$ with a set of N matrices $\{\mathbf{U}^{(n)}\} = \{\mathbf{U}^{(1)}, \mathbf{U}^{(2)}, \ldots, \mathbf{U}^{(N)}\}$ are denoted by

$$\mathcal{Y} \times \{\mathbf{U}^{(n)}\} \triangleq \mathcal{Y} \times_1 \mathbf{U}^{(1)} \times_2 \mathbf{U}^{(2)} \cdots \times_N \mathbf{U}^{(N)}.$$

A tensor $\mathcal{X} \in \mathbb{R}^{I_1 \times I_2 \times \cdots \times I_N}$ is said in Kruskal form if $\mathcal{X} = \sum_{r=1}^{R} \lambda_r\, a_r^{(1)} \circ a_r^{(2)} \circ \cdots \circ a_r^{(N)}$, where "$\circ$" denotes the outer product, $\mathbf{A}^{(n)} = [a_1^{(n)}, a_2^{(n)}, \ldots, a_R^{(n)}] \in \mathbb{R}^{I_n \times R}$ are factor matrices, $a_r^{(n)T} a_r^{(n)} = 1$, for $r = 1, \ldots, R$ and $n = 1, \ldots, N$, and $\lambda_1 \geq \lambda_2 \geq \cdots \geq \lambda_R > 0$.

A tensor $\mathcal{X} \in \mathbb{R}^{I_1 \times I_2 \times \cdots \times I_N}$ has multilinear rank-$(R_1, R_2, \ldots, R_N)$ if $rank(\mathbf{X}_{(n)}) = R_n \leq I_n$ for $n = 1, \ldots, N$, and can be expressed in the Tucker form as

$$\mathcal{X} = \sum_{r_1=1}^{R_1} \sum_{r_2=1}^{R_2} \cdots \sum_{r_n=1}^{R_N} g_{r_1 r_2 \ldots r_N}\, a_{r_1}^{(1)} \circ a_{r_2}^{(2)} \circ \cdots \circ a_{r_N}^{(N)}, \tag{1}$$

where $\mathcal{G} = [g_{r_1 r_2 \ldots r_N}]$, and $\mathbf{A}^{(n)}$ are of full column rank. For compact expression, $[\![\lambda; \{\mathbf{A}^{(n)}\}]\!]$ denotes a Kruskal tensor, where $[\![\mathcal{G}; \{\mathbf{A}^{(n)}\}]\!]$ represents a Tucker tensor.

The main focus of this paper is a block deflation which splits a rank-R CPD into two sub rank-K and rank-$(R - K)$ CPDs. This tensor decomposition is a particular case of the block tensor decomposition [5] but with only two blocks

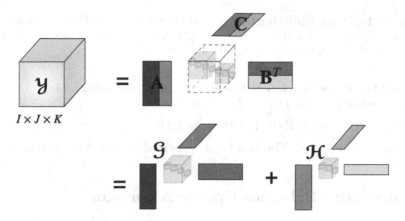

Fig. 1. Rank splitting for the CP decomposition of a rank-R tensor into two multilinear rank-$(K, \ldots, K)$ and rank-$(R - K, \ldots, R - K)$ tensors $\mathcal{G}$ and $\mathcal{H}$.

of multilinear rank-(K, K, K) and rank-$(R - K, R - K, R - K)$ as illustrated in Fig. 1. That is

$$\mathcal{Y} \approx [\![\mathcal{G}; \mathbf{U}^{(1)}, \mathbf{U}^{(2)}, \ldots, \mathbf{U}^{(N)}]\!] + [\![\mathcal{H}; \mathbf{V}^{(1)}, \mathbf{V}^{(2)}, \ldots, \mathbf{V}^{(N)}]\!] + \mathcal{E} \qquad (2)$$

where $\mathbf{U}^{(n)}$ and $\mathbf{V}^{(n)}$ are matrices of size $I_n \times K$ and $I_n \times (R - K)$, respectively. Following this tensor decomposition, decomposition of a rank-R tensor can proceed simultaneously through decompositions of sub-tensors with smaller ranks.

For this kind of tensor decomposition and block tensor deflation, we can use the ALS algorithm [5] or the non-linear least squares (NLS) algorithm [6] developed for the multilinear rank-(L_r, M_r, N_r) block tensor decomposition with two blocks. However, these existing algorithms are expensive due to a large number of parameters of the two core tensors $\mathcal{G}$ and $\mathcal{H}$. The proposed algorithm will estimate only four vectors of length R per dimension whereas the core tensors $\mathcal{G}$ and $\mathcal{H}$ need not be estimated.

We will first introduce an orthogonal normalisation for the block tensor deflation, then state the correctness of the proposed deflation scheme.

Lemma 1 (Orthogonal Normalization for Rank Splitting). *Given a decomposition of $\mathcal{Y}$ as $\mathcal{Y} \approx [\![\mathcal{G}; \mathbf{U}^{(1)}, \mathbf{U}^{(2)}, \ldots, \mathbf{U}^{(N)}]\!] + [\![\mathcal{H}; \mathbf{V}^{(1)}, \mathbf{V}^{(2)}, \ldots, \mathbf{V}^{(N)}]\!]$, where $\mathbf{U}^{(n)} \in \mathbb{R}^{I_n \times (K)}$ and $\mathbf{V}^{(n)} \in \mathbb{R}^{I_n \times (R-K)}$, $K \leq R - K$, one can construct an equivalent decomposition, denoted by tildas, which has the same approximation error, such that*

- $[\![\mathcal{G}; \{\mathbf{U}^{(n)}\}]\!] = [\![\widetilde{\mathcal{G}}; \{\widetilde{\mathbf{U}}^{(n)}\}]\!]$, $[\![\mathcal{H}; \{\mathbf{V}^{(n)}\}]\!] = [\![\widetilde{\mathcal{H}}; \{\widetilde{\mathbf{V}}^{(n)}\}]\!]$
- $\widetilde{\mathbf{U}}^{(n)}$ *and* $\widetilde{\mathbf{V}}^{(n)}$ *are orthogonal, i.e.,* $(\widetilde{\mathbf{U}}^{(n)})^T \widetilde{\mathbf{U}}^{(n)} = \mathbf{I}_K$ *and* $(\widetilde{\mathbf{V}}^{(n)})^T \widetilde{\mathbf{V}}^{(n)} = \mathbf{I}_{R-K}$.
- *and obey conditions* $(\widetilde{\mathbf{U}}^{(n)})^T \widetilde{\mathbf{V}}^{(n)} = [\text{diag}\{\boldsymbol{\sigma}_n\}, \mathbf{0}_{R-2K}]$ *where* $\boldsymbol{\sigma}_n = [\sigma_{n,1}, \ldots, \sigma_{n,K}] \in \mathbb{R}^K$ *and* $0 \leq \sigma_{n,r} < 1$.

Theorem 1 (Rank Splitting). *A rank-R tensor $\mathcal{Y} = [\![\boldsymbol{\beta}; \{\mathbf{B}^{(n)}\}]\!]$ has an exact decomposition $\mathcal{Y} = [\![\mathcal{G}; \mathbf{U}^{(1)}, \ldots, \mathbf{U}^{(N)}]\!] + [\![\mathcal{H}; \mathbf{V}^{(1)}, \ldots, \mathbf{V}^{(N)}]\!]$ as in (2) where $\mathbf{U}^{(n)} \in \mathbb{R}^{I_n \times K}$ and $\mathbf{V}^{(n)} \in \mathbb{R}^{I_n \times (R-K)}$, $K \leq R - K$ and*

- *at least two factor matrices $\mathbf{B}^{(n)} \in \mathbb{R}^{I_n \times R}$ are of full column rank,*
- *$\mathcal{G}$ has multilinear rank-$(K, \ldots, K)$.*

Then $\mathcal{G}$ is a tensor of rank-K and $\mathcal{H}$ of rank $(R - K)$.

Proofs of Lemma 1 and Theorem 1 are provided in the full version of this paper [7].

3   Alternating Subspace Update Algorithm

In this section, we consider order-3 tensors of size $R \times R \times R$. Tensors of larger and unequal sizes should be compressed to this size using the Tucker decomposition [8–10]. We will develop an algorithm for the block tensor deflation which reduces the rank by $K = 2$. For this particular case, the core tensor $\mathcal{G}$ is size of $2 \times 2 \times 2$, and the core tensor $\mathcal{H}$ of size $(R - 2) \times (R - 2 \times (R - 2)$. The factor matrices $\mathbf{U}^{(n)}$ and $\mathbf{V}^{(n)}$ are of size $R \times 2$ and $R \times (R - 2)$, respectively. The rank-2 block deflation has an advantage over the rank-1 tensor deflation when factor matrices have two nearly collinear components.

We denote matrices $\bar{\mathbf{V}}^{(n)} = [\boldsymbol{v}_1^{(n)}, \boldsymbol{v}_2^{(n)}]$ which comprise the first two columns of $\mathbf{V}^{(n)}$, and perform reparameterization of $\mathbf{U}^{(n)}$ as

$$\mathbf{U}^{(n)} = \mathbf{W}^{(n)} \operatorname{diag}(\boldsymbol{\xi}_n) + \bar{\mathbf{V}}^{(n)} \operatorname{diag}(\boldsymbol{\sigma}_n), \tag{3}$$

where $\boldsymbol{\xi}_n = [\xi_{n1}, \xi_{n2}]^T$, $\xi_{nr} = \sqrt{1 - \sigma_{nr}^2}$, and $\mathbf{W}^{(n)} = [\boldsymbol{w}_1^{(n)}, \boldsymbol{w}_2^{(n)}]$ of size $R \times 2$. $[\mathbf{W}^{(n)}, \mathbf{V}^{(n)}]$ are orthonormal matrices of size $R \times R$, i.e., $[\mathbf{W}^{(n)}, \mathbf{V}^{(n)}]^T [\mathbf{W}^{(n)}, \mathbf{V}^{(n)}] = \mathbf{I}_R$.

Consider the following criterion to be minimized,

$$D = \frac{1}{2} \|\mathcal{Y} - \mathcal{G} \times \{\mathbf{U}^{(n)}\} - \mathcal{H} \times \{\mathbf{V}^{(n)}\}\|_F^2. \tag{4}$$

We will later simplify the objective function in (4) by replacing the core tensors by their closed-form expressions and applying the above reparameterization. The objective function will finally depend only on $\mathbf{W}^{(n)}$, $\bar{\mathbf{V}}^{(n)}$ and $\boldsymbol{\sigma}_n$ for $n = 1, 2, 3$.

3.1   Closed-Form Expressions for the Core Tensors

From the model (2) and the cost function D in (4), we can derive closed-form expressions for $\mathcal{H}$ and $\mathcal{G}$ as

$$\mathcal{H} = \mathcal{Y} \times \{\mathbf{V}^{(n)T}\} - \mathcal{G} \times \left\{ \begin{bmatrix} \operatorname{diag}(\boldsymbol{\sigma}_n) \\ \mathbf{0}_{(R-2) \times 2} \end{bmatrix} \right\}, \tag{5}$$

$$\mathcal{G} = \left(\mathcal{Y} \times \{\mathbf{U}^{(n)T}\} - \left(\mathcal{Y} \times \{\bar{\mathbf{V}}^{(n)T}\} \right) \circledast \mathcal{S} \right) \oslash (1 - \mathcal{S} \circledast \mathcal{S}), \tag{6}$$

where $\mathcal{S} = \boldsymbol{\sigma}_1 \circ \boldsymbol{\sigma}_2 \circ \boldsymbol{\sigma}_3$ is a rank-1 tensor of size $2 \times 2 \times 2$, $\circledast$ and $\oslash$ represent the Hadamard (element-wise) product and division, respectively.

We replace $\mathcal{H}$ in the cost function (4) by its closed-form in (5), and rewrite D as

$$D = \frac{1}{2}\|\mathcal{Y} - \mathcal{Y} \times \left\{\mathbf{V}^{(n)}\mathbf{V}^{(n)T}\right\} - \mathcal{G} \times \{\mathbf{U}^{(n)}\} + \mathcal{G} \times \{\bar{\mathbf{V}}^{(n)} \operatorname{diag}(\boldsymbol{\sigma}_n)\}\|_F^2$$

$$= \frac{1}{2}\left(\|\mathcal{Y}\|_F^2 - \|\mathcal{Y} \times \left\{\mathbf{V}^{(n)}\mathbf{V}^{(n)T}\right\}\|_F^2 - \langle\mathcal{G} \circledast (1 - \mathcal{S} \circledast \mathcal{S}), \mathcal{G}\rangle\right). \quad (7)$$

For an index $n \in \{1,2,3\}$, define n_1 and n_2 with $n_1 < n_2$ as its complement in $\{1,2,3\}$, i.e., $\{n, n_1, n_2\} = \{1,2,3\}$. Put $t_{r,s}^{(n)} = \mathcal{Y} \times_{n_1} \boldsymbol{u}_r^{(n_1)T} \times_{n_2} \boldsymbol{u}_s^{(n_2)T}$, $z_{r,s}^{(n)} = \mathcal{Y} \times_{n_1} \boldsymbol{v}_r^{(n_1)T} \times_{n_2} \boldsymbol{v}_s^{(n_2)T}$, and $d_{r,s}^{(n)} = t_{r,s}^{(n)} - z_{r,s}^{(n)} \sigma_{n_1,r} \sigma_{n_2,s}$. The objective function in (7) can be expressed as

$$D = \frac{1}{2}\left(\|\mathcal{Y}\|_F^2 - \|\mathcal{Y} \times \left\{\mathbf{V}^{(n)}\mathbf{V}^{(n)T}\right\}\|_F^2 - \sum_{r_1,r_2,r_3=1}^{2} \frac{(\xi_{1,r_1} \boldsymbol{w}_{r_1}^{(1)T} t_{r_2,r_3}^{(1)} + \sigma_{1,r_1} \boldsymbol{v}_{r_1}^{(1)T} d_{r_2,r_3}^{(1)})^2}{1 - \sigma_{1,r_1}^2 \sigma_{2,r_2}^2 \sigma_{3,r_3}^2}\right). \quad (8)$$

3.2   Estimation of $\boldsymbol{\sigma}_n$

We begin with deriving update rules for $\boldsymbol{\sigma}_1 = [\sigma_{1,1}, \sigma_{1,2}]$. As shown in the cost function in (8), the parameters $\boldsymbol{\sigma}_1$ involve only the third term. In order to estimate $\boldsymbol{\sigma}_1$, we keep other parameters fixed. Then minimization of the cost function (8) leads to maximization of the following function of $\boldsymbol{\sigma}_1$

$$\max_{\sigma_{1,1},\sigma_{1,2}} \sum_{r_1=1}^{2}\sum_{r_2=1}^{2}\sum_{r_3=1}^{2} \frac{(\xi_{1,r_1} \boldsymbol{w}_{r_1}^{(1)T} t_{r_2,r_3}^{(1)} + \sigma_{1,r_1} \boldsymbol{v}_{r_1}^{(1)T} d_{r_2,r_3}^{(1)})^2}{1 - \sigma_{1,r_1}^2 \sigma_{2,r_2}^2 \sigma_{3,r_3}^2}. \quad (9)$$

Each σ_{1,r_1} is found as $\sigma_{1,r_1} = 1/\sqrt{1 + x_{r_1}^2}$ where x_{r_1} is solution to the problem

$$x_{r_1} = \arg\max_{x} \sum_{r_2=1}^{2}\sum_{r_3=1}^{2} \frac{(\alpha_{r_2,r_3} x + \beta_{r_2,r_3})^2}{x^2 + 1 - \sigma_{2,r_2}^2 \sigma_{3,r_3}^2} \quad (10)$$

$\alpha_{r_2,r_3} = \boldsymbol{w}_{r_1}^{(1)T} t_{r_2,r_3}^{(1)}$ and $\beta_{r_2,r_3} = \boldsymbol{v}_{r_1}^{(1)T} d_{r_2,r_3}^{(1)}$. The optimal x_{r_1} is a root of a polynomial of degree-8. The others $\sigma_{n,r}$ can be estimated similarly.

3.3   Estimation of Orthogonal Components $\mathbf{W}^{(n)}$ and $\mathbf{V}^{(n)}$

This section presents update rules which preserve orthogonality constrains on $\mathbf{W}^{(n)}$ and $\mathbf{V}^{(n)}$. Indeed we only need to update $\mathbf{W}^{(n)}$ and the first two column vectors $\bar{\mathbf{V}}^{(n)} = [\boldsymbol{v}_1^{(n)}, \boldsymbol{v}_2^{(n)}]$, whereas the last $(R-4)$ columns $[\boldsymbol{v}_3^{(n)}, \ldots, \boldsymbol{v}_{R-2}^{(n)}]$ are chosen as arbitrary orthogonal complement to $[\mathbf{W}^{(n)}, \bar{\mathbf{V}}^{(n)}]$.

Since $\mathbf{V}^{(n)}\mathbf{V}^{(n)T} = \mathbf{I}_R - \mathbf{W}^{(n)}\mathbf{W}^{(n)T}$, we have

$$\|\mathcal{Y} \times \left\{\mathbf{V}^{(n)}\mathbf{V}^{(n)T}\right\}\|_F^2 = \operatorname{tr}(\boldsymbol{\Phi}_n) - \operatorname{tr}(\mathbf{W}^{(n)T}\boldsymbol{\Phi}_n\mathbf{W}^{(n)}) \quad (11)$$

where $\boldsymbol{\Phi}_n = \mathbf{Y}_{(n)} \left(\bigotimes_{k \neq n} \mathbf{V}^{(n)} \mathbf{V}^{(k)T} \right) \mathbf{Y}_{(n)}^T$ are matrices of size $R \times R$. The cost function in (8) is rewritten as

$$D = \frac{1}{2} \left(\|\mathcal{Y}\|_F^2 - \mathrm{tr}(\boldsymbol{\Phi}_n) + \sum_{r=1}^{2} \boldsymbol{w}_r^{(n)T} \mathbf{Q}_{n,r} \boldsymbol{w}_r^{(n)} - \boldsymbol{v}_r^{(n)T} \mathbf{F}_{n,r} \boldsymbol{v}_r^{(n)} - 2 \boldsymbol{w}_r^{(n)T} \mathbf{K}_{n,r} \boldsymbol{v}_r^{(n)} \right) \quad (12)$$

where

$$\mathbf{Q}_{n,r} = \boldsymbol{\Phi}_n - \xi_{n,r}^2 \sum_{k,l} \frac{\boldsymbol{t}_{k,l}^{(n)} \boldsymbol{t}_{k,l}^{(n)T}}{1 - \sigma_{n,r}^2 \sigma_{n_1,k}^2 \sigma_{n_2,l}^2}, \qquad \mathbf{F}_{n,r} = \sigma_{n,r}^2 \sum_{k,l} \frac{\boldsymbol{d}_{k,l}^{(n)} \boldsymbol{d}_{k,l}^{(n)T}}{1 - \sigma_{n,r}^2 \sigma_{n_1,k}^2 \sigma_{n_2,l}^2}, \quad (13)$$

$$\mathbf{K}_{n,r} = \xi_{n,r} \sigma_{n,r} \sum_{k,l} \frac{\boldsymbol{t}_{k,l}^{(n)} \boldsymbol{d}_{k,l}^{(n)T}}{1 - \sigma_{n,r}^2 \sigma_{n_1,k}^2 \sigma_{n_2,l}^2}. \quad (14)$$

It follows that $\mathbf{W}^{(n)}$ and $\bar{\mathbf{V}}^{(n)}$ are solutions to the following quadratic optimisation

$$\min f(\mathbf{W}^{(n)}, \bar{\mathbf{V}}^{(n)}) = \frac{1}{2} \left(\sum_{r=1}^{2} \boldsymbol{w}_r^{(n)T} \mathbf{Q}_{n,r} \boldsymbol{w}_r^{(n)} - \boldsymbol{v}_r^{(n)T} \mathbf{F}_{n,r} \boldsymbol{v}_r^{(n)} - 2 \boldsymbol{w}_r^{(n)T} \mathbf{K}_{n,r} \boldsymbol{v}_r^{(n)} \right) \quad (15)$$

subject to $[\mathbf{W}^{(n)} \bar{\mathbf{V}}^{(n)}]^T [\mathbf{W}^{(n)} \bar{\mathbf{V}}^{(n)}] = \mathbf{I}_4$.

Following the Crank-Nicholson-like scheme [11], we can update the orthogonal matrices $\mathbf{X}_n = [\mathbf{W}^{(n)}, \bar{\mathbf{V}}^{(n)}]$ with $\mathbf{X}_n^T \mathbf{X}_n = \mathbf{I}_4$ using the following rules

$$\mathbf{X}_n \leftarrow \mathbf{X}_n - 2\tau [\mathbf{G}_f, \mathbf{X}_n] \left(\mathbf{I}_8 + \tau \begin{bmatrix} \mathbf{X}_n^T \mathbf{G}_f & \mathbf{I}_4 \\ -\mathbf{G}_f^T \mathbf{G}_f & -\mathbf{G}_f^T \mathbf{X}_n \end{bmatrix} \right)^{-1} \begin{bmatrix} \mathbf{I}_4 \\ -\mathbf{G}_f^T \mathbf{X}_n \end{bmatrix}, \quad (16)$$

where $\mathbf{G}_f = [\boldsymbol{g}_{f,\boldsymbol{w}_1^{(n)}}, \boldsymbol{g}_{f,\boldsymbol{w}_2^{(n)}}, \boldsymbol{g}_{f,\boldsymbol{v}_1^{(n)}}, \boldsymbol{g}_{f,\boldsymbol{w}_2^{(n)}}]$ of size $R \times 4$ are the first order derivatives of the function $f(\mathbf{W}^{(n)}, \bar{\mathbf{V}}^{(n)})$ with respect to $[\mathbf{W}^{(n)}, \bar{\mathbf{V}}^{(n)}]$

$$\boldsymbol{g}_{f,\boldsymbol{w}_r^{(n)}} = \frac{\partial f}{\partial \boldsymbol{w}_r^{(n)}} = \mathbf{Q}_{n,r} \boldsymbol{w}_r^{(n)} - \mathbf{K}_{n,r} \boldsymbol{v}_r^{(n)}, \quad \boldsymbol{g}_{f,\boldsymbol{v}_r^{(n)}} = \frac{\partial f}{\partial \boldsymbol{v}_r^{(n)}} = -\mathbf{F}_{n,r} \boldsymbol{v}_r^{(n)} - \mathbf{K}_{n,r}^T \boldsymbol{w}_r^{(n)}, \quad (17)$$

and $\boldsymbol{\Gamma}_n = \mathbf{X}_n^T \mathbf{G}_f$ and $\tau > 0$ is a step size chosen using the Barzilai-Borwein method [12].

The most expensive step in the ASU algorithm is computation of the matrices $\boldsymbol{\Phi}_n = \mathbf{Y}_{(n)} \left(\bigotimes_{k \neq n} \mathbf{V}^{(n)} \mathbf{V}^{(k)T} \right) \mathbf{Y}_{(n)}^T$. A naive computation method might cost $\mathcal{O}(R^4)$. We present a more efficient computation which requires a cost of order $\mathcal{O}(R^3)$

$$\boldsymbol{\Phi}_n = \mathbf{Y}_{(n)} \left((\mathbf{I} - \mathbf{W}^{(n_2)} \mathbf{W}^{(n_2)T}) \otimes (\mathbf{I} - \mathbf{W}^{(n_1)} \mathbf{W}^{(n_1)T}) \right) \mathbf{Y}_{(n)}^T$$
$$= \mathbf{Y}_{(n)} \mathbf{Y}_{(n)}^T - \langle \mathcal{Y} \times_{n_1} \mathbf{W}^{(n_1)}, \mathcal{Y} \times_{n_1} \mathbf{W}^{(n_1)} \rangle_{n_1,n_2} - \langle \mathcal{Y} \times_{n_2} \mathbf{W}^{(n_2)}, \mathcal{Y} \times_{n_2} \mathbf{W}^{(n_2)} \rangle_{n_1,n_2}$$
$$- \langle \mathcal{Y} \times_{n_1} \mathbf{W}^{(n_1)} \times_{n_2} \mathbf{W}^{(n_2)}, \mathcal{Y} \times_{n_1} \mathbf{W}^{(n_1)} \times_{n_2} \mathbf{W}^{(n_2)} \rangle_{n_1,n_2}, \quad (18)$$

where $\{n_1 < n_2\} = \{1, 2, 3\} \setminus \{n\}$.

The proposed Alternating Subspace Update (ASU) algorithm is summarized in Algorithm 1.

Algorithm 1. Alternating Subspace Update (ASU)

Input: Data tensor $\mathcal{Y}$: $(R \times R \times R)$ of rank R
Output: A rank-(2,2,2) tensor $[\![\mathcal{G}; \{\mathbf{U}^{(n)}\}]\!]$ and rank-$(R-2, R-2, R-2)$
      tensor $[\![\mathcal{H}; \{\mathbf{V}^{(n)}\}]\!]$
begin

1 | Initialise components $\mathbf{U}^{(n)}$ and $\mathbf{V}^{(n)}$
2 | Orthogonal normalization to $\mathbf{U}^{(n)}$ and $\mathbf{V}^{(n)}$ and compute $\boldsymbol{\sigma}_n = [\sigma_{n,1}, \sigma_{n,2}]^T$
   | and $\mathbf{W}^{(n)}$
   | **repeat**
   | | **for** $n = 1, 2, 3$ **do**
3 | | | **for** $r = 1, 2$ **do** Update $\sigma_{n,r} = \frac{1}{\sqrt{1+x^2}}$ where x is solved as in (10)
4 | | | Compute $\mathbf{G}_f$ as in (17), $\boldsymbol{\Gamma}_n = \mathbf{X}_n^T \mathbf{G}_f$ where $\mathbf{X}_n = [\mathbf{W}^{(n)}, \bar{\mathbf{V}}^{(n)}]$
5 | | | Update $\mathbf{X}_n = [\mathbf{W}^{(n)}, \bar{\mathbf{V}}^{(n)}]$ as in (16)
6 | | | $\mathbf{U}^{(n)} \leftarrow \mathbf{W}^{(n)} \operatorname{diag}(\boldsymbol{\xi}_n) + \bar{\mathbf{V}}^{(n)} \operatorname{diag}(\boldsymbol{\sigma}_n)$
   | **until** *a stopping criterion is met*
7 | **for** $n = 1, \ldots, N$ **do** Select $\mathbf{V}_{3:R-2}^{(n)}$ as an orthogonal complement of
   | $[\mathbf{W}^{(n)}, \bar{\mathbf{V}}^{(n)}]$
8 | Compute output $\mathcal{G}$ and $\mathcal{H}$ as in (6) and (5)

---

4   Simulations

Example 1 *[Decomposition of Small Tensors Admitting the CP Model].* In this
first example, we illustrate the block deflation of tensor of size $R \times R \times R$ and of
rank R where $R = 10, 20, 30$. The weight coefficients λ_r were set to 1, whereas
collinearity degrees between components $\boldsymbol{a}_r^{(n)}$ and $\boldsymbol{a}_s^{(n)}$ for all $r \neq s$ were set to
c in the range $[0, .9]$, $\boldsymbol{a}_r^{(n)T} \boldsymbol{a}_s^{(n)} = c$ and $\boldsymbol{a}_r^{(n)T} \boldsymbol{a}_r^{(n)} = 1$ for all n. We compare the
ASU algorithm with the ALS algorithm [5] for the multilinear rank-(L_r, M_r, N_r)
block tensor decomposition with two blocks. For this problem, one can use the
non-linear least squares (NLS) algorithm [6]. However, as similar to the ALS
algorithm [5], the NLS algorithm needs to estimate two core tensors and full
factor matrices. Hence this algorithm is much more expensive than the ASU
algorithm. Simulations were run on a Macbook-air laptop having 4 GB memory
and a 1.8 GHz core i7. Due to space and time consuming, the ALS [5] was only
ran in simulations for $R = 10$.

The algorithms were initialised by the same values generated using the
Direct Trilinear Decomposition (DTLD) [13]. The algorithms ran until differ-
ences between consecutive approximation errors were small enough, $|\varepsilon_k - \varepsilon_{k+1}| \leq$
$10^{-6} \varepsilon_k$ where $\varepsilon = \|\mathcal{Y} - \hat{\mathcal{Y}}\|_F^2$, or when the number of iterations exceeded 1000.
Performances were assessed through the squared angular error SAE in estima-
tion of components $\boldsymbol{a}^{(n)}$ SAE $= \arccos\left(\frac{\boldsymbol{a}^T \hat{\boldsymbol{a}}}{\|\boldsymbol{a}\|_2 \|\hat{\boldsymbol{a}}\|_2}\right)^2$. There were 100 independent

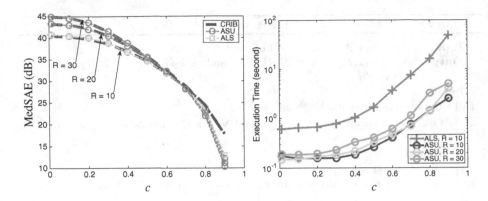

Fig. 2. Comparison of median SAEs and execution times of the ASU and ALS algorithms [5] in decomposition of tensors of size $R \times R \times R$ and rank R where $R = 10$, 20 and 30 for Example 1.

Table 1. Comparison of execution times of the ASU algorithm to extract two components from high rank-R tensors, and those of the CP-FastALS algorithm for Example 2.

	Execution time (second)					
	$c = 0.1$	0.2	0.3	0.4	0.5	0.6
$R = 300$						
ASU	3.81	3.66	3.76	3.82	3.89	3.77
CP-FastALS	530.6	543.5	537.6	537.6	541.9	539.2
$R = 500$						
ASU	38.4	16.7	16.5	16.9	16.8	17.1
CP-FastALS	3658	3672	3679	3693	3678	3669

runs for each rank $R = 10, 20$ and 30. The Gaussian noise was added into the tensor with signal-noise-ratio SNR = 30 dB.

Figure 2 shows median SAE (MedSAE) in dB ($-10 \log_{10} SAE$) obtained by ASU and ALS [5] compared with the Cramér-Rao Induced bound (CRIB) [14] on the squared angular error. The algorithms succeeded in most cases, but failed only when $c = 0.9$. For such difficult scenarios, CRIB on SAE was about 17.8 dB, indicating an angular error of 7.4 degrees between the original and estimated components. We note that in practice, it is hard to estimate a component with CRIB less than 20 dB, i.e., angular error of 5.7 degrees [15].

In Fig. 2, we compare execution times (in second) of algorithms for different ranks. Since the decomposition became more difficult when c was close to 1, running times of algorithms increased as shown in Fig. 2 (right). The ASU algorithm was on average 8 times faster than ALS [5] when $R = 10$.

The results confirmed high speed and accuracy of the proposed ASU algorithm.

Example 2 *[Decomposition of Large-Scale Tensors with High Rank].* This example illustrates an advantage of ASU over existing algorithms for the ordinary CPD in decomposition of large-scale tensors with relatively high rank $R = 300$ and 500. We generated rank-R synthetic tensors of size $R \times R \times R$ as in the previous example. Components $a_r^{(n)}$ and $a_s^{(n)}$ for $r \neq s$ have identical collinearity degrees, i.e., $a_r^{(n)T} a_s^{(n)} = c$ where $c = 0.1, 0.2, \ldots, 0.6$. The Gaussian noise was at SNR $= 30$ dB. Simulations were run on a computer consisted of Intel Xeon 2 processors clocked at 3.33 GHz, 64 GB of main memory. Comparison of execution times of ASU and FastALS [16] is given in Table 1.

5   Conclusions

We have introduced a rank-splitting scheme for CPD, and developed an ASU algorithm for rank-2 block deflation. The algorithm needs to estimate only 4 vectors and two scalars per dimension, and has a computational cost of $\mathcal{O}(R^3)$ for a tensor of size $R \times R \times R$. Further detail and applications of the block tensor deflation are described in the full paper [7]. The algorithm can be extended to higher order tensors, and decomposition with additional constraints. Algorithms for the block tensor deflation are implemented in the Matlab package TENSORBOX which is available online at: http://www.bsp.brain.riken.jp/~phan/tensorbox.php.

References

1. Stegeman, A., Comon, P.: Subtracting a best rank-1 approximation may increase tensor rank. Linear Algebra Appl. **433**(7), 1276–1300 (2010)
2. Phan, A.H., Tichavský, P., Cichocki, A.: Deflation method for CANDECOMP/PARAFAC tensor decomposition. In: IEEE International Conference on Acoustics, Speech and Signal Processing (ICASSP), May 2014, pp. 6736–6740 (2014)
3. Phan, A.H., Tichavský, P., Cichocki, A.: Tensor deflation for CANDECOMP/PARAFAC. Part 1: alternating subspace update algorithm. IEEE Trans. Sig. Process. (2015). (in print)
4. Cichocki, A., Zdunek, R., Phan, A.H., Amari, S.: Nonnegative Matrix and Tensor Factorizations: Applications to Exploratory Multi-way Data Analysis and Blind Source Separation. Wiley, Chichester (2009)
5. De Lathauwer, L., Nion, D.: Decompositions of a higher-order tensor in block terms - part iii: alternating least squares algorithms. SIAM J. Matrix Anal. Appl. 30(3), 1067–1083 (2008). Special issue tensor decompositions and applications
6. Sorber, L., Van Barel, M., De Lathauwer, L.: Structured data fusion. Technical report, ESAT-SISTA, Internal report 13–177 (2013)
7. Phan, A.H., Tichavský, P., Cichocki, A.: Tensor deflation for CANDECOMP/PARAFAC. Part 3: rank splitting. arXiv, CoRR, vol. abs/1506.04971 (2015)
8. De Lathauwer, L., Moor, B.D., Vandewalle, J.: On the best rank-1 and rank-(R1, R2, RN) approximation of higher-order tensors. SIAM J. Matrix Anal. Appl. **21**(4), 1324–1342 (2000)

9. Comon, P., Luciani, X., de Almeida, A.L.F.: Tensor decompositions, alternating least squares and other tales. J. Chemometr. **23**, 393–405 (2009)
10. Phan, A.H., Cichocki, A., Tichavský, P.: On fast algorithms for orthogonal Tucker decomposition. In: IEEE International Conference on Acoustics, Speech and Signal Processing (ICASSP), May 2014, pp. 6766–6770 (2014)
11. Wen, Z., Yin, W.: A feasible method for optimization with orthogonality constraints. Math. Program. **142**, 397–434 (2013)
12. Barzilai, J., Borwein, J.M.: Two-point step size gradient methods. IMA J. Numer. Anal. **8**(1), 141–148 (1988)
13. Sanchez, E., Kowalski, B.: Tensorial resolution: a direct trilinear decomposition. J. Chemometr. **4**, 29–45 (1990)
14. Tichavský, P., Phan, A.H., Koldovský, Z.: Cramér-Rao-induced bounds for CANDECOMP/PARAFAC tensor decomposition. IEEE Trans. Sig. Process. **61**(8), 1986–1997 (2013)
15. Phan, A.H., Tichavský, P., Cichocki, A.: Low complexity damped Gauss-Newton algorithms for CANDECOMP/PARAFAC. SIAM J. Matrix Anal. Appl. **34**(1), 126–147 (2013)
16. Phan, A.H., Tichavský, P., Cichocki, A.: Fast alternating LS algorithms for high order CANDECOMP/PARAFAC tensor factorizations. IEEE Trans. Sig. Process. **61**(19), 4834–4846 (2013)

Some Rank Conditions for the Identifiability of the Sparse Paralind Model

Sebastian Miron$^{(\boxtimes)}$ and David Brie

Centre de Recherche en Automatique de Nancy (CRAN), UMR 7039,
Université de Lorraine, 54506 Vandœuvre, France
sebastian.miron@univ-lorraine.fr

Abstract. In this paper we study the identifiability of the PARALIND model with sparse interaction matrices (*i.e.* S-PARALIND). We provide some theoretical results on how to obtain the sparsest interaction matrices in some particular configurations and when these matrices are unique. These results could be use for the design and analysis of ℓ_0-based decomposition algorithms.

1 Introduction

The PARAFAC [8,14] decomposition of an $\mathcal{X}$ ($I \times J \times K$) 3-way array (or tensor) into sum of R rank-1 tensors is given by $\mathcal{X} = \sum_{r=1}^{R} \mathbf{a}_r \circ \mathbf{b}_r \circ \mathbf{c}_r$, where $\mathbf{a}_r$, $\mathbf{b}_r$ and $\mathbf{c}_r$ are vectors of dimensions I, J and K, respectively, and "$\circ$" denotes the outer vector product. For simplicity, the noise/error term in the model is ignored at this point of the presentation. By regrouping the vectors of the three dimensions (or "modes") of $\mathcal{X}$ into three component matrices $\mathbf{A} = [\mathbf{a}_1 \ldots \mathbf{a}_R]$, $\mathbf{B} = [\mathbf{b}_1 \ldots \mathbf{b}_R]$ and $\mathbf{C} = [\mathbf{c}_1 \ldots \mathbf{c}_R]$, an alternative notation for the PARAFAC decomposition of $\mathcal{X}$ is obtained:

$$\mathcal{X} = [\![\mathbf{A}, \mathbf{B}, \mathbf{C}]\!]. \tag{1}$$

In some applications, prior knowledge on the existence of linear dependencies between the columns of the component matrices is available. This information can be explicitly taken into account by introducing some constraint (or interaction) matrices $\boldsymbol{\Psi}(R_1 \times R)$, $\boldsymbol{\Phi}(R_2 \times R)$, $\boldsymbol{\Omega}(R_3 \times R)$, containing the linear dependency patterns between the columns of $\mathbf{A}$, $\mathbf{B}$, $\mathbf{C}$, respectively. Thus, instead of $[\![\mathbf{A}, \mathbf{B}, \mathbf{C}]\!]$ the decomposition is given by

$$\mathcal{X} = [\![\tilde{\mathbf{A}}\boldsymbol{\Psi}, \tilde{\mathbf{B}}\boldsymbol{\Phi}, \tilde{\mathbf{C}}\boldsymbol{\Omega}]\!], \tag{2}$$

with $\tilde{\mathbf{A}}(I \times R_1)$, $\tilde{\mathbf{B}}(J \times R_2)$ and $\tilde{\mathbf{C}}(K \times R_3)$ full column rank matrices. This type of decomposition was introduced in [6] and previous versions, and named PARALIND[1]. A slightly different version, CONFAC[2], with the constraint matrices

---

[1] PARAllel profiles with LINear Dependencies.
[2] CONstrained FACtor decomposition.

© Springer International Publishing Switzerland 2015
E. Vincent et al. (Eds.): LVA/ICA 2015, LNCS 9237, pp. 41–48, 2015.
DOI: 10.1007/978-3-319-22482-4_5

having canonical vectors as columns, was proposed in [2,3]. A less general framework (involving structured types of linear dependencies), but often highly interpretable, called Block Component Model (BCM) was introduced in [10]. These decompositions proved their usefulness in various domains such as telecommunications [3,17,19,20], spectroscopy [4,5,9] or direction finding [16,21].

In general, the algorithms for fitting the PARALIND model assumes that the constraint matrices are *a priori* known. However, this is not always the case in practice. Moreover, in some real life applications it may be of practical interest to estimate these constraint matrices, as they provide important information on the interactions between the physical mechanisms generating the data. A blind alternating least squares (ALS) estimator for the PARALIND model, referred to as ALS-PARALIND, was proposed in [6]. However, for identifiability reasons (as explained in the next section), the interaction matrices estimated by this approach are highly dependent on the algorithm initialization, which limits their practical utility. To regularize this ill-posed inverse problem, we proposed in [7] to impose sparsity constraints on the interaction matrices, leading to sparse PARALIND (S-PARALIND). These constraints are physically meaningful as they aim at explaining the interactions between the mechanisms generating the data in the simplest way possible. However, no results were given in [7] regarding the identifiability of the S-PARALIND model; the objective of this paper is to shed some light on this aspect.

2　Identifiability of S-Paralind Model

2.1　Preliminaries

A model is said *identifiable* if all its parameters can be *uniquely* estimated from the data, up to some trivial indeterminacies. Thus, in this paper, identifiability can be understood as a uniqueness problem. For example, the PARAFAC model given by (1) is identifiable if the matrices $\mathbf{A}, \mathbf{B}, \mathbf{C}$ can be uniquely estimated from $\mathcal{X}$ up to simultaneous column permutation and column-wise rescaling. The most well-known PARAFAC identifiability condition is due to Kruskal [15] and is based on the Kruskal-rank[3] of the component matrices $\mathbf{A}, \mathbf{B}, \mathbf{C}$. Following [6], identifiability of the PARALIND model is essentially the same as that of the PARAFAC model. If the interaction matrices are fixed and known, identifiability conditions specific to PARALIND can be found in [18]. If these interaction matrices are not known, the identifiability problem can be much more complicated. In particular, it may happen that only some components of the three matrices, or only one matrix (among the three) are identifiable, resulting in the so-called *partial uniqueness* or *uni-mode uniqueness* results. The interested reader is referred to [13] for details.

Let us now assume that the uniqueness of matrix $\mathbf{A}$ is fulfilled and that we aim at estimating the constraint matrix $\mathbf{\Psi}$ together with the full column

---

[3] The Kruskal-rank of a matrix $\mathbf{A}$ (denoted $k_{\mathbf{A}}$) is the maximum number ℓ such that every ℓ columns of $\mathbf{A}$ are linearly independent.

rank matrix $\tilde{\mathbf{A}}$. The identifiability of $\boldsymbol{\Psi}$ and $\tilde{\mathbf{A}}$ comes down to the uniqueness of the bilinear decomposition $\mathbf{A} = \tilde{\mathbf{A}}\boldsymbol{\Psi}$. Without any further constraints, such a decomposition is not unique since an alternative decomposition can be obtained as $\mathbf{A} = \tilde{\mathbf{A}}\boldsymbol{\Psi} = (\tilde{\mathbf{A}}\mathbf{T}^{-1})(\mathbf{T}\boldsymbol{\Psi}) = \tilde{\mathbf{A}}'\boldsymbol{\Psi}'$, for any non-singular matrix $\mathbf{T}$. By imposing sparsity on the constraint matrix $\boldsymbol{\Psi}$ (which should have a minimum number of non-zero entries), we try to explain the rank deficiency of matrix $\mathbf{A}$ by considering the simplest dependency pattern between its columns. This problem has close connection with the problem of dictionary identification using sparse matrix factorization and sparse component analysis, which has been studied in different papers such as [1,11,12]. Basically, in [1,11], the problem is addressed as a $\ell_2 - \ell_0$ optimization problem. Using a geometrical point of view, identifiability conditions are obtained requiring that the size of the training set grows exponentially with the number of atoms. In contrast, the work of [12] addresses the problem as a $\ell_2 - \ell_1$ (non combinatorial) optimization problem. It is shown that the size of training set only needs to grow quadratically with the number of atoms.

All these works consider the problem of overcomplete dictionary recovery. We to stress up the fact that this is one of the main differences with the problem addressed in this paper where only full column rank dictionaries are considered.

The rest of the section aims at giving some answers to the following questions:
- When is the matrix $\tilde{\mathbf{A}}$, yielding the sparsest $\boldsymbol{\Psi}$, a submatrix of $\mathbf{A}$?
- When is the decomposition $\mathbf{A} = \tilde{\mathbf{A}}\boldsymbol{\Psi}$ unique?

Before addressing analytically these problems, let us consider some examples to illustrate the purpose. Let $\mathbf{A}$ be given by

$$\mathbf{A} = [\mathbf{a}_1 \quad \mathbf{a}_2 \quad \mathbf{a}_3 \quad \mathbf{a}_1 + \mathbf{a}_2] = [\mathbf{a}_1 \quad \mathbf{a}_2 \quad \mathbf{a}_3] \begin{bmatrix} 1 & 0 & 0 & 1 \\ 0 & 1 & 0 & 1 \\ 0 & 0 & 1 & 0 \end{bmatrix} \tag{3}$$

$$= [\mathbf{a}_1 + \mathbf{a}_2 \quad \mathbf{a}_1 + \mathbf{a}_3 \quad \mathbf{a}_2 + \mathbf{a}_3] \begin{bmatrix} 1/2 & 1/2 & -1/2 & 1 \\ 1/2 & -1/2 & 1/2 & 0 \\ -1/2 & 1/2 & 1/2 & 0 \end{bmatrix} \tag{4}$$

As illustrated by (3) and (4), it appears that the sparsest matrix $\boldsymbol{\Psi}$ is obtained by selecting R_1 independent columns of $\mathbf{A}$ to form $\tilde{\mathbf{A}}$. It is worth noting that imposing sparsity of $\boldsymbol{\Psi}$ does not ensure the uniqueness of the bilinear decomposition. For example, another possible decomposition of $\mathbf{A}$ is

$$\mathbf{A} = [\mathbf{a}_1 \quad \mathbf{a}_3 \quad \mathbf{a}_1 + \mathbf{a}_2] \begin{bmatrix} 1 & -1 & 0 & 0 \\ 0 & 0 & 1 & 0 \\ 0 & 1 & 0 & 1 \end{bmatrix}. \tag{5}$$

One can see that $\boldsymbol{\Psi}$ matrix in (5) has the same sparsity degree as the one in (3).

2.2 Choosing a Basis of the Column Space of a that Yields the Sparsest $\boldsymbol{\Psi}$

Now we are ready to provide some results on what is the "best" basis, that is the "best" matrix $\tilde{\mathbf{A}}$, for having the sparsest $\boldsymbol{\Psi}$ matrix. However, it is first necessary to introduce some notations. Let $\mathbf{A}$ be a matrix of dimension $(M \times N)$, $M \geq N$ and let $r_{\mathbf{A}} = \mathrm{rank}(\mathbf{A}) \leq N$. We aim at finding a factorization of the matrix $\mathbf{A} = \tilde{\mathbf{A}}\boldsymbol{\Psi}$ where $\tilde{\mathbf{A}}$ is a $(M \times r_{\mathbf{A}})$ (tall) matrix and $\boldsymbol{\Psi}$ is $(r_{\mathbf{A}} \times N)$ (fat) matrix. The considered factorization problem is known to be subject to permutation and scale ambiguities. To remove scale ambiguities we impose to have the maximum value of each column of $\boldsymbol{\Psi}$ equal to 1.

Let us denote the set of admissible bases of the column space of $\mathbf{A}$ by $\mathcal{A} = \{\tilde{\mathbf{A}} \text{ of size } (M \times r_{\mathbf{A}})/\mathrm{span}(\tilde{\mathbf{A}}) = \mathrm{span}(\mathbf{A})\}$. The problem of finding the factorization of $\mathbf{A}$ having the sparsest $\boldsymbol{\Psi}$ can then be formulated as follows:

$$\min_{(\tilde{\mathbf{A}},\boldsymbol{\Psi}),\ \tilde{\mathbf{A}}\in\mathcal{A},\ \mathbf{A}=\tilde{\mathbf{A}}\boldsymbol{\Psi}} \|\boldsymbol{\Psi}\|_0 \tag{6}$$

where $\|\boldsymbol{\Psi}\|_0$ stands for the ℓ_0 pseudo-norm of matrix $\boldsymbol{\Psi}$, that is the number of non-zero entries of $\boldsymbol{\Psi}$.

Proposition 1. *Let* $\mathbf{A} = \tilde{\mathbf{A}}_1\boldsymbol{\Psi}_1 = \tilde{\mathbf{A}}_2\boldsymbol{\Psi}_2$ *where both* $\tilde{\mathbf{A}}_1$ *and* $\tilde{\mathbf{A}}_2$ *are full column-rank matrices in* $\mathcal{A}$ *such that* $\tilde{\mathbf{A}}_1$ *is composed of* $r_{\mathbf{A}} \leq N$ *independent columns of* $\mathbf{A}$ *and* $\tilde{\mathbf{A}}_2$ *is composed of* $r_{\mathbf{A}} \leq N$ *independent linear combination of the columns of* $\mathbf{A}_1$ *which are not proportional to the columns of* $\mathbf{A}$. *If* $r_{\mathbf{A}}$ *satisfies* $r_{\mathbf{A}}^2 - (N+1)r_{\mathbf{A}} + 2N \geq 0$, *then* $\|\boldsymbol{\Psi}_1\|_0 \leq \|\boldsymbol{\Psi}_2\|_0$.

Proof. Let $\mathbf{A} = \tilde{\mathbf{A}}_1\boldsymbol{\Psi}_1 = \tilde{\mathbf{A}}_2\boldsymbol{\Psi}_2$ where both $\tilde{\mathbf{A}}_1$ and $\tilde{\mathbf{A}}_2$ are full column-rank matrices in $\mathcal{A}$. The matrix $\boldsymbol{\Psi}_1$ can be written as $\boldsymbol{\Psi}_1 = [\boldsymbol{\psi}_1(1) \cdots \boldsymbol{\psi}_1(N)]$ and $\boldsymbol{\Psi}_2 = [\boldsymbol{\psi}_2(1) \cdots \boldsymbol{\psi}_2(N)]$. Now we assume that $\tilde{\mathbf{A}}_1$ is composed of $r_{\mathbf{A}} \leq N$ independent columns of $\mathbf{A}$, which, without loss of generality, are assumed to be the first $r_{\mathbf{A}}$ columns: $\boldsymbol{\psi}_1(1), \cdots, \boldsymbol{\psi}_1(r_{\mathbf{A}})$. Thus, the number of non-zero elements of $\boldsymbol{\Psi}_1$ is given by:

$$\|\boldsymbol{\Psi}_1\|_0 = \sum_{i=1}^{N}\|\boldsymbol{\psi}_1(i)\|_0 = \sum_{i=1}^{r_{\mathbf{A}}}\|\boldsymbol{\psi}_1(i)\|_0 + \sum_{i=r_{\mathbf{A}}+1}^{N}\|\boldsymbol{\psi}_1(i)\|_0 = r_{\mathbf{A}} + \sum_{i=r_{\mathbf{A}}+1}^{N}\|\boldsymbol{\psi}_1(i)\|_0$$

As $\forall i = r_{\mathbf{A}} + 1, \cdots, N$, $k_{\mathbf{A}} \leq \|\boldsymbol{\psi}_1(i)\|_0 \leq r_{\mathbf{A}}$, $\|\boldsymbol{\Psi}_1\|_0$ is bounded by: $r_{\mathbf{A}} + (N - r_{\mathbf{A}})k_{\mathbf{A}} \leq \|\boldsymbol{\Psi}_1\|_0 \leq r_{\mathbf{A}} + (N - r_{\mathbf{A}})r_{\mathbf{A}}$. Let us now consider the matrix $\mathbf{A}_2$. Since $\tilde{\mathbf{A}}_2$ is composed of $r_{\mathbf{A}} \leq N$ independent linear combination of the columns of $\mathbf{A}_1$ which are not proportional to the columns of $\mathbf{A}$, we have $\forall i = 1, \cdots, N, \|\boldsymbol{\psi}_2(i)\|_0 \geq 2$ and $\|\boldsymbol{\Psi}_2\|_0 \geq 2N$. Thus, $\|\boldsymbol{\Psi}_1\|_0 \leq \|\boldsymbol{\Psi}_2\|_0$ if $r_{\mathbf{A}} + (N - r_{\mathbf{A}})r_{\mathbf{A}} \leq 2N$, that is: $r_{\mathbf{A}}^2 - (N+1)r_{\mathbf{A}} + 2N \geq 0$.

Remark 1. It can be noticed that $r_{\mathbf{A}}^2 - (N+1)r_{\mathbf{A}} + 2N \geq 0$ is satisfied for all $r_{\mathbf{A}} \leq N \leq 6$. This is no longer true when $N > 6$.

Remark 2. The result of Proposition 1 is based on the worst case scenario since it corresponds to the least favorable case $\|\boldsymbol{\psi}_1(i)\|_0 \leq r_{\mathbf{A}}$. This condition can be relaxed by imposing a more favorable situation such as: $\|\boldsymbol{\psi}_1(i)\|_0 \leq r_{\mathbf{A}} - k$ which results in $r_{\mathbf{A}} + (N - r_{\mathbf{A}})(r_{\mathbf{A}} - k) \leq 2N$, that is $r_{\mathbf{A}}^2 - (N+k+1)r_{\mathbf{A}} + (2+k)N \geq 0$.

Remark 3. In Proposition 1, it is assumed that $\tilde{\mathbf{A}}_2$ does not include any column of $\mathbf{A}$. Let us now examine what happens if $\tilde{\mathbf{A}}_2$ does include a number of $l \leq r_{\mathbf{A}}$ columns of $\mathbf{A}$. In such a case, the sufficient condition for having $\|\mathbf{\Psi}_1\|_0 \leq \|\mathbf{\Psi}_2\|_0$ is $r_{\mathbf{A}}^2 - (N + k + 1)r_{\mathbf{A}} + (2 + k)N \geq l$.

The results corresponding to Proposition 1 and Remark 2 are shown in Fig. 1 for different values of N. The bottom curve corresponds to the case of Proposition 1 ($k = 0$). The other curves are obtained for increasing values of k. For all the cases corresponding to the values of the plotted curves greater that the threshold (set to 0), the solutions obtained by considering independent columns of $\mathbf{A}$ are sparser than those obtained by considering linear combinations of columns of $\mathbf{A}$ which are not proportional to the columns of $\mathbf{A}$. As mentioned in Remark 1, Proposition 1 is always true for $N \leq 6$. The case $N = 6$ is depicted on the left-hand side of Fig. 1. The case of Remark 3, which corresponds to having l columns of $\mathbf{A}$ in $\tilde{\mathbf{A}}_2$, is simply obtained by shifting the threshold to a value equal to l.

To further illustrate these results, we provide next an example in which the condition of Proposition 1 is not fulfilled and for which it is possible to find a sparser decomposition with a basis not consisting of columns of $\mathbf{A}$. Let $\mathbf{A}$ be the following matrix:

$$\mathbf{A} = [\mathbf{a}_1 \quad \mathbf{a}_2 \quad \mathbf{a}_3 \quad \mathbf{a}_4 \quad \mathbf{a}_1 - \mathbf{a}_2 - \mathbf{a}_4 \quad \mathbf{a}_1 + 2\mathbf{a}_2 - \mathbf{a}_3 + \mathbf{a}_4 \quad \mathbf{a}_1 - \mathbf{a}_2 + \mathbf{a}_3 \quad \mathbf{a}_2 - \mathbf{a}_3 - \mathbf{a}_4]$$

$$= [\mathbf{a}_1 \quad \mathbf{a}_2 \quad \mathbf{a}_3 \quad \mathbf{a}_4] \begin{bmatrix} 1 & 0 & 0 & 0 & 1 & 1 & 1 & 0 \\ 0 & 1 & 0 & 0 & -1 & 2 & -1 & 1 \\ 0 & 0 & 1 & 0 & 0 & -1 & 1 & -1 \\ 0 & 0 & 0 & 1 & -1 & 1 & 0 & -1 \end{bmatrix} \tag{7}$$

$$= [(\mathbf{a}_1 + \mathbf{a}_4)/2 \quad (\mathbf{a}_1 - \mathbf{a}_4)/2 \quad \mathbf{a}_2 - (\mathbf{a}_1 - \mathbf{a}_4)/2 \quad \mathbf{a}_3 - \mathbf{a}_2 + (\mathbf{a}_1 - \mathbf{a}_4)/2]$$

$$\begin{bmatrix} 1 & 0 & 0 & 1 & 0 & 0 & 1 & 1 \\ 1 & 1 & 0 & -1 & 1 & 0 & 0 & 0 \\ 0 & 1 & 1 & 0 & -1 & -1 & 0 & 0 \\ 0 & 0 & 1 & 0 & 0 & 1 & 1 & -1 \end{bmatrix}. \tag{8}$$

The number of non-zero elements of the matrix $\mathbf{\Psi}$ corresponding to the first decomposition equals 17 while that of the second is 16. Indeed, in that case $N = 8$ and $r_{\mathbf{A}} = 4$. Thus $r_{\mathbf{A}}^2 - (N + 1)r_{\mathbf{A}} + 2N < 0$, and the sparsest decomposition cannot be guaranteed to correspond to a matrix $\tilde{\mathbf{A}}$ including only columns of $\mathbf{A}$. For example, the second decomposition (8) is sparser than the decomposition (7).

Finally, to conclude this part, we can consider a special case where the dependencies take only the form of collinear loadings. This corresponds to having $k = r_{\mathbf{A}} - 1$ in $r_{\mathbf{A}}^2 - (N + k + 1)r_{\mathbf{A}} + (2 + k)N \geq 0$, yielding $r_{\mathbf{A}} \geq 0$ which is always satisfied. In other words, in the case of collinear loading only, the matrix $\tilde{\mathbf{A}}$ yielding the sparsest solution consists (obviously) in a selection of $r_{\mathbf{A}}$ independent columns of $\mathbf{A}$.

2.3   Uniqueness of the Sparsest Decomposition

In this part we aim at studying the uniqueness of the sparsest decomposition $\mathbf{A} = \tilde{\mathbf{A}}\mathbf{\Psi}$. For all the uniqueness results presented in this part we assume that

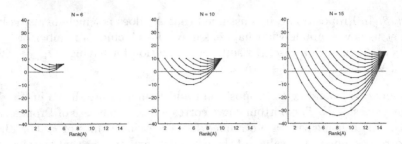

Fig. 1. Illustration of the sparsity properties: the bottom curve corresponds to the case of Proposition 1. The others correspond to the case of Remark 2 for different values of k. As k is increasing, there are much more situations in which the condition $\|\Psi_1\|_0 \leq \|\Psi_2\|_0$ is fulfilled.

the full column rank matrix $\tilde{\mathbf{A}}$ yielding the sparsest solution is a submatrix of $\mathbf{A}$. The uniqueness properties of the case where $\tilde{\mathbf{A}}$ is not a sub matrix of $\mathbf{A}$ are much more difficult to analyze. Thus, the matrix $\mathbf{A}$ can be written as: $\mathbf{A} = [\tilde{\mathbf{A}} \quad \breve{\mathbf{A}}] = [\tilde{\mathbf{a}}_1, \cdots, \tilde{\mathbf{a}}_{r_A}, \breve{\mathbf{a}}_{r_A+1}, \cdots, \breve{\mathbf{a}}_N]$. Defining the matrix $\mathbf{A}^i = [\tilde{\mathbf{A}} \quad \breve{\mathbf{a}}_i]$, its Kruskal-rank satisfies $k_{\mathbf{A}} \leq k_{\mathbf{A}^i} \leq r_{\mathbf{A}}$ and each $\breve{\mathbf{a}}_i$ can be expressed as a linear combination of exactly $k_{\mathbf{A}^i}$ columns of $\tilde{\mathbf{A}}$. In other words : $\breve{\mathbf{a}}_i = \tilde{\mathbf{A}}\psi(i)$ and $\|\psi(i)\|_0 = k_{\mathbf{A}^i}$. We define the set $\tilde{\mathcal{A}} = \{\tilde{\mathbf{A}}/\tilde{\mathbf{A}} \text{ is a submatrix of } \mathbf{A}, \text{span}(\tilde{\mathbf{A}}) = \text{span}(\mathbf{A})\} \subset \mathcal{A}$.

Proposition 2. *Let $\tilde{\mathbf{A}}_1 \neq \tilde{\mathbf{A}}_2$ two matrices of $\tilde{\mathcal{A}}$ and Ψ_1 and Ψ_2 the two matrices satisfying $\mathbf{A} = \tilde{\mathbf{A}}_1\Psi_1 = \tilde{\mathbf{A}}_2\Psi_2$. Then $\|\Psi_1\|_0 \leq \|\Psi_2\|_0$ if and only if $\sum_{r_A+1}^N k_{\mathbf{A}_1^i} \leq \sum_{r_A+1}^N k_{\mathbf{A}_2^i}$.*

Proof. After a proper column permutation, the matrix $\mathbf{A}$ can be written as $\mathbf{A} = [\tilde{\mathbf{A}}_1 \quad \breve{\mathbf{A}}_1]$, thus we have: $\mathbf{A} = \tilde{\mathbf{A}}_1\Psi_1 = [\mathbf{I}_{r_A}, \psi(r_A+1)\cdots\psi(r_N)]$ where $\mathbf{I}_{r_A}$ is the identity matrix of dimension $r_{\mathbf{A}}$. The number of non-zero elements of Ψ_1 is given by $\|\Psi_1\|_0 = r_{\mathbf{A}} + \sum_{r_A+1}^N \|\psi_1(i)\|_0$, which, as shown earlier, is equivalent to $\|\Psi_1\|_0 = r_{\mathbf{A}} + \sum_{r_A+1}^N k_{\mathbf{A}_1^i}$. Similarly we can write $\|\Psi_2\|_0 = r_{\mathbf{A}} + \sum_{r_A+1}^N k_{\mathbf{A}_2^i}$. It follows immediately that $\|\Psi_1\|_0 \leq \|\Psi_2\|_0$ if and only if $\sum_{r_A+1}^N k_{\mathbf{A}_1^i} \leq \sum_{r_A+1}^N k_{\mathbf{A}_2^i}$.

A straightforward extension of Proposition 2 is given by the following proposition which gives the condition for having the sparsest and possibly unique decomposition of $\mathbf{A}$:

Proposition 3. *If $\exists \tilde{\mathbf{A}}_0 \in \tilde{\mathcal{A}}$ such as $\forall \tilde{\mathbf{A}} \in \tilde{\mathcal{A}}, \tilde{\mathbf{A}} \neq \tilde{\mathbf{A}}_0, \sum_{r_A+1}^N k_{\mathbf{A}_0^i} \leq \sum_{r_A+1}^N k_{\mathbf{A}^i}$ then the decomposition $\mathbf{A} = \tilde{\mathbf{A}}_0\Psi_0$ is the sparsest decomposition. If the inequality is strict $(<)$, the sparsest decomposition is unique.*

It should be noted that Proposition 3 does not provide any effective means to find the sparsest decomposition of $\mathbf{A}$. Indeed, finding it would require to find all the possible basis consisting in $r_{\mathbf{A}}$ columns of $\mathbf{A}$ which is actually an NP-complete combinatorial problem.

The result of Proposition 3 can be specialized into the following cases:

- The linear dependencies between the columns of $\mathbf{A}$ are only colinearities. In that case $\tilde{\mathcal{A}}$ only includes a single element since any selection of $r_{\mathbf{A}}$ independent columns of $\mathbf{A}$ will result in the same basis up to scale and permutation. Thus the decomposition is unique.
- $\mathbf{A}$ has a Kruskal-rank equal to its rank *i.e.* $k_{\mathbf{A}} = r_{\mathbf{A}}$. In that case, any selection of $r_{\mathbf{A}}$ columns of $\mathbf{A}$ is a basis and $\forall \tilde{\mathbf{A}} \in \tilde{\mathcal{A}}$, $k_{\mathbf{A}^i} = k_{\mathbf{A}}, \forall i = r_{\mathbf{A}} + 1, \cdots, N$. Thus $\forall \tilde{\mathbf{A}} \in \tilde{\mathcal{A}}$, $\|\boldsymbol{\Psi}\|_0 = (N - r_{\mathbf{A}} + 1)r_{\mathbf{A}}$.

In practice, having a unique sparsest matrix $\boldsymbol{\Psi}$ is not crucial. Indeed, from a PARALIND point of view, having a number of decompositions yielding the same degree of sparsity simply means that all these decompositions are equivalent.

3 Conclusion

In this paper we provided some rank-based results for the identifiability of the PARALIND model with sparse interaction matrices (S-PARALIND). More precisely, we prove a condition that indicates in which cases, choosing the PARALIND loadings between the loadings of the associated PARAFAC decomposition yields the sparsest interaction matrix. These results could be helpful for the design and the analysis of ℓ_0-based algorithms for the decomposition of bilinear/multilinear arrays.

References

1. Aharon, M., Elad, M., Bruckstein, A.: On the uniqueness of overcomplete dictionaries, and a practical way to retrieve them. Linear Algebra Appl. **416**, 48–67 (2006)
2. de Almeida, A.L.F., Favier, G., Mota, J.C.M.: A constrained factor decomposition with application to MIMO antenna systems. IEEE Trans. Signal Process. **56**(6), 2429–2442 (2008)
3. de Almeida, A.L.F., Favier, G., Mota, J.C.M.: Constrained tensor modeling approach to blind multiple-antenna CDMA schemes. IEEE Trans. Signal Process. **56**(6), 2417–2428 (2008)
4. Bahram, M., Bro, R.: A novel strategy for solving matrix effect in three-way data using parallel profiles with linear dependencies. Anal. Chim. Acta **584**(2), 397–402 (2007)
5. Brie, D., Klotz, R., Miron, S., Moussaoui, S., Mustin, C., Ph, B., Grandemange, S.: Joint analysis of flow cytometry data and fluorescence spectra as a non-negative array factorization problem. Chemometr. Intell. Lab. **137**(1), 21–32 (2014)
6. Bro, R., Harshman, R.A., Sidiropoulos, N.D., Lundy, M.E.: Modeling multi-way data with linearly dependent loadings. J. Chemometr. **23**(7–8), 324–340 (2009). special Issue: In Honor of Professor Richard A. Harshman
7. Caland, F., Miron, S., Brie, D., Mustin, C.: A blind sparse approach for estimating constraint matrices in paralind data models. In: Signal Processing Conference (EUSIPCO), 2012 Proceedings of the 20th European, pp. 839–843. Bucharest, Romania (2012)

8. Carroll, J.D., Chang, J.J.: Analysis of individual differences in multidimensional scaling via an N-way generalization of "Eckart-Young" decomposition. Psychometrika **35**(3), 283–319 (1970)

9. Chen, H., Zheng, B., Song, Y.: Comparison of PARAFAC and PARALIND in modeling three-way fluorescence data array with special linear dependences in three modes: a case study in 2-naphthol. Chemometr. Intell. Lab. **25**(1), 20–27 (2011)

10. De Lathauwer, L.: Decompositions of a higher-order tensor in block terms - Part II: Definitions and uniqueness. SIAM J. Matrix Anal. Appl. **30**(3), 1033–1066 (2008)

11. Georgiev, P., Theis, F., Cichocki, A.: Sparse component analysis and blind source separation of underdetermined mixtures. IEEE Trans. Neural Netw. **16**(4), 992–996 (2005)

12. Gribonval, R., Schnass, K.: Dictionary identification - sparse matrix-factorisation via ℓ_1-minimisation. IEEE Trans. Inf. Theory **56**(7), 3523–3539 (2010)

13. Guo, X., Miron, S., Brie, D., Stegeman, A.: Uni-mode and partial uniqueness conditions for CANDECOMP/PARAFAC of three-way arrays with linearly dependent loadings. SIAM J. Matrix Anal. Appl. **33**(1), 111–129 (2012)

14. Harshman, R.A.: Foundations of the PARAFAC procedure: Models and conditions for an 'explanatory' multimodal factor analysis. UCLA Working Papers Phonetics **16**, 1–84 (1970)

15. Kruskal, J.B.: Three-way arrays: Rank and uniqueness of trilinear decompositions, with application to arithmetic complexity and statistics. Linear Algebra Appl. **18**(2), 95–138 (1977)

16. Liu, X., Guang, L., Yang, L., Zhu, H.: PARALIND-based blind joint angle and delay estimation for multipath signals with uniform linear array. EURASIP J. Appl. Signal Process. **2012**(1), 1–13 (2012)

17. Nion, D., De Lathauwer, L.: A block component model-based blind DS-CDMA receiver. IEEE Trans. Signal Process. **56**(11), 5567–5579 (2008)

18. Stegeman, A., de Almeida, A.L.F.: Uniqueness conditions for constrained three-way factor decompositions with linearly dependent loadings. SIAM J. Matrix Anal. Appl. **31**(3), 1469–1499 (2009)

19. Xiaofei, Z., Fei, W., Dazhuan, X.: Blind signal detection algorithm for MIMO-OFDM systems over multipath channel using PARALIND model. IET Commun. **5**(5), 606–611 (2011)

20. Zhang, X., Gao, X., Wang, Z.: Blind paralind multiuser detection for smart antenna CDMA system over multipath fading channel. Prog. Electromagnet. Res. **89**, 23–38 (2009)

21. Zhang, X., Zhou, M., Li, J.: A PARALIND decomposition-based coherent two-dimensional direction of arrival estimation algorithm for acoustic vector-sensor arrays. Sensors **13**(4), 5302–5316 (2013)

Tensors and Latent Variable Models

Mariya Ishteva[✉]

Vrije Universiteit Brussel (VUB), 1050 Brussels, Belgium
mariya.ishteva@vub.ac.be
http://homepages.vub.ac.be/~mishteva

Abstract. In this paper we discuss existing and new connections between latent variable models from machine learning and tensors (multi-way arrays) from multilinear algebra. A few ideas have been developed independently in the two communities. However, there are still many useful but unexplored links and ideas that could be borrowed from one of the communities and used in the other. We will start our discussion from simple concepts such as independent variables and rank-1 matrices and gradually increase the difficulty. The final goal is to connect discrete latent tree graphical models to state of the art tensor decompositions in order to find tractable representations of probability tables of many variables.

Keywords: Latent variable models · Tensor · Low rank

1 Introduction

The goal of this paper is to draw a parallel between some classical concepts and models in probability theory and statistics, and on the other hand notions and decompositions from (multi)linear algebra. Some related analogies can be found in [2,16,20]. Here we put more emphasis on the linear algebra aspects.

We consider discrete random variables, which are variables that take values from a set of possible values, each with an associated probability. We denote these random variables as X_i, $i = 1, 2, \ldots, d$. For simplicity, let us assume that their set of possible values is $\{1, 2, \ldots, n\}$. We denote the probability of a variable X_i taking value j by $P_i(j) = P(X_i = j) \in \mathbb{R}$.[1] The marginal distribution of X_i is then the vector $P_i = \begin{bmatrix} P_i(1) \cdots P_i(n) \end{bmatrix}^\top$, $i = 1, 2, \ldots, d$.

We first discuss the special cases of two and three random variables, in Sects. 2 and 3, respectively. In Sect. 4, we deal with larger numbers of random variables, aiming at tractable representations of their joint probability tables. In the last part of the paper we provide a discussion and the conclusions.

2 Two Random Variables

The joint probability table of two variables X_1 and X_2 can be represented as a matrix $P(X_1, X_2) \in \mathbb{R}^{n \times n}$, denoted here simply by P_{12}. We have $P_{12}(x_1, x_2) =$

---

[1] The values of $P_i(j)$ are actually in the interval $[0, 1]$ but for the purposes of this paper it is easier to think of them as real numbers.

© Springer International Publishing Switzerland 2015
E. Vincent et al. (Eds.): LVA/ICA 2015, LNCS 9237, pp. 49–55, 2015.
DOI: 10.1007/978-3-319-22482-4_6

$P(X_1 = x_1, X_2 = x_2)$. For example, $P_{12}(2,3)$ is then the probability that X_1 and X_2 take values 2 and 3, respectively.

In the special case when the variables are independent, i.e., $X_1 \perp X_2$, we have

$$P_{12}(x_1, x_2) = P_1(x_1)P_2(x_2).$$

In this case, we can represent their joint probability table P_{12} as the outer product of the marginal distributions, i.e.,

$$P_{12} = P_1 P_2^\top.$$

In linear algebra terms, this means that the rank of the matrix P_{12} is 1.

- $X_1, X_2;\quad P(X_1, X_2)$ $P_{12} \in \mathbb{R}^{n \times n}$

- $X_1 \perp X_2$ rank-1 matrix
  $P_{12}(x_1, x_2) = P_1(x_1)P_2(x_2)$

Rank-1 matrices play an important role in linear algebra, as do independent variables in statistics. However, more interesting cases occur when we encounter rank-k matrices. These matrices are generalizations of rank-1 matrices and as such they carry more information. On the other hand, compared to full rank matrices, they can also be more informative since they imply a simplification or structure in some sense.

A rank-k matrix of size $n \times n$, with $k < n$, can be represented as a product of three matrices of dimensions $n \times k$, $k \times k$, and $k \times n$, respectively. One such representation can be obtained from the singular value decomposition (SVD), but there are many other possibilities. For example, in case of probability tables, we can use conditional probability tables (CPT) as factors, as follows

$$P_{12} = P(X_1|H)\, \text{diag}(P(H))\, P(X_2|H)^\top,$$

where H is a new hidden (latent) variable with k states and $\text{diag}(P(H))$ is a diagonal matrix, whose diagonal equals to the vector $P(H)$. Element-wise we have

$$P_{12}(x_1, x_2) = \sum_h P(x_1|h)P(x_2|h)P(h).$$

- $X_1, X_2;\quad P(X_1, X_2)$ $P_{12} \in \mathbb{R}^{n \times n}$

-                                  low-rank matrix
                                                        rank-k matrix, $k < n$

  $P_{12}(x_1, x_2) = \sum_h P(x_1|h)P(x_2|h)P(h)$

3 Three Random Variables

In case of three independent random variables, we have

$$P_{123}(x_1, x_2, x_3) = P_1(x_1)P_2(x_2)P_3(x_3).$$

In (multi)linear algebra terms, the joint probability table P_{123} is then a rank-1 tensor, as it can be represented as the outer product of three vectors, namely the marginal distributions P_1, P_2, and P_3, i.e.,

$$P_{123} = P_1 \circ P_2 \circ P_3,$$

where $\circ$ denotes outer product.

- $X_1, X_2, X_3; \quad P(X_1, X_2, X_3)$ $\qquad\qquad$ $P_{123} \in \mathbb{R}^{n \times n \times n}$

- X_1, X_2, X_3 independent $\qquad\qquad\qquad$ rank-1 tensor
  $P_{123}(x_1, x_2, x_3) = P_1(x_1)P_2(x_2)P_3(x_3)$

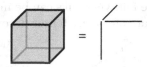

As in the matrix case, low-rank tensors are more general than rank-1 tensors, while being more informative than "full rank" tensors. A low-rank probability table represents the fact that the three observed variables depend on a latent variable with fewer (k) states.

- $X_1, X_2, X_3; \quad P(X_1, X_2, X_3)$ $\qquad\qquad$ $P_{123} \in \mathbb{R}^{n \times n \times n}$

- 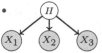 $\qquad\qquad\qquad\qquad\qquad$ rank-k tensor, $k < n$

$P_{123}(x_1, x_2, x_3)$
$= \sum_h P_1(x_1|h)P_2(x_2|h)P_3(x_3|h)P(h)$

As before, the factors of the tensor decomposition can be related to the conditional probability tables $P(X_1|H)$, $P(X_2|H)$, and $P(X_3|H)$ and a diagonal tensor with $P(H)$ on its diagonal. This decomposition is equivalent to the so-called canonical polyadic decomposition [1,10], which is a decomposition of a tensor in rank-1 terms. Element-wise we have

$$P_{123}(x_1, x_2, x_3) = \sum_h P_1(x_1|h)P_2(x_2|h)P_3(x_3|h)P(h).$$

4   Tractable Representations for a Larger Number of Random Variables

We could continue further by adding more random variables and making an analogy between higher-order low-rank tensor decompositions and expressions for the probability table based on conditional probability tables. However, in order to build more realistic models, it could be more useful to consider adding more latent variables as well.

The challenges are to

- Choose an appropriate model,
- Learn the correct structure,
- Estimate the parameters.

The focus of the current paper is mainly on the representation (choice of a model). Some pointers to the remaining two problems are mentioned in Sect. 5.

The following links can be made between some widely used graphical models and state-of-the art tensor decompositions. We elaborate on them below.

- 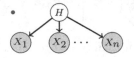   Canonical Polyadic decomposition

-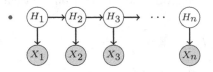

Hidden Markov model   Tensor Train

-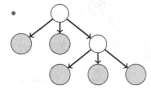

Latent tree model   Hierarchical Tucker

Consider, for example, 10 variables, each having 10 states. Their joint probability table has 10^{10} entries and is intractable. Latent variable models assume that the observed random variables are linked through hidden (latent) variables. Latent tree models further assume that the conditional independence structures are trees and usually that the observed variables depend on the latent ones, the latter having a smaller number of states. Assuming a tree structure leads to a reduction of the number of parameters from exponential to polynomial, while still capturing rich probabilistic dependencies among random variables.

In (multi)linear algebra, analogous decompositions have also been studied. These are the so-called hierarchical decompositions, which are applicable for high-order tensors. The two most popular decompositions are the tensor train [15] and the hierarchical Tucker decomposition [7]. These decompositions allow for a high-order tensor to be decomposed as a product of third-order tensors and second-order tensors (matrices). In this way, a table with 10^{10} elements is decomposed or approximated by less than 10×10^3 parameters. The fact that the latent variables usually have less states than the observed ones translates to having low-rank factors in the hierarchical tensor decompositions, which reduces the number of meaningful parameters further.

The tensor train and hierarchical Tucker decompositions can directly be applied to decompose the given joint probability table if no information about connections between the observed variables is given. On the other hand, if additional information is provided, the hierarchical decompositions can be modified to account for this information. Moreover, latent (tree) variable models can also be used as an inspiration to design more meaningful hierarchical tensor decompositions, by performing the decompositions in accordance with the data, rather than choosing to compute a sequence of factors in the fastest possible way.

5   Further Reading and Discussion

Tensor decompositions are becoming increasingly popular in various research ares. For further information on tensor decompositions and their applications, we refer to the overview papers [3,5,8,12,13], books [4,9,14,17] and the references therein.

The canonical polyadic decomposition avoids the curse of dimensionality and is unique under mild conditions [6], which can be useful for practical applications. However, it is possible that the best approximation of certain (low) rank may not exist. The hierarchical tensor decompositions avoid the curse of dimensionality and have relatively small number of parameters. These decompositions always exist and can be computed using multiple singular value decompositions. For a recent tutorial on breaking the curse of dimensionality, see [19].

In the current paper we concentrated on equivalences in representations between tensors and latent tree graphical models. Some of these links have been used for learning tree structures [11] and estimating (alternative) parameters [18]. Note that due to some rotational invariances of the hierarchical Tucker decomposition, alternative parameters are learned and the low-rank factors in the hierarchical decomposition do not have an interpretation of conditional probability tables. However, the obtained parameters still allow to represent the marginal probability table of the observed variables in a compact and robust way. Imposing constraints, such as nonnegativity, could potentially improve the interpretability of the factors.

6   Conclusions

Real data are often multi-way and modeling them as tensors not only provides a higher-level view, but also has a number of additional advantages. Some tensor decompositions such as (hierarchical) Tucker and tensor train are flexible enough to allow different ranks in different modes. Other decompositions, such as the canonical polyadic decomposition are unique under mild conditions, which can be useful for practical applications. Last but not least, the canonical polyadic decomposition as well as the hierarchical decompositions avoid the curse of dimensionality, allowing for tractable representations of the probability tables of large number of variables.

To simplify the presentation, we assumed that all random variables had the same number of states. The analogies are valid also if this assumptions does not hold true. It is also possible to extend the presented ideas from the discrete to the continuous case. Finally, the fact that the probabilities are always non-negative and smaller than one should also be taken into account.

Acknowledgments. This work was supported in part by the Fund for Scientific Research (FWO-Vlaanderen), by FWO project G.0280.15N, by the Flemish Government (Methusalem), by the Belgian Government through the Inter-university Poles of Attraction (IAP VII) Program (DYSCO II, Dynamical systems, control and optimization, 2012–2017), by the ERC Advanced Grant SNLSID under contract 320378, and by the ERC Starting Grant SLRA under contract 258581. Mariya Ishteva was an FWO Pegasus Marie Curie Fellow.

References

1. Carroll, J., Chang, J.: Analysis of individual differences in multidimensional scaling via an N-way generalization of "Eckart-Young" decomposition. Psychometrika **35**(3), 283–319 (1970)
2. Cichocki, A.: Tensor networks for big data analytics and large-scale optimization problems (2014). arXiv preprint arXiv:1407.3124
3. Cichocki, A., Mandic, D., De Lathauwer, L., Zhou, G., Zhao, Q., Caiafa, C., Phan, H.A.: Tensor decompositions for signal processing applications: from two-way to multiway component analysis. IEEE Sig. Process. Mag. **32**(2), 145–163 (2015)
4. Cichocki, A., Zdunek, R., Phan, A., Amari, S.: Nonnegative Matrix and Tensor Factorizations. Wiley, Chichester (2009)
5. Comon, P.: Tensors: a brief introduction. IEEE Signal Process. Mag. **31**(3), 44–53 (2014)
6. Domanov, I., De Lathauwer, L.: On the uniqueness of the canonical polyadic decomposition of third-order tensors – part II: Uniqueness of the overall decomposition. SIAM J. Matrix Anal. Appl. **34**(3), 876–903 (2013)
7. Grasedyck, L.: Hierarchical singular value decomposition of tensors. SIAM J. Matrix Anal. Appl. **31**(4), 2029–2054 (2010)
8. Grasedyck, L., Kressner, D., Tobler, C.: A literature survey of low-rank tensor approximation techniques. GAMM Mitt. **36**(1), 53–78 (2013)
9. Hackbusch, W.: Tensor Spaces and Numerical Tensor Calculus. Springer Series in Computational Mathematics. Springer, Heidelberg (2012). vol. 42
10. Harshman, R.A.: Foundations of the PARAFAC procedure: model and conditions for an "explanatory" multi-mode factor analysis. UCLA Working Pap. Phonetics **16**(1), 1–84 (1970)

11. Ishteva, M., Song, L., Park, H.: Unfolding latent tree structures using 4th order tensors. In: International Conference on Machine Learning (ICML) (2013)
12. Khoromskij, B.N.: Tensors-structured numerical methods in scientific computing: Survey on recent advances. Chemometr. Intell. Lab. Syst. **110**(1), 1–19 (2012)
13. Kolda, T.G., Bader, B.W.: Tensor decompositions and applications. SIAM Rev. **51**(3), 455–500 (2009)
14. Kroonenberg, P.M.: Applied Multiway Data Analysis. Wiley, New York (2008)
15. Oseledets, I.V.: Tensor-train decomposition. SIAM J. Sci. Comput. **33**, 2295–2317 (2011)
16. Shashua, A.: The applications of tensor factorization in inference, clustering, graph theory, coding and visual representation, 2012. Keynote talk at the 10th International Conference on Latent Variable Analysis and Signal Separation
17. Smilde, A., Bro, R., Geladi, P.: Multi-way Analysis Applications in the Chemical Sciences. Wiley, Chichester (2004)
18. Song, L., Ishteva, M., Parikh, A., Xing, E., Park, H.: Hierarchical tensor decomposition of latent tree graphical models. In: International Conference on Machine Learning (ICML) (2013)
19. Vervliet, N., Debals, O., Sorber, L., De Lathauwer, L.: Breaking the curse of dimensionality using decompositions of incomplete tensors. Sig. Process. Mag. IEEE **31**(5), 71–79 (2014)
20. Yılmaz, Y.K., Cemgil, A.T.: Algorithms for probabilistic latent tensor factorization. Sig. Process. **92**(8), 1853–1863 (2012)

Tensor Factorisation Approach for Separation of Convolutive Complex Communication Signals

Samaneh Kouchaki$^{(\boxtimes)}$ and Saeid Sanei

Faculty of Engineering and Physical Sciences, University of Surrey, Guildford, UK
s.kouchaki@surrey.ac.uk

Abstract. Communication signals such as multiple constellation signals have complex waveforms and may have multipath source reflection and noncircularity problems. Tensor based source separation techniques have become increasingly popular for various applications as they exploit different inherent diversities of the sources. In this paper, a tensor based convolutive source separation algorithm is developed based on PARAFAC2. The optimisation technique is based on the direct model fitting of PARAFAC and augmented statistics. The proposed method is evaluated using simulated data with multiple pathways and various noncircularity levels. Simulation results confirm the superiority of the proposed method over the existing popular techniques.

Keywords: Augmented statistics · Complex valued signals · Convolutive mixtures · Noncircularity level · Phase shift keying · Tensor factorisation

1 Introduction

In a number of applications such as communication signal processing, complex valued signals are recorded at the receiver. Phase shift keying (PSK) and quadrature amplitude modulation (QAM) can be considered as some examples where the multiple constellation signals are represented in complex domain. The recorded complex valued signals usually suffer from multipath including time delay and noise that confirms the necessity of having source separation for complex valued signals.

Among several approaches, complex fastICA as an extension of traditional fastICA has been proposed recently [1]. This method is based on a fixed-point algorithm with local stability considering circular sources. FastICA is an orthogonal separation method that limits the fastICA applications to the instantaneous linear systems.

In the case of complex signals, circularity is a common assumption. It means that the correlation between real and imaginary parts of the complex values is ignored. Therefore, the complex valued mixture has a rotationally invariant probability distribution in the complex plane. This assumption ignores the full statistical information that results in stability problem when the sources are

© Springer International Publishing Switzerland 2015
E. Vincent et al. (Eds.): LVA/ICA 2015, LNCS 9237, pp. 56–63, 2015.
DOI: 10.1007/978-3-319-22482-4_7

noncircular (there is some correlation between real and imaginary parts) as in the complex fastICA approach [6]. To solve this problem, augmented statistics that exploit the complete second-order information by involving the effect of pseudo covariance can improve the separation performance [4,14]. For example, noncircular complex ICA (NCICA) is one of ICA extensions that considers the statistical dependency between the real and imaginary parts [17].

On the other hand, having signals of various amplitude-phase constellations may further complicate the whole process. This problem addressed as convolutive blind source separation (CBSS) is tackled in this article. Similar to the instantaneous case, CBSS aims at receiving the sources from the observed convolutive mixtures. However, CBSS problems are more challenging due to the numbers of channel parameters that should be estimated.

The proposed CBSS techniques are usually classified to frequency and time domain methods [20]. Frequency based methods transfer time domain signal to the frequency domain where the CBSS problem is changed to instantaneous case for each frequency bin. The advantage of such techniques is having multiple independent estimations with fewer parameters at each frequency bin. Parra and Spence [19] suggested a method based on a multiple decorrelation approach. It is quite similar to the other techniques by Kawamoto [11] and Murata et al. [15] for the frequency domain and Wee and Principe [21] method in time domain. The main shortcoming of frequency based techniques is the arbitrary permutation ambiguity.

On the other hand, the shortcoming of time domain techniques is their need for estimation of a larger set of parameters. However, they usually have good performance compared with that of frequency domain methods.

The proposed method based on tensor factorisation which is suitable for separation of convolutive complex signals exploits augmented statistics. We call the method complex augmented tensor factorisation (CATF). Tensor factorisation has become popular in many applications such as chemometrics [2], biomedical [16], antenna array and signal processing [5,13]. A tensor is a multi-way data representation and using it the true underlying structure of the data can be extracted. Tensor factorisation is applicable to both nonstationary and underdetermined cases. The signal that in general changes in time, frequency, and space can be tracked using tensor factorisation. The proposed CATF method is an extension of parallel factor analysis (PARAFAC) or more specifically PARAFAC2 [8,12] algorithm for which the convolutive model is considered. The CATF is tested for various conditions and compared with those of NCICA and Parra's method.

The remainder of the paper is structured as follows: first, our CATF method and the decomposition technique used here are explained. Then, the results of applying the CATF to synthetic data are shown and discussed.

2   PARAFAC and PARAFAC2

PARAFAC [3,8] decomposes a multi-way array into a sum of some rank-one tensors. For instance, each third-way tensor $\underline{\mathbf{X}}$ can be written in matrix form as:

$$\mathbf{X}_k = \mathbf{B} \mathbf{D}_k \mathbf{A}^T, \quad k = 1, 2, ..., K \tag{1}$$

where $(.)^T$ stands for transpose and $\mathbf{X}_k$ is the transposed version of kth frontal slice of $\underline{\mathbf{X}}$. The tensor factors in the first and second modes are respectively $\mathbf{A}$ and $\mathbf{B}$. $\mathbf{D}_k$ is a diagonal matrix with diagonal elements equal to the kth row of the third matrix $\mathbf{C}$. In addition, $\mathbf{E}_k$ is the kth frontal tensor slice error term.

PARAFAC2 is one of PARAFAC extensions and is a common method for source separation as it provides a certain freedom in the shape of each slab. Matrix notation for PARAFAC2 is:

$$\mathbf{X}_k = \mathbf{B}_k \mathbf{D}_k \mathbf{A}^T, \quad k = 1, 2, ..., K \quad \text{subject to} \quad \mathbf{B}_k^H \mathbf{B}_k = \mathbf{\Phi} \tag{2}$$

where $\mathbf{B}_k$ corresponds to the kth frontal tensor slice and $(.)^H$ stands for hermitian or conjugate transpose. Here, the component in the second mode can be different across slices. $\mathbf{\Phi}$ is invariant for all slices to maintain the uniqueness of the solution. Moreover, $\mathbf{B}_k$ can be written $\mathbf{B}_k = \mathbf{P}_k \mathbf{F}$ where $\mathbf{P}_k$ is an orthonormal and $\mathbf{F}$ is an arbitrary matrix:

$$\mathbf{X}_k = \mathbf{P}_k \mathbf{F} \mathbf{D}_k \mathbf{A}^T \tag{3}$$

A direct fitting model suggested by Kiers [12] has been used in this work as follows.

3   Problem Formulation

The following system can be considered as a general formulation for CBSS.

$$x_q(t) = \sum_{i=1}^{N_s} \sum_{\tau=0}^{M-1} s_i(t - \tau) a_{qi}(\tau) + e_q(t), \quad q = 1, 2, ..., N_x \tag{4}$$

where $x_q(t)$ shows the recorded mixture at the time t and qth channel, N_x and N_s indicate respectively the number of sensors and sources, M is the number of time lags, $a_{qi}(\tau)$ indicates the qith element of mixing matrix at a time lag τ, and $e_q(t)$ is the additive noise for qth mixture. The matrix form of Eq. (4) can be shown as:

$$\mathbf{X} = \sum_{\tau=0}^{M-1} \theta_\tau(\mathbf{S}) \mathbf{A}_\tau^T + \mathbf{E}, \theta_\tau(\mathbf{S}) = \mathbf{\Xi}_\tau \mathbf{S} \tag{5}$$

where $\theta_\tau(\mathbf{S})$ is a shift operator that is done by multiplication of a shift matrix $\mathbf{\Xi}_\tau$ by the source matrix [9], $\mathbf{S}$, $\mathbf{E}$, and $\mathbf{X}$ are the source, noise, and recorded matrices respectively, and $\mathbf{A}_\tau$ shows the mixing matrix at time lag τ. The matrices of observed signals, source signals, and noise are all considered as complex valued. In this work, a tensor is built up from our measurements using temporal segmentation with a segment size greater than the expected number of lags. Equation (5) for each data segment can be written as:

$$\mathbf{X}_k = \sum_{\tau=0}^{M-1} \mathbf{\Xi}_\tau \mathbf{S}_k \mathbf{A}_\tau^T + \mathbf{E}_k \text{ for } k = 0, 1, ..., K \tag{6}$$

where K is the number of segments. Here, Eq. (6) can be changed by putting $\mathbf{S}_k = \mathbf{P}_k \mathbf{F} \mathbf{D}_k$ where $\mathbf{P}_k$ is an orthonormal matrix, $\mathbf{D}_k$ is a diagonal matrix, and $\mathbf{F}$ is an arbitrary matrix:

$$\mathbf{X}_k = \sum_{\tau=0}^{M-1} \mathbf{\Xi}_\tau \mathbf{P}_k \mathbf{F} \mathbf{D}_k \mathbf{A}_\tau^T + \mathbf{E}_k, \text{ given that } \mathbf{P}_k^H \mathbf{P}_k = \mathbf{I}_{N_s \times N_s} \tag{7}$$

where $\mathbf{I}_{N_s \times N_s}$ indicates an identity matrix. Obviously, Eq. (7) is a PARAFAC2 extension (it is a PARAFAC2 model if $M = 1$). Hence, the following cost function for CBSS is used in this work.

$$J = ||\mathbf{X}_k - \sum_{\tau=0}^{M-1} \mathbf{\Xi}_\tau \mathbf{P}_k \mathbf{F} \mathbf{D}_k \mathbf{A}_\tau^T||_F^2 \tag{8}$$

where $||.||_F$ stands for the Frobenius norm of a matrix.

To exploit the power difference or the correlation between the data channels, widely linear modelling of the systems has been recently introduced [10]. For this purpose, "augmented" statistics have been established to incorporate the complementary covariance matrices and exploit the complete second-order information [4,14]. In complex domain, the augmented basis matrix is defined as $\mathbf{X}^a = [\mathbf{X}, \mathbf{X}^*]^T \in \mathbb{C}$. Thus, in order to exploit the second-order information, we change the optimisation technique to the following optimisation equation:

$$J = ||\mathbf{X}_k - \sum_{\tau=0}^{M-1} \mathbf{\Xi}_\tau (\mathbf{S}_k \hat{\mathbf{A}}_{1\tau}^T + \mathbf{S}_k^* \hat{\mathbf{A}}_{2\tau}^T)||_F^2 = ||\mathbf{X}_k - \sum_{\tau=0}^{M-1} \mathbf{\Xi}_\tau \mathbf{S}_k^a \mathbf{A}_\tau^T||_F^2 \tag{9}$$

where $\mathbf{S}_k^a = [\mathbf{S}_k, \mathbf{S}_k^*]^T$ and $\mathbf{A}_\tau = [\hat{\mathbf{A}}_{1\tau}^T, \hat{\mathbf{A}}_{2\tau}^T]^T$. Alternating least squares (ALS) optimisation is used here to estimate each model parameter by fixing the rest. After estimating the augmented source and mixing matrices, the original source is obtained using the first part of $\mathbf{S}_k^a$. The detailed parameter optimisation procedure is explained in the following subsection. Based on the above assumption, $\mathbf{P}_k$, $\mathbf{F}$, and $\mathbf{D}_k^a$ are changed to augmented matrices $\mathbf{P}_k^a$, $\mathbf{F}^a$, and $\mathbf{D}_k^a$.

3.1   Estimation of $\mathbf{P}_k^a$

By assuming that all the parameters except for $\mathbf{P}_k^a$ are all known and fixed, the objective function in Eq. (8) can be rewritten as:

$$J_{\mathbf{P}_k^a} = tr(\mathbf{X}_k^H \mathbf{X}_k) + \sum_{\tau=0}^{M-1} tr(\mathbf{A}_\tau^* \mathbf{D}_k^{aH} \mathbf{F}^{aH} \mathbf{F}^a \mathbf{D}_k^a \mathbf{A}_\tau^T)$$

$$- \sum_{\tau=0}^{M-1} tr(2\mathbf{A}_\tau^* \mathbf{D}_k^{aH} \mathbf{F}^{aH} \mathbf{P}_k^{aH} \mathbf{\Xi}_\tau^H \mathbf{X}_k) \tag{10}$$

where $tr(.)$ and $(.)^*$ indicate respectively the matrix trace and conjugate. In Eq. (10) only the last term depends on $\mathbf{P}_k^a$ and the remaining terms are positive semi-definite matrices. Therefore, to estimate $\mathbf{P}_k^a$, we can maximise the following:

$$J_{P_k} = tr(\sum_{\tau=0}^{M-1} \mathbf{F}^a \mathbf{D}_k^a \mathbf{A}_\tau^T \mathbf{X}_k^H \mathbf{\Xi}_\tau \mathbf{P}_k^a) \tag{11}$$

By defining a new variable $\mathbf{Z}_k$ and calculating its SVD, $\mathbf{P}_k^a$s is estimated [7]:

$$\mathbf{Z}_k = \sum_{\tau=0}^{M-1} \mathbf{F}^a \mathbf{D}_k^a \mathbf{A}_\tau^T = \mathbf{U}_k \mathbf{\Sigma}_k \mathbf{V}_k^H \tag{12}$$

where $\mathbf{\Sigma}_k$ is a diagonal matrix with nonnegative diagonal elements and $\mathbf{U}_k$ and $\mathbf{V}_k$ are two orthonormal matrices. Then, $\mathbf{P}_k$s can be obtained as follows:

$$\mathbf{P}_k^a = \mathbf{V}_k \mathbf{U}_k^H, \quad \mathbf{P}_k^{aH} \mathbf{P}_k^a = \mathbf{I}_{2N_s \times 2N_s} \tag{13}$$

3.2   Estimation of $\mathbf{A}_\tau$

To estimate $\mathbf{A}_\tau$, $\sum_{\tau=0}^{M-1}(.)$ is expanded into matrix products as:

$$\mathbf{X}_k = \mathbf{G}_k \mathbf{A},$$

$$\mathbf{G}_k = (\mathbf{\Xi}_0 \mathbf{P}_k^a \mathbf{F}^a \mathbf{D}_k^a, \mathbf{\Xi}_1 \mathbf{P}_k^a \mathbf{F}^a \mathbf{D}_k^a, ..., \mathbf{\Xi}_{M-1} \mathbf{P}_k^a \mathbf{F}^a \mathbf{D}_k^a), \quad \mathbf{A} = \begin{pmatrix} \mathbf{A}_0^T \\ \mathbf{A}_1^T \\ ... \\ \mathbf{A}_{M-1}^T \end{pmatrix} \tag{14}$$

Therefore, by stacking $\mathbf{X}_k$ and $\mathbf{G}_k$, a set of linear equations can be obtained:

$$\hat{\mathbf{X}} = \begin{pmatrix} \mathbf{X}_1 \\ \mathbf{X}_2 \\ ... \\ \mathbf{X}_{K-1} \end{pmatrix}, \quad \hat{\mathbf{G}} = \begin{pmatrix} \mathbf{G}_1 \\ \mathbf{G}_2 \\ ... \\ \mathbf{G}_{K-1} \end{pmatrix}, \quad \hat{\mathbf{X}} = \hat{\mathbf{G}}\mathbf{A} \tag{15}$$

Then, $\mathbf{A}$ can be calculated as:

$$\mathbf{A} = \hat{\mathbf{G}}^\dagger \hat{\mathbf{X}} \tag{16}$$

where $(.)^\dagger$ shows Moore-Penrose pseudo-inverse (pinv) of a complex matrix that can be computed using Gram-Schmidt or SVD factorisation techniques. As a result, $\mathbf{A}_\tau$ is updated using the $\mathbf{A}$ estimation.

3.3   Estimation of $\mathbf{D}_k^a$ and $\mathbf{F}^a$

The following equation can be considered for arbitrary matrices $\mathbf{Y}$, $\mathbf{L}$, and $\mathbf{W}$.

$$vec(\mathbf{YLW}) = (\mathbf{W}^T \otimes \mathbf{Y})vec(\mathbf{L}) \tag{17}$$

where $\otimes$ is Kronecker product and vec is obtained by vectorising the input matrix. Regarding this relation and Eq. (7), we define $\mathbf{Y}_\tau = \mathbf{\Xi}_\tau$, $\mathbf{L}_k = \mathbf{P}_k^a \mathbf{F}^a \mathbf{D}_k^a$, $\mathbf{W}_\tau = \mathbf{A}_\tau^T$. Consequently, $vec(\mathbf{L}_k)$ can be calculated as follows:

$$vec(\mathbf{L}_k) = (\sum_{\tau=0}^{M-1}(\mathbf{A}_\tau \otimes \mathbf{\Xi}_\tau))^\dagger vec(\mathbf{X}_k) \tag{18}$$

Having $\mathbf{L}_k$, we can consider PARAFAC2 factorisation for it, $\mathbf{L}_k = \mathbf{P}_k^a \mathbf{F}^a \mathbf{D}_k^a \mathbf{A}^T$ with $\mathbf{A}$ here approximated as identity matrix. This assumption is not necessarily accurate and for applications to real signals some a priori information about the source locations can be utilised instead. Thus, the normal PARAFAC optimisation technique can be used to estimate $\mathbf{D}_k^a$ and $\mathbf{F}^a$ given that $\mathbf{P}_k^a$ is known:

$$J = ||\mathbf{P}_k^{aT} \mathbf{L}_k - \mathbf{F}^a \mathbf{D}_k^a \mathbf{A}^T||_F^2, \tilde{\mathbf{X}}_k = \mathbf{P}_k^{aT} \mathbf{L}_k, \ \mathbf{A} = \mathbf{I}$$

$$\mathbf{F}^a = \sum_{k=1}^{K} \tilde{\mathbf{X}}_k^T \mathbf{I}^* \mathbf{D}_k^{a*} (\mathbf{G}^\dagger)^T, \mathbf{G} = (\mathbf{I}^H \mathbf{I}) \circ (\mathbf{C}^{aH} \mathbf{C}^a)$$

$$\mathbf{C}^a = \begin{pmatrix} diag(\mathbf{I}^H \tilde{\mathbf{X}}_1 \mathbf{F}^{a*})^T \\ diag(\mathbf{I}^H \tilde{\mathbf{X}}_2 \mathbf{F}^{a*})^T \\ \cdots \\ diag(\mathbf{I}^H \tilde{\mathbf{X}}_K \mathbf{F}^{a*})^T \end{pmatrix} (\mathbf{O}^\dagger)^T, \mathbf{O} = (\mathbf{I}^H \mathbf{I}) \circ (\mathbf{F}^{aH} \mathbf{F}^a) \quad (19)$$

where $\circ$ stands for the element-wise Hadamard product and $diag(.)$ returns a column vector of its input diagonal elements. As explained before, $\mathbf{D}_k$ is achieved from $\mathbf{C}$. Finally, the augmented source matrix can be obtained and the desired source matrix is obtained from the augmented one using the first part of $\mathbf{S}^a$.

4   Experimental Results

To evaluate the performance of the CATF and compare it with the other available methods, a number of experiments were performed using simulated complex data. The simulated scenario consists of a finite number of taps with tunable delays. Each tap signal is modulated in phase or amplitude using a baseband tap-gain function. Then, the transmitted signal or final mixture can be expressed as $\mathbf{x}(t) = \sum_{\tau=0}^{M-1} \mathbf{s}(t-\tau) \mathbf{A}(\tau)^T + \mathbf{e}(t)$ where $\mathbf{x}(t)$ is the complex mixture at a time instant t, $\mathbf{s}(t)$ is the complex input, $\mathbf{A}(\tau)$ is the mixing matrix, $\mathbf{e}(t)$ is additive normalised white Gaussian noise (WGN), and M is the number of time lags. The mixing matrices are selected randomly. Therefore, the sources are convolved with different mixing matrices and summed to form the multichannel mixtures.

We compare the performance of the proposed method with two other benchmark methods; NCICA and Parra's algorithm. Here, the separation of 8PSK sources has been performed for convolutive mixtures and also noncircular sources. Then, the intersymbol interference (ISI) has been estimated as the performance measure (less ISI indicates a better performance).

Different Numbers of Delay Paths: In this part, the performance is obtained from the CATF, NCICA, and Parra's for signals with different numbers of time lags M. M in the proposed method and the number of delay paths that we used to generate the data is considered the same. The results are brought in Fig. 1 for 8PSK which shows the average obtained over 100 runs for each point. Moreover, all the methods are tested using noiseless mixtures.

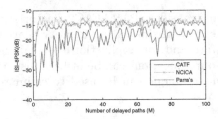

Fig. 1. The effect of having multiple delayed paths on the CATF, NCICA, and Parra's.

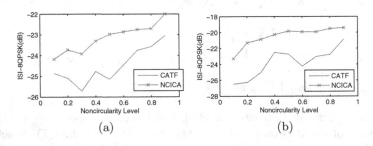

Fig. 2. The effect of noncircularity level on the CATF, NCICA, and Parra's; (a) $M = 1$ and (b) $M = 3$.

As can be seen, when there is no time delays or in instantaneous situation, NCICA performs well and sometimes better than the proposed method. However, as the number of time lags increases, the CATF and Parra's perform better. The main reason is that both the CATF and Parra's methods consider the multipath effect in their formulation. Moreover, in general the performance of the proposed method is better than Parra's.

The Effect of Noncircularity Level: The procedure suggested in [18] is used to generate noncircular sources with different correlation levels. Using this algorithm, the exact correlation or noncircularity level can be set where 0 refers to no correlation and 1 to a perfect correlation. The average results of applying the CATF and NCICA are brought in Fig. 2. The results show that the CATF is more robust against the changes in noncircularity level. Moreover, NCICA works better for $M = 1$ compared with $M > 1$.

5   Conclusions

A convolutive tensor factorisation method based on PARAFAC2 was proposed here for complex valued signals. To improve the performance in the case of noncircular sources, augmented statistics were used. The convolutive model was considered and the parameters were obtained for different time lags. To evaluate the performance of the proposed method, some simulated data with multiple pathways and various noncircularity levels were generated. Moreover, the results were compared with NCICA and Parra's. The simulation results confirm the superiority of the proposed method over the existing benchmarks.

References

1. Bingham, E., Hyvärinen, A.: A fast fixed-point algorithm for independent component analysis of complex valued signals. J. Neural Syst. **10**(01), 1–8 (2000)
2. Bro, R., Kiers, H.A.: A new efficient method for determining the number of components in parafac models. J. Chemometrics **17**(5), 274–286 (2003)
3. Carroll, J.D., Chang, J.J.: Analysis of individual differences in multidimensional scaling via an N-way generalization of 'Eckart-Young' decomposition. Psychometrika **35**(3), 283–319 (1970)
4. Cheong Took, C., Mandic, D.P.: Augmented second-order statistics of quaternion random signals. Signal Process. **91**(2), 214–224 (2011)
5. Comon, P.: Tensor decompositions, state of the art and applications (2009). arXiv preprint arXiv:0905.0454
6. Douglas, S.C.: Fixed-point algorithms for the blind separation of arbitrary complex-valued non-gaussian signal mixtures. EURASIP **2007**, 83 (2007)
7. Green, B.F.: The orthogonal approximation of an oblique structure in factor analysis. Psychometrika **17**(4), 429–440 (1952)
8. Harshman, R.A.: Foundations of the PARAFAC procedure: models and conditions for an explanatory multimodal factor analysis. UCLA Working Pap. Phonetics **16**, 1–84 (1970)
9. Huang, L.K., Manikas, A.: Blind single-user array receiver for MAI cancellation in multipath fading DS-CDMA channels. EUSIPCO Proc. **2**, 647–650 (2000)
10. Javidi, S., Took, C.C., Mandic, D.P.: Fast independent component analysis algorithm for quaternion valued signals. IEEE Trans. Neural Network. **22**(12), 1967–1978 (2011)
11. Kawamoto, M., Matsuoka, K., Ohnishi, N.: A method of blind separation for convolved non-stationary signals. Neurocomputing **22**(1), 157–171 (1998)
12. Kiers, H.A., Ten Berge, J.M., Bro, R.: PARAFAC2-Part I: a direct fitting algorithm for the PARAFAC2 model. J. Chemometrics **13**(3–4), 275–294 (1999)
13. Kolda, T.G., Bader, B.W.: Tensor decompositions and applications. SIAM Rev. **51**(3), 455–500 (2009)
14. Mandic, D.P., Goh, V.S.L.: Complex Valued Nonlinear Adaptive Filters: Noncircularity, Widely Linear and Neural Models, vol. 59. Wiley, New York (2009)
15. Murata, N., Ikeda, S., Ziehe, A.: An approach to blind source separation based on temporal structure of speech signals. Neurocomputing **41**(1), 1–24 (2001)
16. Nazarpour, K., Praamstra, P., Miall, R., Sanei, S.: Steady-state movement related potentials for BCI an exploratory approach in the STF domain. In: Proceedings of BCI (2008)
17. Novey, M., Adali, T.: On extending the complex fastica algorithm to noncircular sources. IEEE Trans. Sig. Process. **56**(5), 2148–2154 (2008)
18. Ollila, E.: On the circularity of a complex random variable. IEEE Sig. Process. Lett. **15**, 841–844 (2008)
19. Parra, L., Spence, C.: Convolutive blind separation of non-stationary sources. IEEE Trans. Speech Audio Process. **8**(3), 320–327 (2000)
20. Pedersen, M.S., Larsen, J., Kjems, U., Parra, L.C.: A survey of convolutive blind source separation methods. Multichannel Speech Process. 1065–1084 (2007)
21. Wee, H., Principe, J.: A criterion for BSS based on simultaneous diagonalization of time correlation matrices. In: NNSP, vol. 97, pp. 496–508 (1997)

Enhanced Tensor Based Semi-blind Estimation Algorithm for Relay-Assisted MIMO Systems

Jianshu Zhang$^{(\boxtimes)}$, Ahmad Nimr, Kristina Naskovska, and Martin Haardt

Communications Research Laboratory, Ilmenau University of Technology,
98684 Ilmenau, Germany
jianshu.zhang@tu-ilmenau.de
http://www.tu-ilmenau.de/crl

Abstract. In this paper we study tensor based semi-blind estimation algorithms for one-way amplify-and-forward relaying systems. By exploiting the tensor structure of the equivalent channel, a bilinear alternating least squares (ALS) algorithm is proposed to jointly estimate the data and the channels at the destination. Sufficient conditions for the unique estimation are derived. The simulation results show that the proposed algorithms outperforms the state of the art algorithms.

1 Introduction

Multiple-input and multiple-output (MIMO) techniques can enhance the performance of relaying networks, e.g., [1]. To fully exploit the MIMO gain, channel knowledge is required at the transmitter and the receiver. Traditionally matrix based solutions are used to estimate MIMO channels, e.g., [5]. Recently, research results in [8,9] have shown that a tensor-based channel estimation method for relaying networks can provide a better estimation accuracy and requires less training overhead compared to the matrix based solution. Due to the use of training sequences, training based estimation methods provide accurate estimation results [8]. But they sacrifice physical resources which can be used for data transmission. Hence, blind or semi-blind estimation techniques, which do not need the transmission of training sequences, become more attractive, e.g., [10,12]. More specifically, in [12] a semi-blind estimation algorithm, which jointly estimates the data and the channels based on the PARAFAC-PARATUCK tensor decomposition of the received signal, is proposed for a two-hop one-way amplify-and-forward (AF) relaying assisted MIMO system. It is shown in [12] that the received signal via the source-to-destination (SD) link satisfies a PARAFAC tensor model, and the received signal via the source-to-relay-to-destination (SRD) link satisfies the PARATUCK2 tensor model [2]. The general PARATUCK2 decomposition requires a diagonal relay amplification matrix and is computed using an alternating least squares (ALS) method. More precisely, a trilinear ALS is proposed in [12]. Moreover, Kruskal's identifiability condition [4] must be satisfied for uniqueness. These stringent conditions have restricted the applicability of the proposed tensor based algorithms in [12].

© Springer International Publishing Switzerland 2015
E. Vincent et al. (Eds.): LVA/ICA 2015, LNCS 9237, pp. 64–72, 2015.
DOI: 10.1007/978-3-319-22482-4_8

In this paper we introduce enhanced tensor based semi-blind algorithms for a joint data and channel estimation in the two-hop AF relaying system. In contrast to [12], our design is applicable even if a full relay amplification matrix is used. Our contribution can be summarized as: (1) A demodulation step is introduced during the estimation of the data symbols to provide well-conditioned data estimates, which can accelerate the converge of the ALS algorithm as compared to [12]. (2) In contrast to the trilinear ALS, bilinear ALS algorithms for jointly estimating the source-to-relay (SR) channel, the relay-to-destination (RD) channel, and the data are proposed. In each iteration, algebraic solutions for computing the SR channel and the RD channel are introduced based on the PARAFAC tensor model. Sufficient conditions for a unique estimation up to scaling ambiguities are derived. The derived conditions are more relaxed compared to Kruskal's condition in [12]. Numerical results show that significant improvements are achieved in terms of channel estimation accuracy and system BER performance using only a few number of iterations.

2   System Model

We consider an AF relaying scenario where one source node communicates with one destination node via a relay assisted network as in [12]. The source, the relay, and the destination have M_S, M_R, and M_D antennas, respectively. We assume that the channel is i.i.d. frequency flat and quasi-static block fading. The matrices $\boldsymbol{H}^{(SD)} \in \mathbb{C}^{M_D \times M_S}$, $\boldsymbol{H}^{(SR)} \in \mathbb{C}^{M_R \times M_S}$, and $\boldsymbol{H}^{(RD)} \in \mathbb{C}^{M_D \times M_R}$ denote the SD channel, the SR channel, and the RD channel, respectively. All the nodes operate in half-duplex modes and thus a complete transmission takes two phases. In the first phase, the source transmits to the relay and the destination. In the second phase, the source is silent and the relay transmits to the destination. To perform a blind estimation, the source transmits data using the Khatri-Rao space-time (KRST) coding scheme proposed in [10]. Assume that N blocks of KRST coded symbols are transmitted and during the n-th block ($n \in \{1, \cdots, N\}$) the transmitted signal is denoted as $\boldsymbol{X}_n = D_n\{\boldsymbol{S}\}\boldsymbol{C}^T \in \mathbb{C}^{M_S \times K}$, where $\boldsymbol{S} \in \mathbb{C}^{N \times M_S}$ denotes the overall data to be transmitted in N blocks, and $\boldsymbol{C} \in \mathbb{C}^{K \times M_S}$ is the known KRST coding matrix, which maps M_S symbols to K time slots, i.e., the spatial code rate is M_S/K [10]. The operation $D_n\{\boldsymbol{A}\}$ creates a diagonal matrix by aligning the elements of the n-th row of $\boldsymbol{A}$ onto its diagonal. Therefore, the n-th block of the received signal via the SD link is given by [12]

$$\boldsymbol{Y}_n^{(SD)} = \boldsymbol{H}^{(SD)} D_n\{\boldsymbol{S}\}\boldsymbol{C}^T + \boldsymbol{V}_n^{(SD)} \in \mathbb{C}^{M_D \times K}, \tag{1}$$

where $\boldsymbol{V}_n^{(SD)} \in \mathbb{C}^{M_D \times K}$ denotes the zero-mean circularly symmetric complex Gaussian (ZMCSCG) noise and each element has a variance of σ_d^2. The received signal after N blocks can be expressed as a three-way tensor with the n-th frontal slice denoted by (1). The resulting PARAFAC model is given by

$$\boldsymbol{\mathcal{Y}}^{(SD)} = \boldsymbol{\mathcal{I}}_{3,M_S} \times_1 \boldsymbol{H}^{(SD)} \times_2 \boldsymbol{C} \times_3 \boldsymbol{S} + \boldsymbol{\mathcal{V}}^{(SD)} \in \mathbb{C}^{M_D \times K \times N} \tag{2}$$

where $\mathcal{I}_{3,M_S}$ is the identity tensor and $\times_i$ denotes the i-mode product [3]. Similarly, the n-th block of the received signal via the SRD link is expressed as [12]

$$Y_n^{(\mathrm{SRD})} = H^{(\mathrm{RD})} G_n H^{(\mathrm{SR})} D_n\{S\} C^{\mathrm{T}} + \bar{V}_n^{(\mathrm{SRD})} \in \mathbb{C}^{M_D \times K} \qquad (3)$$

where $\bar{V}_n^{(\mathrm{SRD})} = H^{(\mathrm{RD})} G_n H^{(\mathrm{SR})} V_n^{(\mathrm{R})} + V_n^{(\mathrm{SRD})}$ denotes the effective noise, where $V_n^{(\mathrm{R})}$ and $V_n^{(\mathrm{SRD})}$ are the ZMCSCG noise at the relay and the destination with a variance of σ_r^2 per element of $V_n^{(\mathrm{R})}$, and $G_n \in \mathbb{C}^{M_R \times M_R}$ is the n-th relay amplification matrix.

Our objective is to develop tensor based semi-blind estimation schemes such that the channels $H^{(\mathrm{SD})}$, $H^{(\mathrm{SR})}$, $H^{(\mathrm{RD})}$, and the data matrix S can be uniquely identified.

3 Enhanced Design in the Direct Link

An estimate of S and $H^{(\mathrm{SD})}$ in the direct link can be obtained via the Khatri-Rao factorization as described in [12]. To this end, the 2-mode unfolding of the tensor $\mathcal{Y}^{(\mathrm{SD})}$ is expressed as

$$\left[\mathcal{Y}^{(\mathrm{SD})}\right]_{(2)} \approx C(H^{(\mathrm{SD})} \diamond S)^{\mathrm{T}}, \qquad (4)$$

where $\approx$ implies that the noise is ignored and $\diamond$ is the Khatri-Rao product. If the coding matrix C has a full column rank, i.e., $K \geq M_S$, by pre-multiplying C^+ on both sides of (4) an estimate of the Khatri-Rao product is obtained as $\hat{Y}^{(\mathrm{SD})} = \left(C^+ \left[\mathcal{Y}^{(\mathrm{SD})}\right]_{(2)}\right)^{\mathrm{T}} \approx H^{(\mathrm{SD})} \diamond S$, where $^+$ is the Moore-Penrose pseudo inverse. The resulting problem is a Khatri-Rao factorization problem, for which a LS solution can be obtained using the singular value decomposition (SVD) [5,6,8,12] and a detailed implementation is also found in [8]. The drawback of this method is that the Khatri-Rao factorization has inherent scaling ambiguities, i.e., one scaling ambiguity per column. A typical way to resolve these ambiguities is to assume that one row of S is known [9,12]. The described estimation procedure so far follows [12] exactly.

The estimates in the SD link and especially the estimated signal matrix $\hat{S}$ will be used to initialize the trilinear ALS based PARATUCK2 decomposition in the SRD link [12]. It is known that the ALS algorithm is sensitive to ill-conditioned matrices due to the inverse operation. We observe that this happens quite often when the algorithms in [12] are applied. To deal with this phenomenon, we propose to demodulate the entries of the data matrix $\hat{S}$. For this purpose, an element-wise hard-decision demodulation technique [7] is used and the output of the demodulation step is denoted as $\hat{S}_{\mathrm{demod}}$. Numerical results show that the matrix $\hat{S}_{\mathrm{demod}}$ is in general well-conditioned. Hence, this demodulation step is applied in both the SD link and the SRD link. Furthermore, we apply a LS based refinement of the channel estimate $\hat{H}^{(\mathrm{SD})}$. This is simply computed by using the 1-mode unfolding of $\mathcal{Y}^{(\mathrm{SD})}$, i.e., $\hat{H}_{\mathrm{enh}}^{(\mathrm{SD})} \approx \left[\mathcal{Y}^{(\mathrm{SD})}\right]_{(1)} ((\hat{S}_{\mathrm{demod}} \diamond C)^{\mathrm{T}})^+$.

4   Bilinear ALS Algorithm for the SRD Link

Let us briefly review the trilinear ALS algorithm proposed in [12]. If the relay amplification matrix G_n in (3) is a diagonal matirx, i.e., $G_n = D_n\{G\}$ and $G \in \mathbb{C}^{N \times M_R}$, the obtained M_D-by-K-by-N tensor by stacking N received blocks satisfies a PARATUCK2 model [2], where (3) is the n-th frontal slice of the corresponding tensor. In our case, the matrices G and C can be designed and we need to estimate S, $H^{(SR)}$, and $H^{(RD)}$ from (3). When two out of three parameters are fixed, (3) can be rewritten as a linear function of the third parameter. Therefore, a trilinear ALS algorithm can be applied and an exact PARATUCK2 decomposition can be achieved. According to Kruskal's condition [4], to ensure the uniqueness of the PARATUCK2 decomposition, i.e., up to scaling ambiguities, it is required that $K \geq M_S$ and $\min(M_S, M_D) \geq M_R$ [12]. To resolve the scaling ambiguities, one row of H^{RD} or one column of H^{SR} needs to be known. Moreover, since S is involved in both the SD link, i.e., (1), and the SRD link, i.e., (3), a better estimation of S and $H^{(SD)}$ might be obtained if (1) and (3) are combined. Depending on whether (1) and (3) are jointly exploited for estimating S or not, the proposed algorithms in [12] are divided into the combined PARAFAC/PARATUCK2 (CPP) method and the sequential PARAFAC/PARATUCK2 (SPP), respectively. In the following, we discuss our proposed approaches based on the CPP method. But the extension to the SPP method is straightforward.

4.1   Bilinear ALS Based Design

If we do not restrict ourselves to the PARATUCK2 model in [2], more flexibilities are obtained, e.g., the use of a full relay amplification matrix. To see this, define $H_n^{(SRD)} = H^{(RD)} G_n H^{(SR)}$, where G_n can be a full matrix in contrast to [12]. Assume that an ALS algorithm is used. Using equation (3), a LS estimate of the effective SRD channel $H_n^{(SRD)}$ at the n-th block in the i-th step is calculated as

$$\hat{H}_{n,i}^{(SRD)} \approx Y_n^{(SRD)} (D_n\{\hat{S}_{i-1}\} C^T)^+ \in \mathbb{C}^{M_D \times M_S}, \tag{5}$$

$\forall n$, where $\hat{S}_{i-1}$ denotes the estimate of S in the $(i-1)$-th step. To obtain $\hat{H}_{n,i}^{(SRD)}$ uniquely, it is sufficient that C has full column rank, i.e., $K \geq M_S$, because in this case $D_n\{\hat{S}_{i-1}\}$ is invertible. This coincides with the requirement in the SD link. Afterwards, the channels $\hat{H}_i^{(RD)}$ and $\hat{H}_i^{(SR)}$ are estimated algebraically from the tensor representation of the equivalent channel $\bar{\mathcal{H}}_i^{(SRD)} = \left[\hat{H}_{1,i}^{(SRD)} \sqcup_3 \hat{H}_{2,i}^{(SRD)} \cdots \sqcup_3 \hat{H}_{N,i}^{(SRD)} \right] \in \mathbb{C}^{M_D \times M_S \times N}$, where $\sqcup_3$ denotes the concatenation of matrices along the third dimension [8]. The proposed estimation methods will be introduced in the sequel. After that, the estimated equivalent channel is reconstructed as $\tilde{H}_{n,i}^{(SRD)} = \hat{H}_i^{(RD)} G_n \hat{H}_i^{(SR)}$. Then by combining (1) and (3), i.e., the CPP concept in [12] is used, a LS estimate of the n-th row of S in the i-th step is determined as

$$\hat{s}_{n,i} \approx \begin{bmatrix} C \diamond \hat{H}^{(\mathrm{SD})} \\ C \diamond \tilde{H}_{n,i}^{(\mathrm{SRD})} \end{bmatrix}^{+} \begin{bmatrix} y_n^{(\mathrm{SD})} \\ y_n^{(\mathrm{SRD})} \end{bmatrix}, \tag{6}$$

$\forall n$, where $y_n^{(\mathrm{SD})} = \mathrm{vec}\{Y_n^{(\mathrm{SD})}\}$ and $y_n^{(\mathrm{SRD})} = \mathrm{vec}\{Y_n^{(\mathrm{SRD})}\}$. The conclusion in [12, Theorem 1] can be still applied to (6). That is, the condition $K \geq M_S$ is sufficient to guarantee that a unique pseudo inversion is obtained in (6). Since in the i-th iteration step only two parameters, i.e., $\hat{H}_{n,i}^{(\mathrm{SRD})}$ and $\hat{s}_{n,i}$, are computed using the ALS algorithm, $\forall n$, we name this algorithm the bilinear ALS method. Due to the nature of an ALS method, the uniqueness is guaranteed up to a diagonal scaling matrix. However, since one row of S is assumed to be known in Sect. 3, the scaling ambiguity is resolved.

In the following we introduce algebraic methods for obtaining $\hat{H}_i^{(\mathrm{RD})}$ and $\hat{H}_i^{(\mathrm{SR})}$ in the i-th iteration. The index i will be dropped for notational simplicity. Moreover, let us define $\mathcal{G} = [G_1 \sqcup_3 G_2 \cdots \sqcup_3 G_N] \in \mathbb{C}^{M_\mathrm{R} \times M_\mathrm{R} \times N}$.

4.2   Khatri-Rao Factorization (KRF) Based Approach

The tensor $\bar{\mathcal{H}}^{(\mathrm{SRD})} \approx \mathcal{G} \times_1 H^{(\mathrm{RD})} \times_2 H^{(\mathrm{SR})^\mathrm{T}}$ satisfies a Tucker2 tensor model. Let $r_{\mathcal{G}}$ be the rank of the tensor $\mathcal{G}$ [8]. Then the PARAFAC decomposition of $\mathcal{G}$ is computed as $\mathcal{G} = \mathcal{I}_{r_{\mathcal{G}}} \times_1 F_1 \times_2 F_2 \times_3 F_3$, where $F_1 \in \mathbb{C}^{M_\mathrm{R} \times r_{\mathcal{G}}}$, $F_2 \in \mathbb{C}^{M_\mathrm{R} \times r_{\mathcal{G}}}$, and $F_3 \in \mathbb{C}^{N \times r_{\mathcal{G}}}$ are the corresponding factor matrices. Inserting this PARAFAC decomposition of $\mathcal{G}$ into the Tucker2 model, we get

$$\bar{\mathcal{H}}^{(\mathrm{SRD})} \approx \mathcal{I}_{3,r_{\mathcal{G}}} \times_1 (H^{(\mathrm{RD})} F_1) \times_2 (H^{(\mathrm{SR})^\mathrm{T}} F_2) \times_3 F_3. \tag{7}$$

Equation (7) is also a PARAFAC decomposition. Note that we can construct F_1, F_2, and F_3 such that $H^{(\mathrm{RD})}$ and $H^{(\mathrm{SR})}$ can be uniquely estimated. By applying the 3-mode unfolding of (7), we obtain

$$\left[\bar{\mathcal{H}}^{(\mathrm{SRD})}\right]_{(3)} \approx F_3 \cdot ((H^{(\mathrm{RD})} F_1) \diamond (H^{(\mathrm{SR})^\mathrm{T}} F_2))^\mathrm{T}. \tag{8}$$

It is straightforward to see that (8) yields a similar Khatri-Rao structure as (4). To estimate $H^{(\mathrm{RD})}$ and $H^{(\mathrm{SR})}$, we first isolate the Khatri-Rao product from (8). For this purpose, it is sufficient that F_3 has a full column rank, i.e., $N \geq r_{\mathcal{G}}$. Then the remaining problem is a Khatri-Rao factorization problem, which can be solved up to a diagonal scaling matrix $\Lambda_\mathrm{f} \in \mathbb{C}^{r_{\mathcal{G}} \times r_{\mathcal{G}}}$. That is, the obtained two factor matrices B_1 and B_2 of the Khatri-Rao factorization can be written as $B_1 = H^{(\mathrm{RD})} F_1 \Lambda_\mathrm{f}$ and $B_2 = H^{(\mathrm{SR})^\mathrm{T}} F_2 \Lambda_\mathrm{f}^{-1}$. To resolve this scaling ambiguity, we still assume that one column of $H^{(\mathrm{RD})}$ or $H^{(\mathrm{SR})}$ is known, e.g., it could be obtained via a training based initial estimate between the relay and the destination [12]. After the scaling ambiguity is resolved, i.e., Λ_f is estimated, the following relationships are valid, i.e., $B_1 \Lambda_\mathrm{f}^{(-1)} = H^{(\mathrm{RD})} F_1$ and $B_2 \Lambda_\mathrm{f} = H^{(\mathrm{SR})^\mathrm{T}} F_2$.

To obtain $H^{(RD)}$ and $H^{(SR)}$ uniquely, we require that F_1 and F_2 have full row rank, i.e., $r_{\mathcal{G}} \geq M_R$. Also, to render the pseudo inversions numerically stable, F_1 and F_2 should have orthogonal rows. Overall, the Khatri-Rao factorization (KRF) based approach can be applied if $N \geq r_{\mathcal{G}} \geq M_R$. To reduce the channel estimation overhead, $r_{\mathcal{G}}$ should be as small as possible. Therefore, we choose $r_{\mathcal{G}} = M_R$. The advantage of this KRF approach is that an algebraic solution is obtained. The disadvantage of it is that the resulting scaling ambiguity is the same as in [12].

Finally, the proposed bilinear algorithm is described in Algorithm 1 and the corresponding sufficient conditions to estimate the channels $H^{(RD)}$, $H^{(SR)}$, and the symbol matrix S in the SRD link without ambiguities are summarized as follows: the identifiablity is guaranteed if C and F_3 have full column rank, and F_1 and F_2 have full row rank, i.e., $K \geq M_S$ and $N \geq M_R$. Note that in contrast to [12] the proposed bilinear ALS algorithm does not impose any restrictions on the number of antennas at the source, the relay, or the destination.

Remark. When G_n is a diagonal matrix, i.e., $G_n = D_n\{G\}$, it is a special case of the PARAFAC based design, i.e., F_1 and F_2 are identity matrices. In such a case the equivalent channel is expressed as $H_n^{(SRD)} \approx H^{(RD)}D_n\{G\}H^{(SR)}$. It has the same structure as (1) and thus an equivalent PARAFAC model is obtained as $\bar{\mathcal{H}}^{(SRD)} \approx \mathcal{I}_{3,M_S} \times_1 H^{(RD)} \times_2 H^{(SR)^T} \times_3 G$. The channels $H^{(RD)}$ and $H^{(RD)}$ can then be obtained by using its 3-mode unfolding $\left[\bar{\mathcal{H}}^{(SRD)}\right]_{(3)} \approx$ $G(H^{(RD)} \diamond H^{(SR)^T})^T$ and the LS Khatri-Rao factorization.

Algorithm 1. A bilinear ALS algorithm for estimating $H^{(RD)}$, $H^{(SR)}$, and S in the SRD link using the CPP method

---

1: **Initialize:** set initial value of $\hat{S}_0$ using the estimate from the SD link, $\hat{H}^{(SD)} = \hat{H}_{enh}^{(SD)}$, $i = 1$, and the threshold value ϵ.

2: **Main step:**

3: **repeat**

4:    Calculate $\hat{H}_{n,i}^{(SRD)}$ using (5).

5:    Estimate $\hat{H}_i^{(RD)}$ and $\hat{H}_i^{(SR)}$ using the KRF approach and obtain $\tilde{H}_{n,i}^{(SRD)} = \hat{H}_i^{(RD)} G_n \hat{H}_i^{(SR)}$.

6:    Compute $\hat{s}_{n,i}$ using (6) and then demodulate $\hat{s}_{n,i}$ using the hard-decision demodulation method [7].

7: **until** $\sum_{n=1}^{N} \|Y_n^{(SRD)} - \hat{H}_i^{(RD)} G_n \hat{H}_i^{(SR)} D_n\{\hat{S}_i\} C^T\| \leq \epsilon$

8: **Output:** $H^{(RD)}$, $H^{(SR)}$, S, and $\hat{H}^{(SD)}$.

---

5   Simulation Results and Concluding Remarks

The proposed bilinear ALS algorithms are evaluated using Monte Carlo simulations. The simulated channels $H^{(SD)}$, $H^{(SR)}$ and $H^{(RD)}$ are uncorrelated Rayleigh fading channels. The transmit power at the source and the relay are

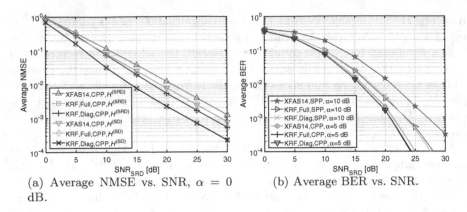

(a) Average NMSE vs. SNR, $\alpha = 0$ dB.

(b) Average BER vs. SNR.

Fig. 1. Comparison of different algorithms. $M_{\mathrm{S}} = M_{\mathrm{D}} = M_{\mathrm{R}} = K = 2$, $N = 4$, $N_f = 40$, and QPSK modulation.

set to unity. The noise power at the relay and the destination are identical, i.e., $\sigma_{\mathrm{d}}^2 = \sigma_{\mathrm{r}}^2 = \sigma_{\mathrm{n}}^2$. The SNR of the relay-assisted link and the direct link are denoted by $\mathrm{SNR}_{\mathrm{SRD}}$ and $\mathrm{SNR}_{\mathrm{SD}}$, respectively. We have $\mathrm{SNR}_{\mathrm{SRD}} = \mathrm{SNR}_{\mathrm{SD}} + \alpha$ [dB] and $\alpha \geq 0$, which is a realistic assumption [12]. The coding matrix C is chosen as a Vandermonde matrix, i.e., $C_{k,m} = e^{\frac{j2\pi(k-1)(m-1)}{M_{\mathrm{S}}}}$, $k \in \{1, \cdots K\}$, and $m \in \{1, \cdots M_{\mathrm{S}}\}$.

The channel estimation accuracy is measured by using the normalized mean squared error (NMSE) criterion. Let $\boldsymbol{H}^{(X)}$ and $\hat{\boldsymbol{H}}^{(X)}$ denote the true channel and the estimated channel, where $X \in \{\mathrm{SD}, \mathrm{SRD}\}$. It implies that for the SRD link, the equivalent channel is evaluated [12]. For each simulation, the NMSE is defined as

$$\mathrm{e}(\boldsymbol{H}^{(X)}, \hat{\boldsymbol{H}}^{(X)}) = \frac{\|\boldsymbol{H}^{(X)} - \hat{\boldsymbol{H}}^{(X)}\|_{\mathrm{F}}^2}{\|\boldsymbol{H}^{(X)}\|_{\mathrm{F}}^2}. \tag{9}$$

The sphere decoder described in [11] is used to decode the modulated signal at the end. The BER performance is determined by transmitting a total number of N_f blocks of KRST coded symbols, where the first N blocks are also used for blind channel estimation. Since the proposed algorithms in [12] outperform the training based solutions in [5,9], we only compare our proposed algorithms to the algorithms in [12], which are denoted as "XFAS14". Note that the XFAS14 algorithms are applicable only if a diagonal relay amplification matrix is used. Moreover, "full" and "diag" stand for whether a full relay amplification matrix or a diagonal relay amplification matrix is used. All the simulation results are averaged over 2000 channel realizations.

Figure 1a and 1b demonstrate the performance of the bilinear ALS algorithms in terms of the NMSE and the BER performance when a CPP or a SPP method is used. When a full relay amplification matrix, $\boldsymbol{F}_1$ and $\boldsymbol{F}_2$ are set to M_{R}-by-M_{R} DFT matrices while $\boldsymbol{F}_3$ is a Vandermonde matrix. When a diagonal relay amplification matrix is used as explained in Remark 1, the matrix $\boldsymbol{G} \in \mathbb{C}^{M_{\mathrm{R}} \times M_{\mathrm{R}}}$

is set to a Vandermonde matrix. As depicted in Fig. 1a, the KRF approach always provides the best channel estimates compared to the XFAS14 algorithms regardless whether a CPP or a SPP procedure is applied. It also provides a better BER performance especially when the SPP method is applied. When the CPP based method is used, the performance gain over the XFAS14 algorithm is not significant due to the fact that the estimate of S from the direct link dominates the performance. Moreover, all the proposed algorithms require only a few iterations compared to the XFAS14 algorithm as shown in Table 1. Finally, we conclude that the proposed bilinear ALS algorithms are computationally more efficient and can provide a better performance compared to the state of the art algorithms in [12]. Moreover, although a full relay amplification matrix contains more free parameters than a diagonal one, to fully exploit this freedom an appropriate design of the relay amplification matrix is desired.

Table 1. Comparison of the average number of required iterations when the SPP method is used, $M_S = M_D = M_R = K = 2$, $N = 4$, and $\alpha = 10$ dB under different values of SNR_{SRD}.

Algorithm	$\text{SNR}_{\text{SRD}} = 0$ dB	$\text{SNR}_{\text{SRD}} = 15$ dB	$\text{SNR}_{\text{SRD}} = 30$ dB
XFAS14	107.2	99.2	55.1
KRF (full)	3.8	2.6	2.0
KRF (diag)	3.8	2.7	2.0

References

1. Chae, C.B., Tang, T., Heath, R.W., Cho, S.: MIMO relaying with linear processing for multiuser transmission in fixed relay networks. IEEE Trans. Signal Process. **56**, 727–738 (2008)
2. Harshman, R.A., Lundy, M.E.: Uniqueness proof for a family of models sharing features of Tucker's three-mode factor analysis and PARAFAC/CANDECOMP. Proc. Psychometrika **61**, 133–154 (1996)
3. Kolda, T.G., Bader, B.W.: Tensor decompositions and applications. SIAM Rev. **51**, 455–500 (2009)
4. Kruskal, J.B.: Three-way arrays: rank and uniqueness of trilinear decompositions, with application to arithmetic complexity and statistics. Linear Algebra Appl. **19**, 95–138 (1977)
5. Lioliou, P., Viberg, M.: Least-squares based channel estimation for MIMO relays. In: Proceedings of the 11th International ITG Workshop on Smart Antennas (WSA), March 2008
6. Liu, S.: Matrix results on the Khatri-Rao and Tracy-Singh products. Elsevier Linear Algebra Appl. **289**(3), 267–277 (1999)
7. Proakis, J.G., Salehi, M.: Digital Communications. McGraw-Hill higher education, New York (2008)
8. Roemer, F., Haardt, M.: Tensor-based channel estimation (TENCE) and iterative refinements for two-way relaying with multiple antennas and spatial reuse. IEEE Trans. Signal Process. **58**, 5720–5735 (2010)

9. Rong, Y., Khandaker, M.R.A., Xiang, Y.: Channel estimation of dual-hop MIMO relay system via parallel factor analysis. IEEE Trans. Wireless Commun. **11**(6), 2224–2233 (2012)
10. Sidiropoulos, N.D., Budampati, R.S.: Khatri-Rao space-time codes. IEEE Trans. Signal Process. **50**, 2396–2407 (2002)
11. Viterbo, E., Boutros, J.: A universal lattice code decoder for fading channels. IEEE Trans. Inf. Theory **45**, 1639–1642 (1999)
12. Ximenes, L.R., Favier, G., de Almeida, A.L.F., Silva, Y.C.B.: PARAFAC-PARATUCK semi-blind receivers for two-hop cooperative MIMO relay systems. IEEE Trans. Signal Process. **62**, 3604–3615 (2014)

Deep Neural Networks for Supervised Speech Separation/Enhancement

Improving Deep Neural Network Based Speech Enhancement in Low SNR Environments

Tian Gao[1]([✉]), Jun Du[1], Yong Xu[1], Cong Liu[2], Li-Rong Dai[1],
and Chin-Hui Lee[3]

[1] University of Science and Technology of China,
Hefei, Anhui, People's Republic of China
{gtian09,xuyong62}@mail.ustc.edu.cn, {jundu,lrdai}@ustc.edu.cn
[2] iFlytek Research, iFlytek Co., Ltd., Hefei, Anhui, People's Republic of China
congliu2@iflytek.com
[3] Georgia Institute of Technology, Atlanta, GA, USA
chl@ece.gatech.edu

Abstract. We propose a joint framework combining speech enhancement (SE) and voice activity detection (VAD) to increase the speech intelligibility in low signal-noise-ratio (SNR) environments. Deep Neural Networks (DNN) have recently been successfully adopted as a regression model in SE. Nonetheless, the performance in harsh environments is not always satisfactory because the noise energy is often dominating in certain speech segments causing speech distortion. Based on the analysis of SNR information at the frame level in the training set, our approach consists of two steps, namely: (1) a DNN-based VAD model is trained to generate frame-level speech/non-speech probabilities; and (2) the final enhanced speech features are obtained by a weighted sum of the estimated clean speech features processed by incorporating VAD information. Experimental results demonstrate that the proposed SE approach effectively improves short-time objective intelligibility (STOI) by 0.161 and perceptual evaluation of speech quality (PESQ) by 0.333 over the already-good SE baseline systems at -5dB SNR of babble noise.

Keywords: Speech enhancement · Low SNR · Deep neural networks · Voice activity detection · Speech intelligibility

1 Introduction

Speech enhancement (SE) has been an open research problem for the past several decades. Many approaches are developed to solve this problem, and they can be classified into two categories, namely unsupervised and supervised methods. As for the unsupervised approaches, there are, spectral subtraction [1], MMSE-based log-spectral amplitude estimator [2] and optimally modified log-MMSE

This work was supported by the National Natural Science Foundation of China under Grants No. 61305002. We would like to thank iFLYTEK Research for providing the training data and DNN training platform.

© Springer International Publishing Switzerland 2015
E. Vincent et al. (Eds.): LVA/ICA 2015, LNCS 9237, pp. 75–82, 2015.
DOI: 10.1007/978-3-319-22482-4_9

estimator [3], etc. However, many assumptions were made during the derivation process of these solutions, and the resulting enhanced speech often suffers from an annoying artifact called musical noise.

Various supervised methods have also been developed in recent years, which have been demonstrated to generate enhanced speech with better quality. Non-negative matrix factorization (NMF) based SE [4] was one of the notable methods. Speech and noise basis were learned from the speech data and noise data, respectively. Then the clean speech could be decomposed given the noisy speech. In [5,6], masking techniques were used to train DNNs for speech separation and recognition. More recently, our proposed DNN-based SE where the DNN was regarded as a regression model to predict the clean log-power spectra (LPS) [7] from the noisy LPS has been successfully applied to noisy speech enhancement [8,9], separation [10] and recognition [11,12].

Figure 1 shows noisy speech mixed with babble noise from the NOISEX-92 [13] corpus at SNR = 0dB along with the corresponding clean speech and frame-level SNR sequence. Speech segment covered by high-energy noises, such as the noted part in Fig. 1, remains difficult to handle. When noise is removed from those segments by conventional DNN approaches, the quality of speech is also severely degraded as it is not easy for a DNN to distinguish in those segments between speech and noise. The noisy speech segments with very weak speech energy are very similar to those pure noise segments in terms of frame-level SNR, which is a challenge for the data-driven approaches using a single DNN. In the frame-level DNN-based SE, local SNR distribution is more meaningful than global (e.g. utterance-level) for learning convergence. From the Fig. 1, we observe that the frame-level SNR values have a high fluctuation from the global SNR at 0dB. This indicates that the training set with a fixed, global SNR is multifarious

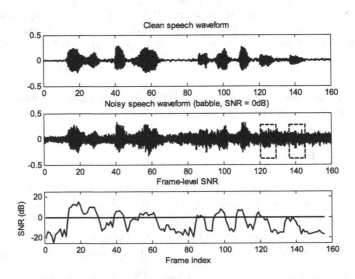

Fig. 1. Illustration of an utterance example in the babble noise environment at SNR = 0dB along with the corresponding clean speech and frame-level SNR sequence.

at frame level especially in low SNR conditions, and it will undoubtedly increase the difficulty of model learning.

In this paper, we propose a combined VAD+SE framework using DNNs in low SNR environments. The main contributions of this paper are summarized as follows: (i) We employ a system with dual outputs of speech features for both target and interference sources in the output layer as our baseline. (ii) We use the speech segments of the multi-condition training set using VAD [14–16] annotations from the corresponding clean speech to train a conservative speech enhancement (denoted as CSE) DNN model to well preserve the weak-energy speech segments in low SNR environments and conservatively remove the pure noise segments. (iii) A DNN-based VAD model is trained for system fusion. Empirical results demonstrate that the proposed framework can significantly improve the performance in low SNR environments.

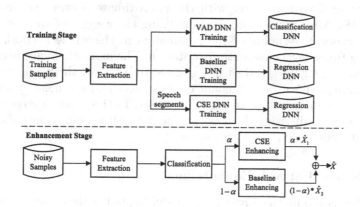

Fig. 2. The proposed system.

2   System Overview

The overall flowchart of the proposed SE system is illustrated in Fig. 2. First, the acoustic features of both clean speech and synthesized noisy speech training data are extracted. Then three DNNs, namely VAD DNN, baseline DNN and CSE DNN, are trained. In the enhancement stage, after feature extraction of the noisy utterance, frame-level soft decision is first given by the DNN-based VAD. To achieve better VAD performance, a long-term smoothing of the multiple DNN outputs with a half-window size τ can be applied. The classification DNN with smoothing is quite similar to the boosted DNN proposed in [17]. Then both the noisy features and speech/non-speech probabilities are presented to CSE and baseline system simultaneously. A fusion is performed with VAD classification probability to obtain the final enhanced speech signals as shown in Fig. 2. α is the probability of speech class, and $(1 - \alpha)$ belong to the non-speech class. $\hat{X}$,

$\hat{X}_1$ and $\hat{X}_2$ are the vectors of final enhanced speech, enhanced speech processed by CSE and by baseline system, respectively. This fusion can smooth the final enhanced speech and improve system performance. The details of both regression and classification DNNs are elaborated in Sect. 3.

3   DNN-based VAD and Speech Enhancement

3.1   DNN-based VAD

DNN for VAD is designed as a classification model where the output refers to the probabilities of two classes. The input to DNN is the noisy LPS features with neighboring frames. The training of this DNN consists of unsupervised pre-training and supervised fine-tuning. The former treats each consecutive pair of layers as a restricted Boltzmann machine (RBM) while the parameters of RBM are trained layer by layer with the approximate contrastive divergence algorithm [18]. After pre-training for initializing the weights of the first several layers, supervised fine-tuning of the parameters in the whole network is performed via a frame-level cross-entropy criterion. The main difference from other DNN approaches, e.g. [17], is the training data. In [17], only three noise types are used for training with a small amount of utterances and the noise types of the test set are the same as those of the training set. In this work, a large training set is formed by synthesizing the noisy speech data with a wide range of additive noises at different SNRs.

3.2   DNN-based Speech Enhancement

In [9], DNN was adopted as a regression model to predict the clean LPS features given the input noisy LPS features with acoustic context. This work improves the framework to predict the clean LPS and noise LPS features simultaneously in the output layer [10]. We believe the estimation of noise LPS will act as a regularization to the clean part. As for the DNN training, we first perform pre-training of a deep generative model with the LPS features of noisy speech by a stacking of multiple RBMs. Then the back-propagation with the MMSE-based objective function between the LPS features of the estimated and the reference (clean speech and noise) is adopted to train the DNN. Another two techniques, namely dropout training and noise-aware training (NAT) can be found in [19]. A stochastic gradient descent algorithm is performed in minibatches with multiple epochs to improve learning convergence as follows,

$$Er = \frac{1}{N} \sum_{n=1}^{N} (\beta \|\hat{X}_n^{\text{clean}} - X_n^{\text{clean}}\|_2^2 + (1 - \beta)\|\hat{X}_n^{\text{noise}} - X_n^{\text{noise}}\|_2^2) \qquad (1)$$

where $\hat{X}_n^{\text{clean}}$ and X_n^{clean} are the nth D-dimensional vectors of estimated and reference clean features, respectively. In the same way, $\hat{X}_n^{\text{noise}}$ and X_n^{noise} are the vectors of estimated and reference noise features. β is used to tune the contribution from the speech part and the noise part. As the noise variance is large and not stable, we mainly focus on the speech part. The second term of Eq. (1) can be considered as a regularization term, which leads to a better generalization capacity for estimating the clean speech. Another benefit from the dual outputs DNN is the estimation of noise can be used in the following ideal ratio mask (IRM) based post-processing module:

$$\widehat{IRM}_n(d) = \sqrt{\frac{\exp(\hat{X}_n^{clean}(d))}{\exp(\hat{X}_n^{clean}(d)) + \exp(\hat{X}_n^{noise}(d))}} \tag{2}$$

Different from [6] where the IRM is directly predicted by a well trained IRM-DNN, the IRM here is estimated by the DNN output for each dimension d, which is used for post-processing as follows

$$\hat{X}_n(d) = \begin{cases} Y_n(d) & \widehat{IRM}_n(d) > \gamma \\ \hat{X}_n^{clean}(d) & \widehat{IRM}_n(d) < \lambda \\ (\hat{X}_n^{clean}(d) + Y_n(d))/2 & otherwise \end{cases} \tag{3}$$

where, $\hat{X}_n$ and Y_n are the vectors of final enhanced speech and noisy speech, respectively. γ and λ are the thresholds to improve the overall performance.

4   Experimental Results and Analysis

4.1   Experimental Setup

In [9], 104 noise types were used as the noise signals for synthesizing the noisy speech training samples. In this study, we add another home-made 200 h real-world noises[1] to handle a wide range of additive noise in the real-world situations. 100 h clean Mandarin data collected by iFlytek were added with the above-mentioned background noises and 5 levels of SNR, at 20dB, 15dB, 10dB, 5dB and 0dB, to build a multi-condition stereo training set. The whole 100-hour training data was used for baseline system and VAD model training. As for VAD training, the frame-level reference labels of each noisy utterance were generated by conventional VAD tool on the corresponding clean utterance. Then, we use the speech segments of the multi-condition training set (about 60 h) for CSE model training. The training method is same with the baseline enhancement subsystem. The final joint DNN based SE system designed for low SNR environments was obtained under the framework illustrated in Fig. 2, denoted as JDNN-SE. Another 200 clean utterances covering 20 males and 17 females were used to construct the test set for each combination of noise types (NOISEX-92 corpus: babble and factory, real-recorded: mess hall and Karaoke Television (KTV)) and SNR levels (−5dB, 0dB, 5dB). All the noises, speakers and texts in test set are different from those in the training set.

---

[1] The noise types are vehicle: bus, train, plane and car; exhibition hall; meeting room; office; emporium; family living room; factory; bus station; mess hall; KTV; musical instruments.

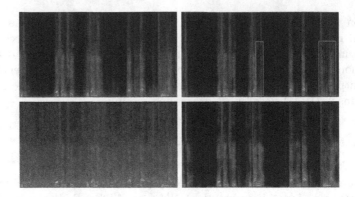

Fig. 3. Four spectrograms of an utterance corrupted by babble noise at 0dB SNR: JDNN-SE system (upper left, PESQ = 2.115), DNN baseline (upper right, PESQ = 1.585), noisy (bottom left, PESQ = 1.602) and clean speech (bottom right, PESQ = 4.5).

For both the regression DNN and classification DNN, sigmoid activation function was used and the number of units in each hidden layer was set to 2048 by default. The mini-batch size N was set to 128. The regularization weighting coefficient β in Eq. (1) was 0.8. γ and λ in Eq. (3) were set to 0.75 and 0.1, respectively. The other tuning parameters of DNN were set according to [19, 20]. The half-window size τ for VAD smoothing was 5. The performance was evaluated using two measures, namely short-time objective intelligibility (STOI) [21] and perceptual evaluation of speech quality (PESQ) [22] measures.

4.2   Results and Analysis

Table 1 gives a performance comparison of different DNN-based SE systems for the four unseen noise environments with different SNRs averaged on the test set. Noisy means the original noisy speech without any processing. The difference between Oracle and JDNN-SE is whether they use clean reference VAD annotations in the enhancement stage. Compared with the noisy speech results, baseline system showed that the speech quality is very poor at SNR = −5dB, and the performance was not satisfactory at SNR = 0dB. Our proposed JDNN-SE system overwhelmed baseline at all SNRs, especially at low SNRs, e.g., 0.333 PESQ improvement and 0.161 STOI improvement at SNR = −5dB in babble noise environment. Finally, the gap between JDNN-SE and Oracle was small compared with that between Baseline and JDNN-SE. This implied that our DNN-based VAD was effective and robust to noise types. Figure 3 presented spectrograms of an utterance. The improved DNN could enhance the speech with less speech distortion, especially at the noisy speech segments which are similar to noise. More results can be found at the demo website[2].

---

[2] http://home.ustc.edu.cn/~gtian09/demos/LowSNR-SEDNN.html.

Table 1. PESQ and STOI comparisons of four DNN-based SE systems averaged on the test sets for the four unseen noise conditions at different SNRs.

Noise type	SNR	PESQ				STOI			
		Noisy	Baseline	JDNN-SE	Oracle	Noisy	Baseline	JDNN-SE	Oracle
Babble	5dB	1.709	2.043	2.248	2.279	0.778	0.795	0.840	0.856
	0dB	1.341	1.307	1.732	1.802	0.678	0.603	0.717	0.758
	−5dB	1.057	0.793	1.126	1.174	0.567	0.396	0.557	0.606
Factory	5dB	1.594	1.990	2.300	2.353	0.778	0.761	0.839	0.861
	0dB	1.233	1.500	1.905	1.951	0.679	0.606	0.745	0.772
	−5dB	0.950	1.030	1.332	1.332	0.571	0.463	0.601	0.627
Mess hall	5dB	1.655	2.048	2.286	2.280	0.787	0.809	0.854	0.863
	0dB	1.311	1.506	1.895	1.894	0.689	0.664	0.761	0.778
	−5dB	1.057	0.927	1.272	1.291	0.579	0.478	0.609	0.633
KTV	5dB	1.885	2.347	2.416	2.403	0.829	0.874	0.885	0.891
	0dB	1.526	1.939	2.077	2.066	0.754	0.796	0.824	0.835
	−5dB	1.198	1.394	1.595	1.619	0.665	0.672	0.726	0.741

5 Conclusion

We have proposed an improved speech enhancement framework to increase speech intelligibility in low SNR environments. In this method, speech and non-speech frames are presented to specific subsystem separately. With frame-level VAD prediction and corresponding soft decision fusion, we obtain the final enhanced speech. The proposed joint DNN based SE system can yield a significant improvement when compared with our baseline, especially in low SNR conditions. As for future work, we will focus on designing multiple DNNs with even more detailed resolution at various frame-level SNRs.

References

1. Boll, S.: Suppression of acoustic noise in speech using spectral subtraction. IEEE Trans. Acoust. Speech Signal Process. **27**(2), 113–120 (1979)
2. Ephraim, Y., Malah, D.: Speech enhancement using a minimum mean-square error log-spectral amplitude estimator. IEEE Trans. Acoust. Speech Signal Process. **33**(2), 443–445 (1985)
3. Cohen, I.: Noise spectrum estimation in adverse environments: improved minima controlled recursive averaging. IEEE Trans. Speech Audio Process. **11**(5), 466–475 (2003)
4. Mohammadiha, N., Smaragdis, P., Leijon, A.: Supervised and unsupervised speech enhancement using nonnegative matrix factorization. IEEE Trans. Audio Speech Lang. Process. **21**(10), 2140–2151 (2013)
5. Wang, Y.X., Wang, D.L.: Towards scaling up classification-based speech separation. IEEE Trans. Audio Speech Lang. Process. **21**(7), 1381–1390 (2013)
6. Narayanan, A., Wang, D.L.: Ideal ratio mask estimation using deep neural networks for robust speech recognition. In: ICASSP, pp. 7092–7096 (2013)

7. Du, J., Huo, Q.: A speech enhancement approach using piecewise linear approximation of an explicit model of environmental distortions. In: INTERSPEECH, pp. 569–572 (2008)
8. Xu, Y., Du, J., Dai, L.-R., Lee, C.-H.: An experimental study on speech enhancement based on deep neural networks. IEEE Signal Process. Lett. 21(1), 65–68 (2014)
9. Xu, Y., Du, J., Dai, L.-R., Lee, C.-H.: A regression approach to speech enhancement based on deep neural networks. IEEE/ACM Trans. Audio Speech Lang. Process. 23(1), 7–19 (2015)
10. Tu, Y.-H., Du, J., Xu, Y., Dai, L.-R., Lee, C.-H.: Speech separation based on improved deep neural networks with dual outputs of speech features for both target and interfering speakers. In: ISCSLP, pp. 250–254 (2014)
11. Du, J., Wang, Q., Gao, T., Xu, Y., Dai, L.-R., Lee, C.-H.: Robust speech recognition with speech enhanced deep neural networks. In: INTERSPEECH, pp. 616–620 (2014)
12. Gao, T., Du, J., Dai, L.-R., Lee, C.-H.: Joint training of front-end and back-end deep neural networks for robust speech recognition. In: ICASSP (2015, accepted)
13. Varga, A., Steeneken, H.J.: Assessment for automatic speech recognition: II. NOISEX-92: a database and an experiment to study the effect of additive noise on speech recognition systems. Speech Commun. 12(3), 247–251 (1993)
14. Sohn, J., Sung, W.: A voice activity detector employing soft decision based noise spectrum adaptation. In: ICASSP, pp. 365–368 (1998)
15. Sohn, J., Kim, N.S., Sung, W.: A statistical model-based voice activity detection. IEEE Signal Process. Lett. 6(1), 1–3 (1999)
16. Zhang, X.-L., Wu, J.: Deep belief networks based voice activity detection. IEEE Trans. Audio Speech Lang. Process. 21(4), 697–710 (2013)
17. Zhang, X.-L., Wang, D.L.: Boosted deep neural networks and multi-resolution cochleagram features for voice activity detection. In: INTERSPEECH, pp. 1534–1538 (2014)
18. Hinton, G.E., Osindero, S., Teh, Y.-W.: A fast learning algorithm for deep belief nets. Neural Comput. 18(7), 1527–1554 (2006)
19. Xu, Y., Du, J., Dai, L.-R., Lee, C.-H.: Dynamic noise aware training for speech enhancement based on deep neural networks. In: INTERSPEECH, pp. 2670–2674 (2014)
20. Hinton, G.E.: A practical guide to training restricted Boltzmann machines. In: Montavon, G., Orr, G.B., Müller, K.-R. (eds.) NN: Tricks of the Trade, 2nd edn. LNCS, vol. 7700, pp. 599–619. Springer, Heidelberg (2012)
21. Taal, C.H., Hendriks, R.C., Heusdens, R., Jensen, J.: A short-time objective intelligibility measure for time-frequency weighted noisy speech. In: ICASSP, pp. 4214–4217 (2010)
22. Rix, A.W., Beerends, J.G., Hollier, M.P., Hekstra, A.P.: Perceptual evaluation of speech quality (PESQ)-a new method for speech quality assessment of telephone networks and codecs. In: ICASSP, pp. 749–752 (2001)

Noise Perturbation Improves Supervised Speech Separation

Jitong Chen[✉], Yuxuan Wang, and DeLiang Wang

The Ohio State University, Columbus, OH 43210, USA
{chenjit,wangyuxu,dwang}@cse.ohio-state.edu

Abstract. Speech separation can be treated as a mask estimation problem where interference-dominant portions are masked in a time-frequency representation of noisy speech. In supervised speech separation, a classifier is typically trained on a mixture set of speech and noise. Improving the generalization of a classifier is challenging, especially when interfering noise is strong and nonstationary. Expansion of a noise through proper perturbation during training exposes the classifier to more noise variations, and hence may improve separation performance. In this study, we examine the effects of three noise perturbations at low signal-to-noise ratios (SNRs). We evaluate speech separation performance in terms of hit minus false-alarm rate and short-time objective intelligibility (STOI). The experimental results show that frequency perturbation performs the best among the three perturbations. In particular, we find that frequency perturbation reduces the error of misclassifying a noise pattern as a speech pattern.

Keywords: Speech separation · Supervised learning · Noise perturbation

1 Introduction

Speech separation is a task of separating target speech from noise interference. Monaural speech separation is proven to be very challenging as it only uses single-microphone recordings, especially in low SNR conditions. One way of dealing with this problem is to apply speech enhancement [6] on a noisy signal, where certain assumptions are made regarding general statistics of the background noise. The speech enhancement approach is usually limited to relatively stationary noises. Looking at the problem from another perspective, computational auditory scene analysis (CASA) exploits perceptual principles to speech separation. In CASA, interference can be reduced by applying masking on a time-frequency (T-F) representation of noisy speech. An ideal mask suppresses noise-dominant T-F units and keeps the speech-dominant T-F units. Therefore, speech separation can be treated as a mask estimation problem where supervised learning is employed to construct the mapping from acoustic features to a mask.

© Springer International Publishing Switzerland 2015
E. Vincent et al. (Eds.): LVA/ICA 2015, LNCS 9237, pp. 83–90, 2015.
DOI: 10.1007/978-3-319-22482-4_10

A binary decision on each T-F unit leads to an estimate of the ideal binary mask (IBM), which is defined as follows.

$$\text{IBM}(t, f) = \begin{cases} 1, & \text{if } \text{SNR}(t, f) > \text{LC} \\ 0, & \text{otherwise} \end{cases} \tag{1}$$

where t denotes time and f frequency. The IBM assigns the value 1 to a T-F unit if its SNR exceeds a local criterion (LC), and 0 otherwise. Therefore, speech separation is translated into a binary classification problem. IBM separation has been shown to improve speech intelligibility in noise for both normal-hearing and hearing-impaired listeners [9,13]. Alternatively, a soft decision on each T-F unit leads to an estimate of the ideal ratio mask (IRM). The IRM is defined below [10].

$$\text{IRM}(t, f) = \left(\frac{10^{(\text{SNR}(t,f)/10)}}{10^{(\text{SNR}(t,f)/10)} + 1} \right)^{\beta} \tag{2}$$

where β is a tunable parameter. A recent study has shown that $\beta = 0.5$ is a good choice for the IRM [15]. In this case, mask estimation becomes a regression problem where the target is the IRM. Ratio masking is shown to lead to slightly better objective intelligibility results than binary masking [15]. In this study, we use the IRM with $\beta = 0.5$ as the learning target.

In supervised speech separation, a training set is typically created by mixing clean speech and noise. When we train and test on a nonstationary noise such as a cafeteria noise, there can be considerable mismatch between training noise segments and test noise segments, especially when the noise resource used for training is restricted. In this study, we aim at expanding the noise resource using noise perturbation to improve the generalization of supervised speech separation. We treat noise expansion as a way to prevent a mask estimator from overfitting the training data. A recent study has shown that speech perturbation improves ASR [7]. However, our study perturbs noise instead of speech since we focus on separating target speech from highly nonstationary noises where the mismatch among noise segments is the major problem.

2 System Overview

To evaluate the effects of noise perturbation, we use a fixed system for mask estimation and compare the quality of estimated masks as well as the resynthesized speech that are derived from the masked T-F representations of noisy speech. As mentioned in Sect. 1, we use the IRM as the learning target. The IRM is computed from the 64-channel cochleagrams of premixed clean speech and noise. A cochleagram is a T-F representation of a signal. We use a 20 ms window and a 10 ms window shift to compute a cochleagram.

We perform IRM estimation using a deep neural network (DNN) and a set of acoustic features. Recent studies have shown that DNN is a strong classifier for ASR [1] and speech separation [16]. As shown in Fig. 1, acoustic features are extracted from a mixture sampled at 16 kHz, and then sent to a DNN for mask

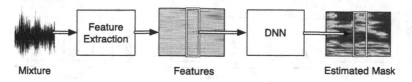

Fig. 1. Diagram of the proposed system.

prediction. To incorporate temporal context and obtain smooth mask estimation, we use 5 frames of features to estimate 5 frames of the IRM [15]. Therefore the output layer of the DNN has 64 × 5 units. Since each frame of the mask is estimated 5 times, we take the average of the 5 estimates. The acoustic features we extract from mixtures are a 59-D complementary feature set (AMS + RASTAPLP + MFCC) [14] combined with 64-D gammatone filterbank (GFB) features. To derive GFB features, an input signal is passed to a 64-channel gammatone filterbank, the response signals are decimated to 100 Hz to form 64-D GFB features.

We use hit minus false-alarm (HIT−FA) rate and short-time objective intelligibility (STOI) score [11] as two criteria for measuring the quality of the estimated IRM and the separated speech respectively. Since HIT−FA is defined for binary masks, we calculate it by binarizing a ratio mask to a binary one, following Eqs. 1 and 2. During the mask conversion, the LC is set to be 5 dB lower than the SNR of a given mixture. Both HIT−FA and STOI are well correlated with human speech intelligibity [8, 11].

3   Noise Perturbation

The goal of noise perturbation is to expand noise segments to cover unseen scenarios so that the overfitting problem is mitigated in supervised speech separation. A recent study has found that three perturbations on speech samples improve ASR performance [7]. These perturbations were used to expand the speech samples by spectral perturbation. The three perturbations are introduced below. Unlike this study, we perturb noise samples instead of perturbing speech samples, as we are dealing with highly nonstationary noises.

3.1   Noise Rate (NR) Perturbation

Speech rate perturbation, a way of speeding up or slowing down speech, is used to expand training utterances during the training of an ASR system. In our study, we extend the method to vary the rate of nonstationary noises. We increase or decrease noise rate by factor γ. When a noise rate is being perturbed, the value of γ is randomly selected from an interval $[\gamma_{min}, 2 - \gamma_{min}]$. The effect of NR perturbation on a spectrogram is shown in Fig. 2a.

3.2   Vocal Tract Length (VTL) Perturbation

VTL perturbation has been used in ASR to cover the variation of vocal tract length among speakers. A recent study suggests that VTL perturbation improves

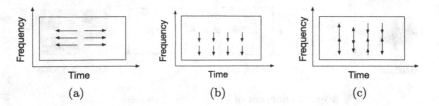

Fig. 2. (a) Illustration of noise rate perturbation. (b) Illustration of vocal tract length perturbation. (c) Illustration of frequency perturbation.

ASR performance [5]. VTL perturbation essentially compresses or stretches the medium and low frequency components of an input signal. We use VTL perturbation as a method of perturbing a noise segment. Specifically, we follow the algorithm in [5] to perturb noise signals:

$$f' = \begin{cases} f\alpha, & \text{if } f \le F_{hi}\frac{min(\alpha,1)}{\alpha} \\ \frac{S}{2} - \frac{\frac{S}{2} - F_{hi}min(\alpha,1)}{\frac{S}{2} - F_{hi}\frac{min(\alpha,1)}{\alpha}}(\frac{S}{2} - f), & \text{otherwise} \end{cases} \tag{3}$$

where f is the original frequency, f' is the mapped frequency, α is the wrapping factor, S is the sampling rate, and F_{hi} controls the cutoff frequency. The effect of VTL perturbation is visualized in Fig. 2b.

3.3   Frequency Perturbation

When frequency perturbation is applied, frequency bands of a spectrogram are randomly shifted upward or downward. We use the method described in [7] to randomly perturb noise samples. Frequency perturbation takes three steps. First, we randomly assign a value to each T-F unit, which is drawn from a uniform distribution.

$$r(f,t) \sim U(-1,1) \tag{4}$$

Then we derive the perturbation factor $\delta(f,t)$ by averaging the assigned values of neighboring T-F units. This averaging step avoids large oscillations in spectrogram.

$$\delta(f,t) = \frac{\lambda}{(2p+1)(2q+1)} \sum_{f'=f-p}^{f+p} \sum_{t'=t-q}^{t+q} r(f',t') \tag{5}$$

where p and q control the smoothness of the perturbation, and λ controls the magnitude of the perturbation. These tunable parameters are decided experimentally. Finally the spectrogram is perturbed as follows.

$$\tilde{S}(f,t) = S(f + \delta(f,t),t) \tag{6}$$

where $S(f,t)$ represents the original spectrogram and $\tilde{S}(f,t)$ is the perturbed spectrogram. Interpolation between neighboring frequencies is used when $\delta(f,t)$ is not an integer. The effect of frequency perturbation is visualized in Fig. 2c.

4 Experimental Results

4.1 Experimental Setup

We use the IEEE corpus recorded by a male speaker [4] and six nonstationary noises from the DEMAND corpus [12] to create mixtures. All signals are sampled at 16 KHz. Note that all recordings of the DEMAND corpus are made with a 16-channel microphone array, we use only one channel of the recordings since this study is on monaural speech separation. We choose six nonstationary noises (each is five-minute long) from the DEMAND corpus, each representing distinct environment: SCAFE noise (recorded in the terrace of a cafe at a public square), DLIVING noise (recorded inside a living room), OMEETING noise (recorded in a meeting room), PCAFETER noise (recorded in a busy office cafeteria), NPARK noise (recorded in a well visited city park) and TMETRO noise (recorded in a subway).

To create a mixture, we mix one IEEE sentence and one noise type at -5 dB SNR. This low SNR is selected with the goal of improving speech intelligibility in mind where there is not much to improve at higher SNRs [3]. The training set uses 600 IEEE sentences and randomly selected segments from the first two minutes of a noise, while the test set uses another 120 IEEE sentences and randomly selected segments from the second two minutes of a noises. Therefore, the test set has different sentences and different noise segments from the training set. We create 50 mixtures for each training sentence by mixing it with 50 randomly selected segments from a given noise, which results in a training set containing 600×50 mixtures. The test set includes 120 mixtures. We train and test using the same noise type and SNR condition.

To perturb a noise segment, we first apply short-time Fourier transform (STFT) to derive noise spectrogram, where a frame length of 20 ms and a frame shift of 10 ms are used. Then we perturb the spectrogram and derive a new noise segment. The parameters of perturbations are selected by using a development set. To evaluate the three noise perturbations, we create five different training sets, each consists of 600×50 mixtures. We train a mask estimator for each training set and evaluate on a fixed test set (i.e. the 120 mixtures created from the original noises). The five training sets are described as follows.

1. Original Noise: All mixtures are created using original noises.
2. NR Perturbation: Half of the mixtures are created from NR perturbed noises, and the other half are from original noises.
3. VTL Perturbation: Half of the mixtures are created from VTL perturbed noises, and the other half are from original noises.
4. Frequency Perturbation: Half of the mixtures are created from frequency perturbed noises, and the other half are from original noises.
5. Combined: Half of the mixtures are created from applying three perturbations altogether, and the other half are from original noises.

As already mentioned, we extract a set of four complementary features (AMS + RASTAPLP + MFCC + GFB) from mixtures. Delta features are appended to

Table 1. HIT−FA rate (in %) for six noises at −5 dB, where FA is shown in parentheses.

Perturbation \ Noise	SCAFE	DLIVING	OMEETING	PCAFETER	NPARK	TMETRO	Average
Original Noise	55 (37)	70 (23)	65 (28)	50 (40)	69 (22)	63 (32)	62 (30)
NR perturbation	64 (24)	77 (15)	72 (18)	60 (26)	77 (12)	72 (21)	70 (19)
VTL Perturbation	64 (24)	76 (16)	71 (19)	60 (27)	78 (10)	72 (21)	70 (20)
Frequency Perturbation	69 (17)	77 (14)	74 (15)	63 (21)	79 (9)	74 (18)	73 (16)
Combined	67 (21)	77 (15)	73 (16)	61 (25)	78 (10)	74 (18)	72 (18)

Table 2. STOI (in %) of separated speech for six noises at −5 dB, where STOI of unprocessed mixtures is shown in parentheses.

Perturbation \ Noise	SCAFE	DLIVING	OMEETING	PCAFETER	NPARK	TMETRO	Average
Original Noise	73.7 (64.1)	87.5 (79.3)	80.0 (67.8)	71.4 (62.5)	80.2 (67.7)	85.9 (77.5)	79.8 (69.8)
NR perturbation	76.5 (64.1)	89.2 (79.3)	82.5 (67.8)	74.1 (62.5)	83.2 (67.7)	87.4 (77.5)	82.1 (69.8)
VTL Perturbation	76.1 (64.1)	88.7 (79.3)	82.2 (67.8)	74.0 (62.5)	83.6 (67.7)	87.2 (77.5)	82.0 (69.8)
Frequency Perturbation	78.2 (64.1)	89.1 (79.3)	83.3 (67.8)	75.1 (62.5)	84.1 (67.7)	87.8 (77.5)	82.9 (69.8)
Combined	77.0 (64.1)	88.6 (79.3)	82.7 (67.8)	74.7 (62.5)	83.8 (67.7)	87.6 (77.5)	82.4 (69.8)

the feature set. A four-hidden-layer DNN is employed to learn the mapping from acoustic features to the IRM. Each hidden layer of the DNN has 1024 rectified linear units [1]. Dropout [1] and adaptive stochastic gradient descent [2] are used to train the DNN.

4.2   Evaluation Results and Comparisons

We evaluate the three perturbations with the five large training sets described in Sect. 4.1. The effects of noise perturbations on speech separation are shown in Tables 1 and 2, in terms of HIT−FA rate and STOI score respectively. The results indicate that all three perturbations lead to better speech separation than the baseline where only the original noises are used. Frequency perturbation performs better than the other two perturbations. Compared to only using the original noises, the frequency perturbed training set on average increases HIT−FA rate and STOI score by 11 % and 3 %, respectively. This indicates that noise perturbation is an effective technique for improving speech separation results. Combining three perturbations, however, does not lead to further improvement over frequency perturbation.

A closer look at Table 1 reveals that the contribution of frequency perturbation lies mainly in the large reduction in FA rate. This means that the problem of misclassifying noise-dominant T-F units as speech-dominant is mitigated. This effect can be illustrated by visualizing the masks estimated from the different training sets and the ground truth mask in Fig. 3a (e.g. around frame 150). When the mask estimator is trained with the original noises, it mistakenly retains the regions where target speech is not present, which can be seen by comparing

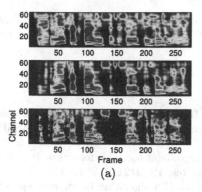
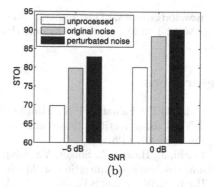

(a)   (b)

Fig. 3. (a) Mask comparison, the top shows a mask estimate using original noise, the middle shows a mask esimate using perturbed noise, the bottom shows the IRM. (b) The effect of frequency perturbation, the average STOI scores (in %) for six noises are shown for unprocessed speech, separated speech using original noise, and separated speech using frequency perturbed noise.

the top and bottom plots of Fig. 3a. Applying frequency perturbation to noises essentially exposes the mask estimator to more noise patterns and results in a more accurate mask estimator, which is shown in the middle plot of Fig. 3a. While HIT−FA rate evaluate the estimated binary masks, STOI directly compares clean speech and the resynthesized speech. As shown in Table 2, frequency perturbation yields higher average STOI scores than using original noises with no perturbation and NR and VTL perturbations.

Finally, to evaluate the effectiveness of frequency perturbation at a higher SNR, we carry out additional experiments at 0 dB input SNRs, where we use the same parameter values as for −5 dB SNR. Figure 3b shows frequency perturbation improves speech separation in terms of STOI in each SNR condition. Also, we find that frequency perturbation remains the most effective among the three perturbations at 0 dB SNR.

5   Concluding Remarks

In this study, we have explored the effects of noise perturbation on supervised monaural speech separation at low SNR levels. Noise perturbation is used to expand training noise to improve generalization of a classifier. We have evaluated three noise perturbations with six nonstationary noises recorded from daily life for speech separation. The three are noise rate, VTL, and frequency perturbations. With perturbed noises, the quality of the estimated ratio mask is improved as the classifier has been exposed to more scenarios of noise interference. In contrast, a mask estimator learned from a training set that only uses original noises tends to make more false alarm errors (i.e. higher FA rate). The experimental results show that frequency perturbation, which randomly perturbs the noise spectrogram along frequency, almost uniformly gives the best speech separation results among the three perturbations examined in terms of HIT−FA rate and STOI score.

Acknowledgments. This research was supported in part by an AFOSR grant (FA9550-12-1-0130), an NIDCD grant (R01 DC012048) and the Ohio Supercomputer Center.

References

1. Dahl, G.E., Sainath, T.N., Hinton, G.E.: Improving deep neural networks for LVCSR using rectified linear units and dropout. In: Proceedings of the ICASSP, pp. 8609–8613 (2013)
2. Duchi, J., Hazan, E., Singer, Y.: Adaptive subgradient methods for online learning and stochastic optimization. J. Mach. Learn. Res. **12**, 2121–2159 (2011)
3. Healy, E.W., Yoho, S.E., Wang, Y., Wang, D.L.: An algorithm to improve speech recognition in noise for hearing-impaired listeners. J. Acoust. Soc. Am. **134**, 3029–3038 (2013)
4. IEEE: IEEE recommended practice for speech quality measurements. IEEE Trans. Audio Electroacoust. 17, 225–246 (1969)
5. Jaitly, N., Hinton, G.E.: Vocal Tract Length Perturbation (VTLP) improves speech recognition. In: Proceedings of the ICML Workshop on Deep Learning for Audio, Speech and Language Processes (2013)
6. Jensen, J., Hendriks, R.C.: Spectral magnitude minimum mean-square error estimation using binary and continuous gain functions. IEEE Trans. Audio, Speech, Lang. Process. **20**, 92–102 (2012)
7. Kanda, N., Takeda, R., Obuchi, Y.: Elastic spectral distortion for low resource speech recognition with deep neural networks. In: Proceedings of the ASRU, pp. 309–314 (2013)
8. Kim, G., Lu, Y., Hu, Y., Loizou, P.C.: An algorithm that improves speech intelligibility in noise for normal-hearing listeners. J. Acoust. Soc. Am. **126**, 1486–1494 (2009)
9. Li, N., Loizou, P.C.: Factors influencing intelligibility of ideal binary-masked speech: Implications for noise reduction. J. Acoust. Soc. Am. **123**, 1673–1682 (2008)
10. Narayanan, A., Wang, D.: Ideal ratio mask estimation using deep neural networks for robust speech recognition. In: Proceedings of the ICASSP, pp. 7092–7096 (2013)
11. Taal, C.H., Hendriks, R.C., Heusdens, R., Jensen, J.: An algorithm for intelligibility prediction of time-frequency weighted noisy speech. IEEE Trans. Audio, Speech, Lang. Process. **19**, 2125–2136 (2011)
12. Thiemann, J., Ito, N., Vincent, E.: The diverse environments multi-channel acoustic noise database: A database of multichannel environmental noise recordings. J. Acoust. Soc. Am. **133**, 3591 (2013)
13. Wang, D.L., Kjems, U., Pedersen, M.S., Boldt, J.B., Lunner, T.: Speech intelligibility in background noise with ideal binary time-frequency masking. J. Acoust. Soc. Am. **125**, 2336–2347 (2009)
14. Wang, Y., Han, K., Wang, D.L.: Exploring monaural features for classification-based speech segregation. IEEE Trans. Audio, Speech, Lang. Process. **21**, 270–279 (2013)
15. Wang, Y., Narayanan, A., Wang, D.L.: On training targets for supervised speech separation. IEEE/ACM Trans. Audio, Speech, Lang. Process. **22**, 1849–1858 (2014)
16. Wang, Y., Wang, D.L.: Towards scaling up classification-based speech separation. IEEE Trans. Audio, Speech, Lang. Process. **21**, 1381–1390 (2013)

Speech Enhancement with LSTM Recurrent Neural Networks and its Application to Noise-Robust ASR

Felix Weninger[1](✉), Hakan Erdogan[2,3], Shinji Watanabe[2],
Emmanuel Vincent[4], Jonathan Le Roux[2], John R. Hershey[2],
and Björn Schuller[5]

[1] Machine Intelligence and Signal Processing Group, TUM, Munich, Germany
weninger@tum.de
[2] Mitsubishi Electric Research Laboratories, Cambridge, MA, USA
[3] Sabanci University, Istanbul, Turkey
[4] Inria, Villers-les-nancy, France
[5] Department of Computing, Imperial College London, London, UK

Abstract. We evaluate some recent developments in recurrent neural network (RNN) based speech enhancement in the light of noise-robust automatic speech recognition (ASR). The proposed framework is based on Long Short-Term Memory (LSTM) RNNs which are discriminatively trained according to an optimal speech reconstruction objective. We demonstrate that LSTM speech enhancement, even when used 'naïvely' as front-end processing, delivers competitive results on the CHiME-2 speech recognition task. Furthermore, simple, feature-level fusion based extensions to the framework are proposed to improve the integration with the ASR back-end. These yield a best result of 13.76 % average word error rate, which is, to our knowledge, the best score to date.

1 Introduction

Supervised training of speech enhancement schemes is becoming increasingly popular especially in the context of single-channel speech enhancement in non-stationary noise [8,16]. There, the source separation problem is formulated as a regression task: determine a time-frequency mask for separating the wanted source, based on acoustic features such as the magnitude spectrogram. Due to their ability to capture the temporal dynamics of speech, RNNs have delivered particularly promising results in the context of regression-based speech enhancement [2,16]. In contrast, the performance of RNN-based speech recognition in noisy conditions is still limited when compared to feedforward deep neural network (DNN) based systems [3,15]. Building on these results, the contributions of this paper are threefold: First, we demonstrate that gains from recent RNN-based speech enhancement methods translate to significant WER improvements. Second, we show a simple, yet very effective method to integrate speech enhancement and recognition by early feature-level fusion in a discriminatively trained DNN acoustic model. Third, we provide a systematic comparison

© Springer International Publishing Switzerland 2015
E. Vincent et al. (Eds.): LVA/ICA 2015, LNCS 9237, pp. 91–99, 2015.
DOI: 10.1007/978-3-319-22482-4_11

of single-channel and two-channel methods, showing that RNN-based single-channel enhancement can yield a recognition performance that is on par with the previous best two-channel system, and at the same time is complementary to two-channel pre-processing.

2   Speech Enhancement Methods

In this work, we consider speech enhancement based on the prediction of time-frequency masks from the magnitude spectrum of a noisy signal. Given an estimated mask $\hat{\mathbf{m}}_t$ for the time frame t, an estimate of the speech magnitudes $|\hat{\mathbf{s}}_t|$ is determined as $|\hat{\mathbf{s}}_t| = \hat{\mathbf{m}}_t \otimes |\mathbf{x}_t|$, where $\mathbf{x}_t$ is the short-term spectrum of the noisy speech and $\otimes$ denotes elementwise multiplication.

In this work, speech separation generally uses the following signal approximation objective, whose minimization maximizes the SNR for the magnitude spectra in each time-frequency bin, and hence directly optimizes for source reconstruction:

$$E^{\mathsf{SA}}(\hat{\mathbf{m}}) = \sum_{f,t} \left(|\hat{s}_{f,t}| - |s_{f,t}| \right)^2 = \sum_{f,t} \left(\hat{m}_{f,t}|x_{f,t}| - |s_{f,t}| \right)^2 . \tag{1}$$

Discriminatively Trained LSTM-DRNN. The above function can be applied to optimize any mask estimation scheme. Here, we consider deep recurrent neural networks (DRNNs), as proposed in [16]. The mask $\hat{\mathbf{m}}_t$ is estimated by the DRNN forward pass, which is defined as follows, for hidden layers $k = 1, \ldots, K - 1$ and time steps $t = 1, \ldots, T$:

$$\mathbf{h}_0^{1,\ldots,K-1} = \mathbf{0}, \tag{2}$$

$$\mathbf{h}_t^0 = |\mathbf{x}_t|, \tag{3}$$

$$\mathbf{h}_t^k = \mathcal{L}(\mathbf{W}^k[\mathbf{h}_t^{k-1}; \mathbf{h}_{t-1}^k; 1]), \tag{4}$$

$$\hat{\mathbf{m}}_t = \sigma(\mathbf{W}^K[\mathbf{h}_t^{K-1}; 1]). \tag{5}$$

Here $\mathcal{L}$ is the LSTM activation function [4], $\mathbf{h}_t^k$ denotes the hidden activations of layer k units at time step t, and σ is the logistic function. The weight matrices $\mathbf{W}^k$, $k = 1, \ldots, K$ are optimized according to (1) by backpropagation through time. There, only the gradient $\partial E^{\mathsf{SA}}/\partial\hat{\mathbf{m}}$ of the objective function with respect to the network output is specific to source separation, whereas the rest of the algorithm is unchanged. Using $\mathcal{L}$ instead of conventional sigmoid or half-wave activation functions helps reducing the vanishing temporal gradient problem of RNNs [5], allowing them to outperform DNNs with static context windows in speech enhancement [16].

Phase-Sensitive Discriminative Training. In [2], it was shown that using a phase-sensitive spectrum approximation (PSA) objective function instead of a magnitude-domain signal approximation (SA) improved source separation performance. The error in the complex short-time spectrum is related to the SNR in the time domain, hence if the network learns to reduce the complex domain error, this would clearly improve the reconstruction SNR. The PSA objective function is given below:

$$E^{\mathsf{PSA}}(\hat{\mathbf{m}}) = \sum_{f,t} |\hat{m}_{f,t} x_{f,t} - s_{f,t}|^2 \tag{6}$$

Note that the network does not predict the phase, but still predicts a masking function. The goal of the complex domain phase-sensitive objective function is to make the network learn to shrink the mask estimates when the noise is high. The exact shrinking amount is the cosine of the angle between the phases of the noisy and clean signals which is known during training but unknown during testing.

Integration of ASR Information. It can be conjectured that adding linguistic information, including word lexica and language models, to the spectro-temporal acoustic information used so far, can help neural network based speech separation. As in [2], we provided such information to the speech separating neural network in the form of additional 'alignment information' vectors appended to each frame's input features. The alignment information we use is derived from the alignment of the one-best decoded transcript at the HMM state-level. Given an active HMM state at a frame, the appended feature is the average of feature vectors that align to that state in the training data. Hence, the additional input has the same dimension as the noisy signal feature. In the results, we denote the neural networks using the additional alignment features as speech state aware (SSA).

Multi-channel Extension. In this work, we always use single-channel input to the neural networks. In case that a multi-channel signal is available, we first perform multi-channel pre-processing (here, delay-and-sum beamforming) prior to single-channel speech separation and recognition. The rationale is that training neural networks on multi-channel input is likely to overfit to the specific microphone placement seen in training, while traditional multi-channel signal processing methods allow for specifying this directly. As a model-based baseline for two-channel source separation, we use multi-channel non-negative matrix factorization (NMF) [9].

3   Experiments and Results

Our methods are evaluated on the corpus of the 2nd CHiME Speech Separation and Recognition Challenge (Track 2: medium vocabulary) [13]. The task is to estimate speech embedded in noisy and reverberant mixtures. Training, development, and test sets of two-channel noisy mixtures along with noise-free reference signals are created from the Wall Street Journal (WSJ-0) corpus of read speech and a corpus of noise recordings. The noise was recorded in a home environment with mostly non-stationary noise sources such as children, household appliances, television, radio, etc. The dry speech recordings are convolved with a time-varying sequence of room impulse responses from the same environment where the noise corpus is recorded. The training set consists of 7 138 utterances at six SNRs from −6 to 9 dB, in steps of 3 dB. The development and test sets consist of 410 and 330 utterances at each of these SNRs, for a total of 2 460 and 1 980 utterances. By construction of the WSJ-0 corpus, our evaluation is speaker-independent. Furthermore, the background noise in the development and test sets is disjoint from the noise in the training set, and a different room impulse response is used to convolve the dry utterances. In the CHiME-2 track 2 setup, the speaker is positioned at an approximate azimuth angle of 0 degrees, i.e., facing the microphone. This means that delay-and-sum beam-forming (BF) corresponds to simply adding the left and right channels. We will consider both BF as well as the left channel as front-ends.

The targets for supervised training according to (1) are derived from the parallel noise-free and multi-condition training sets of the CHiME data. The D(R)NN topology and training parameters were set as in [2,16]. For the NMF-SA baseline, the discriminative objective (1) is optimized as in [17].

Table 1. Speech enhancement results on CHiME-2 database using average of two channels by SDR.

[dB]	SDR (dev)	SDR (eval)						
	Avg	Input SNR [dB]						Avg
Enhancement		−6	−3	0	3	6	9	
BF	0.90	−2.55	−1.12	1.11	2.77	4.47	5.78	1.74
2ch-NMF	4.98	2.75	4.64	5.47	6.53	7.45	8.10	5.82
BF-LSTM-SA	13.19	10.46	11.85	13.40	14.86	16.34	18.07	14.17
BF-LSTM-PSA	13.50	10.97	12.28	13.76	15.13	16.57	18.26	14.49
BF-BLSTM-PSA	13.93	11.30	12.74	14.18	15.46	16.96	18.67	14.88
BF+SSA-BLSTM-PSA	14.11	11.57	12.92	14.33	15.62	17.13	18.81	15.07

3.1   Source Separation Evaluation

Our evaluation measure for speech separation is signal-to-distortion ratio (SDR) [14]. Since results for single-channel systems have already been reported previously [2,16,17], we restrict our evaluation to two-channel systems. In Table 1, we present the results of the same systems when using the channel average as front-end ('beam-forming', BF). Since the reference here is the channel average of the noise-free speech, the noisy baseline is lower than in the single-channel case [16]. We observe that the RNN-based systems outperform the noisy baseline, as well as two-channel NMF by a large margin, and that the gain over the noisy baseline is significantly higher (13.3 dB vs. 12.4 dB) in the two-channel case than in the single-channel case.

Table 2. WER on CHiME-2 database with DNN-HMM acoustic models using stereo training (predicting clean HMM states from noisy data) and sequence discriminative training, using enhanced speech features as input.

Enhancement	WER (dev)	WER (eval)						
	Avg	Input SNR [dB]						Avg
		−6	−3	0	3	6	9	
Single-channel systems								
None	29.39	40.31	30.00	23.37	17.88	15.02	13.86	23.41
NMF-SA [17]	28.38	37.57	28.88	22.23	16.25	14.55	12.63	22.02
LSTM-SA	23.99	30.92	23.26	18.72	14.35	12.85	11.68	18.63
LSTM-PSA	23.72	30.90	22.34	18.77	14.12	12.40	11.34	18.31
BLSTM-PSA	22.87	29.20	23.11	17.11	13.99	11.75	11.26	17.74
SSA-BLSTM-PSA	21.54	28.04	20.03	16.05	13.04	11.38	10.97	**16.58**
Two-channel systems								
BF	25.64	35.55	26.88	21.60	16.61	13.90	12.16	21.12
2ch-NMF	25.13	32.19	23.05	20.04	15.54	13.19	12.72	19.46
BF-LSTM-SA	19.03	24.86	17.65	15.11	11.41	10.20	9.68	14.82
BF-BLSTM-SA	18.35	23.76	17.92	14.48	11.58	9.86	9.19	14.47
BF+SSA-BLSTM-SA	18.41	24.38	16.74	14.80	11.06	9.23	9.32	14.25
BF+SSA-BLSTM-PSA	18.19	23.97	16.81	14.42	11.19	9.64	9.40	**14.24**

3.2 ASR Evaluation

In addition to the source separation measure, we also evaluate the speech separation techniques in terms of word error rate (WER). We use a state-of-the-art ASR setup with discriminatively trained DNN acoustic models. The number of tied HMM states, which are used as DNN targets, is 2,004, and the input feature of the DNN uses 5 left and right context frames of mel filterbank outputs ($40 \times 11 = 440$ dimensions) extracted from noisy and enhanced speech signals. In additional experiments, we also concatenate the noisy and enhanced speech features (i.e., $440 \times 2 = 880$ dimensions) inspired by deep stacking [1] and noise-aware training methods [8,11]. The DNN acoustic models have seven hidden layers, and each layer has 2,048 neurons. Acoustic models are trained with the following steps:

1. Restricted Boltzmann machine based layer-by-layer pretraining.
2. Cross entropy training with reference state alignments. Note that the state alignments are obtained from the Viterbi algorithm of clean signals (the original WSJ0 utterances) so that we can provide correct targets for the DNN [15].
3. Sequence discriminative training. We use the state-level minimum Bayes risk (sMBR) criterion [6] with 5 training iterations, where the lattices were recomputed after the first sMBR iteration [12].

All the experiments use a 5 k closed-vocabulary 3-gram language model.

Table 2 provides the WERs of the development and evaluation sets for each enhancement method. The LSTM methods clearly show an improvement from the baseline (None) and NMF (NMF-SA). Phase-sensitive (LSTM-PSA), bidirectional (BLSTM-PSA), and speech state aware (SSA-BLSTM-PSA) extensions of the LSTM achieve further gains from the standard LSTM (LSTM-SA) by 2.45 % (dev) and 2.05 % (eval) absolute. Similar results are obtained when we use the two-channel systems, and SSA-BLSTM-PSA with the beam-forming inputs (BF+SSA-BLSTM-PSA) finally achieved 18.19 % (dev) and 14.24 % (eval).

Table 3 shows the result of 'deep stacking' (concatenation of the noisy and enhanced features) for the best single/two channel systems in previous results, yielding additional improvements for each system. The final results of 17.87 % (dev) and 13.76 % (eval) are the best reported on this task so far.[1]

Table 3. WER on CHiME-2 database with DNN-HMM acoustic models using stereo training (predicting clean HMM states from noisy data) and sequence discriminative training, using enhanced and noisy speech features as input ('deep stacking').

Enhancement	WER (dev)	WER (eval)						
	Avg	Input SNR [dB]						Avg
		−6	−3	0	3	6	9	
Single-channel systems								
SSA-BLSTM-PSA	19.63	26.34	18.08	14.87	11.43	9.77	9.15	14.94
Two-channel systems								
BF+SSA-BLSTM-PSA	17.87	23.48	17.02	13.71	10.72	8.95	8.67	**13.76**

3.3 Relation Between Speech Recognition and Source Separation Performance

Figure 1 shows the relation of SDR and WER improvements over the single- and two-channel noisy baselines on the test set. Each point corresponds to a

---

[1] The 2nd CHiME challenge regulation forbids the use of parallel data, hence our results are out of competition.

measurement of SDR and WER for the utterances at a single SNR, with a single system shown in Tables 1, 2 and 3, and single-channel results taken from [2, 16, 17]. It can be seen that overall, SDR and WER improvements are significantly correlated (Spearman's rho = .84, $p \ll .001$). It seems that 2ch-NMF (lower left corner) is an outlier, yet we believe this can be explained by the fact that it is not discriminatively trained (unlike the single-channel version used here). Within the single-channel systems, we obtain an even stronger correlation of SDR and WER (Spearman's rho = .92).

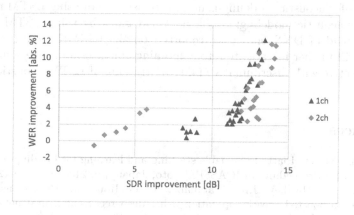

Fig. 1. Relation between improvements in source separation performance (SDR) and word error rate (WER).

4   Conclusions

We have shown that speech separation by recurrent neural networks can be used directly as a front-end for improving the noise robustness of state-of-the-art acoustic models for ASR. A competitive WER result of 14.47 % WER was achieved on the CHiME-2 speech recognition benchmark without deeper integration of source separation and acoustic modeling. This is interesting from a practical point of view, since it allows for a modular design of a noise-robust ASR system, where the same back-end can be used with or without front-end enhancement. Compared to a similar system that uses BF and DNN-based masking as a front-end for a DNN acoustic model [8], we obtain a 20 % relative improvement (from 18.0 %).

Furthermore, by pursuing deeper integration of front-end and back-end by means of two-pass enhancement and decoding, as well as a simple implementation of noise-aware training related to deep stacking, we were able to achieve best results on the CHiME-2 task. Compared to a system using joint training of DNN source separation and acoustic models (DNN-JAT) [8], which achieves a previous best result of 15.4 % WER, we obtain an 11 % relative WER reduction.

Furthermore, our best single-channel system is slightly better (3 % relative) than this previous best two-channel system.

In our results, we observe that back-end WER and front-end SDR are significantly correlated. This is interesting since it stands in contrast to earlier studies which found that SNR and word accuracy gains need not be strongly correlated [7]. However, these studies were carried out on different data and used a different source separation method. It will be highly interesting if, building on these results, one can find sufficient conditions for a good correlation of SNR and WER. Another notable finding is that stacking LSTM networks for source separation with DNNs for acoustic modeling is more promising than using LSTM networks directly for acoustic modeling: In [3], no WER gains by using LSTM acoustic models instead of DNN ones were reported on the CHiME-2 data. In future work, we will further investigate into combining our discriminative source separation objective with discriminative (sMBR) training of LSTM acoustic models as in [10].

References

1. Deng, L., Yu, D., Platt, J.: Scalable stacking and learning for building deep architectures. In: Proceedings of ICASSP, Kyoto, Japan, pp. 2133–2136 (2012)
2. Erdogan, H., Hershey, J.R., Watanabe, S., Le Roux, J.: Phase-sensitive and recognition-boosted speech separation using deep recurrent neural networks. In: Proceedings of ICASSP, Brisbane, Australia (2015)
3. Geiger, J.T., Zhang, Z., Weninger, F., Schuller, B., Rigoll, G.: Robust speech recognition using Long Short-Term Memory recurrent neural networks for hybrid acoustic modelling. In: Proceedings of INTERSPEECH, ISCA, Singapore (2014)
4. Graves, A., Mohamed, A., Hinton, G.: Speech recognition with deep recurrent neural networks. In: Proceedings of ICASSP, Vancouver, Canada, pp. 6645–6649 (2013)
5. Hochreiter, S., Schmidhuber, J.: Long short-term memory. Neural Comput. 9(8), 1735–1780 (1997)
6. Kingsbury, B., Sainath, T.N., Soltau, H.: Scalable minimum Bayes risk training of deep neural network acoustic models using distributed hessian-free optimization. In: Proceedings of ICASSP, Kyoto, Japan (2012)
7. Narayanan, A., Wang, D.L.: The role of binary mask patterns in automatic speech recognition in background noise. J. Acoust. Soc. Am. 133, 3083–3093 (2013)
8. Narayanan, A., Wang, D.L.: Improving robustness of deep neural network acoustic models via speech separation and joint adaptive training. IEEE/ACM Trans. Audio Speech Lang. Process. 23(1), 92–101 (2015)
9. Ozerov, A., Vincent, E., Bimbot, F.: A general flexible framework for the handling of prior information in audio source separation. IEEE Trans. Audio Speech Lang. Process. 20(4), 1118–1133 (2012)
10. Sak, H., Vinyals, O., Heigold, G., Senior, A., McDermott, E., Monga, R., Mao, M.: Sequence discriminative distributed training of long short-term memory recurrent neural networks. In: Proceedings of INTERSPEECH, ISCA, Singapore (2014)
11. Seltzer, M.L., Yu, D., Wang, Y.: An investigation of deep neural networks for noise robust speech recognition. In: Proceedings of ICASSP, Vancouver, Canada, pp. 7398–7402. IEEE (2013)

12. Veselỳ, K., Ghoshal, A., Burget, L., Povey, D.: Sequence-discriminative training of deep neural networks. In: INTERSPEECH, pp. 2345–2349 (2013)
13. Vincent, E., Barker, J., Watanabe, S., Le Roux, J., Nesta, F., Matassoni, M.: The second 'CHiME' speech separation and recognition challenge: datasets, tasks and baselines. In: Proceedings of ICASSP, Vancouver, Canada, pp. 126–130 (2013)
14. Vincent, E., Gribonval, R., Févotte, C.: Performance measurement in blind audio source separation. IEEE Trans. Audio Speech Lang. Process. **14**(4), 1462–1469 (2006)
15. Weng, C., Yu, D., Watanabe, S., Juang, B.H.: Recurrent deep neural networks for robust speech recognition. In: Proceedings of ICASSP, Florence, Italy, pp. 5569–5573 (2014)
16. Weninger, F., Hershey, J.R., Le Roux, J., Schuller, B.: Discriminatively trained recurrent neural networks for single-channel speech separation. In: Proceedings of GlobalSIP, pp. 740–744. IEEE, Atlanta (2014)
17. Weninger, F., Le Roux, J., Hershey, J.R., Watanabe, S.: Discriminative NMF and its application to single-channel source separation. In: Proceedings of INTER-SPEECH, Singapore (2014)

Adaptive Denoising Autoencoders: A Fine-Tuning Scheme to Learn from Test Mixtures

Minje Kim[1](✉) and Paris Smaragdis[2]

[1] Department of Computer Science, University of Illinois at Urbana-Champaign, Champaign, USA
minje@illinois.edu
[2] University of Illinois at Urbana-Champaign, Adobe Research, Champaign, USA
paris@illinois.edu

Abstract. This work aims at a test-time fine-tune scheme to further improve the performance of an already-trained Denoising AutoEncoder (DAE) in the context of semi-supervised audio source separation. Although the state-of-the-art deep learning-based DAEs show sensible denoising performance when the nature of artifacts is known in advance, the scalability of an already-trained network to an unseen signal with an unknown characteristic of deformation is not well studied. To handle this problem, we propose an adaptive fine-tuning scheme where we define a test-time target variables so that a DAE can learn from the newly available sources and the mixing environments in the test mixtures. In the proposed network topology, we stack an AutoEncoder (AE) trained from clean source spectra of interest on top of a DAE trained from a variety of available mixture spectra. Hence, the bottom DAE outputs are used as the input to the top AE, which is to check the purity of the once denoised DAE output. Then, the top AE error is used to fine-tune the bottom DAE during the test phase. Experimental results on audio source separation tasks demonstrate that the proposed fine-tuning technique can further improve the sound quality of a DAE during the test procedure.

Keywords: Deep learning · Deep neural networks · Autoencoders · Speech enhancement · Semi-supervised separation

1 Introduction

Recent advances in the deep learning research greatly improved the single-channel audio source separation performance as well. Most of the time, the Deep Neural Networks (DNN) commonly take a set of frequency coefficients of a short time period of the mixed signal, but there are three different choices for the output. First, the network can produce Ideal Binary Masks (IBM) [11], which are binary labels that tell us whether each frequency coefficient (T-F unit in

© Springer International Publishing Switzerland 2015
E. Vincent et al. (Eds.): LVA/ICA 2015, LNCS 9237, pp. 100–107, 2015.
DOI: 10.1007/978-3-319-22482-4_12

the cochleagram usually) belongs to the interesting source (e.g. speech) or not (e.g. noise). Second, a DNN can generate all the spectra of unmixed sources simultaneously [3]. This kind of models is more difficult to learn due to the higher dimension of the output layer, but they tend to produce even reconstruction qualities for all the sources. Third, a Denoising AutoEncoder (DAE) can take a noisy spectrum, and then outputs its cleaned-up version [5,13]. DAEs are common in deep learning as a feature learning technique, where the inputs are perturbed with some stationary noise [7,10]. In the source separation applications however, a DAE is trained with more realistic acoustic noise types.

There is no good ways for those DNNs for source separation to adapt to an unseen signal, while adapting a half-trained model during the test time has been a common idea in the source separation research in the name of *semi-supervised separation* [1]. It has a merit especially when an established separation model cannot efficiently represent a test signal with some unknown sources in it. To handle this problem, the semi-supervised separation systems build a part of the model for the desired source in advance, and then train the rest of the model from the residual of the test signal.

In this work we propose to vertically stack a pair of a DAE and an AutoEncoder (AE) as a source separation system that adapts to the unknown characteristics of the test signals. First, we train a DAE with an available set of noisy spectra and their corresponding clean spectra as the input and the output, respectively. However, we also consider the fact that the noisy spectra for training might not be diverse enough to cover all the variation of deformation that can happen in the real world, such as different types and levels of additive noise. As in the semi-supervised separation scenario, we improve this imperfectly trained DAE by fine-tuning it to minimize the test-time error we newly define. To this end, we set up another AE that is dedicated to produce a clean spectrum if its input is the clean spectra of a target source as well, which we also call a *purity checker*. It gives us a lower error if its input is clean and higher otherwise. During the separation phase, we first denoise the input using the bottom DAE. Then, we check on the purity of the DAE output by feeding it an input to the top AE. In this way, instead of a single path feedforward for the denoising job, we measure the quality of the once denoised spectrum and backpropagate the error of the top AE to fine-tune the bottom DAE.

We show that the proposed method can sensibly improve Signal-to-Interference Ratio (SIR), while sacrificing Signal-to-Artifact Ratio (SAR) a little. Therefore, there is a point where we get better Signal-to-Distortion Ratio (SDR).

2   Related Work

This section introduces semi-supervised NMF models and two-stage approaches, which are conceptually and structurally similar to our work, respectively.

2.1   Semi-supervised Source Separation

In the semi-supervised source separation methods, Nonnegative Matrix Factorization [4], or Probabilistic Latent Component Analysis (PLCA) [6] as an audio

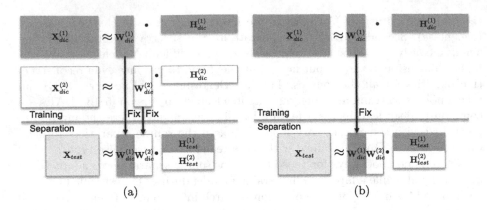

Fig. 1. (a) Supervised separation (b) Semi-supervised separation.

analogy of topic modeling, is a popular tool to discover latent structures of a mixed signal[1]. Figure 1 compares the two different strategies. In (a) we assume that a set of magnitudes of Fourier spectra, e.g. $\mathbf{X}_{dic}^{(1)}$ from clean speech, is available for the NMF algorithm to train the source specific bases, or a dictionary, $\mathbf{W}_{dic}^{(1)}$ and their temporal activations, $\mathbf{H}_{dic}^{(1)}$. As a fully supervised case, we also learn the second source's dictionary $\mathbf{W}_{dic}^{(2)}$ (e.g. "babble" noise). For the separation, we fix the two dictionaries during the final NMF learning, while both sets of their activations $\mathbf{H}_{test}^{(1)}$ and $\mathbf{H}_{test}^{(2)}$ are learned to best describe the unseen test input $\mathbf{X}_{test}$. Finally, we recover the source by multiplying its corresponding dictionary matrix and activations, e.g. $\mathbf{W}_{dic}^{(1)}\mathbf{H}_{test}^{(1)}$ for the first source.

In the semi-supervised scenario on the other hand, we assume that only a part of the sources is known (the first source is known in Fig. 1(b), while the other sources are not). However, by fixing only the first part of the dictionary matrix with the trained one $\mathbf{W}_{dic}^{(1)}$, and learning the other part $\mathbf{W}_{dic}^{(2)}$ along with the activations from the test spectra, we can still recover the test mixture in a semi-supervised fashion. This approach is particularly useful if we are not sure about the quality of the second source's training set, or if it does not exist at all.

2.2 Two-Stage Approaches

In the two-stage approaches [12], an additional NMF approximation helps improve the masking-based DNN results. First, a two-stage system uses its usual DNN-based T-F masking module as a front-end to denois the noisy speech signal. In the second stage of the system, NMF is employed to further improve the estimated speech signal to discover a lower-rank approximation of the denoised speech. The NMF part of the system works, because its job is to re-synthesize the speech estimate with a fixed set of clean NMF bases.

---

[1] In this section we use terminologies from NMF-based models without the loss of generality in the other latent variable models.

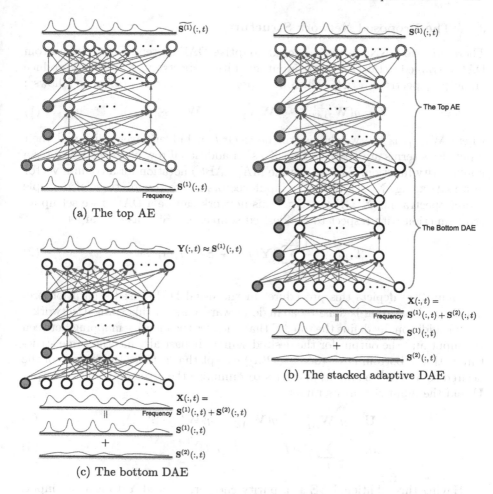

(a) The top AE

(b) The stacked adaptive DAE

(c) The bottom DAE

Fig. 2. The proposed system and DNNs as its building blocks.

If we consider NMF as an AE with a single hidden layer, we can expand the two-stage approaches by replacing the NMF module with a deeper AE trained from the clean source. However, the series of runs of DNN and NMF do not guarantee the adaptation we seek in this work, since the second stage merely works as a post-processor, and there is no chance to fine-tune the main DNN-based separation module.

3   The Proposed Adaptive Denoising Autoencoders

We propose the adaptive source separation system in this section. First, we will present the network structure and the test-time error function we define in Sect. 3.1. Then, we provide the details about the network settings in Sect. 3.2.

3.1   The Proposed Network Structure

There are two DNNs in the proposed adaptive DAE system. First, the bottom DAE is trained to take a mixture input magnitude spectrum $|\mathbf{X}(:,t)|$ and produce an output spectrum $\mathbf{Y}(:,t)$, which we express in the matrix notation as follows[2]:

$$\mathbf{Y} = g(\mathbf{W}_{DAE}^{L+1} \cdots g(\mathbf{W}_{DAE}^2 \cdot g(\mathbf{W}_{DAE}^1 \cdot \mathbf{X}))), \tag{1}$$

where $\mathbf{W}_{DAE}^l$ is the weight matrix between l-th hidden layer units and their input. Bias terms are absolved into $\mathbf{X}$ as an additional row of 1's. $g(\cdot)$ denotes a nonlinearity function. We also use the MATLAB® notation for a column vector in a matrix, e.g. $\mathbf{X}(:,t)$. Since the input vectors $\mathbf{X}$ are the mixtures of multiple source spectra, i.e. $\mathbf{X} = \mathbf{S}^{(1)} + \mathbf{S}^{(2)}$, this network can be a DAE if we set up an error function with respect to the desired source, e.g. $\mathbf{S}^{(1)}$, as the target,

$$\mathcal{E}_{DAE} = \frac{1}{2} \sum_{f,t} \left(\mathbf{Y}(f,t) - \mathbf{S}^{(1)}(f,t) \right)^2. \tag{2}$$

Figure 2(c) depicts this procedure. In the usual DAE-based separation scenario, the separation is done by a single forward pass on this trained network.

We additionally define the top AE that encodes the identity mapping between the input and the output for the desired source. It basically has the same structure with the bottom DAE (See Fig. 2(a)) except that it takes the spectra of the desired source $\mathbf{S}^{(1)}$ as input, and tries to minimize the error between its output $\mathbf{U}$ and the input $\mathbf{S}^{(1)}$ as its target:

$$\mathbf{U} = g(\mathbf{W}_{AE}^{L+1} \cdots g(\mathbf{W}_{AE}^2 \cdot g(\mathbf{W}_{AE}^1 \cdot \mathbf{S}^{(1)}))), \tag{3}$$

$$\mathcal{E}_{AE} = \frac{1}{2} \sum_{f,t} \left(\mathbf{U}(f,t) - \mathbf{S}^{(1)}(f,t) \right)^2. \tag{4}$$

Having this additional AE as a purity checker, we feed $\mathbf{Y}$ to it as an input. It will give us a smaller AE error $\mathcal{E}_{AE}$ if its input $\mathbf{Y}$ has produced a smaller DAE error $\mathcal{E}_{DAE}$ than bigger. It means the bottom DAE did its job well. On the other hand, we should expect a bigger $\mathcal{E}_{AE}$ value if $\mathbf{Y}$ was significantly different from $\mathbf{S}^{(1)}$. Using this concept, we can judge the quality of $\mathbf{Y}$ by checking on $\mathcal{E}_{AE}$ during the test-time without any help from the ground truth. Moreover, we can fine-tune the bottom DAE so that it can further reduce the error between its output $\mathbf{Y}$ and the top AE output $\mathbf{U}$,

$$\mathcal{E}_{AE} = \frac{1}{2} \sum_{f,t} \left(\mathbf{U}(f,t) - \mathbf{Y}(f,t) \right)^2, \tag{5}$$

which is nothing but the AE error. At every epoch, we backpropagate this error to update the bottom DAE to better separate the unseen mixture probably

---

[2] We drop the absolute function $|\cdot|$ from now on for brevity, but the readers should be aware that mixing in the time domain does not hold in the magnitude Fourier domain.

with unfamiliar sources and mixing environment. Finally, we construct a stacked adaptive DAE as in Fig. 2(b), whose bottom and top parts are the DAE to be fine-tuned and the AE as a purity checker, respectively.

3.2   The Proposed Network Setting

For both AEs, we train dropout networks [9] with dropout rates 0 % for the input and 20 % for the other hidden units. We use momentum with parameter 0.95. The main optimization is done by Stochastic Gradient Descent (SGD) with initial step size set to be 10^{-6}. Weights are bound to be between -1 to 1. Once we train both AEs, we first feedforward the new test mixture in the bottom DAE. With its output $\mathbf{Y}$, we do another feedforward in the top AE to get $\mathbf{U}$. Then, we calculate the AE error as in (5) to fine-tune the bottom DAE. We found that the same optimization setting works well for this fine-tuning job, too. The activation is a modified maxout function as suggested in [5],

$$g(x) = \begin{cases} x & \text{if } x \geq \epsilon \\ \frac{-\epsilon}{x-1-\epsilon} & \text{if } x < \epsilon. \end{cases} \tag{6}$$

4   Numerical Experiments

4.1   Speech and Noise

We train the bottom DAE with 10 random female speakers from TIMIT training data, each of which has 10 utterances. Thy are mixed with four different noise types used in [5], i.e. "Babble", "Airport", "Train" and "Subway." Eventually, there are 400 noisy utterances for training, which amount to 80,864 frames after short-time Fourier transform with 1024 pt of the frame size and a 75 % overlap. A square-root of Hann window is used for both analysis and synthesis. With the same setting, we also train an AE, but on the clean speech spectra as its input and target. Both networks have two hidden layers with 2048 hidden units per each layer. As for the test signals, we randomly choose 5 female speakers from the TIMIT test part, and add them up with eight different noise types: "Piano", "Drill", "Bus", "Birds", "Computer keyboard", "Frogs", "Machinegun", and "Street" (400 noisy utterances). We try two input Signal-to-Noise Ratio (SNR) choices: 0 and –5dB. Note that since the networks are trained on 0dB mixtures only, the –5dB test mixtures are more difficult for them to separate. Finally, we either use the spectrum as it is or after concatenating five consecutive frames to test the network with inputs with temporal structures.

The bottom DAE is not perfect, because we deliberately chose different kinds of test noise. Fine-tuning the bottom DAE is to reduce the top AE error in (5). By doing so, we get better DAE outputs $\mathbf{Y}$, in the sense of reducing the level of interfering sources, but in the meantime the recovered speech can also lose some of its energy. Figure 3 shows that the average SIR values increase as we keep fine-tuning, where 0-th epoch means no fine-tuning has been done. Since the SAR decreases more slowly, there is a better SDR value after several epochs inall cases.

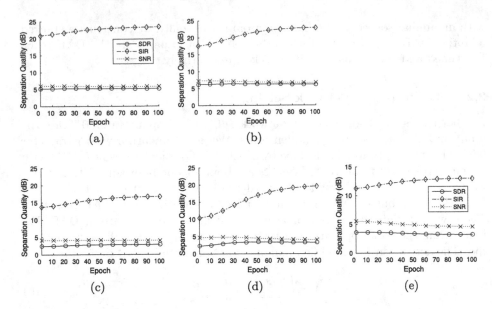

Fig. 3. The separation performance of speech enhancement experiments with different settings (a) Frame-by-frame; input SNR 0dB (b) 5 frames; input SNR 0dB (c) Frame-by-frame; input SNR –5dB (d) 5 frames; input SNR –5dB. (e) The performance on singing voice separation task.

In (d), where the test mixture was –5dB and the input is the vectorized 5 frames, both SIR and SDR improvements are most significant.

4.2   Singing Voice Separation

MIR-1K is a dataset with a thousand karaoke clips played by 19 amateur singers [2]. We followed the basic setting in [8] where only 175 clips from two singers are allowed to be used as the training set, while there are 825 test clips available. We consider the voice part as our desired source, and train the networks with three hidden layers and 2048 units per a layer. Three frames are vectorized to form an input. This lack of training data makes the top AE less reliable. However, we can still see in Fig. 3(e) that the proposed fine-tuning scheme can improve the average SIR values for the test samples.

5   Conclusion

We developed an adaptive source separation system, which consists of a bottom DAE and a top AE for the main separation module and a purity checker for the fine-tuning job, respectively. Although a good target variables are not usually available for the use during the test-time separation, we found that this additional well-trained AE on the clean spectra of the desired source can provide the main separation module with some alternative quality measurements.

References

1. Duan, Z., Mysore, G.J., Smaragdis, P.: Online PLCA for real-time semi-supervised source separation. In: Theis, F., Cichocki, A., Yeredor, A., Zibulevsky, M. (eds.) LVA/ICA 2012. LNCS, vol. 7191, pp. 34–41. Springer, Heidelberg (2012)
2. Hsu, C.L., Jang, J.S.: On the improvement of singing voice separation for monaural recordings using the MIR-1K dataset. IEEE Trans. Audio, Speech, Lang. Process. **18**(2), 310–319 (2010)
3. Huang, P., Kim, M., Hasegawa-Johnson, M., Smaragdis, P.: Deep learning for monaural speech separation. In: Proceedings of the IEEE International Conference on Acoustics, Speech, and Signal Processing (ICASSP), May 2014
4. Lee, D.D., Seung, H.S.: Algorithms for non-negative matrix factorization. In: Advances in Neural Information Processing Systems (NIPS), vol. 13. MIT Press (2001)
5. Liu, D., Smaragdis, P., Kim, M.: Experiments on deep learning for speech denoising. In: Proceedings of the Annual Conference of the International Speech Communication Association (Interspeech), Sep 2014
6. Raj, B., Smaragdis, P.: Latent variable decomposition of spectrograms for single channel speaker separation. In: Proceedings of the IEEE Workshop on Applications of Signal Processing to Audio and Acoustics, pp. 17–20 (2005)
7. Salakhutdinov, R., Hinton, G.: Semantic hashing. Int. J. Approximate Reasoning **50**(7), 969–978 (2009)
8. Sprechmann, P., Bronstein, A., Sapiro, G.: Real-time online singing voice separation from monaural recordings using robust low-rank modeling. In: Proceedings of the International Conference on Music Information Retrieval (ISMIR) (2012)
9. Srivastava, N., Hinton, G., Krizhevsky, A., Sutskever, I., Salakhutdinov, R.: Dropout: A simple way to prevent neural networks from overfitting. J. Mach. Learn. Res. **15**(1), 1929–1958 (2014)
10. Vincent, P., Larochelle, H., Bengio, Y., Manzagol, P.A.: Extracting and composing robust features with denoising autoencoders. In: Proceedings of the International Conference on Machine Learning (ICML), pp. 1096–1103 (2008)
11. Wang, Y., Wang, D.L.: Towards scaling up classification-based speech separation. IEEE Trans. Audio, Speech Lang. Process. **21**(7), 1381–1390 (2013)
12. Williamson, D.S., Wang, Y., Wang, D.L.: Reconstruction techniques for improving the perceptual quality of binary masked speech. J. Acoust. Soc. Am. **136**, 892–902 (2014)
13. Xu, Y., Du, J., Dai, L.R., Lee, C.H.: An experimental study on speech enhancement based on deep neural networks. IEEE Signal Process. Lett. **21**(1), 65–68 (2014)

Joint Analysis of Multiple Datasets, Data Fusion, and Related Topics

Joint Independent Subspace Analysis:
A Quasi-Newton Algorithm

Dana Lahat$^{(\boxtimes)}$ and Christian Jutten

GIPSA-Lab, UMR CNRS 5216, Grenoble Campus,
BP46, 38402 Saint-Martin-d'Hères, France
Dana.Lahat@gipsa-lab.grenoble-inp.fr

Abstract. In this paper, we present a quasi-Newton (QN) algorithm for joint independent subspace analysis (JISA). JISA is a recently proposed generalization of independent vector analysis (IVA). JISA extends classical blind source separation (BSS) to jointly resolve several BSS problems by exploiting statistical dependence between latent sources across mixtures, as well as relaxing the assumption of statistical independence within each mixture. Algebraically, JISA based on second-order statistics amounts to coupled block diagonalization of a set of covariance and cross-covariance matrices, as well as block diagonalization of a single permuted covariance matrix. The proposed QN algorithm achieves asymptotically the minimal mean square error (MMSE) in the separation of multidimensional Gaussian components. Numerical experiments demonstrate convergence and source separation properties of the proposed algorithm.

Keywords: Blind source separation · Independent vector analysis · Independent subspace analysis · Joint block diagonalization

1 Introduction

In this paper, we present a new algorithm for joint independent subspace analysis (JISA) [1]. JISA is a blind source separation (BSS) framework inspired by two recently-proposed extensions to BSS that until recently have been dealt with only separately: (1) relaxing the constraint that latent sources within a set of measurements must be statistically independent, sometimes termed independent subspace analysis (ISA) [2–4], and (2) solving several classical BSS problems simultaneously by exploiting statistical dependencies between latent sources across sets of measurements, a model often known as independent vector analysis (IVA) [5]. JISA provides a new flexible way to exploit links between different datasets, and thus has the potential to be useful to data fusion. The JISA model, and a relative gradient (RG) algorithm that achieves optimal separation in terms of minimal mean square error (MMSE) in the presence of noiseless Gaussian data, were first presented in [1]. A gradient algorithm that performs JISA based

This work is supported by the project CHESS, 2012-ERC-AdG-320684. GIPSA-Lab is a partner of the LabEx PERSYVAL-Lab (ANR–11-LABX-0025).

E. Vincent et al. (Eds.): LVA/ICA 2015, LNCS 9237, pp. 111–118, 2015.
DOI: 10.1007/978-3-319-22482-4_13

on the multivariate Laplace distribution has recently been proposed [6]. The main contribution of this paper is a Newton-based algorithm that achieves optimal separation for Gaussian noise-free data and converges in a much smaller number of iterations than its RG counterpart [1].

Consider T observations of K vectors $\mathbf{x}^{[k]}(t)$, modelled as

$$\mathbf{x}^{[k]}(t) = \mathbf{A}^{[k]}\mathbf{s}^{[k]}(t) \quad 1 \leq t \leq T,\, 1 \leq k \leq K, \tag{1}$$

where $\mathbf{A}^{[k]}$ are $M \times M$ invertible matrices that may be different $\forall k$, and $\mathbf{x}^{[k]}(t)$ and $\mathbf{s}^{[k]}(t)$ are $M \times 1$ vectors. Given the partition $\mathbf{s}^{[k]}(t) = [\mathbf{s}_1^{[k]\mathsf{T}}(t), \ldots, \mathbf{s}_N^{[k]\mathsf{T}}(t)]^\mathsf{T}$, where $\mathbf{s}_i^{[k]}(t)$ are $m_i \times 1$ vectors, $m_i \geq 1$, $\sum_{i=1}^N m_i = M$, $N \leq M$, $\cdot^\mathsf{T}$ denotes transpose, and the probability density function (pdf) of each random process $\mathbf{s}_i^{[k]}(t)$ irreducible in the sense that it cannot be factorized into a product of non-trivial pdfs, then each mixture (1) represents a single ISA [2–4] problem. The model that we define as JISA corresponds to linking several such standalone ISA problems via the assumption that the elements of the $n_i \times 1$ vector $\mathbf{s}_i(t) = [\mathbf{s}_i^{[1]\mathsf{T}}(t), \ldots, \mathbf{s}_i^{[K]\mathsf{T}}(t)]^\mathsf{T}$, where $n_i = K m_i$, are statistically *dependent*, whereas the pairs $(\mathbf{s}_i(t), \mathbf{s}_j(t))$ are statistically *independent* $\forall i \neq j \in \{1, \ldots, N\}$. JISA can be regarded as generalizing IVA since in IVA, $m_i = 1\ \forall i$ (which implies $N = M$).

In the rest of this paper, we focus on JISA using second-order statistics (SOS). In this case, further insights can be obtained by rewriting (1) as

$$\mathbf{x}(t) = \mathbf{A}\mathbf{s}(t) \tag{2}$$

where $\mathbf{s}(t) = [\mathbf{s}^{[1]\mathsf{T}}(t), \ldots, \mathbf{s}^{[K]\mathsf{T}}(t)]^\mathsf{T}$ and $\mathbf{x}(t) = [\mathbf{x}^{[1]\mathsf{T}}(t), \ldots, \mathbf{x}^{[K]\mathsf{T}}(t)]^\mathsf{T}$ are $L \times 1$ vectors, $L = KM$, $\mathbf{A} = \oplus_{k=1}^K \mathbf{A}^{[k]} \in \mathcal{B}_\mathbf{k}$ is a matrix direct sum, $\mathbf{k} = M\mathbf{1}_K$ the block-pattern of $\mathbf{A}$, and $\mathcal{B}_\mathbf{b}$ denotes the subspace of all invertible block-diagonal matrices with block-pattern $\mathbf{b}$. With these notations, $\widetilde{\mathbf{s}}(t) = \mathbf{\Phi}\mathbf{s}(t)$, where $\widetilde{\mathbf{s}}(t) = [\mathbf{s}_1^\mathsf{T}(t), \ldots, \mathbf{s}_N^\mathsf{T}(t)]^\mathsf{T}$ and $\mathbf{\Phi}$ is the corresponding $L \times L$ permutation matrix. Assuming sample independence $\forall t \neq t'$, the model assumptions imply that $\widetilde{\mathbf{S}} \triangleq E\{\widetilde{\mathbf{s}}(t)\widetilde{\mathbf{s}}^\mathsf{T}(t)\} = \begin{bmatrix} \mathbf{S}_{11} & 0 & 0 \\ 0 & \ddots & 0 \\ 0 & & \mathbf{S}_{NN} \end{bmatrix} = \oplus_{i=1}^N \mathbf{S}_{ii} \in \mathcal{B}_\mathbf{n}$, where $\widetilde{\mathbf{S}}$ is an $L \times L$ block-diagonal matrix with block-pattern $\mathbf{n} = [n_1, \ldots, n_N]^\mathsf{T}$ and $\widetilde{\mathbf{S}} = \mathbf{\Phi}\mathbf{S}\mathbf{\Phi}^\mathsf{T} \in \mathcal{B}_\mathbf{n}$. The linear model (2) implies that $\mathbf{X} = \mathbf{A}\mathbf{S}\mathbf{A}^\mathsf{T}$ where $\mathbf{S} = E\{\mathbf{s}(t)\mathbf{s}^\mathsf{T}(t)\}$ and $\mathbf{X} = E\{\mathbf{x}(t)\mathbf{x}^\mathsf{T}(t)\}$. In the sequel, we assume that all $\mathbf{S}_{ii}$ are invertible and do not contain values fixed to zero. Typical structures of some of these matrices are depicted in Fig. 1.

Figure of Merit: The above partition of $\mathbf{s}^{[k]}(t)$ induces a corresponding partition in the mixing matrices: $\mathbf{A}^{[k]} = [\mathbf{A}_1^{[k]} | \cdots | \mathbf{A}_N^{[k]}]$ with $\mathbf{A}_i^{[k]}$ the ith $M \times m_i$ column-block of $\mathbf{A}^{[k]}$. The multiplicative model (1) may now be rewritten as a sum of $N \leq M$ *multidimensional components*: $\mathbf{x}^{[k]}(t) = \sum_{i=1}^N \mathbf{x}_i^{[k]}(t)$ where the ith $M \times 1$ component $\mathbf{x}_i^{[k]}(t)$ is defined as $\mathbf{x}_i^{[k]}(t) = \mathbf{A}_i^{[k]}\mathbf{s}_i^{[k]}(t)$. In a blind context, the component vector $\mathbf{x}_i^{[k]}(t)$ is better defined than the source vector $\mathbf{s}_i^{[k]}(t)$. Indeed, for any invertible $m_i \times m_i$ matrix $\mathbf{Z}_{ii}^{[k]}$, it is impossible to discriminate between the representation of a component $\mathbf{x}_i^{[k]}(t)$ by the pair $(\mathbf{A}_i^{[k]}, \mathbf{s}_i^{[k]}(t))$ and $(\mathbf{A}_i^{[k]}\mathbf{Z}_{ii}^{-[k]}, \mathbf{Z}_{ii}^{[k]}\mathbf{s}_i^{[k]}(t))$, where $\mathbf{Z}_{ii}^{-[k]}$ denotes $(\mathbf{Z}_{ii}^{[k]})^{-1}$. This means that only the column space of $\mathbf{A}_i^{[k]}$, $\text{span}(\mathbf{A}_i^{[k]})$, can be blindly identified. Therefore, JISA is in fact a (joint) subspace estimation problem. Given $\mathbf{m} = [m_1, \ldots, m_N]$ and the set of observations $\mathcal{X} = \{\mathbf{x}^{[k]}(t)\}_{k=1,t=1}^{K,\ T}$, the problem that we define as JISA can thus be stated as estimating $\mathcal{A} = \{\mathbf{A}^{[k]}\}_{k=1}^K$ such that the components $\mathbf{x}_1(t), \ldots, \mathbf{x}_N(t)$, where $\mathbf{x}_i(t) = [\mathbf{x}_i^{[1]\mathsf{T}}(t), \ldots, \mathbf{x}_i^{[K]\mathsf{T}}(t)]^{\mathsf{T}}$, are as independent as possible. In the sequel, we set up a simple statistical model that, via its likelihood function, yields a quantitative measure of independence. Accordingly, we define the figure of merit as the mean square error (MSE) in the estimation of $\mathbf{x}_i(t)$,

$$\widehat{\mathrm{MSE}}_i = \frac{1}{\sigma_i^2} \frac{1}{T} \sum_{t=1}^T \|\widehat{\mathbf{x}}_i(t) - \mathbf{x}_i(t)\|^2, \tag{3}$$

where $\sigma_i^2 = E\{\|\mathbf{x}_i(t)\|^2\}$. For Gaussian data, estimates of $\mathbf{x}_i(t)$ obtained from matrices that are maximum likelihood (ML) estimates of $\mathcal{A}$ achieve asymptotically (i.e., $T \to \infty$) the MMSE [7].

Likelihood and Contrast Function: In the following, we consider a Gaussian model in which $\mathbf{s}_i(t) \sim \mathcal{N}(\mathbf{0}_{n_i \times 1}, \mathbf{S}_{ii})$ are mutually independent samples $\forall t \neq t'$. The log-likelihood for the model just described is [1] $\log p(\mathcal{X}; \mathcal{A}, \mathbf{S}) = -TD(\boldsymbol{\Phi}\mathbf{A}^{-1}\overline{\mathbf{X}}\mathbf{A}^{-\mathsf{T}}\boldsymbol{\Phi}^{\mathsf{T}}, \widetilde{\mathbf{S}}) - \kappa$, where $\mathcal{A} = \{\mathbf{A}^{[k]}\}_{k=1}^K$, and $\overline{\mathbf{X}} = \frac{1}{T}\sum_{t=1}^T \mathbf{x}(t)\mathbf{x}^{\mathsf{T}}(t)$ is the empirical counterpart of $\mathbf{X}$. The term $\kappa = \frac{T}{2}(\log\det(2\pi\overline{\mathbf{X}}) + L)$ is irrelevant to the maximization of the likelihood since it depends only on the data and not on the parameters. The scalar $D(\mathbf{R}_1, \mathbf{R}_2) = \frac{1}{2}(\text{tr}\{\mathbf{R}_1\mathbf{R}_2^{-1}\} - \log\det(\mathbf{R}_1\mathbf{R}_2^{-1}) - M)$, defined for any two $M \times M$ symmetric positive-definite matrices $\mathbf{R}_1$ and $\mathbf{R}_2$, is the Kullback-Leibler divergence (KLD) between the distributions $\mathcal{N}(\mathbf{0}, \mathbf{R}_1)$ and $\mathcal{N}(\mathbf{0}, \mathbf{R}_2)$ [8]. Given the block-diagonal structure of $\widetilde{\mathbf{S}}$, its ML estimate is [1] $\widetilde{\mathbf{S}}^{\mathrm{ML}} = \text{bdiag}_n\{\boldsymbol{\Phi}\mathbf{A}^{-1}\overline{\mathbf{X}}\mathbf{A}^{-\mathsf{T}}\boldsymbol{\Phi}^{\mathsf{T}}\}$. We can now write $\max_{\mathbf{S}} \log p(\mathcal{X}; \mathcal{A}, \mathbf{S}) = -TC(\mathbf{A}) + \kappa$, where in the latter we have defined the *contrast function*

$$C(\mathbf{A}) = D(\boldsymbol{\Phi}\mathbf{A}^{-1}\overline{\mathbf{X}}\mathbf{A}^{-\mathsf{T}}\boldsymbol{\Phi}^{\mathsf{T}}, \text{bdiag}_n\{\boldsymbol{\Phi}\mathbf{A}^{-1}\overline{\mathbf{X}}\mathbf{A}^{-\mathsf{T}}\boldsymbol{\Phi}^{\mathsf{T}}\}). \tag{4}$$

It holds that $D(\mathbf{R}, \text{bdiag}_b\{\mathbf{R}\}) \geq 0$ with equality if and only if (iff) $\mathbf{R} \in \mathcal{B}_b$. Hence, for any positive-definite matrix $\mathbf{R}$, $D(\mathbf{R}, \text{bdiag}_b\{\mathbf{R}\})$ is a measure of the block-diagonality of $\mathbf{R}$. Therefore, minimizing[1] the contrast function (4) amounts to *(approximate) block diagonalization* of $\overline{\mathbf{X}}$ by a permuted block-diagonal matrix $\boldsymbol{\Phi}\mathbf{A}^{-1}$. The RG of (4), $\nabla C(\mathbf{A}) = \text{bdiag}_k\{\boldsymbol{\Phi}^{\mathsf{T}}\text{bdiag}_n^{-1}\{\boldsymbol{\Phi}\mathbf{A}^{-1}\overline{\mathbf{X}}\mathbf{A}^{-\mathsf{T}}\boldsymbol{\Phi}^{\mathsf{T}}\}\boldsymbol{\Phi}\mathbf{A}^{-1}\overline{\mathbf{X}}\mathbf{A}^{-\mathsf{T}}\} - \mathbf{I}$, where $\text{bdiag}_m^{-1}\{\cdot\}$ stands for $(\text{bdiag}_m\{\cdot\})^{-1}$, was derived in [1]. Matrices that satisfy $\nabla C(\mathbf{A}) = \mathbf{0}$ are ML estimates of $\mathcal{A}$. This is the basis for the RG algorithm [1].

2   Derivation of the Approximate Hessian

The derivation of the Hessian is based on a relative perturbation of each $\mathbf{A}^{[k]}$ by $\widehat{\mathbf{A}}^{[k]} = \mathbf{A}^{[k]}(\mathbf{I}_M + \boldsymbol{\mathcal{E}}^{[k]})^{-1}\boldsymbol{\Lambda}^{[k]}$, where the $M \times M$ matrix $\boldsymbol{\mathcal{E}}^{[k]}$ reflects the *relative*

---

[1] We assume that an optimum exists.

change in $\mathbf{A}^{[k]}$, up to a scale ambiguity that is represented by the arbitrary invertible matrix $\Lambda^{[k]} \in \mathcal{B}_{\mathbf{m}}$. It can be shown [7] that the MSE (3) is invariant to $\Lambda^{[k]}$. The first-order expansion of the equations that satisfy $\nabla C(\mathbf{A}) = \mathbf{0}$ can be rewritten [7], for each pair $i \neq j$, as $\mathbf{e} = -\mathcal{H}^{-1}\mathbf{g} + \Omega(\frac{1}{T})$, where $\mathbf{e}$ and $\mathbf{g}$ are $2Km_im_j \times 1$ vectors, $\mathbf{e} = [\mathbf{e}_{ij}^\mathsf{T}\ \mathbf{e}_{ji}^\mathsf{T}]^\mathsf{T}$, $\mathbf{e}_{ij} = [\text{vec}^\mathsf{T}\{\mathcal{E}_{ij}^{[1]}\}\ \cdots\ \text{vec}^\mathsf{T}\{\mathcal{E}_{ij}^{[K]}\}]^\mathsf{T}$, $\mathbf{g} = [\mathbf{g}_{ij}^\mathsf{T}\ \mathbf{g}_{ji}^\mathsf{T}]^\mathsf{T}$,

$\mathbf{g}_{ij} = [\text{vec}^\mathsf{T}\{[\mathbf{S}_{ii}^{-1}\overline{\mathbf{S}}_{ij}]_{11}\}\ \cdots\ \text{vec}^\mathsf{T}\{[\mathbf{S}_{ii}^{-1}\overline{\mathbf{S}}_{ij}]_{KK}\}]^\mathsf{T}$, $\mathcal{H} = \begin{bmatrix} \mathbf{S}_{jj} \odot \mathbf{S}_{ii}^{-1} & \mathbf{I}_K \otimes \mathcal{T}_{m_j, m_i} \\ \mathbf{I}_K \otimes \mathcal{T}_{m_i, m_j} & \mathbf{S}_{ii} \odot \mathbf{S}_{jj}^{-1} \end{bmatrix}$

is a $2Km_im_j \times 2Km_im_j$ matrix that we assume invertible, $\otimes$ is the Kronecker product, and $\mathbf{S}_{jj} \odot \mathbf{S}_{ii}^{-1} = \begin{bmatrix} \mathbf{S}_{jj}^{[1,1]} \otimes [\mathbf{S}_{ii}^{-1}]_{11} & \cdots & \mathbf{S}_{jj}^{[1,K]} \otimes [\mathbf{S}_{ii}^{-1}]_{11} \\ \vdots & & \vdots \\ \mathbf{S}_{jj}^{[K,1]} \otimes [\mathbf{S}_{ii}^{-1}]_{K1} & \cdots & \mathbf{S}_{jj}^{[K,K]} \otimes [\mathbf{S}_{ii}^{-1}]_{KK} \end{bmatrix}$ is a $Km_im_j \times$

Km_im_j matrix whose (k,l)th block according to the partition $m_im_j\mathbf{1}_K$ is $\mathbf{S}_{jj}^{[k,l]} \otimes [\mathbf{S}_{ii}^{-1}]_{kl}$. The commutation matrix $\mathcal{T}_{P,Q} \in \mathbb{R}^{PQ \times PQ}$ is such that $\text{vec}\{\mathbf{M}^\mathsf{T}\} = \mathcal{T}_{P,Q}\text{vec}\{\mathbf{M}\}$ for any $\mathbf{M} \in \mathbb{R}^{P \times Q}$. $\Omega(f)$ stands for zero-mean stochastic terms whose standard deviation is proportional to f, or to higher powers thereof.

3   Algorithm

The approximation of the Hessian gives rise to a quasi-Newton (QN) algorithm, which is given in pseudocode in Algorithm 1. In line 5 of Algorithm 1 we introduce the operator $\text{vecbd}_{\alpha \times \beta}\{\mathbf{X}\} \triangleq [\text{vec}^\mathsf{T}\{\mathbf{X}_{11}\}\ \ldots\ \text{vec}^\mathsf{T}\{\mathbf{X}_{KK}\}]^\mathsf{T}$, where $\text{vecbd}_{\alpha \times \beta}\{\mathbf{X}\}$ is a vector that consists only of the (vectorized) entries of the main-diagonal blocks of matrix $\mathbf{X}$. Matrices $\mathbf{X}_{kk}$ are the blocks on the main diagonal of $\mathbf{X}$ where the rows of $\mathbf{X}$ are partitioned according to α and the columns by β. The difference from the RG algorithm is in lines 5–8, see [1, Algorithm 2].

4   Numerical Validation

In this section, we explore some numerical properties of the QN algorithm and validate its proper functioning. In all the following numerical experiments, the real positive definite matrices $\mathbf{S}_{ii}$ are generated as $\mathbf{S}_{ii} = \text{diag}^{-\frac{1}{2}}\{\mathbf{U}\Lambda\mathbf{U}^\mathsf{T}\}\mathbf{U}\Lambda\mathbf{U}^\mathsf{T}$ $\text{diag}^{-\frac{1}{2}}\{\mathbf{U}\Lambda\mathbf{U}^\mathsf{T}\}$, where $\mathbf{U}\Lambda\mathbf{V}^\mathsf{T}$ is the singular value decomposition (SVD) of a $Km_i \times Km_i$ matrix whose independent and identically distributed (i.i.d.) entries $\sim \mathcal{N}(0,1)$. The corresponding samples are created by right-multiplying the transpose of the Cholesky factorization of $\widetilde{\mathbf{S}}_{ii}$ with $Km_i \times T$ i.i.d. $\sim \mathcal{N}(0,1)$ numbers. $\mathbf{A}^{[k]}$ is arbitrary and thus, for simplicity, fixed to $\mathbf{I}$. In order to allow varying degrees of initialization, $\mathbf{A}_{\text{init}}^{[k]} = p\Upsilon + (1-p)\mathbf{I}$, $0 \leq p \leq 1$, where $p = 1$ implies fully random initialization. The entries of Υ are $\sim \mathcal{N}(0,1)$ i.i.d. and drawn anew for each new $\mathbf{A}_{\text{init}}^{[k]}$. The stopping threshold is set to 10^{-6}, and $T = 10^4$. In order to emphasize the difference between JISA and IVA, at each Monte Carlo (MC) trial, the algorithm is run twice on the same data, in two modes: in the first mode, the input parameter $\mathbf{m}$ (Algorithm 1 line 1) reflects the true data properties. In the second mode, the input $\mathbf{m}$ is set to $\mathbf{1}_{M \times 1}$. The latter amounts to

Algorithm 1. An iterative Newton-based algorithm for JISA

1: **function** JISA($\overline{\mathbf{X}}$, $\mathbf{\Phi}$, $\mathbf{A}_{\text{init}}$, $\mathbf{m}$, threshold, K)
2:     $\mathbf{A} \leftarrow \mathbf{A}_{\text{init}}$, $\mathbf{R} \leftarrow \overline{\mathbf{X}}$ ▷ initialization
3:     **while** $\|\nabla C\| >$ threshold **do**
4:         **for** i=2:N, j=1:i−1 **do**   ▷ Sweep on $(i, j \neq i)$
5:             $\mathbf{g} \leftarrow \begin{bmatrix} \text{vecbd}_{m_i 1_K \times m_j 1_K} \{\mathbf{R}_{ii}^{-1}\mathbf{R}_{ij}\} \\ \text{vecbd}_{m_j 1_K \times m_i 1_K} \{\mathbf{R}_{jj}^{-1}\mathbf{R}_{ji}\} \end{bmatrix}$
6:             $\mathcal{H} \leftarrow \begin{bmatrix} \mathbf{R}_{jj} \odot \mathbf{R}_{ii}^{-1} & \mathbf{I}_K \otimes \mathcal{T}_{m_j,m_i} \\ \mathbf{I}_K \otimes \mathcal{T}_{m_i,m_j} & \mathbf{R}_{ii} \odot \mathbf{R}_{jj}^{-1} \end{bmatrix}$
7:             Evaluate $\mathcal{E}_{ij}^{[k]}, \mathcal{E}_{ji}^{[k]}$, k=1,...,K
8:             Set $\{\mathcal{E}_{ij}^{[k]}, \mathcal{E}_{ji}^{[k]}\}_{k=1}^{K}$ in $\mathcal{E} = \oplus_{k=1}^{K} \mathcal{E}^{[k]}$
9:             $\mathbf{T} \leftarrow \mathbf{I} + \mathcal{E}$ ▷ $\mathcal{E}_{ii}^{[k]} = 0$
10:            $\mathbf{R} \leftarrow \mathbf{T}^{-1}\mathbf{R}\mathbf{T}^{-\intercal}$
11:            $\mathbf{A} \leftarrow \mathbf{A}\mathbf{T}$ ▷ For output only
12:        **end for**
13:        $\nabla C \leftarrow \text{bdiag}_{\mathbf{k}}\{\mathbf{\Phi}^{\intercal} \text{bdiag}_{\mathbf{n}}^{-1}\{\mathbf{\Phi}\mathbf{A}^{-1}\mathbf{R}\mathbf{A}^{-\intercal}\mathbf{\Phi}^{\intercal}\}\mathbf{\Phi}\mathbf{A}^{-1}\mathbf{R}\mathbf{A}^{-\intercal}\mathbf{\Phi}^{\intercal}\mathbf{\Phi}\} - \mathbf{I}$
14:    **end while**
15:    **return A**
16: **end function**

assuming that each $Km_i \times Km_i$ (irreducible, by definition) block on the diagonal of $\widetilde{\mathbf{S}}$ *can* be further reduced into m_i smaller blocks of dimension $K \times K$, i.e., ignoring the true subspace structure of the data. We denote this approach "mismodeling" [9].

4.1   Sensitivity to Initizalization and Number of Iterations

In order to focus here on the initialization, we avoid finite sample size errors by using data that can be exactly block-diagonalized. Figure 1 depicts typical such data, as well as outputs of QN, in mismodeling and correct model scenarios, for two types of initialization: mild ($p = 0.2$, Fig. 1(d)–(e)), and fully random ($p = 1$, Fig. 1(f)–(i)). A key observation is that JISA is sensitive to initialization: compare Fig. 1(e) with (g). In the latter, a fully random initialization results in an inability to recover the block structure. On the other hand, due to the convexity of the mismodeled algorithm (see [10]), it minimizes properly the *mismodeled* contrast function $\forall p$. However, minimizing the mismodeled contrast function does not imply separation: there is still need to cluster the blocks, as shows Fig. 1(h). In Fig. 1(i), we cluster by simple enumeration on all $M!$ possibilities. This is definitely not a viable approach. Further discussion of this topic is beyond the scope of this paper.

We now compare the QN and RG algorithms in terms of number of iterations. Both RG and QN minimize the same contrast function (4) and thus achieve the same MSE, up to numerical precision. In the experiments that follow, only $\mathbf{A}_{\text{init}}$ varies at each of the 100 MC trials, while $\mathbf{A}$, $\mathbf{S}$ and $\overline{\mathbf{S}}$ (the empirical counterpart of $\mathbf{S}$) remain fixed. In this example, $\mathbf{m} = [1, 2, 4]^{\intercal}$, $K = 4$. Figure 2 validates that

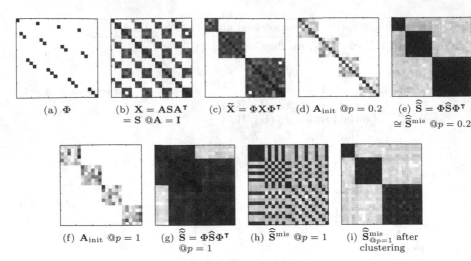

Fig. 1. Typical matrices and output of the QN algorithm, on error-free data ($\mathbf{X}$ is input to the algorithm). $\mathbf{A} = \mathbf{I}$, $\mathbf{m} = [1, 2, 3]^\mathsf{T}$, $K = 4$. In (d)–(e), $p = 0.2$. In (f)–(i), $p = 1$. In (e), the blocks are reconstructed properly both for JISA and mismodeling. Figure (g) is a typical case where a fully random initialization prohibits the proper reconstruction of the blocks in JISA. Figure (h)–(i) reflect the output in a mismodeling scenario before and after correcting the global permutation, respectively. False colours: in (b,c,e–i) we depict $\log 10 | \cdot |$ in order to enhance small numerical features. White=zero.

indeed, the QN algorithm improves over RG in terms of number of iterations. In addition, these results reflect the fact that in mismodeling, the algorithm is trying to block-diagonalize $\widetilde{\mathbf{S}}$ into smaller blocks than is actually possible and thus doing unnecessary work. These results conform with previous ones [1,11]. In this scenario, we set $p = 0.15$. This value guaranteed proper convergence to the correct minimum of the contrast function in all trials. For larger values of p, the number of iterations in the RG becomes prohibitive. Hence, in this respect, the advantage of the QN approach over RG is clear.

4.2 Component Separation

The component separation quality of the QN algorithm, quantified by its MSE, is illustrated in Fig. 3. In the following experiment, we run mutliple trials for fixed $\mathbf{S}$, $\mathbf{A}$, and $\mathbf{A}_{\text{init}}$. Only $\overline{\mathbf{S}}$ varies. For each trial we evaluate the normalized empirical MSE (3). As in the previous experiments, we compare JISA with its mismodeling counterpart. We set $\mathbf{m} = [6, 5, 1]^\mathsf{T}$, $K = 5$. Here, $M = 12$ which is prohibitive for enumeration (recall Fig. 1(h)–(i)) and thus we use $p = 0.2$ in order to have a good chance that the output is automatically clustered properly into the correct N blocks, before evaluating the MSE (3). In fact, the convexity of the "mismodeled" variant, mentioned in Sect. 4.1, no longer holds as m_i largely diverges from 1. In order to overcome this *for the error-analysis validation purpose only*, we choose a more strict initialization strategy in which in the first attempt $\mathbf{A}_{\text{init}}$ is taken from the output of the JISA run, and if the empirical

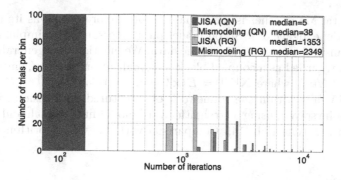

Fig. 2. Histogram of number of iterations in QN and RG, on the same data. Init with mild perturbation ($p = 0.15$): all runs converged properly. Logarithmic X-axis.

MSE indicates large errors, new $\mathbf{A}_{\text{init}}$ are generated according to the original procedure until no permutation issues are detected. Figure 3 illustrates our results. Subplot i corresponds to component i. The mean and standard deviation (std) of $\widehat{\text{MSE}}_i$ are written above the corresponding sublot, together with the theoretically predicted MSE for the JISA scenario [7]. The histograms represent 200 MC trials. The averaged MSE and its theoretical prediction for JISA are marked on the histograms. Figure 3 shows good fit between the predicted and empirical values. It also shows improved MSE from using the correct multidimensional model, including for the component with $m_i = 1$, as expected.

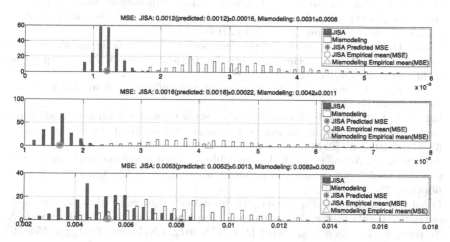

Fig. 3. Component separation. Histogram of the normalized empirical MSE for correct and mismodeling scenarios. Subplots correspond to components with dimensions 6, 5 and 1, respectively. $K = 5$, 200 trials.

Concluding Remarks: In this paper, we introduced a new Newton-based algorithm for JISA that achieves asymptotically optimal performance for Gaussian noise-free data. Many other issues remain to be explored, such as its numerical complexity, dependence of MSE on model parameters, efficient implementation, and behaviour in the presence of real-life data. We mention that generalizing [12], JISA can be regarded as a coupled block diagonalization problem, since $\mathbf{X}^{[k,l]} = \mathbf{A}^{[k]}\mathbf{S}^{[k,l]}\mathbf{A}^{[l]\mathsf{T}}\ \forall k,l$, where $\mathbf{X}^{[k,l]} = E\{\mathbf{x}^{[k]}(t)\mathbf{x}^{[l]\mathsf{T}}(t)\}$ and $\mathbf{S}^{[k,l]} \in \mathcal{B}_{\mathbf{m}}$. Consequently, JISA falls within the framework of structured data fusion (SDF) [13] and thus can be solved, using a straightforward model-fit approach and a Euclidean norm, using Tensorlab [14]. Comparison with the QN algorithm is left for future work.

References

1. Lahat, D., Jutten, C.: Joint blind source separation of multidimensional components: model and algorithm. In: Proceedings of EUSIPCO, Lisbon, Portugal, September 2014, pp. 1417–1421 (2014)
2. Comon, P.: Supervised classification, a probabilistic approach. In: Proceedings of ESANN, Brussels, Belgium, April 1995, pp. 111–128 (1995)
3. De Lathauwer, L., De Moor, B., Vandewalle, J.: Fetal electrocardiogram extraction by source subspace separation. In: Proceedings of IEEE SP/ATHOS Workshop on HOS, Girona, Spain, June 1995, pp. 134–138 (1995)
4. Cardoso, J.-F.: Multidimensional independent component analysis. In: Proceedings of ICASSP, Seattle, WA, May 1998, vol. 4, pp. 1941–1944 (1998)
5. Kim, T.-S., Eltoft, T., Lee, T.-W.: Independent vector analysis: an extension of ICA to multivariate components. In: Rosca, J.P., Erdogmus, D., Príncipe, J.C., Haykin, S. (eds.) ICA 2006. LNCS, vol. 3889, pp. 165–172. Springer, Heidelberg (2006)
6. Silva, R.F., Plis, S., Adalı, T., Calhoun, V.D.: Multidataset independent subspace analysis extends independent vector analysis. In: Proceedings of ICIP, Paris, France, October 2014, pp. 2864–2868 (2014)
7. Lahat, D., Jutten, C.: Joint independent subspace analysis using second-order statistics. GIPSA-Lab, Technical report hal-01132297, March 2015. https://hal.archives-ouvertes.fr/hal-01132297
8. Kullback, S., Leibler, R.A.: On information and sufficiency. Ann. Math. Statist. **22**(1), 79–86 (1951)
9. Lahat, D., Cardoso, J.-F., Messer, H.: Blind separation of multidimensional components via subspace decomposition: performance analysis. IEEE Trans. Sig. Process. **62**(11), 2894–2905 (2014)
10. Anderson, M., Adalı, T., Li, X.-L.: Joint blind source separation with multivariate gaussian model: algorithms and performance analysis. IEEE Trans. Sig. Process. **60**(4), 1672–1683 (2012)
11. Lahat, D., Cardoso, J.-F., Messer, H.: Joint block diagonalization algorithms for optimal separation of multidimensional components. In: Theis, F., Cichocki, A., Yeredor, A., Zibulevsky, M. (eds.) LVA/ICA 2012. LNCS, vol. 7191, pp. 155–162. Springer, Heidelberg (2012)
12. Li, X.-L., Adalı, T., Anderson, M.: Joint blind source separation by generalized joint diagonalization of cumulant matrices. Sig. Process. **91**(10), 2314–2322 (2011)
13. Sorber, L., Van Barel, M., De Lathauwer, L.: Structured data fusion. IEEE J. Sel. Top. Sig. Process. **9**(4), 586–600 (2015)
14. Sorber, L., Van Barel, M., De Lathauwer, L.: Tensorlab v2.0, January 2014. http://www.tensorlab.net/

Joint Decompositions with Flexible Couplings

Rodrigo Cabral Farias$^{(\boxtimes)}$, Jérémy Emile Cohen, Christian Jutten,
and Pierre Comon

GIPSA-Lab, UMR CNRS 5216, Grenoble Campus,
38400 Saint Martin d'Hères, France
rodrigo.cabral-farias@gipsa-lab.grenoble-inp.fr

Abstract. A Bayesian framework is proposed to define flexible coupling
models for joint decompositions of data sets. Under this framework, a
solution to the joint decomposition can be cast in terms of a maximum
a posteriori estimator. Examples of joint posterior distributions are provided, including general Gaussian priors and non Gaussian coupling priors. Then simulations are reported and show the effectiveness of this
approach to fuse information from data sets, which are inherently of different size due to different time resolution of the measurement devices.

Keywords: Tensor decompositions · Coupled decompositions · Data
fusion · Multimodal data · Heterogeneous data

1 Introduction

In domains such as brain imaging, metabolomics and link prediction, various
data gathering devices are used to collect information on some underlying phenomena. Since no device has a complete view of the phenomena, data fusion
can be used to blend the different views provided by each device, thus allowing
a broader understanding. It is then not surprising that multimodal data fusion
has become an important topic in these domains [2,8,18].

One way of defining a framework for heterogeneous data fusion is to state
it as a problem of latent variable analysis. Variations of the hidden variables
are supposed to explain most of the variations in the data sets. Since the latter
are considered to be different views of the same phenomena, part of the hidden
variables are related to each other. We can say that the latent models are coupled through subsets of their variables. By exploiting this coupling in the joint
estimation of the latent models, we expect that the information from one data
set will help in the estimation of the latent variables related to the other.

Although the framework described above dates back to the coupled (or
linked) tensor model described in [9], it was repopularized recently in [15], where

R.C. Farias—This research was supported in part by the ERC Grants AdG-2013-
320594 "DECODA" (R. Cabral Farias, J. E. Cohen and P. Comon) and AdG-
2012-320684 "CHESS" (Christian Jutten). This paper is a short version of a report
available online at: https://hal.archives-ouvertes.fr/hal-01135920/.

© Springer International Publishing Switzerland 2015
E. Vincent et al. (Eds.): LVA/ICA 2015, LNCS 9237, pp. 119–126, 2015.
DOI: 10.1007/978-3-319-22482-4_14

the problem of joint matrix factorization was considered under the constraint that one of the factors is shared by all matrices. In both cases the coupling occurs through equality constraints on latent factors. Following the work in [15], the framework of coupled tensor decompositions was revisited in [5,12]. Variations on this framework, such as tensor-matrix factorizations [1,3] and more general latent models [16] have also been proposed. Uniqueness and algorithm issues for the exact coupled tensor decomposition are addressed in [17], while algorithms are developed in [8,19] for general cost functions.

A more flexible model of coupling has been proposed in [2]. Instead of considering equality constraints for the entire factors of a tensor model, only a few components are constrained. In [14], the problem of coupled nonnegative matrix factorization is considered and a flexible coupling is proposed by assuming that the shared components are similar in L_1 or L_2 sense, and not equal. In this paper, we propose a generalization of flexible models for joint decompositions using a Bayesian approach. We subsequently give some examples of non trivial couplings, namely coupled factors that do not have the same size due to different sampling rates. Techniques of [2] or [14] cannot deal with this type of coupling.

The following notation will be used: scalars and vectors are denoted by lower case x and bold lower case $\boldsymbol{x}$ letters respectively. Matrices are denoted by upper case bold letters $\boldsymbol{X}$, while tensors by calligraphic letters $\mathcal{X}$. Elements of a given array are indicated by subscripts $\mathcal{X}_{ijk}$. Vectorization of parameters is indicated by vec($\cdot$). The Kronecker product of two matrices $\boldsymbol{X}$ and $\boldsymbol{Y}$ is denoted by $\boldsymbol{X} \boxtimes \boldsymbol{Y}$, while the Khatri-Rao product (column-wise Kronecker product) by $\boldsymbol{X} \odot \boldsymbol{Y}$. Left and right pseudo-inverses are denoted with superscript (†) on the left and on the right respectively.

2   Coupled Decompositions: A Bayesian Approach

Consider two arrays of measurements, $\mathcal{Y}$ and $\mathcal{Y}'$, which can be tensors of possibly different orders and dimensions. Arrays $\mathcal{Y}$ and $\mathcal{Y}'$ are related to two parametric models characterized by parameter vectors $\boldsymbol{\theta}$ and $\boldsymbol{\theta}'$, respectively.

For instance, if $\mathcal{Y}$ is a matrix (a second order tensor) to be diagonalized, the model can be the SVD $\mathcal{Y} = \boldsymbol{U}\boldsymbol{\Sigma}\boldsymbol{V}^H$, so that $\boldsymbol{\theta} = $ vec($\boldsymbol{U}, \boldsymbol{\Sigma}, \boldsymbol{V}$). If $\mathcal{Y}'$ is a third order tensor, its Canonical Polyadic (CP) decomposition [7,11] writes $\mathcal{Y}' = (\boldsymbol{A}', \boldsymbol{B}', \boldsymbol{C}')$, meaning that $\mathcal{Y}'_{ijk} = \sum_{r=1}^{R} A'_{ir}B'_{jr}C'_{kr}$, and $\boldsymbol{\theta}' = $ vec($[\boldsymbol{A}'^\top, \boldsymbol{B}'^\top, \boldsymbol{C}'^\top]^\top$).

In the case where $\boldsymbol{\theta}$ and $\boldsymbol{\theta}'$ are not coupled, they can be obtained (non uniquely in the matrix case) by processing the data arrays separately. On the other hand, if they are coupled, then the data needs to be processed jointly (and parameters are then uniquely estimated). We can consider that the coupling between $\boldsymbol{\theta}$ and $\boldsymbol{\theta}'$ is flexible, for example, we could have $\boldsymbol{V} \approx \boldsymbol{B}$, or even $\boldsymbol{V} \approx \boldsymbol{W}\boldsymbol{B}$ for a transformation matrix $\boldsymbol{W}$ that is known only approximately. To formalize this we assume that the pair $\boldsymbol{\theta}$, $\boldsymbol{\theta}'$ is random and that we have at our disposal a joint probability distribution $p(\boldsymbol{\theta}, \boldsymbol{\theta}')$.

Maximum a posteriori (MAP) estimator: the approximation setting under the MAP criterion becomes[1]

$$\arg\max_{\theta,\theta'} p(\theta,\theta'|\mathcal{Y},\mathcal{Y}') = \arg\max_{\theta,\theta'} p(\theta,\theta',\mathcal{Y},\mathcal{Y}') = \arg\min_{\theta,\theta'} \Upsilon(\theta,\theta'), \quad (1)$$

where $\Upsilon(\theta,\theta') = -\log p(\theta,\theta',\mathcal{Y},\mathcal{Y}')$. Conditioning on the parameters leads to a cost function that can be decomposed in a joint data likelihood term plus a term involving the coupling:

$$\Upsilon(\theta,\theta') = -\log p(\mathcal{Y},\mathcal{Y}'|\theta,\theta') - \log p(\theta,\theta'). \quad (2)$$

We assume conditional independence, that is $p(\mathcal{Y}|\mathcal{Y}',\theta,\theta') = p(\mathcal{Y}|\theta)$ and $p(\mathcal{Y}'|\mathcal{Y},\theta,\theta') = p(\mathcal{Y}'|\theta')$. In addition, we assume that the likelihoods $p(\mathcal{Y}|\theta)$ and $p(\mathcal{Y}'|\theta')$ and the joint prior, *e.g.* $p(V,B)$ or one of its conditional distributions, *e.g.* $p(V|B)$ or $p(B|V)$, are known. Then, we get the simplified cost:

$$\Upsilon(\theta,\theta') = -\log p(\mathcal{Y}|\theta) - \log p(\mathcal{Y}'|\theta') - \log p(\theta,\theta'). \quad (3)$$

where the first two terms are data related, and the last term is a penalization given by the coupling prior.

3   Examples

In what follows we consider that the parametric models underlying the data arrays are two CP models (A, B, C) and (A', B', C') with dimensions I, J, K and I', J', K' and rank (*i.e* number of matrix columns) R and R' respectively. We consider that the coupling occurs between matrices C and C', and exploit this framework with two different illustrating examples in Sects. 3.1 and 3.2.

3.1   Joint Gaussian Modeling

A general joint Gaussian model comprising coupled and uncoupled variables is given by the following expression:

$$M\left[\theta^\top \theta'^\top\right]^\top = \Sigma u + \mu, \quad (4)$$

where M is a matrix defining the structural relations between variables, u is a white Gaussian vector with zero mean and unit variances, Σ is a diagonal matrix of standard deviations and μ is a constant vector. Observe that a condition for the pair (θ,θ') to define a joint Gaussian vector is the left invertibility of M. Under this condition we have $[\theta^\top \theta'^\top]^\top \sim \mathcal{N}\{^\dagger M\mu, \Gamma\}$, where $\Gamma = {}^\dagger M^\top \Sigma^2 {}^\dagger M$ is the covariance matrix of the joint vector. The MAP objective function is

$$\Upsilon(\theta,\theta') = -\log p(\mathcal{Y}|\theta) - \log p(\mathcal{Y}'|\theta') + \{[\theta^\top \theta'^\top] - \mu^\top {}^\dagger M^\top\} \Gamma^{-1} \{[\theta^\top \theta'^\top]^\top - {}^\dagger M\mu\} \quad (5)$$

---

[1] We could also consider a minimum mean squared error setting but then we would need to evaluate $p(\mathcal{Y},\mathcal{Y}')$, which is normally cumbersome.

Sampling a Continuous Function: an important problem in multimodal data fusion is related to sampling. Different measurement devices have different sampling frequencies or even different non uniform sampling grids. In some situations, the continuous functions being measured can be approximated by a common function $c(t)$. For two sampled vectors c and c', their relation with the continuous function can be obtained with an interpolation kernel (see the general description in [4])

$$c(t) \approx \sum_{k=1}^{K} c_k h(t, t_k) \approx \sum_{k'=1}^{K'} c'_k h'(t, t'_k), \tag{6}$$

for some kernels $h(\cdot, \cdot)$ and $h'(\cdot, \cdot)$ and for sampling times $\{t_k, k \in 1, \cdots, K\}$ and $\{t_{k'}, k' \in 1, \cdots, K'\}$. Therefore, we can impose a new common sampling grid of size L where both interpolations should match. This leads to the linear relations $Hc \approx H'c'$, where $H_{lk} = h(t_l, t_k)$, $H'_{lk'} = h(t_l, t_{k'})$ and $\{t_l, l \in 1, \cdots, L\}$. As a consequence, in the coupled CP models, when C and C' have different dimensions due to different samplings, the coupling can be rewritten in the joint Gaussian setting as

$$\begin{bmatrix} 0 \ \mathrm{diag}(H) \ -\mathrm{diag}(H') \ 0 \end{bmatrix} \begin{bmatrix} \cdots \ \mathrm{vec}(C)^\top \ \mathrm{vec}(C')^\top \ \cdots \end{bmatrix}^\top = \Sigma u, \tag{7}$$

where $\mathrm{diag}(H)$ is a block diagonal matrix with repetitions of H on the diagonal.

3.2 Non Gaussian Conditional Coupling

Non trivial couplings between the factors C and C' can be considered by assuming that the coupling is given by a non Gaussian conditional distribution $p(C|C')$. When $C_{ij} > 0$ and $C'_{ij} > 0$, an additive Gaussian random coupling may not be the best option, since to ensure positiveness the support of the additive term has to depend on the values of C', which is not realistic. Therefore, other alternatives naturally ensuring positiveness can be considered, for example a Tweedie's conditional distribution [10]. Special cases of this distribution are the Poisson, Gamma and inverse-Gaussian distributions (the Gaussian distribution is a limit special case). In general, the PDF of the Tweedie's distribution has no analytical form, thus we cannot directly use it to write down a coupling term in the MAP objective function. However, if we consider that the coupling between $C_{ij} > 0$ and $C' > 0$ is strong (dispersion δ is small), then a saddle point approximation can be used [10]

$$p(C_{ij}|C'_{ij}) \approx (2\pi\delta^2 C_{ij}^\beta)^{-1/2} \exp[-d_\beta(C_{i,j}|C'_{i,j})/\delta^2], \tag{8}$$

where β is a shape parameter ($\beta = 1, 2, 3$ for the Poisson, Gamma and inverse-Gaussian distributions respectively) and d_β is the beta divergence [20]

$$d_\beta(C_{i,j}|C'_{i,j}) = C_{i,j}'^{1-\beta}/[(1-\beta)(2-\beta)] \left[C_{i,j}^{2-\beta} C_{i,j}'^{\beta-1} - C_{ij}(2-\beta) + C'_{ij}(1-\beta) \right]. \tag{9}$$

Under this conditional distribution the coupling term in the MAP objective becomes $\sum_{ij} \left[(\beta/2) \log(C_{ij}) + (1/\delta) d_\beta(C_{i,j}|C'_{i,j}) \right]$.

4 Gaussian Problem with Flat Priors and Alternating Least Squares (ALS)

From now on we will focus on the specific case of joint Gaussian modeling. Note that the joint decomposition can be found by minimizing (5). This can be done, for example, using an all-at-once minimization procedure such as a gradient algorithm.

If we consider that the priors are flat and that the coupling is of the form $G\mathrm{vec}(C) - G'\mathrm{vec}(C') = (I/\sigma_c^2)u$, where G and G' are two coupling matrices, I is the identity matrix, σ_c is a standard deviation related to the coupling intensity and u is a white Gaussian vector. Then the coupling term becomes a quadratic term on $\mathrm{vec}(C)$ and $\mathrm{vec}(C')$. Moreover, if we suppose that the two tensors $\mathcal{Y}$ and $\mathcal{Y}'$ are measured each with i.i.d. Gaussian noise with respective standard deviations σ_n and σ_n'. Then the objective function to be minimized is

$$\Upsilon = \frac{1}{\sigma_n^2}\|\mathcal{Y} - (A,B,C)\|_F^2 + \frac{1}{\sigma_n'^2}\|\mathcal{Y}' - (A',B',C')\|_F^2 + \frac{1}{\sigma_c^2}\|G\mathrm{vec}(C) - G'\mathrm{vec}(C')\|_F^2 \tag{10}$$

To minimize this function we can use an easy to implement algorithm, such as the alternating least squares (ALS) algorithm. Observe that the alternating updates for the uncoupled factors are simply the standard ALS steps for CP approximation, while for the coupled factors the updates are the solution of a joint least squares problem with a coupling term. The ALS procedure is the following, at iteration k:

$$\begin{aligned}
\hat{A}_k &= Y_{(1)}(\hat{C}_{k-1} \odot \hat{B}_{k-1})^\dagger & \hat{A}_k' &= Y'_{(1)}(\hat{C}'_{k-1} \odot \hat{B}'_{k-1})^\dagger, \\
\hat{B}_k &= Y_{(2)}(\hat{C}_{k-1} \odot \hat{A}_k)^\dagger & \hat{B}_k' &= Y'_{(2)}(\hat{C}'_{k-1} \odot \hat{A}_k')^\dagger,
\end{aligned} \tag{11}$$

$$\begin{bmatrix} \mathrm{vec}(\hat{C}_k) \\ \mathrm{vec}(\hat{C}_k') \end{bmatrix} = \begin{bmatrix} \frac{(F^\top F)\boxtimes I}{\sigma_n^2} + \frac{G^\top G}{\sigma_c^2} & -\frac{G^\top G'}{\sigma_c^2} \\ -\frac{G'^\top G}{\sigma_c^2} & \frac{(F'^\top F')\boxtimes I}{\sigma_n'^2} + \frac{G'^\top G'}{\sigma_c^2} \end{bmatrix}^\dagger \begin{bmatrix} (F\boxtimes I)\mathrm{vec}\left(Y_{(1)}\right) \\ (F'\boxtimes I)\mathrm{vec}\left(Y_{(2)}\right) \end{bmatrix}$$

where $F = \hat{B}_k \odot \hat{A}_k$, $F' = \hat{B}_k' \odot \hat{A}_k'$.

5 Simulations

To show the effects of the flexible coupling of two CP models on approximation performance, we apply the ALS algorithm presented in Sect. 4 to the problem of coupling two CP models with different sizes through interpolation. We consider $I = I' = J = J' = 10$ and different sizes on the third mode of the CP models $K = 37$ and $K' = 53$. We suppose that the components of C and C' are sampled versions of the same underlying continuous functions, however, the sampling periods to obtain the factors are different so that the elements in the factors cannot be similar, at least, in most of the points. Since the factors are not similar, we cannot apply the direct coupling model and we must use an interpolation approach as explained at the end of Sect. 3.

In this example we consider functions which are band limited and periodic. For an odd number of samples, the interpolation kernel is given by the Dirichlet kernel [13]

$$\boldsymbol{H}_{lk} = \frac{\sin\{K\pi[(l-1)T_i - (k-1)T]/[(L-1)T_i]\}}{K\sin\{\pi[(l-1)T_i - (k-1)T]/[(L-1)T_i]\}}, \tag{12}$$

where T is the original sampling period and T_i is the sampling period of the interpolation. As a consequence we have $\boldsymbol{G} = \boldsymbol{I} \boxtimes \boldsymbol{H}$ and $\boldsymbol{G}' = \boldsymbol{I} \boxtimes \boldsymbol{H}'$.

We simulate two random CP models with $R = R' = 3$. The components of the $\boldsymbol{C}$ factors are generated by sampling $c_r(t) = \sum_{i=1}^{3} \gamma_{ir} \sin(2\pi f_i t)$ where γ_{ir} are generated randomly and $f_1 = 2$, $f_2 = 2.5$, $f_3 = 3.5$. The sampling periods are $T = 1/9$ and $T' = 1/13$. An example of continuous-time component with its sampled points on different grids is shown in Fig. 1.

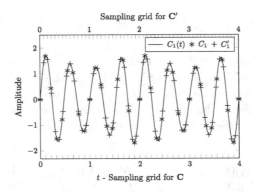

Fig. 1. Underlying continuous function $c_1(t)$ for the first components of the $\boldsymbol{C}$ factors and their corresponding sampled versions $\boldsymbol{c}_1$ and $\boldsymbol{c}_1'$ on different sampling grids.

We fix $\sigma_n' = 0.1$, while σ_n varies from 0.001 to 0.1, so that the ratio σ_n'/σ_n varies in the interval $[0.1, 10]$. Since the signals are band limited and since we observe them over a finite duration, interpolation can only approximate the continuous signal and it is necessary to set a nonzero σ_c even if the continuous signals are the same for both data sets. We set $L = 100$, $\sigma_c = 0.01$. The two CP models are first approximated separately (disregarding the coupling); in this case an all-at-once conjugate gradient algorithm is used. After convergence of the algorithm, the columns of the resulting factors are permuted so that the components in the coupling match. The permuted factors are then used to initialize the ALS procedure described in Sect. 4. We simulate 50 times this procedure with different noise realizations and we evaluate the total mean squared error (MSE) on the coupled factors. The results are shown in Fig. 2. When the noise ratio increases, the total MSE for the uncoupled approach increases sharply, while the coupled approach has a smooth increase. This shows that even though the coupled factors are not similar, information can still be exchanged between them though the interpolation.

We also investigated the case when the number of interpolation points for the coupling L is varied from 5 to 70. Both data arrays are noisy $\sigma_n = \sigma_n' = 0.1$

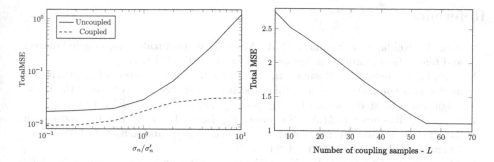

Fig. 2. Total MSE for the estimation of the C factors for different noise levels ratios σ_n/σ'_n. The CP models are coupled through interpolation.

Fig. 3. Total MSE for the estimation of the C factors as a function of the number of interpolation samples.

we set $\sigma_c = 0.001$ in the ALS algorithm. The total MSE is shown in Fig. 3. Note that only the sampling time points $t = \{0, 1, 2, 3, 4\}$ coincide in the original factors. Thus in a standard coupling approach only these points can be coupled and the total MSE that we obtain is the first point in the curve. By increasing the number of interpolation points the information exchanged within the model is larger and MSE decreases almost linearly. Above $L = 53$, only a small quantity of information can be exchanged because this is the maximum resolution present in the data and the total MSE curve becomes flat.

6   Conclusions

Since the expression of a phenomenon can be different in different data sets, it is clear that the link between factorizations of the data sets must be somehow flexible. To give a meaning to this flexibility we have proposed a Bayesian setting for factor couplings. Under this setting, we can model not only trivial flexible links between factors, but also joint Gaussian models and nonnegative similarity models. As an example of application of this framework to multimodal data fusion, we have presented the problem of fusing two data sets in which one dimension is different due to different sampling rates. Although the factors are almost completely different, the underlying hypothesis that they come from the same continuous-time function allows to exchange information between the data sets in an interpolated domain. A deeper analysis of the influence of noise will be reported in a longer paper [6].

In the simulation example, we have focused on a joint Gaussian modeling for the couplings; in future works we shall concentrate on non Gaussian couplings such as the Tweedie's coupling for nonnegative variables presented only briefly here. Moreover, since the CP approximation problem is an estimation problem, we can evaluate the Cramér-Rao bounds (CRB) for the coupled problem so that we can assess approximately the estimation performance without resorting to extensive simulation. In the hard coupling case a constrained CRB can be considered, while in the fully Bayesian and Bayesian with flat priors cases, the Bayesian and hybrid CRB can be considered.

References

1. Acar, E., Kolda, T.G., Dunlavy, D.M.: All-at-once optimization for coupled matrix and tensor factorizations. arXiv preprint arXiv:1105.3422 (2011)
2. Acar, E., Lawaetz, A.J., Rasmussen, M.A., Bro, R.: Structure-revealing data fusion model with applications in metabolomics. In: Conference Proceedings of the IEEE Engineering in Medicine and Biology Society, pp. 6023–6026. IEEE (2013)
3. Acar, E., Rasmussen, M.A., Savorani, F., Næs, T., Bro, R.: Understanding data fusion within the framework of coupled matrix and tensor factorizations. Chemometr. Intell. Lab. **129**, 53–63 (2013)
4. Aldroubi, A., Gröchenig, K.: Nonuniform sampling and reconstruction in shift-invariant spaces. SIAM Rev. **43**(4), 585–620 (2001)
5. Banerjee, A., Basu, S., Merugu, S.: Multi-way clustering on relation graphs. In: SDM, vol. 7, pp. 225–334. SIAM (2007)
6. Cabral Farias, R., Cohen, J.E., Comon, P.: Exploring multimodal data fusion through joint decompositions with flexible couplings. arXiv preprint arXiv:1505.07717 (2015)
7. Comon, P., Luciani, X., De Almeida, A.L.F.: Tensor decompositions, alternating least squares and other tales. J. Chemometr. **23**(7–8), 393–405 (2009)
8. Ermiş, B., Acar, E., Cemgil, A.T.: Link prediction via generalized coupled tensor factorisation. arXiv preprint arXiv:1208.6231 (2012)
9. Harshman, R.A., Lundy, M.E.: Data preprocessing and the extended PARAFAC model, pp. 216–284. Research Methods for Multimode Data Analysis (1984)
10. Jørgensen, B.: The theory of exponential dispersion models and analysis of deviance. Number 51 in Monografias de matematica. CNPQ, IMPA (Brazil) (1992)
11. Kolda, T.G., Bader, B.W.: Tensor decompositions and applications. SIAM Rev. **51**(3), 455–500 (2009)
12. Lin, Y.-R., Sun, J., Castro, P., Konuru, R., Sundaram, H., Kelliher, A.: Metafac: community discovery via relational hypergraph factorization. In: Proceedings of the ACM SIGKDD Conference, pp. 527–536. ACM (2009)
13. Margolis, E., Eldar, Y.C.: Nonuniform sampling of periodic bandlimited signals. IEEE Trans. Sig. Process. **56**(7), 2728–2745 (2008)
14. Seichepine, N., Essid, S., Fevotte, C., Cappé, O.: Soft nonnegative matrix co-factorization. IEEE Trans. Sig. Process. **62**(22), 5940–5949 (2014)
15. Singh, A.P., Gordon, G.J.: Relational learning via collective matrix factorization. In: Proceedings of the 14th ACM SIGKDD Conference, pp. 650–658. ACM (2008)
16. Sorber, L., Van Barel, M., De Lathauwer, L.: Structured data fusion. Technical report, KU Leuven (Belgium), pp. 13–177 (2013)
17. Sørensen, M., Domanov, I., Nion, D., De Lathauwer, L.: Coupled canonical polyadic decompositions and (coupled) decompositions in multilinear rank-$(L_{r,n}, L_{r,n}, 1)$ terms. Technical report, KU Leuven (Belgium) (2013)
18. Sui, J., Adali, T., Yu, Q., Chen, J., Calhoun, V.D.: A review of multivariate methods for multimodal fusion of brain imaging data. J. Neurosci. Meth. **204**(1), 68–81 (2012)
19. Yılmaz, K.Y., Cemgil, A.T., Simsekli, U.: Generalised coupled tensor factorisation. In: Advances Neural Information Processing Systems, pp. 2151–2159 (2011)
20. Yilmaz, Y.K., Cemgil, A.T.: Alpha/beta divergences and tweedie models. arXiv preprint arXiv:1209.4280 (2012)

Learning Coupled Embedding Using MultiView Diffusion Maps

Ofir Lindenbaum, Arie Yeredor$^{(\boxtimes)}$, and Moshe Salhov

Tel Aviv University, P.O. Box 39040, 69978 Tel-aviv, Israel
ofirlin@gmail.com, arie@eng.tau.ac.il, moshebar-s@013.net

Abstract. In this study we consider learning a reduced dimensionality representation from datasets obtained under multiple views. Such multiple views of datasets can be obtained, for example, when the same underlying process is observed using several different modalities, or measured with different instrumentation. Our goal is to effectively exploit the availability of such multiple views for various purposes, such as non-linear embedding, manifold learning, spectral clustering, anomaly detection and non-linear system identification. Our proposed method exploits the intrinsic relation within each view, as well as the mutual relations between views. We do this by defining a cross-view model, in which an implied Random Walk process between objects is restrained to hop between the different views. Our method is robust to scaling of each dataset, and is insensitive to small structural changes in the data. Within this framework, we define new diffusion distances and analyze the spectra of the implied kernels.

Keywords: Dimensionality reduction · Manifold learning · Diffusion maps · Multiview

1 Introduction

High dimensional big datasets are difficult to analyze as is. The challenge is to identify the essential features. Unsupervised dimensionality reduction methods, aim to find a lower dimensional representation based on the geometry of the analyzed dataset. A dimensionality reduction methodology reduces the complexity of processing while preserving the coherence of the original data such that clustering, classification, manifold learning and many other data analysis tasks can be applied effectively in the reduced space.

The problem of learning from two views has been studied in the field of spectral clustering. Some approaches addressing this problem are Bilinear Models [1], Partial Least Squares [2] and Canonical Correlation Analysis [3]. These methods are powerful for learning the relation between the different views but do not provide insights separately or together into the low dimensional geometry or structure of each view. Recently, a few kernel based methods (e.g. [4]) proposed a model of co-regularizing kernels in both views in a way which resembles joint

© Springer International Publishing Switzerland 2015
E. Vincent et al. (Eds.): LVA/ICA 2015, LNCS 9237, pp. 127–134, 2015.
DOI: 10.1007/978-3-319-22482-4_15

diagonalization, by searching for an orthogonal transformation which maximizes the diagonal terms of the kernel matrices obtained from all views. A mixture of Markov chains is proposed in [5] to model the multiple views in order to apply spectral clustering. A way to incorporate multiple given metrics for the same data using a cross diffusion process is presented in [6–9], however the applicability of the suggested approach is limited to clustering tasks only. An iterative algorithm for spectral clustering is proposed in [10]. The idea is to iteratively modify each view using the representation of the other view. The derivation of two manifolds from the same dataset (i.e. two views) is described in [11]. This approach is similar to Canonical Correlation Analysis [3] that seeks a linear transformation which maximizes the correlation among the views. A similar approach is proposed in [12,13]. It suggests data modeling that uses a bipartite graph and then, based on the 'minimum-disagreement' algorithm, partitions the dataset. This approach attempts to minimize the clusters' disagreement between multiple views. In [14] joint diagonalization is used to extend the diffusion framework for multiple modalities.

In this work we present a framework based on De Sa's construction [12], and show that the approach is a special case of a more general diffusion based process. We build and analyze a new framework which generalizes the random walk model while using multiple views. Our proposed method utilizes the intrinsic relation within each view, as well as the mutual relations between views. The multiview is achieved by defining a cross diffusion process in which a specially structured Random Walk is imposed between the various views. The multiview method is robust to scalings of each dataset and it is insensitive to small structural changes in the datasets. Within this framework, we define new diffusion distances and study the spectral decomposition of the new kernel. The framework provides coupled low dimensional embbedings for coupled views.

2   Multiview Diffusion Maps

Problem Formulation: Consider two sets of observations $X = \{x_1, x_2, x_3, ..., x_M\} \in \mathbb{R}^{D_1}$ and $Y = \{y_1, y_2, y_3, ..., y_M\} \in \mathbb{R}^{D_2}$, which are views with bijective correspondence, sampled from the same physical process. The goal is to find a lower dimensional representation for each view that preserves the interactions of data points (in some sense) within and between the views X and Y.

2.1   Multiview Dimensionality Reduction

We begin by generalizing the Diffusion Maps [15] framework for handling a multiview scenario. Our goal is to impose a random walk model using the local connectivities of data points within both views. Our way to incorporate the conectivities is by restraining the random walker to "hop" between views in each step. The first step in this construction is to choose symmetrical, positive-semi-definite kernel functions, one for each view: $\mathcal{K}^x : X \times X \longrightarrow R$ and $\mathcal{K}^y : Y \times Y \longrightarrow R$. A common choice is a Gaussian kernel. These kernels should capture the local intrinsic geometry of each view and neglect the global geometry. By using $\mathcal{K}^x$ and $\mathcal{K}^y$, we form a large row-stochastic matrix of size $2M \times 2M$, as follows; First we compute a matrix product $\boldsymbol{K}^z = \boldsymbol{K}^x \boldsymbol{K}^y$ between the kernels $\mathcal{K}^x$ and $\mathcal{K}^y$, such that (for Gaussian kernels)

$$K_{i,j}^z = \sum_m K_{i,m}^x K_{m,j}^y = \sum_m e^{-\frac{||\boldsymbol{x}_i - \boldsymbol{x}_m||^2}{2\sigma_x^2}} e^{-\frac{||\boldsymbol{y}_m - \boldsymbol{y}_j||^2}{2\sigma_y^2}} \tag{1}$$

(where σ_x and σ_y are the "scaling" parameters of the two kernels), then, the multiview generalized kernel is formed by constructing the following matrix

$$\widehat{\boldsymbol{K}} = \begin{bmatrix} \boldsymbol{0}_{M \times M} & \boldsymbol{K}^z \\ (\boldsymbol{K}^z)^T & \boldsymbol{0}_{M \times M} \end{bmatrix}. \tag{2}$$

Finally, by using the diagonal matrix $\widehat{\boldsymbol{D}}$, $\widehat{D}_{i,i} = \sum_j \widehat{K}_{i,j}$, we compute the normalized row-stochastic matrix

$$\widehat{\boldsymbol{P}} = \widehat{\boldsymbol{D}}^{-1} \widehat{\boldsymbol{K}}, \quad \widehat{P}_{i,j} = \frac{\widehat{K}_{i,j}}{D_{ii}}, \tag{3}$$

that describes the probability matrix of a specially constrained Markov random walk between the data points of $\boldsymbol{X}$ and $\boldsymbol{Y}$. The block anti diagonal form of $\widehat{\boldsymbol{K}}$ is symmetric, and the normalized version $\widehat{\boldsymbol{P}}$ provides a probabilistic interpretation to the construction (explained in 2.3).

2.2   Alternative Multiview Approaches

We briefly mention two alternative methods for incorporating views, which we shall only use as references for comparisons in the experimental evaluations.

Kernel Product (KP): Multiplying the probability matrix element wise $\boldsymbol{K}^\circ \triangleq \boldsymbol{K}^x \circ \boldsymbol{K}^y$, $K_{ij}^\circ \triangleq K_{ij}^x \cdot K_{ij}^y$ and then normalizing by the sum of the rows, resulting in a row stochastic matrix. This kernel corresponds to the approach in [15].

Kernel Sum (KS): Defining the sum kernel $\boldsymbol{K}^+ \triangleq \boldsymbol{K}^x + \boldsymbol{K}^y$. Normalizing the sum kernel by the sum of rows, to compute $\boldsymbol{P}^+$. This random walk sums the step probabilities from each view. This approach is proposed in [5].

2.3   Probabilistic Interpretation of $\widehat{\boldsymbol{P}}$

Under our proposed construction, the elements $[\widehat{\boldsymbol{P}}^t]_{i,j} = \widehat{p}_t(\boldsymbol{x}_i, \boldsymbol{x}_j)$ denote (for each $i, j \in [1, M]$) the transition probability from node $\boldsymbol{x}_i$ to node $\boldsymbol{x}_j$ in t time steps "hopping" between the views $\boldsymbol{X}$ and $\boldsymbol{Y}$ in each step. Note that due to the block-anti-diagonal structure of $\widehat{\boldsymbol{K}}$ (and $\widehat{\boldsymbol{P}}$), for odd values of t, this probability is zero. However, for even values of t, this probability is nonzero and describes an even time transition from view $\boldsymbol{X}$ through view $\boldsymbol{Y}$ and back to $\boldsymbol{X}$. In the same way, $[\widehat{\boldsymbol{P}}^t]_{i+M,j+M} = \widehat{p}_t(\boldsymbol{y}_i, \boldsymbol{y}_j)$ denotes the transition probability from node $\boldsymbol{y}_i$ to node $\boldsymbol{y}_j$ $(i, j \in [1, M])$ in t time steps. Likewise, $[\widehat{P}^t]_{i,j+M} = \widehat{p}_t(\boldsymbol{x}_i, \boldsymbol{y}_j)$ denotes the transition probability from node $\boldsymbol{x}_i$ to node $\boldsymbol{y}_j$ $(i, j \in [1, M])$ in t time steps. Note that this probability is nonzero only for odd values of t. This probability takes into consideration all the various possibilities of crossing from node $\boldsymbol{x}_i$ to node $\boldsymbol{y}_j$ by propagating in view 1 and then in view 2.

2.4 Spectral Decomposition

The matrix $\widehat{\boldsymbol{P}}$ is algebraically similar to a symmetric matrix $\widehat{\boldsymbol{P}}_s = \widehat{\boldsymbol{D}}^{1/2}\widehat{\boldsymbol{P}}\widehat{\boldsymbol{D}}^{-1/2} = \widehat{\boldsymbol{D}}^{-1/2}\widehat{\boldsymbol{K}}\widehat{\boldsymbol{D}}^{-1/2}$. Therefore, both $\widehat{\boldsymbol{P}}$ and $\widehat{\boldsymbol{P}}_s$ share the same set of eigenvalues $\{\lambda_m\}$. Due to the symmetry of the matrix $\widehat{\boldsymbol{P}}_s$, it has a set of $2M$ real eigenvalues $\{\lambda_i\}_{i=0}^{2M-1} \in \mathbb{R}$ and corresponding real orthonormal eigenvectors $\{\boldsymbol{\pi}_m\}_{m=0}^{2M-1} \in \mathbb{R}^{2M}$, thus, $\widehat{\boldsymbol{P}}_s = \boldsymbol{\Pi}\boldsymbol{\Lambda}\boldsymbol{\Pi}^T$. By denoting $\boldsymbol{\Psi} \overset{\triangle}{=} \widehat{\boldsymbol{D}^{-1/2}}\boldsymbol{\Pi}$ and $\boldsymbol{\Phi} \overset{\triangle}{=} \widehat{\boldsymbol{D}}^{1/2}\boldsymbol{\Pi}$, we conclude that the set $\{\boldsymbol{\psi}_m, \boldsymbol{\phi}_m\}_{m=0}^{2M-1} \in \mathbb{R}^{2M}$ denotes the right and left eigenvectors of $\widehat{\boldsymbol{P}} = \boldsymbol{\Psi}\boldsymbol{\Lambda}\boldsymbol{\Phi}^T$, respectively, satisfying the bi-orthonormality property $\boldsymbol{\psi}_i^T\boldsymbol{\phi}_j = \delta_{ij}$. In the sequel, we use the matrix $\widehat{\boldsymbol{P}}_s$ to simplify the analysis.

Although we have formed a large matrix of size $2M \times 2M$, no computational complexity is added by using our construction relative to using the construction of a single DM-based view for each view. The spectral decomposition of $\widehat{\boldsymbol{P}}_s$ can be computed using a Singular Value Decomposition (SVD) of the $M \times M$ matrix $\bar{\boldsymbol{K}}^z = \boldsymbol{D}^{rows-1/2}\boldsymbol{K}^z\boldsymbol{D}^{cols-1/2}$, where $D_{i,i}^{rows} = \sum_{j=1}^{M} K_{i,j}^z$ and $D_{j,j}^{cols} = \sum_{i=1}^{M} K_{i,j}^z$ are diagonal matrices. Theorem 1 enables us to form the eigenvectors of $\widehat{\boldsymbol{P}}$ as a concatenation of the singular vectors of $\bar{\boldsymbol{K}}^z$:

Theorem 1. *By using the left and right singular vectors of $\bar{\boldsymbol{K}}^z = \boldsymbol{V}\boldsymbol{\Sigma}\boldsymbol{U}^T$, the eigenvectors and the eigenvalues of $\widehat{\boldsymbol{P}}_s$ are given explicitly by*

$$\boldsymbol{\Pi} = \frac{1}{\sqrt{2}}\begin{bmatrix} \boldsymbol{V} & \boldsymbol{V} \\ \boldsymbol{U} & -\boldsymbol{U} \end{bmatrix}, \boldsymbol{\Lambda} = \begin{bmatrix} \boldsymbol{\Sigma} & \boldsymbol{0}_{M\times M} \\ \boldsymbol{0}_{M\times M} & -\boldsymbol{\Sigma} \end{bmatrix}. \tag{4}$$

A similar theorem and proof is given in [16,17].

2.5 Multiview Diffusion Distance

Tasks such as classification, clustering or system identification require a measure for the intrinsic connectivity between data points. This type of measure is only satisfied locally by the Euclidian distance in the high dimensional ambient space. The multiview diffusion kernel (defined in Sect. 2.1) provides indication about all the small local connections between data points. The row stochastic matrix $\widehat{\boldsymbol{P}}^t$ incorporates all possibilities for transition in t time steps between data points while hopping between both views. For a fixed value of $t > 0$, two data points are *intrinsically similar* if the conditional distributions $\widehat{p}_t(\boldsymbol{x}_i, :) = [\widehat{\boldsymbol{P}}^t]_{i,:}$ and $\widehat{p}_t(\boldsymbol{x}_j, :) = [\widehat{\boldsymbol{P}}^t]_{j,:}$ are similar. This type of similarity measure indicates that the points $\boldsymbol{x}_i$ and $\boldsymbol{x}_j$ are similarly connected to several mutual points. Thus, they are connected by a geometrical path.

Based on this observation, by expanding the single view construction given by [15], we define the weighted inner view diffusion distances to be

$$\mathcal{D}_t{}^2(\boldsymbol{x}_i, \boldsymbol{x}_j) = \sum_{k=1}^{2M} \frac{([\widehat{\boldsymbol{P}}^t]_{i,k} - [\widehat{\boldsymbol{P}}^t]_{j,k})^2}{\phi_o(k)} = ||(\boldsymbol{e}_i - \boldsymbol{e}_j)^T \widehat{\boldsymbol{P}}^t||^2_{\widehat{D}^{-1}} \tag{5}$$

where $\boldsymbol{e}_i$ is the i-th column of an $2M \times 2M$ identity matrix, $\boldsymbol{\phi}_0$ is the first left eigenvector of $\widehat{\boldsymbol{P}}$, its k-th element is $\phi_0(k) = \widehat{D}_{k,k}$. $|| \cdot ||^2_W$ denotes the weighted norm, which for a vector ξ and a positive-definite weight matrix $\mathbf{W}$ is defined as $||\xi||^2_{\mathbf{W}} = \xi^T \mathbf{W}\xi$. Similarly,

$$\mathcal{D}_t{}^2(\boldsymbol{y}_i, \boldsymbol{y}_j) = \sum_{k=1}^{2M} \frac{([\widehat{\boldsymbol{P}}^t]_{M+i,k} - [\widehat{\boldsymbol{P}}^t]_{M+j,k})^2}{\phi_o(k)}. \tag{6}$$

The main advantage of these distances ((5) and (6)) is that they can be expressed in terms of eigenvalues and eigenvectors of the matrix $\widehat{\boldsymbol{P}}$. This insight allows us to use a representation (defined in Sect. 2.6) in which the induced Euclidean distance is proportional to the diffusion distances (defined in (5) and (6)).

Theorem 2. *The inner view diffusion distances ((5) and (6)) are equal to:*

$$\mathcal{D}_t{}^2(\boldsymbol{x}_i, \boldsymbol{x}_j) = 2 \cdot \sum_{\ell=1}^{M-1} \lambda_\ell^{2t}(\psi_\ell(i) - \psi_\ell(j))^2, \tag{7}$$

$$\mathcal{D}_t{}^2(\boldsymbol{y}_i, \boldsymbol{y}_j) = 2 \cdot \sum_{\ell=1}^{M-1} \lambda_\ell^{2t}(\psi_\ell(M+i) - \psi_\ell(M+j))^2. \tag{8}$$

Proof. We prove for one view (with proper adjustments, the proof applies to the second view as well).

Note first that we can express $\widehat{\boldsymbol{P}}^t \widehat{\boldsymbol{D}}^{-1}(\widehat{\boldsymbol{P}}^t)^T$ as $\widehat{\boldsymbol{P}}^t \widehat{\boldsymbol{D}}^{-1}(\widehat{\boldsymbol{P}}^t)^T = \boldsymbol{\Psi}\boldsymbol{\Lambda}^t \boldsymbol{\Phi}^T \widehat{\boldsymbol{D}}^{-1}$ $\boldsymbol{\Phi}\boldsymbol{\Lambda}^t \boldsymbol{\Psi}^T = \boldsymbol{\Psi}\boldsymbol{\Lambda}^{2t}\boldsymbol{\Psi}^T$ (since $\boldsymbol{\Phi}^T \widehat{\boldsymbol{D}}^{-1}\boldsymbol{\Phi} = \boldsymbol{\Pi}^T \boldsymbol{\Pi} = \mathbf{I}$), and therefore $\mathcal{D}_t{}^2(\boldsymbol{x}_i, \boldsymbol{x}_j) =$ $||(\boldsymbol{e}_i - \boldsymbol{e}_j)^T \widehat{\boldsymbol{P}}^t||^2_{\widehat{D}^{-1}} = (\boldsymbol{e}_i - \boldsymbol{e}_j)^T \widehat{\boldsymbol{P}}^t \widehat{\boldsymbol{D}}^{-1}(\widehat{\boldsymbol{P}}^t)^T(\boldsymbol{e}_i - \boldsymbol{e}_j) = (\boldsymbol{e}_i - \boldsymbol{e}_j)^T \boldsymbol{\Psi}\boldsymbol{\Lambda}^{2t}\boldsymbol{\Psi}^T(\boldsymbol{e}_i - \boldsymbol{e}_j) = \sum_{\ell=0}^{2M-1} \lambda_\ell^{2t}(\psi_\ell(i) - \psi_\ell(j))^2 = 2\sum_{\ell=1}^{M-1} \lambda_\ell^{2t}(\psi_\ell(i) - \psi_\ell(j))^2$.

The last equality is due to the repetitive form of $\widehat{\boldsymbol{D}}, \boldsymbol{\Pi}$ (therefore of $\boldsymbol{\Psi}$) and $\boldsymbol{\Lambda}$, as described in (4); $\ell = 0$ is excluded from the sum due to the property of $\boldsymbol{\psi}_0 = 1$ (an all-ones vector), which holds for all stochastic matrices.

2.6   Multiview Data Parametrization

Tasks such as classification, clustering or regression in a sampled high-dimensional feature space are considered to be computationally expensive. In addition, the performance of these tasks is highly dependent on the distance measure. As explained above, distance measures in the original ambient space could be meaningless in many real life situations. Interpreting Theorem 2 in terms of the Euclidean distance, enables us to define two mappings for X and Y by using the right eigenvectors weighted by λ_i^t. A representation for X is

$$\boldsymbol{\Psi}_t(\boldsymbol{x}_i) : \boldsymbol{x}_i \longmapsto \left[\lambda_1^t \psi_1(i), ..., \lambda_{M-1}^t \psi_{M-1}(i)\right]^T \in \mathbb{R}^{M-1} \tag{9}$$

and a representation for Y is

$$\boldsymbol{\Psi}_t(\boldsymbol{y}_i) : \boldsymbol{y}_i \longmapsto \left[\lambda_1^t \psi_1(M+i), ..., \lambda_{M-1}^t \psi_{M-1}(M+i)\right]^T \in \mathbb{R}^{M-1}. \tag{10}$$

These mappings capture the intrinsic geometry of both views as well as a mutual relation between the views. Our framework is based on the decaying properties of the Gaussian kernel, as studied in [15]. These properties enable us to approximate (7) and (8) by neglecting all the eigenvalues that are smaller than some (small) δ. Therefore we can compute a low-dimensional mapping such that

$$\widehat{\boldsymbol{\Psi}}_t^r(\boldsymbol{x}_i) : \boldsymbol{x}_i \longmapsto \left[\lambda_1^t \psi_1(i), ..., \lambda_{r-1}^t \psi_{r-1}(i)\right]^T \in \mathbb{R}^{r-1}. \tag{11}$$

This mapping of dimension $r-1$ provides a low dimensional space in which tasks such as clustering or classification are more computationally efficient and more precise than when the computation takes place in the original ambient space.

3   Experimental Results

3.1   Coupled Manifold Learning

In this section, we examine the embedding extracted using our method and compare it to the KP approach (Sect. 2.2). We generate two manifolds with a common underlying open circular structure. The manifolds were generated by applying a 3 dimensional function to 1000 data points spread linearly within the following lines $a_i \rightarrow [0, 2\pi]$, $b_i = a_i + 0.5\pi \mod 2\pi$, where $i = 1, 2, 3, ..., 1000$. We use the following functions to generate the datasets for View-I and View-II:

$$\boldsymbol{X} = \begin{bmatrix} 4\cos(0.9a_i) + 0.3\cos(20a_i) \\ 4\sin(0.9a_i) + 0.3\sin(20a_i) \\ 0.1(6.3a_i^2 - a_i^3) \end{bmatrix}, \boldsymbol{Y} = \begin{bmatrix} 4\cos(0.9b_i) + 0.3\cos(20b_i) \\ 4\sin(0.9b_i) + 0.3\sin(20b_i) \\ 0.1(6.3b_i - b_i^2) \end{bmatrix} \tag{12}$$

The 3-dimensional manifolds are presented in Fig. 1.

The standard diffusion mapping (KP)- separates the manifold to a horseshoe and a point as shown in Fig. 2-bottom. This structure does not represent any of the original structures, nor does it reveal the underlying parameters (a_i, b_i). On

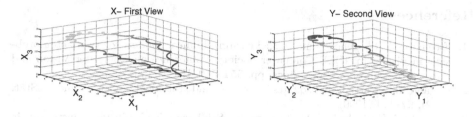

Fig. 1. Manifolds in both views. Both manifolds have some circular structure governed by the angle, colored by the points index.

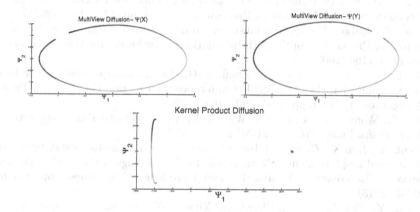

Fig. 2. Top- the mappings computed using our proposed parametrization (9 and 10). Bottom- a standard two dimensional diffusion mapping, computed using the concatenation vector from both views (corresponding to kernel K°).

the other hand, our embedding captures the two structures, one for each view. As shown in Fig. 2-top, one structure represents the angle of a_i (with the gap at the "end", due to the multiplication by 0.9 in (12)) while the other represents the angle of b_i (with the inflicted gap in the "middle"). The Euclidean distance in the new spaces preserves the mutual relations between points based on the geometrical relation in both views. Moreover both manifolds are in the same coordinate system, and this is a strong advantage as it enables to compare the manifolds in the lower dimensional space.

4   Conclusions

In this paper, we presented a framework for MultiView dimensionality reduction. The method enables to extract simultaneous embeddings from coupled views. The framework is useful for various real life machine learning tasks.

A few open issues arise from this work, such as reduction of the sensitivity to the density of points, or computing the spectral properties of the proposed kernel.

References

1. Freeman, W.T., Tenenbaum, J.B.: Learning bilinear models for two-factor problems in vision. In: IEEE Computer Society Conference on Computer Vision and Pattern Recognition, Proceedings, pp. 554–560. IEEE (1997)
2. Helland, I.S.: Partial least squares regression and statistical models. Scand. J. Stat. **17**(2), 97–114 (1990)
3. Chaudhuri, K., Kakade, S.M., Livescu, K., Sridharan, K.: Multi-view clustering via canonical correlation analysis. In: Proceedings of the 26th Annual International Conference on Machine Learning, pp. 129–136. ACM (2009)
4. Kumar, A., Rai, P., Daume, H.: Co-regularized multi-view spectral clustering. In: Advances in Neural Information Processing Systems, pp. 1413–1421 (2011)
5. Zhou, D., Burges, C.: Spectral clustering and transductive learning with multiple views. In: Proceedings of the 24th International Conference on Machine Learning, pp. 1159–1166 (2007)
6. Wang, B., Jiang, J., Wang, W., Zhou, Z.-H., Tu, Z.: Unsupervised metric fusion by cross diffusion. In: 2012 IEEE Conference on Computer Vision and Pattern Recognition (CVPR), pp. 2997–3004. IEEE (2012)
7. Bai, X., Wang, B., Yao, C., Liu, W., Tu, Z.: Co-transduction for shape retrieval. IEEE Trans. Image Process. **21**(5), 2747–2757 (2012)
8. Wang, Y., Lin, X., Zhang, Q.: Towards metric fusion on multi-view data: a cross-view based graph random walk approach. In: Proceedings of the 22nd ACM International Conference on Information and Knowledge Management, pp. 805–810. ACM (2013)
9. Wang, Y., Pei, J., Lin, X., Zhang, Q., Zhang, W.: An iterative fusion approach to graph-based semi-supervised learning from multiple views. In: Tseng, V.S., Ho, T.B., Zhou, Z.-H., Chen, A.L.P., Kao, H.-Y. (eds.) PAKDD 2014, Part II. LNCS, vol. 8444, pp. 162–173. Springer, Heidelberg (2014)
10. Kumar, A., Daumé, H.: A co-training approach for multi-view spectral clustering. In: Proceedings of the 28th International Conference on Machine Learning (ICML-11), pp. 393–400 (2011)
11. Boots, B., Gordon, G.: Two-manifold problems with applications to nonlinear system identification (2012)
12. de Sa, V.R.: Spectral clustering with two views. In: ICML Workshop on Learning with Multiple Views (2005)
13. De Sa, V.R., Gallagher, P.W., Lewis, J.M., Malave, V.L.: Multi-view kernel construction. Mach. Learn. **79**(1–2), 47–71 (2010)
14. Eynard, D., Kovnatsky, A., Bronstein, M.M., Glashoff, K., Bronstein, A.M.: Multimodal manifold analysis by simultaneous diagonalization of laplacians (2012)
15. Coifman, R.R., Lafon, S.: Diffusion maps. Appl. Comput. Harmonic Anal. **21**, 5–30 (2006)
16. Zha, H., He, X., Ding, C., Simon, H., Gu, M.: Bipartite graph partitioning and data clustering. In: Proceedings of the Tenth International Conference on Information and Knowledge Management, pp. 25–32. ACM (2001)
17. Dhillon, I.S.: Co-clustering documents and words using bipartite spectral graph partitioning. In: Proceedings of the Seventh ACM SIGKDD International Conference on Knowledge Discovery and Data Mining, pp. 269–274. ACM (2001)

Extraction of Temporal Patterns in Multi-rate and Multi-modal Datasets

Antoine Liutkus[1]([⊠]), Umut Şimşekli[2], and A. Taylan Cemgil[2]

[1] Inria, Speech processing team, LORIA, Universit de Lorraine,
Villers-ls-nancy, France
antoine.liutkus@inria.fr
[2] Department of Computer Engineering, Boğaziçi University, İstanbul, Turkey
{umut.simsekli,taylan.cemgil}@boun.edu.tr

Abstract. We focus on the problem of analyzing corpora composed of irregularly sampled (multi-rate) heterogeneous temporal data. We propose a novel convolutive multi-rate factorization model for extracting multi-modal patterns from such multi-rate data. Our model builds up on previously proposed multi-view (coupled) nonnegative matrix factorization techniques, and extends them by accounting for heterogeneous sample rates and enabling the patterns to have a duration. We illustrate the proposed methodology on the joint study of audiovisual data for speech analysis.

Keywords: Coupled factorization · Multi-rate data analysis

1 Introduction

The last decade has witnessed a rapid growth in the size of available data. Thanks to the current technological infrastructure, massive amounts of data are continuously produced and the cost of storing this massive data gets cheaper everyday. This growth in the size of the data has brought new scientific challenges.

One major challenge is handling the data-heterogeneity. Data are often collected in different modalities (e.g., audio, video, text, etc.) at different time instances. Combining different but related data can improve estimation and prediction performance drastically, provided the different modes of the data contain sufficiently rich information and a proper model is established for jointly modeling these modes.

Various research fields have focused on the data-heterogeneity problem, such as transfer learning [9], multiple-kernel learning [3], and coupled factorizations [14]. Each of these fields has different application-specific objectives (such as increasing classification acuracy or separation performance) and therefore approach the problem from slightly different perspectives. A common theme in these works is modeling a collection of observed matrices $\{V_n\}_{n=1}^N$ by using a factorization model:

$$V_n(l,t) \approx \hat{V}_n(l,t) = \sum_k W_n(l,k)H(k,t). \tag{1}$$

© Springer International Publishing Switzerland 2015
E. Vincent et al. (Eds.): LVA/ICA 2015, LNCS 9237, pp. 135–142, 2015.
DOI: 10.1007/978-3-319-22482-4_16

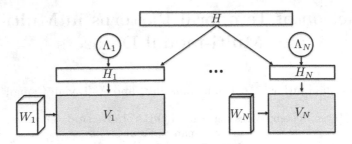

Fig. 1. Illustration of the proposed model (MULTICONV). The blocks represent the matrices and the tensors that appear in the model. The shaded blocks are observed, whereas the other ones are latent. The arrows visualize the dependency structure.

Here, V_n denotes the different modes of data, where each V_n is modeled as the product of a dictionary matrix W_n and an activation matrix H. In this modeling strategy, each mode n has its own dictionary W_n but their corresponding activations are shared among all modes, making the overall model coupled.

When different modes of the data contain temporal information, alternative factorization models can be proposed [2,12]. In this study, we will consider the non-negative matrix factor deconvolution (NMFD) model [12], where the temporal information is incorporated through convolution:

$$V_n(l,t) \approx \hat{V}_n(l,t) = \sum_{k,p} W_n(l,k,p)H(k,p-t). \qquad (2)$$

Here, the dictionary tensors W_n have temporal axis (p) that enables the dictionaries to encapsulate temporal information.

This modeling strategy has yielded many practical applications, when there is only one observed matrix (i.e., $N = 1$). However, when there are multiple observed matrices, this model requires all modalities V_n to be synchronized temporally. However, in practice, different modes of the data are often collected with different technologies. Therefore, they are usually sampled at different sampling rates, which we call 'multi-rate' data. In this study, we propose a novel convolutive factorization model that is able to model multi-rate multi-modal data. In the sequel, we will describe the model in detail and present a practical inference algorithm to estimate the parameters of the model. We illustrate the proposed method on the joint decomposition of audiovisual data for speech analysis.

2   The MULTICONV Model

In this section, we describe our model in detail. We assume that we observe N matrices $\{V_n\}_{n=1}^{N}$ with nonnegative entries, each one of them being of size $L_n \times T_n$, where L_n and T_n are the dimensions of each sample and the number of samples for modality n, respectively. For instance, V_1 can be the magnitude or power spectrogram of audio data, where each column might contain the spectrum of a single audio frame, and V_2 can be video data where each column contains the vectorized version of an image. Our objective is to jointly model different modalities $\{V_n\}_n$ when their sampling rates are different, yielding possibly different number of samples T_n. Without loss of generality, let us assume that the modalities are sorted by decreasing number of samples, so that $T_1 \geq \cdots \geq T_N$. Finally, let $T_0 \geq T_1$ be an arbitrary integer, corresponding to the number of samples in some absolute *time reference*, where sampling is regular and achieved at a high precision. For concision, a sample index t of modality n will be written $t \in \mathbb{T}_n$. Note that, even though we assume all the observed data to be matrices, it is straightforward to extend the model where any V_n can be a tensor.

We will model each V_n by using an NMFD model. In order to accurately model patterns with a temporal structure, they will be taken as lasting P_n samples in modality n. A typical choice for P_n in the case of constant sampling rates is to enforce patterns to have the same absolute duration through different modalities, picking an arbitrary P_0 as the *absolute duration* of the patterns, and then choosing:

$$\forall n, P_n = \left\lceil \frac{P_0 T_n}{T_0} \right\rceil, \tag{3}$$

where $\lceil \cdot \rceil$ is the ceiling function.

The first important issue we face is to establish a temporal correspondence between the samples observed through the different modalities. The difficulty on this point is that not only the different sampling rates may be different, they may also be varying over time or even be irregular. In full generality, we introduce a *link tensor* Λ_n for each modality, of dimension $T_n \times P_n \times T_0 \times T_0$. In essence, $\Lambda_n(t, p, \tau, \tau')$ is high whenever time instants t and $t - p$ in $\mathbb{T}_n$ correspond to reference samples τ and $\tau - \tau'$ in $\mathbb{T}_0$. For instance, assume that all sampling frequencies are constant and equal. Then, we can pick $\Lambda_n(t, p, \tau, \tau') = \delta(t, \tau)\delta(p, \tau')$ with $\delta(t, t') = 1$ iff $t = t'$ and 0 otherwise. If sampling frequencies f_n are constant but unequal, we can for instance pick:

$$\Lambda_n(t, p, \tau, \tau') \propto \delta(\lfloor f_1 t \rfloor, \lfloor f_n \tau \rfloor)\delta(\lfloor f_1 p \rfloor, \lfloor f_n \tau' \rfloor), \quad \forall n \tag{4}$$

where $\lfloor \cdot \rceil$ is the rounding function and $\propto$ denotes equality up to a normalizing constant. Indeed, we assume in the sequel that Λ_n is normalized so that:

$$\forall t, p \in \mathbb{T}_n, \sum_{\tau, \tau' \in \mathbb{T}_0} \Lambda_n(t, p, \tau, \tau') = 1, \quad \forall n \tag{5}$$

With the link tensors Λ_n in hand, we can describe the actual model that decomposes the observations as the sum of only a few multi-modal patterns. For this purpose, we introduce a latent activation matrix $H(k, t)$, with fixed dimension $K \times T_0$, i.e. with the resolution of the reference time line $\mathbb{T}_0$. Finally, observation V_n is modeled as the superposition of the K patterns, activated over time through convolution, given as follows:

$$V_n\left(l,t\right) \approx \hat{V}_n\left(l,t\right) = \sum_{k=0}^{K-1}\sum_{p=0}^{P_n-1} W_n\left(l,k,p\right) \underbrace{\sum_{\tau,\tau'\in \mathbb{T}_0} \Lambda_n\left(t,p,\tau,\tau'\right) H\left(k,\tau-\tau'\right)}_{H_n\left(k,t-p\right)}.$$

(6)

where H_n is the temporally adjusted activations for each modality V_n. Figure 1 illustrates the model.

3 Inference

Once we observe $\{V_n\}_n$, our aim is to estimate the parameters $\Theta = \{\{W_n\}_n, H\}$ and thus to find the multi-modal patterns as well as the way they are activated over time that best permits to account for the observed data $\{V_n\}_n$. For this purpose, we choose the parameters that minimize a cost function $C\left(\Theta\right)$, which is taken as the sum of a data-fit $J_V\left(\Theta\right)$ and a regularization term $\Psi_H\left(\Theta\right)$ for the activations H:

$$\hat{\Theta} = \operatorname*{argmin}_{\Theta} C\left(\Theta\right) = J_V\left(\Theta\right) + \Psi_H\left(\Theta\right).$$

(7)

In our setup, the data-fit criterion in (7) is taken as the sum over all the data of a scalar (element-wise) cost-function:

$$J_V\left(\Theta\right) = \sum_n \lambda_n \left[\sum_{l,t}^{L_n,T_n} d_n\left(V_n\left(l,t\right)\|\hat{V}_n\left(l,t\right)\right)\right],$$

where $d_n\left(v\|\hat{v}\right)$ is the particular cost function used for modality n and $\lambda_n > 0$ is a scalar indicating the global importance of a good fit for modality n. It assesses the similarity between one of the elements v from the observation and the corresponding element $\hat{v}$ from the model (6).

We allow for the cost-function to differ from one modality to another, mainly because observation noise may have strongly different physical origins depending on the modality. In this work, we will assume that d_n belongs to the family of β-divergences, that is defined as follows:

$$d_\beta(v\|\hat{v}) = \frac{v^\beta}{\beta(\beta-1)} - \frac{v\hat{v}^{\beta-1}}{\beta-1} + \frac{\hat{v}^\beta}{\beta}$$

(8)

This divergence is the squared Euclidean distance for $\beta = 2$, and it can be extended by continuity at $\beta = 1$ and $\beta = 0$ to coincide with the Kullback-Leibler and Itakura-Saito divergences, respectively.

The term $\Psi_H\left(\Theta\right)$ in (7) permits to enforce some additional constraints concerning the activations H. In our context, *sparsity* is relevant and means that we expect most activations to be close to 0, and only occasionally to bear a significant magnitude. Sparse regularization for NMF has been the topic of many studies [4,7,11] and here, we pick the ℓ_1 norm over H as a sparsity-enforcing criterion [5].

3.1   Multiplicative Updates

To update the parameters Θ so as to minimize a given cost function $C(\Theta)$, such as $C = J_V + \Psi_H$, we adopt a Majoration-Equalization approach, through Multiplicative Updates (MU) that was first presented in [6] and whose proofs for convergence were recently given in [1] for the β-divergence J_V with $\beta \in [0, 2]$. The MU methodology may be described as follows. First, we compute the derivative of $C(\Theta)$ with respect to any one Θ_i of the parameters and then express it as the difference of two nonnegative terms:

$$\frac{\partial C}{\partial \Theta_i}(\theta) = \underbrace{G_+^i(\theta)}_{\geq 0} - \underbrace{G_-^i(\theta)}_{\geq 0}. \tag{9}$$

In our case, this is easily done for the data fit and regularization functions J_V and Ψ_H chosen here. Then, instead of adopting a classical gradient descent for Θ_i, we update it multiplicatively through:

$$\Theta_i \leftarrow \Theta_i \frac{G_-^i(\Theta_i)}{G_+^i(\Theta_i)}. \tag{10}$$

When applied to $C = J_V + \Psi_H$, (10) leads to the following multiplicative updates for the parameter Θ_i:

$$\Theta_i \leftarrow \Theta_i \frac{\sum_{n,l,t} \lambda_n \hat{V}_n(l,t)^{\beta_n-2} V_n(l,t) \frac{\partial \hat{V}_n(l,t)}{\partial \Theta_i} + \nabla_{\Theta_i}^- \Psi_H(\Theta)}{\sum_{n,l,t} \lambda_n \hat{V}_n(l,t)^{\beta_n-1} \frac{\partial \hat{V}_n(l,t)}{\partial \Theta_i} + \nabla_{\Theta_i}^+ \Psi_H(\Theta)}. \tag{11}$$

These update rules are applied iteratively until convergence.

4   Experiments

4.1   Dataset and Experimental Setup

In this paper, we apply the MULTICONV model to the extraction of multi-modal patterns in the MRI-TIMIT database [8]. This corpus features real time Magnetic Resonance Imaging (rtMRI) data along with the corresponding audio, for 10 different speakers, 5 males and 5 females, each one of them recorded while uttering 460 sentences from the MOCHA-TIMIT database [13].

This corpus hence consists of synchronized rtMRI and audio recordings. The rtMRI (video) has an image resolution of 68×68, with a sampling rate of 23 frames per second. Each image corresponds to a mid-sagittal slice of a speaker. The corresponding audio is sampled at $20\,\mathrm{kHz}$. For analysis, the audio was split into frames of 128 samples ($6.5\,\mathrm{ms}$), with an overlap of 50% between adjacent frames. The resulting Short-Term Fourier Transform frame rate312 frames per second. One of the excerpt of the database is depicted on Fig. 2.

Our objective here is to extract meaningful articulatory audio-visual patterns, or *primitives*, from the MRI-TIMIT database by using the MULTICONV

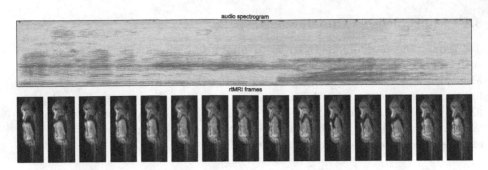

Fig. 2. Excerpt of the MRI-TIMIT database. Speaker F3, "It was easy for us". Audio is sampled at 20 kHz, rtMRI at 23 frames per second.

model. In this respect, the current study goes further in the direction undertaken by the pioneering work presented in [10], that focused on the same objective, but exploited the rtMRI modality only. Furthermore, while [10] also integrated sparsity constraints in the activations, these constraints were slightly different than those considered here and do not lead to straightforward multiplicative updates.

4.2   Results

The MULTICONV model was fitted on the MRI-TIMIT data for speaker F1 (sentences 1–25), using the Kullback-Leibler divergence both for the audio and rtMRI data. We chose to estimate $K = 25$ patterns having a duration of about 380 ms. After two hundred iterations, we recover both the patterns and their underlying activation vectors. For the audio modality, the template is a spectrogram, while it is a short video for the rtMRI modality. Nine patterns are displayed in Fig. 3, where the average of the rtMRI modality is represented, for conciseness.

Interestingly, the joint convolutive modeling clearly isolates some parts of the rtMRI data, thus automatically locating the main places of articulation. In Fig. 3 for instance, we clearly see that the lips are identified as moving together, thus being important parts of the same pattern. However, the main interest of the MULTICONV model is to also automatically relate these places of articulation with a corresponding audio spectrogram, even if the sampling frequencies of these modalities are very different.

Notwithstanding the interest of performing such an unsupervised analysis of multi-modal data, the qualitative use of these results by professional linguists is made difficult by the lack of phonological information. Indeed, it seems natural to associate each pattern to a phoneme, and learning the MULTICONV model would then amount to estimating the best rtMRI and associated spectrogram for each phoneme. To achieve this, we simply need to make use of a transcription and adapt the model so that the activation of each pattern-phoneme is zero except at the beginning of all occurrences of the phoneme. We leave this for future work.

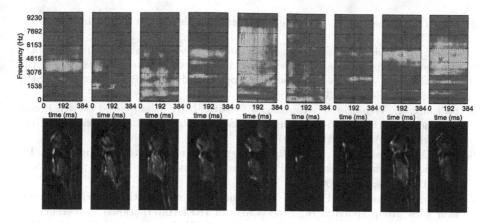

Fig. 3. Nine multi-modal patterns learned with the MULTICONV model on the MRI-TIMIT dataset.

5 Conclusion

Multi-view data analysis is concerned with corpora composed of heterogeneous items from different modalities, such as audio, video, images or text. In most studies, either all modalities are assumed perfectly synchronized or the phenomenon under study is intrinsically non-temporal, such as images or textual documents. In those cases, a joint analysis in effect often boils down to data concatenation. In this paper, we have proposed the MULTICONV model, to extract multi-modal patterns from the joint analysis of data-streams that exhibit different or even non-constant frame-rates, and that capture different aspects of the same phenomenon. This model builds on previously proposed multi-view Nonnegative Matrix Factorization techniques (NMF), but significantly extends them by both accounting for heterogeneous sample rates and by enabling patterns to have a duration, which proves fundamental in the study of datasets that are relative to temporal phenomena. In practice, we propose a convolutive multi-rate NMF model, where temporal patterns are activated simultaneously over the different modalities through a shared underlying activation stream. The multi-rate problem is addressed by the incorporation of a reference time scale, which subsumes many different sampling scenarios and permits to bind the modalities together. We illustrated the proposed methodology with a preliminary analysis of an audio-visual corpus of speech data.

References

1. Févotte, C., Idier, J.: Algorithms for nonnegative matrix factorization with the beta-divergence. Neural Comput. **23**(9), 2421–2456 (2011)
2. Févotte, C., Le Roux, J., Hershey, J.R.: Non-negative dynamical system with application to speech and audio. In: 2013 IEEE International Conference on Acoustics, Speech and Signal Processing (ICASSP), pp. 3158–3162 (2013)

3. Gönen, M., Alpaydın, E.: Multiple kernel learning algorithms. J. Mach. Learn. Res. **12**, 2211–2268 (2011)
4. Joder, C., Weninger, F., Virette, D., Schuller, B.: A comparative study on sparsity penalties for nmf-based speech separation: Beyond lp-norms. In: 2013 IEEE International Conference on Acoustics, Speech and Signal Processing (ICASSP), pp. 858–862. IEEE (2013)
5. Le Roux, J., Weninger, F., Hershey, J.: Sparse NMF - half-baked or well done? Technical report TR2015-023, Mitsubishi Electric Research Laboratories (MERL), Cambridge, MA, USA, March 2015
6. Lee, D.D., Seung, H.S.: Algorithms for non-negative matrix factorization. In: Advances in Neural Information Processing Systems (NIPS), vol. 13, pp. 556–562. The MIT Press, April 2001
7. Lefèvre, A., Bach, F., Févotte, C.: Itakura-saito nonnegative matrix factorization with group sparsity. In: Proceedings of IEEE International Conference on Acoustics, Speech and Signal Processing (ICASSP). Prague, Czech Republic, May 2011
8. Narayanan, S., Bresch, E., Ghosh, P., Goldstein, L., Katsamanis, A., Kim, Y., Lammert, A., Proctor, M., Ramanarayanan, V., Zhu, Y.: A multimodal real-time mri articulatory corpus for speech research. In: INTERSPEECH, pp. 837–840 (2011)
9. Pan, S.J., Yang, Q.: A survey on transfer learning. IEEE Trans. Knowl. Data Eng. **22**, 1664–1677 (2010)
10. Ramanarayanan, V., Katsamanis, A., Narayanan, S.: Automatic data-driven learning of articulatory primitives from real-time MRI data using convolutive NMF with sparseness constraints. In: INTERSPEECH, pp. 61–64. ISCA (2011)
11. Smaragdis, P., Shashanka, M., Raj, B., Mysore, G.J.: Probabilistic factorization of non-negative data with entropic co-occurrence constraints. In: Adali, T., Jutten, C., Romano, J.M.T., Barros, A.K. (eds.) ICA 2009. LNCS, vol. 5441, pp. 330–337. Springer, Heidelberg (2009)
12. Smaragdis, P.: Non-negative matrix factor deconvolution; extraction of multiple sound sources from monophonic inputs. In: Puntonet, C.G., Prieto, A.G. (eds.) ICA 2004. LNCS, vol. 3195, pp. 494–499. Springer, Heidelberg (2004)
13. Wrench, A.: A multi-channel/multi-speaker articulatory database for continuous speech recognition research. Phonus. **5**, 1–13 (2000)
14. Yilmaz, Y.K., Cemgil, A.T., Simsekli, U.: Generalised coupled tensor factorisation. In: NIPS (2011)

Audio-Visual Speech-Turn Detection and Tracking

Israel D. Gebru, Silèye Ba, Georgios Evangelidis, and Radu Horaud$^{(\boxtimes)}$

INRIA Grenoble Rhône-Alpes, Montbonnot Saint-Martin, France
radu.horaud@inria.fr

Abstract. Speaker diarization is an important component of multi-party dialog systems in order to assign speech-signal segments among participants. Diarization may well be viewed as the problem of detecting and tracking speech turns. It is proposed to address this problem by modeling the spatial coincidence of visual and auditory observations and by combining this coincidence model with a dynamic Bayesian formulation that tracks the identity of the active speaker. Speech-turn tracking is formulated as a latent-variable temporal graphical model and an exact inference algorithm is proposed. We describe in detail an audio-visual discriminative observation model as well as a state-transition model. We also describe an implementation of a full system composed of multi-person visual tracking, sound-source localization and the proposed online diarization technique. Finally we show that the proposed method yields promising results with two challenging scenarios that were carefully recorded and annotated.

Keywords: Speaker diarization · Audio-visual fusion · Sound-source localization · Multi-person tracking · Temporal graphical models

1 Introduction

In human-computer interaction (HCI) and human-robot interaction (HRI) it is often necessary to solve the multi-party dialog problem. For example, if two or more persons are engaged in a discussion, one important task to be solved, prior to automatic speech recognition (ASR) and natural language processing (NLP), is to correctly assign speech segments among the participants. In the speech and language processing literatures, this problem is often referred to as speaker diarization and a number of methods has been recently proposed to solve this problem, e.g., [1]. When only auditory data are available, the task is very difficult because of the inherent ambiguity of mixed acoustic signals captured by the microphones. An interesting alternative consists in fusing auditory and visual data. The two modalities provide complementary information and hence

Support from EU-FP7 ERC AdG VHIA (#340113) and STREP EARS (#609645) is greatly acknowledged.

© Springer International Publishing Switzerland 2015
E. Vincent et al. (Eds.): LVA/ICA 2015, LNCS 9237, pp. 143–151, 2015.
DOI: 10.1007/978-3-319-22482-4_17

audio-visual approaches to speaker diarization are likely to be more robust than audio-only approaches.

An audio-visual diarization method was recently proposed [7] where the hidden (latent) discrete variables represent the speaker identity and the speaker visibility at time t. The main limitation of [7] as well as of other audio-visual approaches reviewed in [1] is that these methods require the recognition of frontal faces and characterization of lip motions. Indeed, audio-visual association is often solved using the temporal correlation between facial features and audio features [8].

More generally, audio-visual association for speaker diarization can be achieved on the premise that a speech signal *coincides* with a person that is visible and that emits a sound. This coincidence must occur both in space and time. In formal dialogs, diarization is facilitated by the fact that the participants talk sequentially, that there is a short silence between speech turns, and that the participants face the cameras and are static or remain seated. In these cases, audio-visual association based on temporal coincidence seems to provide satisfactory results, e.g., [5]. In informal settings, which are very common particularly in HRI, the situation is much more complex. The perceived audio signals are corrupted by environmental noise, reverberations, and several people may occasionally speak simultaneously. People wander around, turn their heads away from the sensors, and come in and out of the fields of view of the cameras.

These problems were addressed by several authors in different ways. For example, [4] proposes a multi-speaker tracker using approximate inference implemented with a Markov chain Monte Carlo particle filter (MCMC-PF). [6] uses a 3D visual tracker, based on MCMC-PF as well, to estimate the positions and velocities of the participants which are then passed to a blind source separation method. This provides a proof of concept benchmark for moving speakers. MCMC-PF tracking cannot easily handle a varying number of speakers. Moreover, the reported experiments in both [4,6] are carried out with a microphone array and several cameras to guarantee that frontal views of the speakers are permanently visible. They do not specifically address speaker diarization which is a difficult problem in its own right.

In this paper it is proposed to enforce spatial coincidence into diarization. We consider a setup consisting of participants that are engaged in a multi-party dialog while they are allowed to move and to turn their attention away from the cameras. We propose to combine an online multi-person visual tracker [2], with a voice activity detector [9], and a sound-source localizer [3]. Assuming that the image and audio sequences are synchronized, we propose to group auditory features and visual features on the premise that they share a common location if they are generated by the same speaker. We define a speech-turn latent variable and we devise an online tracker such that at each frame t the identity of the active-speaker is estimated. We propose a discriminative observation model that evaluates the posterior probability of speech turns, conditioned by the outputs of a multi-person visual tracker, a sound-source localizer, and a voice activity detector. We also propose a dynamic model that allows to estimate the transition probabilities, from $t - 1$ to t, of the speech-turn variable. The proposed online speech-turn tracking method uses an efficient exact inference algorithm.

The remainder of this paper is organized as follows. Section 2 formally describes the proposed exact inference method; Sect. 2.1 describes the audio-visual discriminative observation model, Sect. 2.2 describes the proposed transition probabilities model. Section 3 describes implementation details and experiments. Finally, Sect. 4 draws some conclusions.

2   A Graphical Model for Tracking Speech Turns

We start by introducing some notations and definitions. Upper-case letters denote random variables while lower-case letters denote realizations of random variables. We consider an image sequence that is synchronized with an audio sequence and let t denote the temporal index of both visual and audio frames. There are at most N visual observations at frame t, $\mathbf{X}_t = (\boldsymbol{X}_{t1}, \ldots, \boldsymbol{X}_{tn}, \ldots, \boldsymbol{X}_{tN}) \in \mathbb{R}^{2 \times N}$, where the random variable $\boldsymbol{X}_{tn}$ corresponds to the location of person n in image t. Then, a multi-person tracker, e.g., [2] (Sect. 3) provides a time series of N image locations, namely $\mathbf{X}_{1:t} = \{\mathbf{X}_1, \ldots, \mathbf{X}_t\}$ and associated *visual-presence binary masks* $\mathbf{V}_{1:t}$, namely variable V_{tn} associated with $\boldsymbol{X}_{tn}$ such that $V_{tn} = 1$ if person n is present in image t and 0 otherwise. Hence $N_t = \sum_n V_{tn}$ denotes the number of persons that are present (observed) at t. We also consider an audio-source localizer that provides the azimuth and elevation of the dominant sound source at each audio frame t, e.g., [3] (Sect. 3). The sound-source location can then be mapped onto the image plane, such that an azimuth-elevation pair of observations is transformed into an image location modeled by a random variable $\boldsymbol{Y}_t \in \mathbb{R}^2$. Sound-source localization (SSL) together with voice activity detection (VAD) provide a time series of sound locations $\mathbf{Y}_{1:t} = \{\boldsymbol{Y}_1, \ldots, \boldsymbol{Y}_t\}$ and associated *speech-activity binary masks* $\mathbf{A}_{1:t} = \{A_1, \ldots, A_t\}$, such that $A_t = 1$ if there is an active audio source at frame t and 0 otherwise.

The objective is to track the active speaker which amounts to associate over time the audio activity (if any) with one of the tracked persons. This is also referred to as audio-visual speaker diarization, e.g., [7] which is addressed below in the framework of temporal graphical models; A time-series of discrete latent variables is introduced, $\mathbf{S}_{1:t} = \{S_1, \ldots, S_t\}$ such that $S_t = n, n \in \{1, 2, \ldots, N\}$ if person n is both observed and speaks at frame t, and $S_t = 0$ if none of the visible persons speaks at frame t. Notice that $S_t = 0$ encompasses two cases: first, there is an active sound-source at t ($A_t = 1$) but its location cannot be associated with one of the visible persons and second, there is no active sound-source at t ($A_t = 0$). The active-speaker or, equivalently, speech-turn tracker can be formulated as a maximum a posteriori (MAP) estimation problem:

$$\hat{s}_t = \underset{s_t}{\operatorname{argmax}}\, P(S_t = s_t | \boldsymbol{x}_{1:t}, \boldsymbol{v}_{1:t}, \boldsymbol{y}_{1:t}, \boldsymbol{a}_{1:t}). \tag{1}$$

The posterior probability (1) can be written as:

$$P(S_t = s_t | \boldsymbol{u}_{1:t}) = \frac{P(\boldsymbol{u}_t | S_t = s_t, \boldsymbol{u}_{1:t-1}) P(S_t = s_t | \boldsymbol{u}_{1:t-1})}{P(\boldsymbol{u}_t | \boldsymbol{u}_{1:t-1})}, \tag{2}$$

where we used the notation $\boldsymbol{u}_t = (\boldsymbol{x}_t, \boldsymbol{v}_t, \boldsymbol{y}_t, a_t)$. This can be further developed as:

$$P(S_t = s_t | \boldsymbol{u}_{1:t}) = \frac{P(\boldsymbol{u}_t | S_t = s_t) \sum\limits_{i=0}^{N} P(S_t = s_t | S_{t-1} = i) P(S_{t-1} = i | \boldsymbol{u}_{1:t-1})}{\sum\limits_{j=0}^{N} (P(\boldsymbol{u}_t | S_t = j)(\sum\limits_{i=0}^{N} P(S_t = j | S_{t-1} = i) P(S_{t-1} = i | \boldsymbol{u}_{1:t-1})))}$$

(3)

The evaluation of this recursive relationship requires the observed likelihood $P(\boldsymbol{u}_t | S_t = s_t)$ and the transition probabilities $P(S_t = j | S_{t-1} = i)$. Because the number of persons that are simultaneously present is small (3–5 persons), the exact evaluation of (3) is tractable and hence the MAP estimator (1) is straightforward.

2.1 Audio-Visual Observation Model

One crucial feature of the proposed model is its ability to robustly associate the acoustic activity at time t with a person. The generative model that is proposed below assigns the audio activity, if any, to a person, or to nobody. In this context, the state variable S_t plays the role of an assignment variable in a mixture model. If a sound-source is active at time t, $(A_t = 1)$ its location $\boldsymbol{y}_t$ is assumed to be drawn from the following Gaussian/uniform mixture:

$$P(\boldsymbol{y}_t | \boldsymbol{x}_t, \boldsymbol{v}_t, A_t = 1; \boldsymbol{\theta}_t) = \sum_{n=1}^{N} \pi_{tn} v_{tn} \mathcal{N}(\boldsymbol{y}_t | \boldsymbol{x}_{tn}, \boldsymbol{\Sigma}_{tn}) + \pi_{t0} \mathcal{U}(\beta_t),$$

(4)

where $\boldsymbol{\theta}_t = (\{\pi_{tn}\}_{n=0}^{N}, \{\boldsymbol{\Sigma}_{tn}\}_{n=1}^{N}, \beta_t)$ denotes the set of model parameters, namely the priors $\pi_{tn} = P(S_t = n)$, $\pi_{t0} + \sum_{n=1}^{N} v_{tn} \pi_{tn} = 1$, the 2×2 covariance matrices $\boldsymbol{\Sigma}_{tn}$, and a parameter β_t that characterizes the outlier component of the mixture, namely a uniform distribution. The posterior probability of a sound-source to be associated with the n-th visible person writes:

$$P(S_t = n | \boldsymbol{y}_t, \boldsymbol{x}_t, \boldsymbol{v}_t, A_t = 1; \boldsymbol{\theta}_t) = \frac{\pi_{tn} v_{tn} \mathcal{N}(\boldsymbol{y}_t | \boldsymbol{x}_{tn}, \boldsymbol{\Sigma}_{tn})}{\sum_{k=1}^{N} \pi_{tk} v_{tk} \mathcal{N}(\boldsymbol{y}_t | \boldsymbol{x}_{tk}, \boldsymbol{\Sigma}_{tk}) + \pi_{t0} \mathcal{U}(\beta_t)}.$$

(5)

We can also write the posterior probability that a sound source is not associated with a visible person, either because it corresponds to a sound emitted by a non visible person or emitted by another type of source, i.e., the posterior of the uniform component of the mixture:

$$P(S_t = 0 | \boldsymbol{y}_t, \boldsymbol{x}_t, \boldsymbol{v}_t, A_t = 1; \boldsymbol{\theta}_t) = \frac{\pi_{t0} \mathcal{U}(\beta_t)}{\sum_{k=1}^{N} \pi_{tk} v_{tk} \mathcal{N}(\boldsymbol{y}_t | \boldsymbol{x}_{tk}, \boldsymbol{\Sigma}_{tk}) + \pi_{t0} \mathcal{U}(\beta_t)}.$$

(6)

If there is no audio activity at time t $(A_t = 0)$, the posterior can be evaluated with the following formula, where r is a small positive scalar, e.g., $r = 0.2$:

$$P(S_t = 0 | \boldsymbol{y}_t, \boldsymbol{x}_t, \boldsymbol{v}_t, A_t = 0; r) = \begin{cases} r/N_t & \text{if } 1 \leq n \leq N \\ 1 - r & \text{if } n = 0. \end{cases}$$

(7)

Finally, by assuming a uniform distribution over the priors of visible person n ($v_{tn} = 1$), i.e., $\pi_{t0} = \pi_{tn} = 1/(N_t + 1)$, and by remarking that the observed-data likelihood $P(\boldsymbol{y}_t, \boldsymbol{x}_t, \boldsymbol{v}_t, a_t)$ does not depend on S_t, we obtain the following observation model:

$$P(\boldsymbol{y}_t, \boldsymbol{x}_t, \boldsymbol{v}_t, a_t | S_t = n) \propto P(S_t = n | \boldsymbol{y}_t, \boldsymbol{x}_t, \boldsymbol{v}_t, a_t). \tag{8}$$

2.2   State Transition Model

The state transition probabilities, $p(\boldsymbol{S}_t = j | \boldsymbol{S}_{t-1} = i)$, provide a temporal model for tracking speech turns. Several cases need be considered based on the presence/absence of persons and on their speaking status (for convenience and without loss of generality we set $v_{t0} = 1$):

$$p(\boldsymbol{S}_t = j | \boldsymbol{S}_{t-1} = i) = \begin{cases} p_s & \text{if } i = j \text{ and } v_{t-1i} = v_{ti} = 1 \\ (1 - p_s)/N_t & \text{if } i \neq j \text{ and } v_{t-1i} = v_{tj} = 1 \\ 0 & \text{if } v_{t-1i} = v_{t-1j} = 1 \text{ and } v_{tj} = 0 \\ 1/N_t & \text{if } v_{t-1i} = 1, v_{ti} = 0 \text{ and } v_{tj} = 1 \\ 1/N & \text{if } v_{t-1i} = 0 \text{ and } v_{ti} = 0. \end{cases} \tag{9}$$

The first case of (9) defines the self-transition probability, p_s, e.g., $p_s = 0.8$, of person i present at both $t-1$ and t. The second case defines the transition probability from person i present at $t-1$ to another person j present at t. The third case simply forbids transitions from person i present at $t-1$ to person j present at $t-1$ but not present at t. The fourth case defines the transition probability from person i present at $t-1$ but not present at t, to a person j present at t. The fifth case defines the transition probability from person i not present at $t-1$ to person j that is not present at t. These five cases can be grouped in a compact way to yield the state transition probability matrix ($\delta_{ij} = 1$ if $i = j$ and 0 otherwise):

$$p(\boldsymbol{S}_t = j | \boldsymbol{S}_{t-1} = i) = v_{t-1i} v_{tj} \left(p_s \delta_{ij} + \frac{(1 - p_s)(1 - \delta_{ij})}{N_t} + \frac{1 - v_{ti}}{N_t} \right) + \frac{1 - v_{t-1i}}{N}. \tag{10}$$

One may easily verify that $\sum_{j=1}^N p(\boldsymbol{S}_t = j | \boldsymbol{S}_{t-1} = i) = 1$.

3   Implementation and Experiments

As already outlined, the speech-turn tracking method that we propose in this paper may well be viewed as a speaker diarization process – track several persons in parallel, estimate their auditory status, and assign a speech segment to the dominant speaker. Unlike existing audio-visual diarization approaches, which only consider the *temporal coincidence* of the two modalities and which assume that the participants are always visible by the cameras, the proposed method enforces *spatial coincidence* such that it can deal with participants that are temporarily occluded, or who come in and out of the field of view of the cameras.

Unfortunately there are no publicly available datasets that would allow us to test the robustness of our method in the presence of moving/occluding participants and to compare it with other methods. The only datasets currently available correspond to formal meetings where the participants are seated and are permanently facing the cameras. Benchmarking against existing approaches was not possible because other methods do not cope with the audio-visual alignment issue.

Therefore we recorded our own data, gathered with two microphones and one camera [3]. The two modalities are synchronized such that video frames are temporally aligned with audio frames. Hence the frame index t is shared by the two modalities. We gathered two scenarios, the *counting* scenario (Fig. 1) and the *chat* scenario (Fig. 2). The videos are recorded at 25 FPS while the audio signals are sampled at 48000 Hz. Hence a video frame is 40 ms long. To ensure temporal synchronization between the two modalities, we define 40 ms audio frames in the following way. An audio frame is composed of several 64 ms windows shifted by 8 ms. Hence, a 40 ms audio frame is composed of 5 consecutive windows that partially overlap. The *counting* sequence has 500 frames (20 s) while the *chat* sequence has 850 frames (34 s).

We briefly describe the multi-person tracking and sound-source localization techniques used to gather values for the observed auditory and visual variables, i.e., Sect. 2.1. Among the visual tracking methods that are currently available, we chose the multi-person tracker of [2]. This method has several advantages, namely (i) it robustly handles fragmented tracks, which are due to occlusions or to unreliable detections, and (ii) it performs online discriminative learning

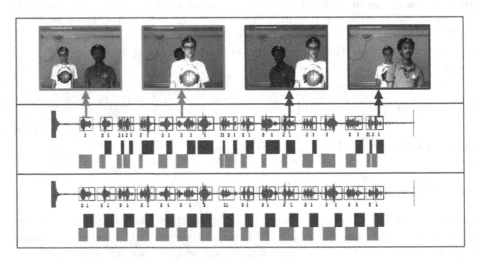

Fig. 1. The *counting* sequence involves two moving persons that occasionally occlude each other (top). Diarization results (middle) are also illustrated with a color diagram. Ground-truth diarization (bottom); notice that there is a systematic overlap between the two speech signals (Color figure online).

to handle similar appearances of different persons. The multi-person tracker provides values of the visual observation variables $\mathbf{X}_{1:t}$ and associated *visual-presence binary masks* $\mathbf{V}_{1:t}$, as explained in detail in Sect. 2.

Sound localization consists in finding the direction of arrival (DOA) of an acoustic signal from multi-microphone recordings. We adopted the method of [3] that estimates the DOA of a sound with two degrees of freedom (azimuth and elevation) using a binaural acoustic dummy head. A prominent advantage of this method over other DOA methods is that it provides a built-in mechanism for directly mapping a binaural feature vector associated with an audio frame, onto an image location. Hence a DOA associated with a sound source in expressed in pixel coordinates. In practice, the STFT is applied to 64 ms windows of the left and right microphone signals and a complex-valued binaural feature vector is built for each window using the ILD (interaural level difference) and IPD (interaural phase difference). Then we apply the method of [3] to a short spectrogram, composed of five consecutive windows (roughly corresponding to a frame), to estimate a DOA for each audio frame. In combination with a voice activity detector (VAD), this provides a time series of realizations of both the sound location variables $\mathbf{Y}_{1:t}$ and the associated *speech-activity binary masks* $\mathbf{A}_{1:t}$, as detailed in Sect. 2.

These auditory and visual observations are used to evaluate the likelihoods (8) and transition probabilities (9) which in turn are plugged into (3) to estimate

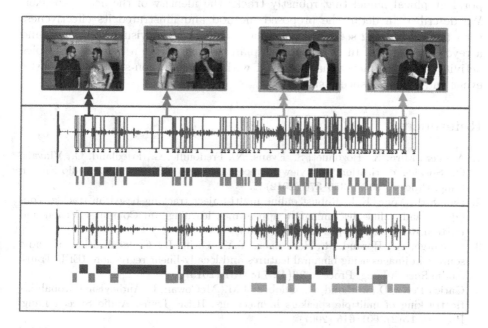

Fig. 2. The *chat* sequence involves two then three moving persons that take speech turns and that occasionally occlude each other (top). Diarization results (middle) and ground-truth (bottom); notice that in this case there is no speech overlap.

the posteriors of the speech-status variable given the observations, $P(S_t = s_t | \boldsymbol{u}_{1:t})$. In all our experiments we used the following numerical values for the model's free parameters: $r = 0.2, p_s = 0.8, \boldsymbol{\Sigma} = \text{Diag}\,[60, 120], \beta = 300000$. The proposed value of p_s achieves a good compromise between either assigning speech to the current speaker or jumping to another person. The parameters $\boldsymbol{\Sigma}$ and β are expressed in pixels. The value of β corresponds to a uniform distribution over an image of 640×480 pixels. The method yields 75 % correct results for the *counting* sequence and 64 % correct results for the *chat* sequence.

4   Conclusions

The paper addressed the problem of speaker diarization using auditory and visual data gathered with two microphones and one camera. Recent work in audio-visual diarization has capitalized on temporal coincidence of the two modalities, e.g., [1,7]. In contrast, we propose a speech-turn detection and tracking method that enforces spatial coincidence, namely it materializes that a sound-source and associated visual-object should have the same spatial location. Consequently, it is possible to perform speaker localization by detecting persons in an image, localizing a sound source, mapping the sound-source location onto the image and associating the source location with one of the persons that are present in the image. Moreover, this process can be plugged into a latent-variable temporal graphical model that robustly tracks the identity of the active speaker. We described in detail the proposed method and illustrated its effectiveness with two challenging scenarios involving moving people, visual occlusions, and a reverberant room. In the future we plan to incorporate a more robust voice activity detector/tracker that is robust with respect to non-stationary acoustic event and to mixed speech signals.

References

1. Anguera Miro, X., Bozonnet, S., Evans, N., Fredouille, C., Friedland, G., Vinyals, O.: Speaker diarization: A review of recent research. IEEE Trans. Audio Speech Lang. Process. **20**(2), 356–370 (2012)
2. Bae, S.H., Yoon, K.J.: Robust online multi-object tracking based on tracklet confidence and online discriminative appearance learning. In: Computer Vision and Pattern Recognition, pp. 1218–1225 (2014)
3. Deleforge, A., Horaud, R., Schechner, Y.Y., Girin, L.: Co-localization of audio sources in images using binaural features and locally-linear regression. IEEE Trans. Audio Speech Lang. Process. **23**(4), 718–731 (2015)
4. Gatica-Perez, D., Lathoud, G., Odobez, J.M., McCowan, I.: Audiovisual probabilistic tracking of multiple speakers in meetings. IEEE Trans. Audio Speech Lang. Process. **15**(2), 601–616 (2007)
5. Kidron, E., Schechner, Y.Y., Elad, M.: Cross-modal localization via sparsity. IEEE Trans. Signal Process. **55**(4), 1390–1404 (2007)
6. Naqvi, S., Yu, M., Chambers, J.: A multimodal approach to blind source separation of moving sources. IEEE J. Sel. Top. Signal Process. **4**(5), 895–910 (2010)

7. Noulas, A., Englebienne, G., Krose, B.J.A.: Multimodal speaker diarization. IEEE Trans. Pattern Anal. Mach. Intell. **34**(1), 79–93 (2012)
8. Potamianos, G., Neti, C., Gravier, G., Garg, A., Senior, A.W.: Recent advances in the automatic recognition of audiovisual speech. Proc. IEEE **91**(9), 1306–1326 (2003)
9. Sohn, J., Kim, N.S., Sung, W.: A statistical model-based voice activity detection. IEEE Signal Process. Lett. **6**(1), 1–3 (1999)

Advances in Nonlinear Blind
Source Separation

An Overview of Blind Source Separation Methods for Linear-Quadratic and Post-nonlinear Mixtures

Yannick Deville[1]($\boxtimes$) and Leonardo Tomazeli Duarte[2]

[1] Université de Toulouse, UPS-CNRS-OMP,
IRAP (Institut de Recherche en Astrophysique et Planétologie),
14 avenue Edouard Belin, 31400 Toulouse, France
yannick.deville@irap.omp.eu
[2] School of Applied Sciences, University of Campinas (UNICAMP),
Limeira, Brazil
leonardo.duarte@fca.unicamp.br

Abstract. Whereas most blind source separation (BSS) and blind mixture identification (BMI) investigations concern linear mixtures (instantaneous or not), various recent works extended BSS and BMI to nonlinear mixing models. They especially focused on two types of models, namely linear-quadratic ones (including their bilinear and quadratic versions, and some polynomial extensions) and post-nonlinear ones. These works are particularly motivated by the associated application fields, which include remote sensing, processing of scanned images (show-through effect) and design of smart chemical and gas sensor arrays. In this paper, we provide an overview of the above two types of mixing models and of the associated BSS and/or BMI methods and applications.

Keywords: Blind source separation · Blind mixture identification · Linear-quadratic mixing model · Post-nonlinear mixing model · Survey

1 Introduction

Blind source separation (BSS) methods aim at estimating a set of source signals from a set of observed signals which are mixtures of these source signals [17]. It has been shown that, if the mixing function applied to the source signals is completely unknown, the BSS problem (or its ICA solution) leads to unacceptable indeterminacies. Therefore, in most investigations the mixing function is requested to belong to a known class and only the values of its parameters are to be estimated. Many of these works are restricted to the simplest class of mixtures, namely linear ones (instantaneous or not) [17]. However, various more advanced studies dealing with *nonlinear* mixtures have also been reported. Two

This work was performed in a project jointly supported by CNRS (France) and FAPESP (Brazil). L.T. Duarte thanks CNPq (Brazil) for funding his research.

© Springer International Publishing Switzerland 2015
E. Vincent et al. (Eds.): LVA/ICA 2015, LNCS 9237, pp. 155–167, 2015.
DOI: 10.1007/978-3-319-22482-4_18

nonlinear mixing models have especially been considered. The first one consists of linear-quadratic (LQ) mixtures (and, to some extent, polynomial ones). It forms a natural extension of linear mixtures, by also including second-order (and possibly higher-order) terms. It may thus first be seen as a generic model, to be used as an approximation (truncated polynomial series) of various, possibly unknown, models faced in practical applications. Moreover, LQ mixing has been shown to *actually* occur in some applications. It has thus mainly been used for unmixing of remote sensing data [32,40,46–49], processing of scanned images involving the show-through effect [5,27,45,50] and analysis of gas sensor array data [10]. The other main nonlinear mixing model is the post-nonlinear (PNL) one. In this case, the mixing process comprises an initial linear mixing stage followed by a set of component-wise nonlinear functions. Therefore, such a model is useful in applications where the first stage of the mixing process is of linear nature but the sensors then exhibit a nonlinear response, due to saturation or more complex nonlinear transducer phenomena. The main field of application for PNL models is the design of smart chemical sensor arrays [11,12,25]. PNL models were also applied in the context of remote sensing data [6].

In this paper, we provide an overview of the two above-defined nonlinear mixing models and associated BSS and/or blind mixture identification (BMI) methods reported so far. We first define both mixing models in Sect. 2. We then present BSS/BMI methods for LQ mixtures in Sect. 3, and methods for PNL mixtures in Sect. 4. To conclude, related topics are briefly discussed in Sect. 5.

2   Considered Nonlinear Mixing Models

Considering continuous-valued signals which depend on a discrete variable n, the scalar form of the LQ (memoryless, or instantaneous) mixing model reads

$$x_i(n) = \sum_{j=1}^{M} a_{ij} s_j(n) + \sum_{j=1}^{M} \sum_{k=j}^{M} b_{ijk} s_j(n) s_k(n) \qquad \forall\, i \in \{1, \ldots, P\} \quad (1)$$

where $x_i(n)$ are the values of the P observed mixed signals for the sample index n and $s_j(n)$ are the values of the M unknown source signals which yield these observations, whereas a_{ij} and b_{ijk} are respectively the linear and quadratic mixing coefficients (with unknown values in the blind case) which define the considered source-to-observation transform. The specific version of this model which contains no second-order auto-terms (i.e. $b_{ijk} = 0$ when $k = j$) is called the bilinear mixing model. It corresponds to replacing the second sum in (1) by $\sum_{j=1}^{M-1} \sum_{k=j+1}^{M}$ (additional constant terms are considered in [5]). Similarly, the quadratic version of this model is obtained when all coefficients a_{ij} are zero.

A first matrix form of that model (1) reads

$$x(n) = As(n) + Bp(n) \qquad (2)$$

where the source and observation vectors are

$$s(n) = [s_1(n), \ldots, s_M(n)]^T, \qquad x(n) = [x_1(n), \ldots, x_P(n)]^T, \qquad (3)$$

where T stands for transpose and matrix A consists of the mixing coefficients a_{ij}. The column vector $p(n)$ is composed of all source products $s_j(n)s_k(n)$ of (1), i.e. with $1 \leq j \leq k \leq M$, arranged in a fixed, arbitrarily selected, order (see e.g. [49] for the natural order). The matrix B is composed of all entries b_{ijk} arranged so that i is the row index of B and the columns of B are indexed by (j,k) and arranged in the same order as the source products $s_j(n)s_k(n)$ in $p(n)$.

An even more compact model may be derived by stacking row-wise the vectors $s(n)$ and $p(n)$ of sources and source products in an extended vector

$$\tilde{s}(n) = \begin{bmatrix} s(n) \\ p(n) \end{bmatrix} \tag{4}$$

whereas the corresponding matrices A and B are stacked column-wise in an extended matrix

$$\tilde{A} = [A\ B]. \tag{5}$$

The LQ mixing model (2) then yields

$$x(n) = \tilde{A}\tilde{s}(n). \tag{6}$$

A third matrix-form model may eventually be derived by stacking column-wise all available signal samples, with n ranging from 1 to N, in the matrices

$$\tilde{S} = [\tilde{s}(1), \dots, \tilde{s}(N)], \quad X = [x(1), \dots, x(N)]. \tag{7}$$

The single-sample model (6) thus yields its overall matrix version

$$X = \tilde{A}\tilde{S}. \tag{8}$$

Some LQ BSS methods are based on the "original sources" $s_1(n), \dots, s_M(n)$ contained in $s(n)$, whereas other methods are based on the signals contained in $\tilde{s}(n)$, which are called the "extended sources" hereafter.

We also consider a second class of nonlinear mixing models known as post-nonlinear (PNL) models, in which each observed mixture corresponds to a univariate nonlinear function of a linear mixture of the sources. In its scalar form, the PNL model is given by

$$x_i(n) = f_i\left(\sum_{j=1}^{M} a_{ij}s_j(n)\right) \qquad \forall\, i \in \{1, \cdots, P\} \tag{9}$$

where a_{ij} and $f_i(\cdot)$ denote the linear mixing coefficients and the univariate nonlinear functions, respectively. In the blind case, both a_{ij} and $f_i(\cdot)$ are unknown. However, one often assumes that $f_i(\cdot)$ are strictly monotonic functions and, thus, admit inverse functions.

3   BSS/BMI Methods For Linear-Quadratic Mixtures

3.1   Independent Component Analysis (ICA) and Statistical Methods

(A) Methods for i.i.d Sources: Various LQ BSS methods were developed for i.i.d and mutually statistically independent sources, by **exploiting the mutual independence of the outputs of a separating system**. A first class of such methods is intended for the version of the LQ model which is determined with respect to the *original* sources, i.e. such that $P = M$. Their separating systems are nonlinear recurrent networks, which were described e.g. in [20,36] (and then extended e.g. in [14,20,50], including to much broader classes of nonlinear mixtures than LQ ones). The first of these LQ BSS methods [36] may be seen as an LQ extension of the linear Hérault-Jutten method, since it adapts the parameters of the above nonlinear recurrent networks so as to achieve an approximation of output independence, more precisely so as to cancel the (3,1) and (2,1) centered output cross-moments. The second reported method [51] is the LQ extension of Comon's linear approach, since it adapts the above parameters so as to minimize the mutual information of the network outputs, thus completely ensuring output independence.

The above recurrent separating systems are attractive because they only require one to know the analytical expression of the mixing model. On the contrary, direct structures require one to know the analytical expression of the *inverse* of the mixing model, which cannot be derived for *nonlinear* mixing models, except in simple situations such as bilinear models with 2 original sources [5,36]. However, nonlinear recurrent structures may yield some limitations: (i) they may be unstable at equilibrium points of interest or they may even lead to chaotic behavior [20] (see also [19–21] for extended networks which solve such problems; such networks have therefore also been used in [7] as original tools for solving nonlinear equations), (ii) they may have spurious equilibrium points and (iii) they require one to iteratively compute each output vector.

The situation becomes simpler when the number of observations can be increased up to the number of *extended* sources in $\tilde{s}(n)$. The mixing model is then determined and linear with respect to these extended sources, as shown by (6) or (8). Especially, performing M *linear* combinations of all these observations, with adequate coefficient values, then makes it possible to restore all M original sources. These coefficients may be adapted so as to enforce the statistical independence of the restored signals. In [26], this is achieved by a two-stage procedure based on minimizing the mutual information of these restored signals.

Still for mutually statistically independent and i.i.d sources, other reported LQ BSS and extended methods are based on **estimating the mixing model** (thus firstly achieving BMI). The first approach is based on maximizing the likelihood of the observations [15,37,38,50]. These investigations deal with determined mixtures of original sources and use the above-defined recurrent separating networks. The link between BSS methods based on likelihood maximization and mutual information minimization was established in [22] for *nonlinear* mixtures, including LQ ones.

The second reported approach for estimating the mixing model consists in starting from the relationships which define the observed signals with respect to the source signals and mixing coefficients, and deriving resulting expressions of some cumulants or moments of the observed signals with respect to those of the source signals and to the mixing coefficients. Solving these equations for known (estimated) values of the observation cumulants or moments then especially yields the values of the mixing coefficients (up to some indeterminacies). This approach was applied to quadratic mixtures in [16]. Also using cumulants, a quite different BMI method was proposed in [42] for complex-valued sources.

Finally, other LQ BSS methods for i.i.d sources **jointly estimate the sources** *and* **mixing model**, whereas the above-defined methods put more emphasis on *one* of these two types of unknowns of the BSS/BMI problem. This joint approach especially includes LQ Bayesian methods [24, 27]. Unlike above-described approaches, Bayesian methods do not explicitly use a separating system and thus avoid the associated potential issues, mainly for determined mixtures of original sources. However, the development of the use of Bayesian methods is limited by the complexity of their implementation and their high computational cost, as compared with the most popular ICA-related methods.

(B) Methods for Non-i.i.d Sources: Other LQ BSS methods have been developed by considering non-i.i.d random source signals and exploiting their autocorrelation (when each source is not independently distributed for different samples n) and/or their non-stationarity (when each source is not identically distributed for different samples). Both properties were used in the extension of the above likelihood-based method proposed in [39]. Similarly, the above-mentioned Bayesian approach [24, 27] has been applied to autocorrelated sources. A BMI method for LQ mixtures of autocorrelated and mutually independent source signals was also proposed in [1], using a joint diagonalization of a set of observation correlation matrices. Using similar tools, a method for extracting source products was also presented in [33] for uncorrelated sources with distinct autocorrelations.

3.2   Extensions of Nonnegative Matrix Factorization (NMF)

Although the considered observations (1) are nonlinear mixtures of the original sources, the reformulated mixing model (8) shows that they are linear mixtures of the associated extended sources and that, from the point of view of the latter sources, they follow the matrix-form mixing model encountered in *linear* NMF. When $\tilde{A}$ and $\tilde{S}$ (and thus X) are nonnegative, this allows one to develop LQ BSS/BMI methods for jointly estimating them by extending linear NMF methods, especially by adapting (estimates of) $\tilde{A}$ and the linear part of $\tilde{S}$ so as to minimize the Frobenius norm $||X - \tilde{A}\tilde{S}||_F$. Resulting gradient-based and/or Hessian-based algorithms have more complex forms than for linear mixtures, because they also involve derivatives of the second-order terms of $\tilde{S}$ with respect to the original sources. Several such methods are detailed in [47] (which also addresses polynomial mixtures) and [49] (the application of a standard NMF algorithm to the *extended* sources is also considered in [46,47]). A similar approach, dedicated to the case when both the sources and mixing coefficients follow bilinear models, is described in [32]. Besides, [45] deals with a very specific configuration involving 2 mixtures of 2 original sources, with the same quadratic contribution in both mixtures.

3.3   Sparse Component Analysis (SCA)

For linear mixtures, two major principles used in the literature for performing SCA may be briefly defined as follows. The first one consists in minimizing a sparsity-based cost function, such as the L0 norm of an "error term". The second one consists in taking advantage of zones (i.e. adjacent samples) in the sources where only one source is active, i.e. non-zero. These two principles have been extended to LQ mixtures, respectively in [28] and [18]. Besides, [40] describes an approach which also takes advantage of small parts of the observed data where only one "contribution" is non-zero, more precisely pixels which correspond to only one pure material (i.e. pure pixels) in the considered application to unmixing of remote sensing spectra. However, the proposed criterion is only guaranteed to yield a *necessary* condition for detecting pure pixels.

4   BSS Methods for Post-nonlinear Mixtures

There are basically two approaches to develop BSS methods for the mixing model expressed in (9). In the first one, which will be referred to as the *joint approach*, the nonlinear functions $f_i(\cdot)$ and the mixing matrix are jointly counterbalanced by means of a single criterion. Alternatively, in the *two-stage approach*, an initial stage aims at estimating the nonlinear functions $f_i(\cdot)$ or their inverses. Once these functions are estimated, the second stage simply becomes a linear BSS problem. In the sequel, we shall review methods for the joint and the two-stage approaches.

4.1   Joint Approaches

As in linear BSS, most of the works in PNL separation consider the determined case ($P = M$) and an adaptation framework based on ICA. An important reason for that comes from the theoretical results ensuring ICA-separability in PNL models. Indeed, the seminal work of Taleb and Jutten [59] showed that, by considering the following mirrored version of (9) as separating system

$$y_i(n) = \sum_{j=1}^{P} w_{ij} g_j(x_j(n)) \quad \forall\, i \in \{1, \cdots, P\}, \tag{10}$$

where w_{ij} and $g_j(\cdot)$ denote, respectively, the separating coefficients and the inverting functions, the recovery of independent components $y_1(n), \cdots, y_P(n)$ leads to source separation under conditions very close to those established for the linear case. Eventually, other works [3, 8, 61] extended [59] by providing more rigorous and less restrictive proofs.

By relying on the ICA-separability of PNL models, several works, starting from [59], proposed strategies to jointly adapt w_{ij} and $g_j(\cdot)$ by minimizing measures of statistical dependence. Most of these works considered a gradient-based framework for minimizing the mutual information [4, 9, 59] in the case of i.i.d sources. However, in [43], a joint method based on mutual information was set up to deal with non-i.i.d sources. It is also worth mentioning that alternative criteria of statistical dependence can be considered, as, for instance, in [2].

In the mutual information approach, a first issue is the estimation of the score functions of $y_i(n)$, which was addressed by several works of Pham (see [53], for instance) — these works dealt with general BSS models but were fundamental to many PNL algorithms. A second issue is the risk of local convergence to non-separating minima. In order to overcome this problem, several works proposed learning algorithms based on meta-heuristics [23, 56]. Finally, another issue that must be handled in joint PNL ICA algorithms is the parametrization of the inverting functions $g_j(\cdot)$. Indeed, since the ICA-separability results for (10) require bijective pairs $f_j(\cdot)$ and $g_j(\cdot)$, one aims at defining parametric functions $g_j(\cdot)$ that are bijective but flexible enough to compensate $f_j(\cdot)$. Possible solutions considered monotonic polynomials [23], splines [57], functions based on quantiles [54] and monotonic neural networks [31].

As in linear and LQ mixtures, an important class of joint PNL methods are obtained by formulating BSS as a Bayesian estimation problem. A Bayesian approach provides a natural framework to take into account prior information that can be expressed through a probabilistic modeling. On the other hand, the challenging aspect here is related to the practical resolution of the resulting inference problem. In [35], the authors introduced a variational learning scheme in order to perform inference. Alternatively, in [25], a Markov chain Monte Carlo (MCMC) strategy was set up to deal with a special class of PNL models that arises in chemical sensing applications. Such an approach has allowed the incorporation of non-negative priors for the sources and the mixing coefficients [25]. More recently, an MCMC-based Bayesian method was also proposed, but now for a special class of PNL models related to hyperspectral imagery [6]. A Bayesian approach was also considered for dealing with the underdetermined case $(P < M)$ [63].

4.2   Two-Stage Approaches

The first two-stage PNL approach [8] addressed the separation of two bounded sources under geometrical arguments. Indeed, since the scatter plot of bounded PNL mixtures presents nonlinear borders, [8] proposed to identify $g_j(\cdot)$ by recovering signals $g_1(x_1)$, $g_2(x_2)$ that provide linear borders in the scatter plot. A similar geometrical approach was proposed in [52] and was able to deal with the case of more than two sources.

Another idea to identify $g_j(\cdot)$ is to exploit prior information related to the sparsity of the sources. This idea is similar to the geometrical approach—indeed, when the sources are sparse, it becomes easier to identify the borders associated with the scatter plot of the mixtures. For instance, in [55], the authors proposed a SCA-based scheme to identify $g_j(\cdot)$ by assuming that there are, for each source $s_i(n)$, a temporal (or spatial) zone in which only $s_i(n)$ is active — such an idea is similar to that previously described for LQ mixtures. Similar SCA schemes were also developed for dealing with the case of underdetermined PNL mixtures [60, 62] — here, of course, the resulting linear BSS problem is more challenging than the determined linear BSS problem.

The underlying criteria for inverting $g_j(\cdot)$ in the two-stage methods presented so far are based on a joint processing of the mixtures. Alternatively, there are two-stage methods that process each mixture in a separate fashion — this approach will be referred to as *independent two-stage methods*. In this case, the resulting method thus comprises P independent executions of an algorithm that blindly compensates each $f_i(\cdot)$ followed by the application of a linear BSS method.

The first independent PNL two-stage methods make use of a well-known property involving probability distributions and nonlinear functions: it is possible to blindly estimate a univariate random variable that underwent a nonlinear distortion by setting up a nonlinear compensating function that provides a new random variable having the same probability distribution as that of the original random variable. This idea of matching the probability distributions of the input and its estimated version was firstly applied in signal processing by White [64].

In the context of PNL methods, it is possible to adapt the strategy proposed in [64] by observing that, after the linear mixing stage, the signals tend to Gaussian variables — this is a consequence of the central limit theorem. Moreover, due to the action of $f_j(\cdot)$, the observations $x_j(n)$ have non-Gaussian distributions. Therefore, a natural idea to counterbalance $f_j(\cdot)$ is to adapt $g_j(\cdot)$ so that its output becomes again Gaussian. Implementations of this strategy can be found in [58, 66, 67]. Interestingly, these Gaussianization-based methods provide better results as the number of sources increases, since the hypothesis of Gaussian linear mixtures is more realistic as P grows. However, even for a small number of sources they can provide at least an initial approximation of $g_j(\cdot)$ [58].

Alternative independent two-stage methods were proposed by taking into account other prior information than the gaussianity of the linear mixtures. For instance, [29] introduced a novel method that is tailored to the case of bandlimited sources. More recently, by considering the assumption that the sources admit a sparse representation in a known domain, [30] extended [29] and introduced a method for blind compensation of nonlinear functions that can be directly applied to PNL separation problems. Note that, differently from the above-discussed PNL methods based on sparsity priors, the introduced method in [30] operates in an independent fashion.

5   Conclusion

In this overview, we especially focused on practical BSS/BMI methods intended for two major types of nonlinear mixtures. Due to space limitations, we hereafter first only briefly mention closely related topics, i.e. the case of finite-alphabet sources [13], the invertibility of the considered mixing models (see e.g. [36]) the extension of LQ mixtures to polynomial ones (see e.g. [13,47,65]), the separability of these models with given separation principles, such as ICA (see [5] for bounded sources), the approaches based on non-blind and semi-blind BSS methods (see e.g. [20,34]). There are also interesting works that deal with PNL models and were not discussed in this overview paper. For instance, some effort has been put on the case of convolutive PNL mixtures [9,41] and on the problem of blind source extraction in PNL mixtures [44]. Finally, let us stress that other types of nonlinear mixing models have also recently been considered in the literature. All this shows that nonlinear BSS is currently a quite active research field, that we plan to present in more detail in a future publication.

References

1. Abed-Meraim, K., Belouchrani, A., Hua, Y.: Blind identification of a linear-quadratic mixture of independent components based on joint diagonalization procedure. In: Proceedings of ICASSP 1996, Atlanta, GA, 7–10 May 1996, vol. 5, pp. 2718–2721 (1996)
2. Achard, S., Pham, D.T., Jutten, C.: Quadratic dependence measure for nonlinear blind source separation. In: Proceedings of the Fourth International Workshop on Independent Component Analysis and Blind Signal Separation, ICA 2003 (2003)
3. Achard, S., Jutten, C.: Identifiability of post-nonlinear mixtures. IEEE Sig. Process. Lett. 12(5), 423–426 (2005)
4. Achard, S., Pham, D.T., Jutten, C.: Criteria based on mutual information minimization for blind source separation in post nonlinear mixtures. Sig. Process. 85, 965–974 (2005)
5. Almeida, M.S.C., Almeida, L.B.: Nonlinear separation of show-through image mixtures using a physical model trained with ICA. Sig. Process. 92, 872–884 (2012)
6. Altmann, Y., Dobigeon, N., Tourneret, J.Y.: Unsupervised post-nonlinear unmixing of hyperspectral images using a Hamiltonian Monte Carlo algorithm. IEEE Trans. Image Process. 23(6), 2663–2675 (2014)
7. Ando, R.A., Duarte, L.T., Soriano, D.C., Attux, R., Suyama, R., Deville, Y., Jutten, C.: Recurrent source separation structures as iterative methods for solving nonlinear equation systems. In: Proceedings of XXX Simposio Brasileiro de Telecomunicacoes, SBrT 2012, Brasilia, Brazil, 13–16 September 2012
8. Babaie-Zadeh, M., Jutten, C., Nayebi, K.: A geometric approach for separating post non-linear mixtures. In: Proceedings of the European Signal Processing Conference (EUSIPCO) (2002)
9. Babaie-Zadeh, M.: On blind source separation in convolutive and nonlinear mixtures. Ph.D. thesis, Institut National Polytechnique de Grenoble (2002)
10. Bedoya, G.: Non-linear blind signal separation for chemical solid-state sensor arrays. Ph.D. thesis, Universitat Politecnica de Catalunya (2006)
11. Bermejo, S., Jutten, C., Cabestany, J.: ISFET source separation: foundations and techniques. Sens. Actuators B 113, 222–233 (2006)
12. Bermejo, S.: A post-non-linear source separation algorithm for bounded magnitude sources and its application to ISFETs. Neurocomputing 148, 477–486 (2015)
13. Castella, M.: Inversion of polynomial systems and separation of nonlinear mixtures of finite-alphabet sources. IEEE Trans. Sig. Process. 56, 3905–3917 (2008)

14. Chaouchi, C., Deville, Y., Hosseini, S.: Nonlinear source separation: a quadratic recurrent inversion structure. In: Proceedings of ECMS 2009, Arrasate-Mondragon, Spain, 8–10 July 2009, pp. 91–98 (2009)
15. Chaouchi, C., Deville, Y., Hosseini, S.: Nonlinear source separation: a maximum likelihood approach for quadratic mixtures. In: Proceedings of MaxEnt 2010, Chamonix, France, 4–9 July 2010
16. Chaouchi, C., Deville, Y., Hosseini, S.: Cumulant-based estimation of quadratic mixture parameters for blind source separation. In: Proceedings of EUSIPCO 2010, Aalborg, Denmark, 23–27 August 2010, pp. 1826–1830 (2010)
17. Comon, P., Jutten, C. (eds.): Handbook of Blind Source Separation. Independent Component Analysis and Applications. Academic Press, Oxford (2010)
18. Deville, Y., Hosseini, S.: Blind identification and separation methods for linear-quadratic mixtures and/or linearly independent non-stationary signals. In: Proceedings of ISSPA 2007, Sharjah, United Arab Emirates, 12–15 February 2007
19. Deville, Y., Hosseini, S.K.: Stable higher-order recurrent neural network structures for nonlinear blind source separation. In: Davies, M.E., James, C.J., Abdallah, S.A., Plumbley, M.D. (eds.) ICA 2007. LNCS, vol. 4666, pp. 161–168. Springer, Heidelberg (2007). ISSN 0302-9743
20. Deville, Y., Hosseini, S.: Recurrent networks for separating extractable-target nonlinear mixtures. Part I: non-blind configurations. Sig. Proc. **89**, 378–393 (2009)
21. Deville, Y., Hosseini, S.: Blind operation of a recurrent neural network for linear-quadratic source separation: fixed points, stabilization and adaptation scheme. In: Vigneron, V., Zarzoso, V., Moreau, E., Gribonval, R., Vincent, E. (eds.) LVA/ICA 2010. LNCS, vol. 6365, pp. 237–244. Springer, Heidelberg (2010)
22. Deville, Y., Hosseini, S., Deville, A.: Effect of indirect dependencies on maximum likelihood and information theoretic blind source separation for nonlinear mixtures. Sig. Process. **91**, 793–800 (2011)
23. Duarte, L.T., Suyama, R., de Faissol Attux, R.R., Von Zuben, F.J., Romano, J.M.T.: Blind source separation of post-nonlinear mixtures using evolutionary computation and order statistics. In: Rosca, J.P., Erdogmus, D., Príncipe, J.C., Haykin, S. (eds.) ICA 2006. LNCS, vol. 3889, pp. 66–73. Springer, Heidelberg (2006)
24. Duarte, L.T., Jutten, C., Moussaoui, S.: Bayesian source separation of linear-quadratic and linear mixtures through a MCMC method. In: Proceedings of IEEE MLSP, Grenoble, France, 2–4 September 2009
25. Duarte, L.T., Jutten, C., Moussaoui, S.: A bayesian nonlinear source separation method for smart ion-selective electrode arrays. IEEE Sens. J. **9**(12), 1763–1771 (2009)
26. Duarte, L.T., Suyama, R., Attux, R., Deville, Y., Romano, J.M.T., Jutten, C.: Blind source separation of overdetermined linear-quadratic mixtures. In: Vigneron, V., Zarzoso, V., Moreau, E., Gribonval, R., Vincent, E. (eds.) LVA/ICA 2010. LNCS, vol. 6365, pp. 263–270. Springer, Heidelberg (2010)
27. Duarte, L.T., Jutten, C., Moussaoui, S.: Bayesian source separation of linear and linear-quadratic mixtures using truncated priors. J. Sig. Process. Syst. **65**, 311–323 (2011)
28. Duarte, L.T., Ando, R.A., Attux, R., Deville, Y., Jutten, C.: Separation of sparse signals in overdetermined linear-quadratic mixtures. In: Theis, F., Cichocki, A., Yeredor, A., Zibulevsky, M. (eds.) LVA/ICA 2012. LNCS, vol. 7191, pp. 239–246. Springer, Heidelberg (2012)
29. Duarte, L.T., Suyama, R., Attux, R.R.F., Rivet, B., Jutten, C., Romano, J.M.T.: Blind compensation of nonlinear distortions: application to source separation of post-nonlinear mixtures. IEEE Trans. Sig. Process. **60**, 5832–5844 (2012)

30. Duarte, L.T., Suyama, R., Attux, R., Romano, J.M.T., Jutten, C.: A sparsity-based method for blind compensation of a memoryless nonlinear distortion: application to ion-selective electrodes. IEEE Sens. J. **15**(4), 2054–2061 (2015)
31. Duarte, L.T., Pereira, F.O., Attux, R., Suyama, R., Romano, J.M.T.: Source separation in post-nonlinear mixtures by means of monotonic networks. In: Vincent, E. et al. (eds.) LVA/ICA 2015. LNCS, vol. 9237, pp. 176–183. Springer, Heidelberg (2015)
32. Eches, O., Guillaume, M.: A bilinear-bilinear nonnegative matrix factorization method for hyperspectral unmixing. IEEE Geosci. Remote Sens. Lett. **11**, 778–782 (2014)
33. Georgiev, P.: Blind source separation of bilinearly mixed signals. In: Proceedings of ICA 2001, San Diego, USA, pp. 328–331 (2001)
34. Halimi, A., Altmann, Y., Dobigeon, N., Tourneret, J.-Y.: Nonlinear unmixing of hyperspectral images using a generalized bilinear model. IEEE Trans. Geosci. Remote Sens. **49**, 4153–4162 (2011)
35. Honkela, A., Valpola, H., Ilin, A., Karhunen, J.: Blind separation of nonlinear mixtures by variational Bayesian learning. Digital Sig. Process. **17**(5), 914–934 (2007)
36. Hosseini, S., Deville, Y.: Blind separation of linear-quadratic mixtures of real sources using a recurrent structure. In: Mira, J., Álvarez, J.R. (eds.) IWANN 2003. LNCS, vol. 2687, pp. 241–248. Springer, Heidelberg (2003)
37. Hosseini, S., Deville, Y.: Blind maximum likelihood separation of a linear-quadratic mixture. In: Puntonet, C.G., Prieto, A.G. (eds.) ICA 2004. LNCS, vol. 3195, pp. 694–701. Springer, Heidelberg (2004). Erratum: http://arxiv.org/abs/1001.0863
38. Hosseini, S., Deville, Y.: Recurrent networks for separating extractable-target nonlinear mixtures. Part II: blind configurations. Sig. Process. **93**, 671–683 (2013)
39. Hosseini, S., Deville, Y.: Blind separation of parametric nonlinear mixtures of possibly autocorrelated and non-stationary sources. IEEE Trans. Sig. Process. **62**, 6521–6533 (2014)
40. Jarboui, L., Hosseini, S., Deville, Y., Guidara, R., Ben Hamida, A.: A new unsupervised method for hyperspectral image unmixing using a linear-quadratic model. In: Proceedings of ATSIP 2014, Sousse, Tunisia, 17–19 March 2014, pp. 423–428 (2014)
41. Koutras, A.: Blind separation of non-linear convolved speech mixtures. In: Proceedings of the IEEE International Conference on Acoustics, Speech, and Signal Processing (ICASSP), vol. 1, pp. I-913–I-916 (2002)
42. Krob, M., Benidir, M.: Blind identification of a linear-quadratic model using higher-order statistics. In: Proceedings of ICASSP 1993, Minneapolis, USA, 27–30 April 1993, pp. 440–443 (1993)
43. Larue, A., Jutten, C., Hosseini, S.: Markovian source separation in post-nonlinear mixtures. In: Puntonet, C.G., Prieto, A.G. (eds.) ICA 2004. LNCS, vol. 3195, pp. 702–709. Springer, Heidelberg (2004)
44. Leong, W.Y., Liu, W., Mandic, D.P.: Blind source extraction: standard approaches and extensions to noisy and post-nonlinear mixing. Neurocomputing **71**, 2344–2355 (2008)
45. Liu, Q., Wang, W.: Show-through removal for scanned images using non-linear NMF with adaptive smoothing. In: Proceedings of ChinaSIP, Beijing, pp. 650–654 (2013)
46. Meganem, I., Déliot, P., Briottet, X., Deville, Y., Hosseini, S.: Physical modelling and non-linear unmixing method for urban hyperspectral images. In: Proceedings of WHISPERS 2011, Lisbon, Portugal, 6–9 June 2011

47. Meganem, I., Deville, Y., Hosseini, S., Déliot, P., Briottet, X., Duarte, L.T.: Linear-quadratic and polynomial non-negative matrix factorization; application to spectral unmixing. In: Proceedings of EUSIPCO 2011, Barcelona, Spain, 29 August–2 September 2011

48. Meganem, I., Déliot, P., Briottet, X., Deville, Y., Hosseini, S.: Linear-quadratic mixing model for reflectances in urban environments. IEEE Trans. Geosci. Remote Sens. **52**, 544–558 (2014)

49. Meganem, I., Deville, Y., Hosseini, S., Déliot, P., Briottet, X.: Linear-quadratic blind source separation using NMF to unmix urban hyperspectral images. IEEE Trans. Sig. Process. **62**, 1822–1833 (2014)

50. Merrikh-Bayat, F., Babaie-Zadeh, M., Jutten, C.: Linear-quadratic blind source separating structure for removing show-through in scanned documents. IJDAR **14**, 319–333 (2011)

51. Mokhtari, F., Babaie-Zadeh, M., Jutten, C.: Blind separation of bilinear mixtures using mutual information minimization. In: Proceedings of IEEE MLSP, Grenoble, France, 2–4 September 2009

52. Nguyen, T.V., Patra, J.C., Emmanuel, S.: gpICA: a novel nonlinear ICA algorithm using geometric linearization. EURASIP J. Adav. Sig. Process. **2007**, 1–12 (2007)

53. Pham, D.T.: Fast algorithms for mutual information based independent component analysis. IEEE Trans. Sig. Process. **52**(10), 2690–2700 (2004)

54. Pham, D.T.: Flexible parametrization of postnonlinear mixtures model in blind sources separation. IEEE Sig. Process. Lett. **11**(6), 533–536 (2004)

55. Puigt, M., Griffin, A., Mouchtaris, A.: Post-nonlinear speech mixture identification using single-source temporal zones and curve clustering. In: Proceedings of the European Signal Processing Conference (EUSIPCO), pp. 1844–1848 (2011)

56. Rojas, F., Puntonet, C.G., Rodríguez-Álvarez, M., Rojas, I., Martín-Clemente, R.: Blind source separation in post-nonlinear mixtures using competitive learning, simulated annealing, and a genetic algorithm. IEEE Trans. Syst. Man Cybern. Part C: Appl. Rev. **34**(4), 407–416 (2004)

57. Solazzi, M., Uncini, A.: Spline neural networks for blind separation of post-nonlinear-linear mixtures. IEEE Trans. Circuits Syst. I: Regul. Pap. **51**(4), 817–829 (2004)

58. Solé-Casals, J., Jutten, C., Pham, D.T.: Fast approximation of nonlinearities for improving inversion algorithms of PNL mixtures and Wiener systems. Sig. Process. **85**, 1780–1786 (2005)

59. Taleb, A., Jutten, C.: Source separation in post-nonlinear mixtures. IEEE Trans. Sig. Process. **47**(10), 2807–2820 (1999)

60. Theis, F.J., Amari, S.: Postnonlinear overcomplete blind source separation using sparse sources. In: Puntonet, C.G., Prieto, A.G. (eds.) ICA 2004. LNCS, vol. 3195, pp. 718–725. Springer, Heidelberg (2004)

61. Theis, F.J., Gruber, P.: On model identifiability in analytic postnonlinear ICA. Neurocomputing **64**, 223–234 (2005)

62. Vaerenbergh, S.V., Santamaria, I.: A spectral clustering approach to underdetermined postnonlinear blind source separation of sparse sources. IEEE Trans. Neural Netw. **17**(3), 811–814 (2006)

63. Wei, C., Woo, W., Dlay, S.: Nonlinear underdetermined blind signal separation using bayesian neural network approach. Digital Sig. Process. **17**, 50–68 (2007)

64. White, S.A.: Restoration of nonlinearly distorted audio by histogram equalization. J. Audio Eng. Soc. **30**(11), 828–832 (1982)

65. Zeng, T.-J., Feng, Q.-Y., Yuan, X.-H., Ma, H.-B.: The multi-component signal model and learning algorithm of blind source separation. In: Proceedings of Progress in Electromagnetics Research Symposium, Taipei, 25–28 March 2013, pp. 565–569 (2013)
66. Zhang, K., Chan, L.W.: Extended gaussianization method for blind separation of post-nonlinear mixtures. Neural Comput. **17**, 425–452 (2005)
67. Ziehe, A., Kawanabe, M., Harmeling, S., Müller, K.R.: Blind separation of post-nonlinear mixtures using gaussianizing transformations and temporal decorrelation. J. Mach. Learn. Res. **4**, 1319–1338 (2003)

Nonlinear Sparse Component Analysis with a Reference: Variable Selection in Genomics and Proteomics

Ivica Kopriva[1]([⊠]), Sanja Kapitanović[2], and Tamara Čačev[2]

[1] Division of Laser and Atomic R&D, Ruđer Bošković Institute,
Bijenička Cesta 54, 10000 Zagreb, Croatia
Ivica.Kopriva@irb.hr
[2] Division of Molecular Medicine, Ruđer Bošković Institute,
Bijenička Cesta 54, 10000 Zagreb, Croatia
{Sanja.Kapitanovic,Tamara.Cacev}@irb.hr

Abstract. Many scenarios occurring in genomics and proteomics involve small number of labeled data and large number of variables. To create prediction models robust to overfitting variable selection is necessary. We propose variable selection method using nonlinear sparse component analysis with a reference representing either negative (healthy) or positive (cancer) class. Thereby, component comprised of cancer related variables is automatically inferred from the geometry of nonlinear mixture model with a reference. Proposed method is compared with 3 supervised and 2 unsupervised variable selection methods on two-class problems using 2 genomic and 2 proteomic datasets. Obtained results, which include analysis of biological relevance of selected genes, are comparable with those achieved by supervised methods. Thus, proposed method can possibly perform better on unseen data of the same cancer type.

Keywords: Variable selection · Nonlinear mixture model · Empirical kernel maps · Sparse component analysis

1 Introduction

Data acquired by microarray gene expression profiling technology [1, 2] or mass spectrometry [3, 4], present "large p, small n" problem: large number of variables (genes or mass-to-charge, m/z, ratios) and small number of labeled (diagnosed) gene or protein expressions. They correspond with the mixtures in blind source separation (BSS) vocabulary while variables correspond with samples in BSS vocabulary. In described scenario learned prediction models adapt to training data (overfitt) and not generalize well on unseen data of the same cancer type [5, 6]. Improvement of predictor performance is enabled by variable selection [6, 7]. This implies selection of small number of variables that discriminate well between cancer and healthy subjects. Here we propose unsupervised variable selection method that performs blind sparseness constrained decomposition of each mixture independently according to implicit, empirical kernel map (EKM)-based [8], nonlinear mixture model. The model is comprised of a test mixture and a reference mixture representing positive (cancer) class.

© Springer International Publishing Switzerland 2015
E. Vincent et al. (Eds.): LVA/ICA 2015, LNCS 9237, pp. 168–175, 2015.
DOI: 10.1007/978-3-319-22482-4_19

Proposed method takes into account biological diversity of mixtures as well as nonlinear nature of the interaction among variables (genes) within the components present in mixtures [9]. Reference mixture enables automatic selection of component within test mixture that is comprised of cancer related variables. Since no label information is used selected cancer related components can be used both for biomarker identification studies as well as for training prediction models. As opposed to that, variable selection based on standard BSS methods, [10–14], use whole dataset for decomposition. Afterwards, one component composed of cancer related variables is selected by using label information. That enables selected component to be used for biomarker identification studies but prevents it to be used for training predictive models (otherwise label information would be used twice). The method proposed herein is nonlinear generalization of the mixture dependent linear model with a reference [15] as well as generalization of mixture dependent nonlinear model with a reference that is based on approximate explicit feature maps (EFM) [16]. Implicit nonlinear mapping is performed variable-wise yielding nonlinear mixture model with the same number of variables and "increased" number of mixtures. Sparse component analysis (SCA) is performed on nonlinearly mapped mixture. Afterwards, variables in cancer related components are ranked by their mixture-wise variance. That yields index set used to access variables in the original input space. They are used to learn two-class support vector machine (SVM) predictive model [17]. We compare proposed method with 3 supervised variable selection methods [18, 19] and 2 unsupervised methods [15, 16]. The methods were compared on 2 well-known cancer types in genomics: colon cancer [1] and prostate cancer [2], as well as on 2 well-known cancer types in proteomics: ovarian cancer [3] and prostate cancer [4]. Furthermore, analysis of biological relevance of selected genes in colon cancer experiment is also provided. We describe proposed method in Sect. 2. Results of comparative performance analysis are described in Sect. 3. Conclusions are proposed in Sect. 4.

2   Method

Let us assume that N mixtures are stored in rows of data matrix $\mathbf{X} \in \mathbb{R}^{N \times K}$, whereas each mixtures is further comprised of K variables. We also assume that N mixtures have diagnoses (label): $\mathbf{x}_n \in \mathbb{R}^{1 \times K}, y_n \in \{1, -1\}$, $n = 1,..., N$, where 1 stands for positive (cancer) and -1 stands for negative (healthy) mixture. Within this paper we assume that mixtures are normalized such that: $-1 \leq x_{nk} \leq 1 \, \forall n = 1, ..., N$ $k = 1, ..., K$. Matrix factorization methods such as principal component analysis, independent component analysis, SCA and/or nonnegative matrix factorization assume linear mixture model: $\mathbf{X} = \mathbf{AS}$, where $\mathbf{A} \in \mathbb{R}_{0+}^{N \times M}$, $\mathbf{S} \in \mathbb{R}^{M \times K}$ and M stands for an unknown number of components imprinted in mixtures. Each component is represented by a row vector of matrix $\mathbf{S}$, that is: $\mathbf{s}_m \in \mathbb{R}^{1 \times K}$, $m = 1,..., M$. Column vectors of matrix $\mathbf{A}$: $\mathbf{a}_m \in \mathbb{R}^{N \times 1}$, $m = 1,..., M$, represent concentration profiles of the corresponding components. To infer component comprised of disease relevant variables label information is used by methods such as [10, 11]. That prevents usage of selected component for training prediction models. This limitation has been addressed in [15] by formulating mixture dependent linear model with a reference. Herein, as in [16], we assume nonlinear model of a mixture:

$$\begin{bmatrix} x_{ref,k} \\ x_{nk} \end{bmatrix} = f_n(\mathbf{s}_{k;n}) \quad n = 1, \ldots, N; \; k = 1, \ldots, K \tag{1}$$

where $f_n : \mathbb{R}^{M_n} \to \mathbb{R}^2$ is an unknown mixture dependent nonlinear function that maps M_n-dimensional vector of variables $\mathbf{s}_{k;n} \in \mathbb{R}^{M_n \times 1}$ to 2-dimensional observation vector. Thereby, first element of the observation vector belongs to the reference mixture and second element to the test mixture. Herein, we assume that reference mixture represents positive (cancer) class. It can be selected by an expert or, as it was the case herein, can be obtained by averaging all the mixtures belonging to positive class. We propose EKM for implicit (kernel-based) mapping of (1). We repeat definition 2.15 from [8]:

Definition 1. For a given set of patterns $\{\mathbf{v}_d \in \mathbb{R}^{N \times 1}\}_{d=1}^{D} \subset \mathbf{X}$, $D \in \mathbb{N}$, we call $\psi : \mathbb{R}^N \to \mathbb{R}^D$, where $\psi : \mathbf{x}_{nk} \mapsto [\kappa(\mathbf{v}_1, \mathbf{x}_{nk}), \ldots, \kappa(\mathbf{v}_D, \mathbf{x}_{nk})]^T$, $\forall k = 1, \ldots, K$, the EKM with respect to basis $\mathbf{V} := \{\mathbf{v}_d\}_{d=1}^{D}$.

Thereby, $\mathbf{x}_{nk} = [x_{ref,k} \; x_{nk}]^T$ is defined in (1). The basis $\mathbf{V}$ has to satisfy:

$$span\{\mathbf{v}_d\}_{d=1}^{D} \approx span\{\mathbf{x}_{nk}\}_{k=1}^{K} \tag{2}$$

To estimate $\mathbf{V}$ we have used *k-means* algorithm to cluster empirical set of patterns (samples) $\{\mathbf{x}_{nk}\}_{k=1}^{K}$ in predefined number of D cluster centroids (basis vectors). If $\mathbf{V}$ satisfies (2) then obviously $\mathbf{V} \cup [1 \; 0]^T$ satisfies (2) as well. Hence, EKM $\psi(\mathbf{x}_{nk})$ is obtained by projecting EFM $\phi(\mathbf{x}_{nk})$ associated with kernel $\kappa(\circ, \mathbf{x}_{nk})$ on a $(D+1)$-dimensional subspace in mapping induced space spanned by $\left\{\phi(\mathbf{v}_d) \in \mathbb{R}^{\bar{D}}\right\}_{d=1}^{D+1}$:

$$\begin{aligned} \psi(\mathbf{x}_{nk}) &= [\phi(\mathbf{v}_1) \; \ldots \; \phi(\mathbf{v}_D) \; \phi(\mathbf{v}_{D+1})]^T \phi(\mathbf{x}_{nk}) \\ &= [\kappa(\mathbf{x}_{nk}, \mathbf{v}_1) \; \ldots \; \kappa(\mathbf{x}_{nk}, \mathbf{v}_D) \; \kappa(\mathbf{x}_{nk}, \mathbf{v}_{D+1})]^T \quad \forall k = 1, \ldots, K \end{aligned} \tag{3}$$

where $\mathbf{v}_{D+1} \in \mathbb{R}_{0+}^{2 \times 1} = [1 \; 0]^T$. We now define mixture dependent linear model in EKM-induced space:

$$\psi\begin{pmatrix} x_{ref,k} \\ x_{nk} \end{pmatrix} \approx \bar{\mathbf{A}}_n \bar{\mathbf{s}}_{k;n} \quad k = 1, \ldots, K \tag{4}$$

where $\bar{\mathbf{A}}_n \in \mathbb{R}_{0+}^{D+1 \times M_n}$, $\bar{\mathbf{s}}_{k;n} \in \mathbb{R}^{M_n \times 1}$ and M_n stands for mixture dependent number of components. The key observation regarding nonlinear model (3)/(4) is that, for suitably chosen kernel, $\kappa(\mathbf{x}_{nk}, \mathbf{v}_{D+1})$ it becomes a function of the reference mixture $x_{ref,k}$ only. As an example, for $\kappa(\mathbf{x}_{nk}, \mathbf{v}_{D+1}) = \exp(|\langle \mathbf{x}_{nk}, \mathbf{v}_{D+1} \rangle|/\sigma^2) = \exp(x_{ref,k}/\sigma^2)$. For Gaussian kernel it applies: $\kappa(\mathbf{x}_{nk}, \mathbf{v}_{D+1}) = \exp(-x_{nk}^2/\sigma^2)\exp(-\sigma^2)\exp\left(\left(2x_{ref,k} - x_{ref,k}^2\right)/\sigma^2\right)$.

Under assumption $-1 \le x_{nk} \le 1$ and $\sigma^2 > x_{nk}^2$ the first part is approximately 1 and the last part $\exp\big((2x_{ref,k} - 1)/\sigma^2\big)$. Thus, $\kappa(\mathbf{x}_{nk}, \mathbf{v}_{D+1}) \approx \exp\big((2x_{ref,k} - 1)/\sigma^2\big)$. Hence, we can express $\psi(\mathbf{x}_{nk})$ in standard Euclidean basis $\{\mathbf{e}_d\}_{d=1}^{D+1}$:

$$\psi\begin{pmatrix} x_{ref,k} \\ \mathbf{x}_{nk} \end{pmatrix} = \kappa(\mathbf{x}_{nk}, \mathbf{v}_1)\mathbf{e}_1 + \ldots + \kappa(\mathbf{x}_{nk}, \mathbf{v}_D)\mathbf{e}_D + f(x_{ref,k})\mathbf{e}_{D+1} \qquad (5)$$

Representation (5) enables automatic selection of component $\bar{\mathbf{s}}_{m*}$, $m^* \in \{1,.., M_n\}$ comprised of cancer relevant variables. $\bar{\mathbf{s}}_{m*}$ is associated with the mixing vector that closes the smallest angle with the axis $\mathbf{e}_{D+1}$ that represents cancer class. Cosine of the angle between mixing vector $\bar{\mathbf{a}}_{m;n}$ and $\mathbf{e}_{D+1}$ s obtained as:

$$\cos \angle(\bar{\mathbf{a}}_{m;n}, \mathbf{e}_{D+1}) = \langle \bar{\mathbf{a}}_{m;n}, \mathbf{e}_{D+1}\rangle / \|\bar{\mathbf{a}}_{m;n}\| \qquad (6)$$

Thus index of component composed of cancer relevant variables is obtained as:

$$m^* = \arg\max_m \cos \angle(\bar{\mathbf{a}}_{m;n}, \mathbf{e}_{D+1}) \qquad (7)$$

When each mixture is decomposed according to (4), components comprised of cancer relevant variables are stored row-wise in a matrix $\bar{\mathbf{S}}_{cancer} \in \mathbb{R}^{N \times K}$. Variables (columns of $\bar{\mathbf{S}}_{cancer}$) are then ranked by their variance across the mixture dimension yielding $\bar{\mathbf{S}}_{cancer}^{ranked} \in \mathbb{R}^{N \times K}$. Let us denote by I a corresponding index set. Variables ranked in the original mixture space are obtained by indexing each mixture by I, that is: $\mathbf{x}_n^{ranked} = \mathbf{x}_n(I)$, $n = 1,\ldots, N$. Mixtures with ranked variables form rows of the matrix $\mathbf{X}^{ranked} \in \mathbb{R}^{N \times K}$. That, when paired with the vector of labels $\mathbf{y}$, is used to learn SVM prediction model.

Decomposition of the linear mixture model (4) is performed enforcing sparseness of the components $\bar{\mathbf{s}}_{m;n}$, $m = 1, \ldots, M_n$. That is because sparse components are comprised of few dominantly expressed variables and that can be good indicator of a disease. Method used to solve, in principle, underdetermined BSS problem (4) estimates mixing matrix $\bar{\mathbf{A}}_n$ first by using the separable NMF algorithm [20] with a MATAB code available at: https://sites.google.com/site/nicolasgillis/publications. The important characteristic of the method [20] is that there are no free parameters to be tuned or defined a priori. The unknown number of components M_n is also estimated automatically and is limited above by $D + 1$. Thus, by cross-validating D we implicitly cross-validate M_n as well. After $\bar{\mathbf{A}}_n$ is estimated, $\bar{\mathbf{S}}_n$ is estimated by solving sparseness constrained optimization problem:

$$\hat{\bar{\mathbf{S}}}_n = \min_{\bar{\mathbf{S}}_n} \left\{ \frac{1}{2}\left\| \hat{\bar{\mathbf{A}}}_n \bar{\mathbf{S}}_n - \psi\begin{pmatrix} \mathbf{x}_{ref} \\ \mathbf{x}_n \end{pmatrix} \right\|_F^2 + \lambda \|\bar{\mathbf{S}}_n\|_1 \right\} \qquad (8)$$

where the hat sign denotes an estimate of the true (but unknown) quantity, λ is regularization parameter and $\|\bar{\mathbf{S}}_n\|_1$ denotes ℓ_1-norm of $\bar{\mathbf{S}}_n$. To solve (8), we have used the

iterative shrinkage thresholding (IST) method [21] with a MATLAB code at: http://ie.
technion.ac.il/Home/Users/becka.html. Sparsity of the solution is controlled by the
parameter λ. There is a maximal value of λ (denoted by λ_{max} here) above which the
solution of the problem (8) is equal to zero. Thus, in the experiments reported below
the value of λ has been selected by cross-validation with respect to λ_{max}.

3 Experiments

Proposed approach is compared with supervised variable selection methods: maximum
mutual information minimal redundancy (MIMR) method [18] and HITTON_PC and
HITTON_MB [19] methods. We also report results for linear [15] and EFM-based
nonlinear [16] counterparts of proposed method. Gene Expression Model Selector
(GEMS) software system [22], has been used for 10-fold cross-validation based
learning of SVM-based diagnostic models with polynomial and Gaussian kernels. The
system is available at: http://www.gems-system.org/. HITON_PC and HITON_MB
algorithms are implemented in GEMS software system while implementation of the
MIMR algorithm is available at MATLAB File Exchange. Order D of the EKM in
(3) has been cross-validated in the range: $D \in \{5, 10, 15, 20, 25, 30\}$. Regularization
constant λ in (8) has been cross-validated in the range: $\lambda \in \{0.05, 0.1, 0.2, 0.3, 0.4, 0.
5\} \times \lambda_{max}$. Methods were compared on 2 cancer types in genomics: colon cancer [1]
and prostate cancer [2], as well as on 2 cancer types in proteomics: ovarian cancer [3]
and prostate cancer [4]. The number of cancer vs. normal mixtures is for 4 datasets
given in respective order as: 40/22, 52/50, 100/100 and 69/63. The number of variables
in each dataset is in respective order given as: 2000, 10509, 15152 and 15154. For each
dataset we report in Table 1 result achieved by: proposed method, the best result
achieved by one of 3 supervised methods and results achieved by [15, 16]. Due to the

Table 1. Classification accuracy and number of selected variables.

Dataset	Proposed method	Supervised method	[16]	[15]
1. Prostate cancer	91.18 % / 12 genes ($D = 20$, $\lambda = 0.3$, Gauss kernel).	MIMR: **98.08 %** / **10 genes**	91.27 % / 38 genes	94.27 % / 477 genes.
2. Colon cancer	**93.57 %** / 20 genes ($D = 20$, $\lambda = 0.3$, Gauss kernel).	HITON_MB: 93.33 % / **4 genes**	91.91 % / 24 genes	90.48 % / 30 genes, $\lambda = 0.05$.
3. Ovarian cancer	94.5 % / 47 m/z lines ($D = 20$, $\lambda = 0.35$, Exp. kernel).	HITON_PC: **99.5 %** / **7 m/z lines**	93 % / 7 m/z lines	82 % / 25 m/z lines, $\lambda = 0.2$.
4. Prostate cancer	94.61 % / 27 m/z lines ($D = 20$, $\lambda = 0.35$, Exp. kernel).	MIMR: **100 %** / **9 m/z lines**	94.06 % / 14 m/z lines	94.01 % / 85 m/z lines, $\lambda = 0.2$.

lack of space we do not report details on parameters of the SVM classifiers. For each of 4 datasets, proposed method achieves result that is worse than but comparable with the result of supervised algorithm and better than its linear and EFM-based nonlinear unsupervised counterparts [15, 16]. Since reported results are achieved with small number of variables the probability of overfitting is reduced. Thus, it is reasonable to expect that performance on unseen data of the same cancer type by proposed unsupervised method will be better than the one achieved with supervised algorithms.

Colon cancer data are available at: http://genomic-pubs.princeton.edu/oncology/affydata/index.html. Prostate cancer genomic data are available at: http://www.gems-system.org/. Ovarian and prostate cancer proteomic data are available at: http://home.ccr.cancer.gov/ncifdaproteomics/ppatterns.asp. To comply with principle of reproducible research software which implements proposed algorithm, datasets used and results presented in Table 1 are available at: http://www.lair.irb.hr/ikopriva/Data/HRZZ/data/LVA_2015.zip.

We also provide brief biological interpretation of genes selected by proposed method in the colon cancer experiment [1]. The majority of genes selected by the proposed algorithm have been previously associated with tumorgenesis. For instance, expression of genes encoding ribosomal proteins (RPS9, RPS18, RPS29, RPS24, RPLP1, RPL30) has been known to increase in tumors as a result of uncontrolled cell proliferation which is one of the key hallmarks of cancer. In addition, several previous microarray studies have reported an increase in mRNA expression of ribosomal genes in solid tumors including colorectal cancer [23]. Several genes which were found to be differentially expressed by our algorithm like IGHG3, FTL, GAPDH and UBC encode proteins involved in cellular metabolism and bioenergetics and have previously been associated with cancer [24, 25]. This is not surprising since changes in metabolic processes are often observed in tumor cells. For instance altered GAPDH expression has been reported in breast, gastric, liver, lung as well as colorectal cancer [26]. Laminin receptor 1 (RPSA) and actin (ACTB), two other genes detected by our algorithm, are involved in wide spectrum of cellular functions including the maintenance of cellular structure as well as adhesion and motility [26]. When specifically colorectal cancer is considered, S100A6 has previously been associated with this type of cancer [27]. In addition, the role of Thymosin beta-4 in cell proliferation, growth and migration has been previously established and its overexpression has been reported during the different stages of colorectal carcinogenesis [28].

4 Conclusion

Since it requires little prior knowledge unsupervised decomposition of a set of mixtures into additive combination of components is of particular importance in addressing overfitting problem. Herein, we have proposed an unsupervised approach for variable selection by decomposing each mixture individually into sparse components according to nonlinear kernel-based model of a mixture, whereas decomposition is performed with respect to a reference mixture that represents positive (cancer) class. That enables selection of cancer related components automatically and, afterwards, their use for either biomarker identification studies or learning diagnostic models. It is conjectured that outlined properties of proposed method enabled competitive diagnostic accuracy with small number of variables on cancer related human gene and protein expression datasets. While proposed method is developed for binary (two-class) problems its extension for multi-category classification problems is aimed for the future work.

Acknowledgments. Work of I. Kopriva has been partially supported through the FP7-REGPOT-2012-2013-1, Grant Agreement Number 316289 – InnoMol and partially through the Grant 9.01/232 funded by Croatian Science Foundation.

References

1. Alon, U., et al.: Broad patterns of gene expression revealed by clustering analysis of tumor and normal colon tissues probed by oligonucleotide arrays. Proc. Natl. Acad. Sci. U.S.A. **96**, 6745–6750 (1999)
2. Singh, D., et al.: Gene expression correlates of clinical prostate cancer behavior. Cancer Cell **1**, 203–209 (2002)
3. Petricoin, E.F., et al.: Use of proteomic patterns in serum to identify ovarian cancer. Lancet **359**, 572–577 (2002)
4. Petricoin, E.F., et al.: Serum proteomic patterns for detection of prostate cancer. J. Natl. Canc. Inst. **94**, 1576–1578 (2002)
5. Guyon, I., et al.: Gene selection for cancer classification using support vector machines. Mach. Learn. **46**, 389–422 (2002)
6. Statnikov, A., et al.: A comprehensive evaluation of multicategory classification methods for microarray gene expression cancer diagnosis. Bioinformatics **21**, 631–643 (2005)
7. Guyon, I., Elisseeff, A.: An introduction to variable and feature selection. J. Mach. Learn. Res. **3**, 1157–1182 (2002)
8. Schölkopf, B., Smola, A.: Learning with Kernels. The MIT Press, Cambridge (2002)
9. Yuh, C.H., Bolouri, H., Davidson, E.H.: Genomic *cis*-regulatory logic: experimental and computational analysis of a sea urchin gene. Science **279**, 1896–1902 (1998)
10. Lee, S.I., Batzoglou, S.: Application of independent component analysis to microarrays. Genome Biol. **4**, R76 (2003)
11. Schachtner, R., et al.: Knowledge-based gene expression classification via matrix factorization. Bioinformatics **24**, 1688–1697 (2008)
12. Stadtlthanner, K., et al.: Hybridizing sparse component analysis with genetic algorithms for microarray analysis. Neurocomputing **71**, 2356–2376 (2008)
13. Gao, Y., Church, G.: Improving molecular cancer class discovery through sparse non-negative matrix factorization. Bioinformatics **21**, 3970–3975 (2005)
14. Kim, H., Park, H.: Sparse non-negative matrix factorizations via alternating non-negativity-constrained least squares for microarray data analysis. Bioinformatics **23**, 1495–1502 (2007)
15. Kopriva, I., Filipović, M.: A mixture model with a reference-based automatic selection of components for disease classification from protein and/or gene expression levels. BMC Bioinformatics **12**, 496 (2011)
16. Kopriva, I.: A Nonlinear Mixture Model Based Unsupervised Variable Selection in Genomics and Proteomics. In: Bioinformatics 2015 – 6th International Conference on Bioinformatics Models, Methods and Algorithms, pp. 85–92, Scitepress (2015)
17. Vapnik, V.: Statistical Learning Theory. Wiley-Interscience, New York (1998)

18. Brown, G.: A new perspective for information theoretic feature selection. J. Mach. Learn. Res. **5**, 49–56 (2009)
19. Aliferis, C.F., et al.: Local causal and markov blanket induction for causal discovery and feature selection for classification - Part I: algorithms and empirical evaluation. J. Mach. Learn. Res. **11**, 171–234 (2010)
20. Gillis, N., Vavasis, S.A.: Fast and robust recursive algorithms for separable nonnegative matrix factorization. IEEE Trans. Pattern Anal. Mach. Intell. **36**, 698–714 (2014)
21. Beck, A., Teboulle, M.: A fast iterative shrinkage-thresholding algorithm for linear inverse problems. SIAM J. Imag. Sci. **2**, 183–202 (2009)
22. Statnikov, A., et al.: GEMS: A system for automated cancer diagnosis and biomarker discovery from microarray gene expression data. Int. J. Med. Inf. **74**, 491–503 (2003)
23. Artero-Castro, A., et al.: Rplp1 bypasses replicative senescence and contributes to transformation. Exp. Cell Res. **315**, 1372–1383 (2009)
24. Bin Amer, S.M., et al.: Gene expression profiling in women with breast cancer in a Saudi population. Saudi Med. J. **29**, 507–513 (2008)
25. Alkhateeb, A.A., Connor, J.R.: The significance of ferritin in cancer: anti-oxidation, inflammation and tumorigenesis. Biochim. Biophys. Acta **1836**, 245–254 (2013)
26. Guo, C., Liu, S., Sun, M.Z.: Novel insight into the role of GAPDH playing in tumor. Clin. Transl. Oncol. **15**, 167–172 (2013)
27. Leśniak, W., Słomnicki, Ł.P., Filipek, A.: S100A6 - new facts and features. Biochem Biophys Res Commun. **390**, 1087–1092 (2009)
28. Sribenja, S., et al.: Roles and mechanisms of β-thymosins in cell migration and cancer metastasis: an update. Cancer Invest. **31**, 103–110 (2013)

Source Separation in Post-nonlinear Mixtures by Means of Monotonic Networks

Leonardo Tomazeli Duarte[1]($\boxtimes$), Filipe de Oliveira Pereira[2], Romis Attux[2], Ricardo Suyama[3], and João M.T. Romano[2]

[1] School of Applied Sciences, University of Campinas (UNICAMP), Limeira, Brazil
leonardo.duarte@fca.unicamp.br
[2] School of Electrical and Computer Engineering,
University of Campinas (UNICAMP), Campinas, Brazil
lipepereira@gmail.com, attux@dca.fee.unicamp.br,
romano@decom.fee.unicamp.br
[3] Engineering, Modeling and Applied Social Sciences Center,
Federal University of ABC (UFABC), Santo André, Brazil
ricardo.suyama@ufabc.edu.br

Abstract. In this work, we investigate the use of monotonic neural networks as compensating functions in the context of source separation of post-nonlinear (PNL) mixtures. We first provide a numerical example that illustrates the importance of having bijective nonlinear compensating functions in PNL models. Then, we propose a separation framework in which a monotonic neural network is considered in the first stage of the PNL separating system. Finally, numerical experiments are performed to assess the proposed framework.

Keywords: Source separation · Nonlinear mixtures · Post-nonlinear mixtures · Monotonic networks

1 Introduction

Blind source separation (BSS) in nonlinear models is still a challenging topic [1–3]. Differently from the linear case, it is difficult to develop a general separation framework — comprising a separation structure and a separation criterion — for nonlinear models. Therefore, research in this area has been focusing on particular classes of nonlinear mixing models that are relevant in practical applications and also interesting from a theoretical point of view.

Among the classes of nonlinear models studied so far, post-nonlinear (PNL) models have been receiving considerable attention since the work of Taleb and Jutten [4]. A PNL mixing model is composed of two stages. In the first one, the sources are submitted to a standard linear mixing process. Then, in the second stage, each linear mixture undergoes a memoryless nonlinear (and bijective) distortion and the outputs of this process correspond to the observed mixtures.

The authors would like to thank FAPESP and CNPq for funding this research.

E. Vincent et al. (Eds.): LVA/ICA 2015, LNCS 9237, pp. 176–183, 2015.
DOI: 10.1007/978-3-319-22482-4_20

The usual PNL separating system has also two stages but they are disposed in a reverse order with respect to the mixing system.

A first interesting feature in PNL models is related to their practical relevance. Indeed, PNL models suit well applications in which the mixing process is linear but the sensing or transducer mechanism is nonlinear. For instance, such a situation is typical in audio signal processing, due to the existence of amplifying stages that may operate in nonlinear regions [4]. Another example of application can be found in electrochemical sensor arrays for measuring ionic activities [5]. In this case, the nonlinear character of the mixtures can be explained by the Nernst equation [6].

Besides their practical relevance, PNL models present a key feature for BSS: they are separable in the sense of independent component analysis (ICA) [4,7]. In other words, assuming that the sources can be modeled as statistically independent random variables, it is possible to blindly recover the original sources by setting a PNL separating system that provides independent signals. This property has opened the way for the development of BSS algorithms based, for instance, on the mutual information minimization principle [4].

The necessary conditions for ICA-separability in PNL models are similar to those of established for linear mixtures (non-Gaussian sources and invertible mixing matrix) [8]. However, there is an additional requirement: the nonlinear compensating functions placed at the first stage of the separating system must also be bijective — otherwise, one may recover independent components that are still mixed versions of the sources. From this requirement, there arises an important aspect related to finding nonlinear compensating functions that are bijective and yet flexible enough to deal with the nonlinear distortions. So far in the literature, this problem has been mainly tackled by defining constrained polynomial functions [9] and spline-based approximations [10].

In the present work, we shall investigate the application of artificial neural networks that are monotonic by construction as compensating functions of a PNL separating system. Our motivation comes from the existence of monotonic networks that exhibit the universal approximation property [11] for bijective functions — in particular, our analysis will focus on the network proposed in [12]. The rest of the paper is organized as follows. In Sect. 2 we provide a mathematical description of the PNL model and a numerical example to illustrate the importance of having bijective compensating functions. Then, in Sect. 3, we introduce a PNL/BSS paradigm based on a monotonic network and on an ICA criterion. In Sect. 4, numerical experiments are performed. Finally, we present our conclusions in Sect. 5.

2   Post-nonlinear Models

In order to define the BSS problem in the context of PNL mixtures, let us denote the N sources to be recovered by the vector $\mathbf{s}(n) = [s_1(n)\ s_2(n) \ldots s_N(n)]^T$. In this work, only the determined case (equal number of sources and mixtures) is considered. In a PNL mixing model, the N mixtures, denoted by $\mathbf{x}(n) = [x_1(n)$

$x_2(n) \ldots x_N(n)]^T$, are given by

$$\mathbf{x}(n) = \mathbf{f}\ (\mathbf{z}(n)) = \mathbf{f}\ (\mathbf{A}\mathbf{s}(n)), \tag{1}$$

where $\mathbf{A}$ is the $N \times N$ mixing matrix and $\mathbf{f}(\cdot) = [f_1(\cdot)\ f_2(\cdot) \ \cdots \ f_N(\cdot)]^T$ denotes the set of nonlinear component-wise functions, which are bijective.

In the usual PNL separating system, the estimated sources, denoted by $\mathbf{y}(n) = [y_1(n)\ y_2(n) \ \ldots \ y_N(n)]^T$, are given by

$$\mathbf{y}(n) = \mathbf{W}\mathbf{e}(n) = \mathbf{W}\mathbf{g}\ (\mathbf{x}(n)), \tag{2}$$

where $\mathbf{W}$ is the separating matrix and $\mathbf{g}(\cdot) = [g_1(\cdot)\ g_2(\cdot) \ \cdots \ g_N(\cdot)]^T$ represents the compensating functions. The adjustment of $\mathbf{W}$ and $\mathbf{g}(\cdot)$ by means of ICA can be formulated as the following optimization problem

$$\min_{\mathbf{W}, \mathbf{g}(\cdot)}\ J(\mathbf{y}(n)), \tag{3}$$

where $J(\mathbf{y}(n))$ is a cost function that attains its minimum value if, and only if, the signals comprised in $\mathbf{y}(n)$ are statistically independent. Very often, $J(\mathbf{y}(n))$ is defined as the mutual information [13].

The ICA framework presented in (3) is meaningful in BSS when the ICA-separability property holds, that is, when the minimization of (3) leads to a solution in which $g_i \circ f_i$ (for every i) is linear and $\mathbf{W} = \mathbf{P}\mathbf{D}\mathbf{A}^{-1}$, where $\mathbf{P}$ and $\mathbf{D}$ denote a permutation and a diagonal matrix, respectively. In short, ICA-separability in PNL models requires three conditions: (1) There exists at most one Gaussian source; (2) $\mathbf{A}$ is invertible; and (3) all functions $g_i(\cdot)$ must be bijective. In the sequel, we shall illustrate by means of a numerical example the risks when the third condition is violated.

2.1   The Requirement of Monotonicity in PNL Models: A Numerical Example

Let us consider a PNL source separation problem in which $N = 2$ and both distortions are given by $x_i(n) = f_i(z_i(n)) = \sqrt[3]{z_i(n)}$. Both sources follow a uniform distribution between $[-1, 1]$, as shown in Fig. 1(a). BSS is performed via mutual information minimization, as described in (3). Moreover, we consider two different configurations of compensating functions $g_i(\cdot)$. In the first one, the following polynomial of degree 4 was defined as compensating function for both mixtures:

$$e_i(n) = w_{i1}x_i(n) + w_{i2}x_i^2(n) + w_{i3}x_i^3(n) + w_{i4}x_i^4(n), \tag{4}$$

where w_{ij} denotes the parametrization of $g_i(\cdot)$. In the second configuration, we adopt the following polynomial compensating functions

$$e_i(n) = w_{i1}x_i(n) + w_{i3}x_i^3(n) + w_{i5}x_i^5(n), \tag{5}$$

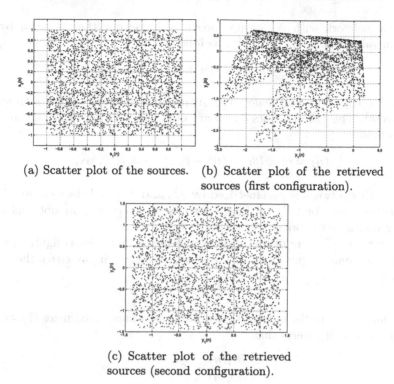

(a) Scatter plot of the sources.   (b) Scatter plot of the retrieved sources (first configuration).

(c) Scatter plot of the retrieved sources (second configuration).

Fig. 1. PNL source separating using polynomial compensating functions.

where $w_{ij} \geq 0$. Therefore, in the second configuration, the compensating function is invertible. Note that, in both configurations, the polynomial function can perfectly compensate $f_i(\cdot)$ when $w_{i3} \neq 0$ and the other parameters are null.

In order to carry out (3), we applied the solution proposed in [9]. The scatter plots of the retrieved sources for both configurations are shown in Fig. 1. It is worth noticing that when the first configuration (given by (4)) is considered, the estimated sources do not correspond to the actual ones although their mutual information is very close to zero. In other words, independent components are retrieved but they are still mixed versions of the sources. Conversely, when the bijectivity condition is imposed (second configuration), source separation is achieved.

3   A PNL Separation Method Based on a Monotonic Neural Network as Compensating Function

In view of the importance of having bijective compensating functions, we here propose a PNL separation framework based on a univariate version of the monotonic neural network introduced in [12]. This neural network is built upon on minimum and maximum operators. More precisely, given an input $x_i(n)$ this monotonic network with N_g groups provides an output that is given by the minimum value among these groups, as follows

$$e_i(n) = \min_k r_k(l_{ik}(n)), \ k = 1, \ldots, N_g, \tag{6}$$

where $r_k(l_{ik}(n))$ and $l_{ik}(n)$ denote the output and the input of the k-th group, respectively[1]. The group inputs $l_{ik}(n)$ are obtained by a set of N_h linear functions, as follows

$$l_{ik}(n) = \max_j \left(\theta_{k,j} x_i(n) - \beta_{k,j} \right), \ j = 1, \ldots, N_h. \tag{7}$$

Therefore, this network is parametrized by $\theta_{k,j}$ and $\beta_{k,j}$ and the total number of parameters is given by $2 N_g N_h$. In order to ensure increasing monotonicity, the following restriction is taken into account: $\theta_{k,j} > 0$.

An interesting feature of the adopted network is that it is straightforward to calculate the input-output derivative. Indeed, for a given input $x_i(n)$, the output is given by

$$e_i(n) = \theta_{\hat{k},\hat{j}} x_i(n) - \beta_{\hat{k},\hat{j}}, \tag{8}$$

where $\hat{k}$ and $\hat{j}$ denote the indexes that minimizes (6) and maximizes (7), respectively. Therefore, it asserts that

$$\frac{\partial e_i(n)}{\partial x_i(n)} = \theta_{\hat{k},\hat{j}}. \tag{9}$$

It is worth noticing that such a network can be seen as a practical way to implement a piecewise linear function. Moreover, the presence of the maximum and minimum operators allows one to approximate nonlinear functions that present both convex and concave regions [12].

3.1   Separation Criterion

Having defined the monotonic network that will be used as compensating function $g_i(\cdot)$, we can now define an ICA separation criterion. As in [4,9], the adjustment of the separating system parameters will be conducted so as to minimize the mutual information between the retrieved sources, which is given by:

$$I(\mathbf{y}(n)) = \sum_{i=1}^{N} H(y_i(n)) - H(\mathbf{y}(n)), \tag{10}$$

where $H(\cdot)$ denotes Shannon's differential entropy [13]. In view of (2), Eq. (10) can be written as

$$I(\mathbf{y}(n)) = \sum_{i=1}^{N} H(y_i(n)) - H(\mathbf{x}(n)) - \log |\det \mathbf{W}| - E \left\{ \sum_{i=1}^{N} \log |g_i'(x_i(n))| \right\}. \tag{11}$$

---

[1] We here keep the index i, which is related to the mixtures, and also the sample index n.

Given that the entropy term $H(\mathbf{x}(n))$ is constant with respect to the separating system and that $I(\mathbf{y}(n)) \geq 0$, with equality achieved if, and only if, $\mathbf{y}(n)$ are mutually independent, a natural ICA separation framework is given by:

$$\min_{\mathbf{W}, \theta_{k,j}, \beta_{k,j}} \sum_{i=1}^{N} H(y_i(n)) - \log |\det \mathbf{W}| - E\left\{ \sum_{i=1}^{N} \log |g_i'(x_i(n))| \right\}. \qquad (12)$$

In this work, the marginal entropy terms $H(y_i(n))$ are calculated by means of the estimator based on order statistics presented in [9]. The last term of (12) can be directly calculated from Eq. (9).

3.2   Optimization Method

In order to minimize (12), we consider the algorithm opt-aiNet [3]. This method performs a population-based search and is particularly useful to deal with cost functions that present local optima and/or are not continuous. More information on the opt-aiNet algorithm and its application to BSS of PNL mixtures can be found in [9].

4   Numerical Experiments

In order to assess the proposed framework, we conduct a set of numerical experiments in two distinct scenarios. In both cases, there are $N = 2$ uniformly distributed sources (between -0.5 and 0.5), the mixing matrix is given by $\mathbf{A} = [1\ 0.5; 0.5\ 1]$ and the number of samples is 1000. In the first scenario, the nonlinear distortions are given by $f_i(x_i(n)) = \tanh(3x_i(n))$ and, in the second one, they correspond to the inverse hyperbolic tangent function $f_i(x_i(n)) = \arctan(1.3x_i(n))$. The adopted monotonic neural network has $N_g = 8$ groups and $N_h = 8$ linear functions per group. We also compare the proposed framework with the PNL method based on monotonic polynomial compensation functions [9] — we have chosen polynomial functions of degree 7.

In Fig. 2(a) and (d), we show the scatter plots of the retrieved sources (first scenario) by the proposed method and by the strategy based on monotonic polynomials. Moreover, we also provide in Fig. 2(b) and (c) the mappings $g \circ f$ obtained by the proposed method, whereas Fig. 2(e) and (f) show the mappings $g \circ f$ obtained by the polynomial-based solution. The obtained results indicate that both approaches were able to achieve fair estimations of the sources (the polynomial-based method performed better in this case).

When the distortions are modeled as inverse hyperbolic tangent functions (second scenario), the solution based on polynomial compensating does not provide a good result. This can be seen in Fig. 2(j), which depicts the scatter plot of the retrieved sources, and in Fig. 2(k) and (l) (mappings $g_i \circ f_i$). Conversely, the proposed framework achieved again a fair solution (see Fig. 2(g), (h) and (i)). The bad performance of the polynomial-based PNL method in this scenario can be attributed to the structural difficulty in inverting an inverse hyperbolic tangent function through a polynomial function.

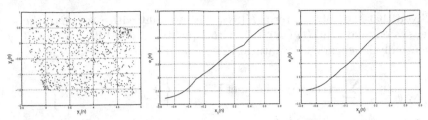

(a) First scenario: scatter plot of the retrieved sources (monotonic network).

(b) First scenario: $g_1 \circ f_1$ (monotonic network).

(c) First scenario: $g_2 \circ f_2$ (monotonic network).

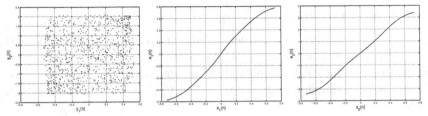

(d) First scenario: scatter plot of the retrieved sources (monotonic polynomial).

(e) First scenario: $g_1 \circ f_1$ (monotonic polynomial).

(f) First scenario: $g_2 \circ f_2$ (monotonic polynomial).

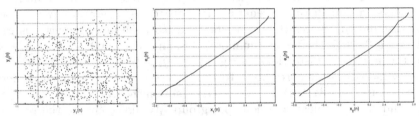

(g) Second scenario: scatter plot of the retrieved sources (monotonic network).

(h) Second scenario: $g_1 \circ f_1$ (monotonic network).

(i) Second scenario: $g_2 \circ f_2$ (monotonic network).

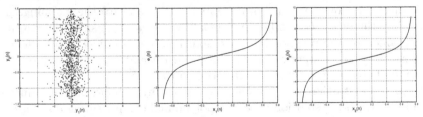

(j) Second scenario: scatter plot of the retrieved sources (monotonic polynomial).

(k) Second scenario: $g_1 \circ f_1$ (monotonic polynomial).

(l) Second scenario: $g_2 \circ f_2$ (monotonic polynomial).

Fig. 2. Numerical experiments for the two scenarios.

5   Conclusions

In this work, we introduced a PNL source separation method based on a monotonic neural network. Our main motivation was to develop a PNL separating system that was flexible enough to deal with a large class of nonlinear distortions yet respecting the requirement of bijectivity needed for ICA-separability. Numerical experiments showed that the adopted neural network could indeed deal with two distinct nonlinear functions. Future works include the derivation of alternative optimization methods to tackle (12).

References

1. Comon, P., Jutten, C. (eds.): Handbook of Blind Source Separation: Independent Component Analysis and Applications. Academic Press, New York (2010)
2. Deville, Y., Duarte, L.T.: An overview of blind source separation methods for linear-quadratic and post-nonlinear mixtures. In: Submitted to the 12th International Conference on Latent Variable Analysis and Signal Separation (LVA/ICA) (2015)
3. Romano, J.M.T., Attux, R.R.F., Cavalcante, C.C., Suyama, R.: Unsupervised Signal Processing: Channel Equalization and Source Separation. CRC Press, Boca Raton (2011)
4. Taleb, A., Jutten, C.: Source separation in post-nonlinear mixtures. IEEE Trans. Signal Process. **47**(10), 2807–2820 (1999)
5. Duarte, L.T., Jutten, C., Moussaoui, S.: A Bayesian nonlinear source separation method for smart ion-selective electrode arrays. IEEE Sens. J. **9**(12), 1763–1771 (2009)
6. Gründler, P.: Chemical Sensors: An Introduction for Scientists and Engineers. Springer, Heidelberg (2007)
7. Achard, S., Jutten, C.: Identifiability of post-nonlinear mixtures. IEEE Signal Process. Lett. **12**(5), 423–426 (2005)
8. Comon, P.: Independent component analysis, a new concept? Sig. Process. **36**, 287–314 (1994)
9. Duarte, L.T., Suyama, R., de Faissol Attux, R.R., Von Zuben, F.J., Romano, J.M.T.: Blind source separation of post-nonlinear mixtures using evolutionary computation and order statistics. In: Rosca, J.P., Erdogmus, D., Príncipe, J.C., Haykin, S. (eds.) ICA 2006. LNCS, vol. 3889, pp. 66–73. Springer, Heidelberg (2006)
10. Pham, D.T.: Flexible parametrization of postnonlinear mixtures model in blind sources separation. IEEE Signal Process. Lett. **11**(6), 533–536 (2004)
11. Haykin, S.: Neural Networks: A Comprehensive Foundation. Prentice-Hall, New York (1998)
12. Sill, J.: Monotonic networks. In: Proceedings of the Conference on Advances in Neural Information Processing Systems (NIPS) (1998)
13. Cover, T.M., Thomas, J.A.: Elements of Information Theory. Wiley-Interscience, New York (1991)

From Blind Quantum Source Separation to Blind Quantum Process Tomography

Yannick Deville[1](✉) and Alain Deville[2]

[1] Université de Toulouse, UPS-CNRS-OMP IRAP (Institut de Recherche
en Astrophysique et Planétologie), 14 avenue Edouard Belin, 31400 Toulouse, France
yannick.deville@irap.omp.eu
[2] Aix-Marseille Université, CNRS, IM2NP UMR 7334, 13397 Marseille, France
alain.deville@univ-amu.fr

Abstract. We here extend the field of blind (i.e. unsupervised) quantum computation into two directions. On the one hand, we introduce a new class of blind quantum source separation (BQSS) methods, which perform quantum/classical data conversion by means of spin component measurements, followed by classical processing. They differ from our previous class of classical-processing BQSS methods by using extended types of measurements (three directions, possibly different for the considered two spins), which yield a more complete nonlinear mixing model. This allows us (i) to develop a new disentanglement-based separation procedure, which requires a much lower number of source values for adaptation and (ii) to restore a larger set of sources. On the other hand, these extended measurements motivate us to introduce a new research field, namely *Blind* Quantum Process Tomography, which may be seen both as the blind extension of its existing non-blind version and as the quantum extension of classical blind identification of mixing systems.

Keywords: Blind quantum system identification and inversion · Nonlinear mixture · Disentanglement-based separation principle · Unsupervised unmixing · Multidirectional measurements of spin components

1 Prior Work and Problem Statement

Source Separation (SS), also called signal separation, is a generic Information Processing (IP) problem, where the inverting block of a separating system eventually receives signals, which are mixtures of source signals that it does not know, and aims at recovering these source signals only from their known mixtures [2]. In the ideal case, the separating system initially completely knows the mixing function. On the contrary, in many applications, this system initially knows which class the mixing function belongs to, but does not know its parameter values. This system therefore contains an adapting block which is initially used to tune the parameter values of the inverting block so that the latter block achieves the inverse of the mixing function (possiby up to some indeterminacies). This adapting block thus typically aims at estimating the parameter values of

© Springer International Publishing Switzerland 2015
E. Vincent et al. (Eds.): LVA/ICA 2015, LNCS 9237, pp. 184–192, 2015.
DOI: 10.1007/978-3-319-22482-4_21

the mixing function. The SS problem is thus closely linked to the Mixture Identification (MI) problem (see e.g. [1], and [2] pp. 65–66), which may be seen as a multiple-input multiple-output system identification problem.

The above initial adaptation/identification may be performed in two modes. In the less demanding, i.e. non-blind, mode, the adapting block receives both a set of known source values and the associated values of the mixed signals. The more challenging mode corresponds to the *Blind* (or unsupervised) Source Separation (BSS) [2] and associated Blind Mixture Identification (BMI) problems. In this blind mode, the adapting block only receives values of the mixed signals. The sources have unknown values, but they are requested to possess some known properties (e.g. they are mutually statistically independent for ICA methods).

Until recently, all (B)SS and BMI investigations were performed in a "classical", i.e. non-quantum, framework. Independently from them, another growing field within the overall IP domain is Quantum Information Processing (QIP) [7]. QIP is closely related to Quantum Physics (QP). It uses abstract representations of systems whose behavior is requested to obey the laws of QP. This already made it possible to develop new and powerful IP methods, which manipulate the states of so-called quantum bits, or qubits.

In 2007, we bridged the gap between classical (B)SS and QIP/QP, by introducing a new field, namely Quantum Source (or Signal) Separation (QSS) [3]. The QSS problem consists in restoring the information contained in individual *quantum* source signals, eventually only using the mixtures (in SS terms [4]) of states of these qubits which result from their undesired coupling. We especially developed two main classes of Blind (i.e. unsupervised) QSS (BQSS) methods for qubits implemented as spins 1/2. Briefly, in the first class (see e.g. [3–5]), we first perform a quantum/classical conversion by using monodirectional spin component measurements and then process the resulting data with classical means. In the second class (see e.g. [6]), we only use quantum processing means in the inverting block, whereas the adapting block preferably performs bidirectional spin component measurements and then classical processing.

In this paper, we first introduce a new mixing model (Sect. 2), defined by our already used spin coupling model and an *extended* set of spin component measurements (along three directions, that may moreover be different for the two spins). We then present major extensions of the above quantum-source processing methods, into two directions. On the one hand, we introduce a new class of BQSS methods (Sects. 3 and 4), which use classical processing after the quantum/classical conversion performed by the above measurements, as in our first class of methods, but which take advantage of this new set of measurements to achieve additional capabilities. On the other hand, these capabilities motivate us to explicitly introduce a new research field in Sect. 5, namely *Blind* (or unsupervised) Quantum Process Tomography (BQPT). This field may be seen both as the quantum counterpart of classical BMI when applied to separately initialized qubits as in this paper, and as the blind extension of the field of (non-blind) QPT, previously developed in the framework of QIP [7]. Conclusions are drawn from this investigation in Sect. 6.

2   New Mixing Model

2.1   Heisenberg Quantum Coupling

As stated above, qubits are used instead of classical bits for performing computations in the field of QIP [7]. In our previous papers (e.g. [3,4]), we first detailed the required concepts for a single qubit and then presented the type of coupling between two qubits that is involved in the "mixing model", in (B)SS terms, of our investigation. We hereafter summarize the major aspects of that discussion, which are required in the current paper.

A qubit with index i considered at a given time t_0 has a quantum state. If this state is pure, it belongs to a two-dimensional space $\mathcal{E}_i$ and may be expressed as

$$|\psi_i(t_0)\rangle = \alpha_i|+\rangle + \beta_i|-\rangle \tag{1}$$

in the basis of $\mathcal{E}_i$ defined by the two orthonormal vectors that we hereafter denote $|+\rangle$ and $|-\rangle$, whereas α_i and β_i are two complex-valued coefficients constrained to meet the condition

$$|\alpha_i|^2 + |\beta_i|^2 = 1 \tag{2}$$

which expresses that the state $|\psi_i(t_0)\rangle$ is normalized.

In the QSS configuration studied in this paper, we first consider a system composed of two qubits, called "qubit 1" and "qubit 2" hereafter, at a given time t_0. This system has a quantum state. If this state is pure, it belongs to the four-dimensional space $\mathcal{E}$ defined as the tensor product (denoted $\otimes$) of the spaces $\mathcal{E}_1$ and $\mathcal{E}_2$ respectively associated with qubits 1 and 2, i.e. $\mathcal{E} = \mathcal{E}_1 \otimes \mathcal{E}_2$. The considered basis of $\mathcal{E}$ is composed of the four orthonormal vectors $|++\rangle, |+-\rangle, |-+\rangle, |--\rangle$, where e.g. $|+-\rangle$ is an abbreviation for $|+\rangle \otimes |-\rangle$, with $|+\rangle$ corresponding to qubit 1 and $|-\rangle$ corresponding to qubit 2. Any pure state of the above two-qubit system may then be expressed as

$$|\psi(t_0)\rangle = c_1(t_0)|++\rangle + c_2(t_0)|+-\rangle + c_3(t_0)|-+\rangle + c_4(t_0)|--\rangle \tag{3}$$

and has unit norm. In particular, we study the case when the two qubits are separately initialized, with states defined by (1) respectively with $i = 1$ and $i = 2$. Then

$$|\psi(t_0)\rangle = |\psi_1(t_0)\rangle \otimes |\psi_2(t_0)\rangle \tag{4}$$

$$= \alpha_1\alpha_2|++\rangle + \alpha_1\beta_2|+-\rangle + \beta_1\alpha_2|-+\rangle + \beta_1\beta_2|--\rangle. \tag{5}$$

Moreover, we consider the case when the two qubits correspond to two electron or nuclear spins 1/2, called "spin 1" and "spin 2", which have undesired coupling after they have been initialized according to (4). The considered coupling is based on the Heisenberg model with a cylindrical-symmetry axis collinear to Oz, the direction common to the applied magnetic field and to our first chosen quantization axis [4]. Due to that coupling, and for negligible coupling with the environment, the state of the above system at any time $t > t_0$ reads [4]

$$|\psi(t)\rangle = c_1(t)|++\rangle + c_2(t)|+-\rangle + c_3(t)|-+\rangle + c_4(t)|--\rangle, \tag{6}$$

with

$$c_1(t) = \alpha_1 \alpha_2 e^{-i\omega_{1,1}(t-t_0)} \tag{7}$$

$$c_2(t) = \frac{1}{2}\left[(\alpha_1\beta_2 + \beta_1\alpha_2)e^{-i\omega_{1,0}(t-t_0)} + (\alpha_1\beta_2 - \beta_1\alpha_2)e^{-i\omega_{0,0}(t-t_0)}\right] \tag{8}$$

$$c_3(t) = \frac{1}{2}\left[(\alpha_1\beta_2 + \beta_1\alpha_2)e^{-i\omega_{1,0}(t-t_0)} - (\alpha_1\beta_2 - \beta_1\alpha_2)e^{-i\omega_{0,0}(t-t_0)}\right] \tag{9}$$

$$c_4(t) = \beta_1\beta_2 e^{-i\omega_{1,-1}(t-t_0)}. \tag{10}$$

where all four (angular) frequencies $\omega_{k,l}$ are unknown in practical, i.e. non-ideal, configurations (see also Sect. 5 for more details about this physical model).

2.2 Extended Quantum/Classical Conversion

Classical-form data may be derived from the above coupled state $|\psi(t)\rangle$ by measuring the components of the considered two spins along given directions. In our first class of QSS methods [3–5], we only used measurements along Oz for both spins. We explained that this couple of measured spin components has four possible values only, namely $(+\frac{1}{2},+\frac{1}{2})$, $(+\frac{1}{2},-\frac{1}{2})$, $(-\frac{1}{2},+\frac{1}{2})$ and $(-\frac{1}{2},-\frac{1}{2})$ (in normalized units), with respective probabilities

$$p_{1zz} = |c_1(t)|^2, \quad p_{2zz} = |c_2(t)|^2, \quad p_{3zz} = |c_3(t)|^2, \quad p_{4zz} = |c_4(t)|^2. \tag{11}$$

These probabilities may be estimated by the sample frequencies of the associated measured values, using the Repeated Write Read (RWR) procedure that we proposed. This allowed us to derive a *nonlinear* mixing model, where the mixed signals are three of these (estimated) probabilities and the sources are the two moduli $|\alpha_1|$ and $|\alpha_2|$ and a single combination of the four phases of α_i and β_i.

In this paper, we extend this nonlinear mixing model by also performing other types of measurements for $|\psi(t)\rangle$ (for additional initializations of the qubits). More precisely, we first consider the case when one again measures the Oz component of spin 1, but now the Ox component of spin 2. QP then tells us that such measurements again yield the same four possible values as above and that, in particular, the probabilities of $(+\frac{1}{2},+\frac{1}{2})$, and $(-\frac{1}{2},+\frac{1}{2})$ respectively read

$$p_{1zx} = \frac{1}{2}|c_1(t) + c_2(t)|^2 \quad \text{and} \quad p_{3zx} = \frac{1}{2}|c_3(t) + c_4(t)|^2. \tag{12}$$

Similarly, the probabilities of the above two values when measuring the Oz and Oy components respectively of spins 1 and 2 read

$$p_{1zy} = \frac{1}{2}|c_1(t) - ic_2(t)|^2 \quad \text{and} \quad p_{3zy} = \frac{1}{2}|c_3(t) - ic_4(t)|^2, \tag{13}$$

and the probabilities for getting the value $(+\frac{1}{2},+\frac{1}{2})$, for the two couples of directions (Ox, Oz) and (Oy, Oz) for spins 1 and 2, are respectively

$$p_{1xz} = \frac{1}{2}|c_1(t) + c_3(t)|^2 \quad \text{and} \quad p_{1yz} = \frac{1}{2}|c_1(t) - ic_3(t)|^2. \tag{14}$$

The output of the mixing stage of the considered QSS configuration consists of all probabilities in (11)–(14) (more precisely of their *estimates* derived from our RWR procedure). It is sent to the input of the inverting block of the separating system defined in the next section.

3   Inverting Block of Separating System

We hereafter present the five steps of the operation of the classical-processing inverting block of the proposed separating system, respectively calling "Case 1" and "Case 2" the ideal and blind (Q)SS configurations defined in Sect. 1.

In both Cases, **Step 1** consists in restoring (estimates of: this is not stated everywhere below) the coefficients $c_j(t)$, with $j \in \{1, \ldots, 4\}$, from the (estimates of) probabilities derived in Sect. 2.2. To this end, we use the polar representation $c_j(t) = \rho_j e^{i\xi_j}$ of these coefficients. All their moduli ρ_j are directly derived from (11). Using (11), (12) and $c_j(t) = \rho_j e^{i\xi_j}$, it may then be shown that

$$\cos(\xi_1 - \xi_2) = \frac{2p_{1zx} - p_{1zz} - p_{2zz}}{2\sqrt{p_{1zz}p_{2zz}}}, \qquad \cos(\xi_3 - \xi_4) = \frac{2p_{3zx} - p_{3zz} - p_{4zz}}{2\sqrt{p_{3zz}p_{4zz}}}. \tag{15}$$

The sines of the above phase differences may then similarly be derived by using (13) instead of (12). Finally, using (14) instead, one obtains the cosine and sine of $(\xi_1 - \xi_3)$. All differences between the four phases ξ_j are thus known (modulo 2π). Moreover, a quantum state (here (6)) is only defined up to a phase factor. One may therefore arbitrarily fix one of the above phases ξ_j (e.g. to 0). As an overall result, we thus know all phases ξ_j and moduli ρ_j, i.e. all coefficients $c_j(t)$.

Keeping in mind that these restored versions of $c_1(t)$ to $c_4(t)$ here meet (7)–(10), we then process them so as to derive successive transformed versions of this set of four coefficients, which progressively bring us back to the original, or source, data defined by (1) and (4). The four transformed coefficients obtained at the output of each processing step n with $n = 2$ to 4 are denoted as c_{jn}, with $j \in \{1, \ldots, 4\}$. **Step 2** then consists in reducing its input coefficients $c_j(t)$ to expressions which only depend on a single frequency $\omega_{k,l}$. To this end, both in Cases 1 and 2, we keep $c_{12} = c_1(t)$ and $c_{42} = c_4(t)$, while respectively setting c_{22} and c_{32} to the sum and difference of $c_2(t)$ and $c_3(t)$, moreover rescaled so that the coefficients c_{j2} form a unit-norm vector, as in (3) and (6). This yields

$$c_{22} = \frac{1}{\sqrt{2}}[c_2(t) + c_3(t)] = \frac{1}{\sqrt{2}}(\alpha_1\beta_2 + \beta_1\alpha_2)e^{-i\omega_{1,0}(t-t_0)} \tag{16}$$

$$c_{32} = \frac{1}{\sqrt{2}}[c_2(t) - c_3(t)] = \frac{1}{\sqrt{2}}(\alpha_1\beta_2 - \beta_1\alpha_2)e^{-i\omega_{0,0}(t-t_0)}. \tag{17}$$

Step 3 then aims at compensating for the phase factors $e^{-i\omega_{k,l}(t-t_0)}$ in the above c_{j2}. This is achieved by setting $c_{j3} = c_{j2} \times e^{i\gamma_j}$, with $j \in \{1, \ldots, 4\}$, which yields

$$c_{13} = \alpha_1\alpha_2 e^{i\delta_1}, \qquad c_{23} = \frac{1}{\sqrt{2}}(\alpha_1\beta_2 + \beta_1\alpha_2)e^{i\delta_2} \tag{18}$$

$$c_{33} = \frac{1}{\sqrt{2}}(\alpha_1\beta_2 - \beta_1\alpha_2)e^{i\delta_3}, \qquad c_{43} = \beta_1\beta_2 e^{i\delta_4} \tag{19}$$

with

$$\delta_j = \gamma_j - \omega_{k,l}(t - t_0). \tag{20}$$

In Case 2, all parameters γ_j are adapted as explained in Sect. 4, because all $\omega_{k,l}$ are unknown. In Case 1, all parameters γ_j are set to the known values $\omega_{k,l}(t-t_0)$. All phase factors $e^{i\delta_j}$ thus disappear in (18)–(19). In **Step 4**, we reduce the above coefficients to a single product of α_i and/or β_j parameters in Case 1. To this end, we use the same approach as in Step 2, i.e. in both Cases we keep $c_{14} = c_{13}$, $c_{44} = c_{43}$ and we set

$$c_{24} = \frac{1}{\sqrt{2}}[c_{23} + c_{33}], \qquad c_{34} = \frac{1}{\sqrt{2}}[c_{23} - c_{33}]. \tag{21}$$

In Case 1, this yields

$$c_{14} = \alpha_1\alpha_2, \qquad c_{24} = \alpha_1\beta_2, \qquad c_{34} = \beta_1\alpha_2, \qquad c_{44} = \beta_1\beta_2. \tag{22}$$

Step 5 then aims at deriving all source parameters α_i and β_i (i.e. a larger set of sources than in our previous classical-processing BQSS methods) from all coefficients c_{j4}. This is relevant only if these coefficients correspond to a non-entangled quantum state, i.e. a tensor product such as (4), e.g. as in Case 1. In the latter case, one computes the moduli of the outputs of this step e.g. as $\sqrt{|c_{14}|^2 + |c_{24}|^2}$ because (22) and (2) show that this yields $|\alpha_1|$ in Case 1. Moreover, one of the four phases of the parameters α_i and β_i, say $\arg(\alpha_1)$, may be arbitrarily selected. Then combining (22) with the polar expressions of c_{j4}, α_i and β_i e.g. yields $\arg(\beta_1) = \arg(\alpha_1) + \arg(c_{34}) - \arg(c_{14})$ (modulo 2π). The calculations for the other source parameters are similar and therefore skipped.

4   Interpretation and Adaptation of the Separating system

The inverting block of the separating system that we developed in our QSS method proposed in [6] only uses quantum states and quantum processing means. It is thus quite different from the inverting block depicted in Sect. 3, which receives classical-form data (which have quantum properties, however) and processes them with classical means. Yet, it may be shown that (i) the classical-form coefficients $c_j(t)$ here restored in Step 1 are those of the quantum state at the input of the inverting block of [6], again up to estimation errors, and (ii) the quantum processing achieved in that block of [6] is governed by the same equations as in Steps 2 to 4 above, although they are expressed in a quite differ-ent way in [6]. Processing Steps 2 to 4 of the block of Sect. 3 may therefore be considered as a new classical-processing counterpart of the quantum-processing block of [6] (but they here receive an approximate version of coefficients $c_j(t)$).

To blindly adapt the parameters γ_j of the inverting block of Sect. 3, we then propose a procedure which is partly related to the one introduced in [6]. We here take advantage of the availability of the complex-valued coefficients c_{j4} in classical form. On the contrary, in [6] their counterpart is only available in quantum form, which required us to develop an adaptation criterion based on related real-valued probabilities. The new adaptation procedure proposed here consists in tuning all γ_j so as to enforce the quantum disentanglement condition

$$c_{14}c_{44} = c_{24}c_{34} \tag{23}$$

for (at least) two (non-redundant [6]) source states (5). QP calculations skipped here show that condition (23) above implies the probability-based separation conditions (17) and (24) of [6]. As shown in [6], the latter conditions themselves entail

$$\delta_3 - \delta_2 = m\pi, \qquad \delta_1 + \delta_4 = 2\delta_2 + 2k\pi, \tag{24}$$

where m and k are arbitrary integers, and these conditions ensure separability, so that they here force the coefficients c_{j4} to become equal to those in (5), up to some permutation and phase indeterminacies. This approach thus yields a new classical-processing BQSS method, which only requires a very limited number of source states for adaptation, by using the disentanglement condition (23). On the contrary, our previous, statistical, methods related to ICA [5] need hundreds to thousands of (also repeatedly prepared) source states.

5 Blind Quantum Process Tomography

The considered cylindrical-symmetry Heisenberg quantum coupling model was initially defined by the corresponding Hamiltonian (see e.g. [4]). We showed that this yields the coupled state expression in (6)–(10), moreover with

$$\omega_{1,1} = \frac{1}{\hbar}\left[GB - \frac{J_z}{2}\right], \qquad \omega_{1,0} = \frac{1}{\hbar}\left[-J_{xy} + \frac{J_z}{2}\right] \tag{25}$$

$$\omega_{0,0} = \frac{1}{\hbar}\left[J_{xy} + \frac{J_z}{2}\right], \qquad \omega_{1,-1} = \frac{1}{\hbar}\left[-GB - \frac{J_z}{2}\right]. \tag{26}$$

In these expressions, $\hbar$ is the reduced Planck constant and $G = g\mu_e$, where g is the principal value of the considered isotropic $\overline{\overline{g}}$ tensor and the constant μ_e is the Bohr magneton [4]. The value of g may be experimentally determined. B is the magnitude of the applied magnetic field, which can be known thanks to measurements. J_{xy} and J_z are the principal values of the exchange tensor, which are unknown in practice. The frequencies $\omega_{k,l}$ are thus unknown.

QPT, mentioned in Sect. 1, is a generic, therefore complex, procedure for identifying the behavior of a quantum system by applying known input states (thus in the non-blind mode) to this system and measuring its corresponding outputs. We here aim at developing an extension of QPT tailored to the Heisenberg Hamiltonian and operating in the *blind* mode, i.e. with unknown input states. To this end, we analyze the BMI capabilities of the adaptation procedure proposed in Sect. 4. Eqs. (24), (20) and (25)–(26) then yield

$$J_{xy} = \frac{\hbar}{2(t - t_0)}(\gamma_3 - \gamma_2 - m\pi), \qquad J_z = \frac{\hbar}{2(t - t_0)}(\gamma_2 + \gamma_3 - \gamma_1 - \gamma_4 + 2k\pi - m\pi). \tag{27}$$

The only unknown values of the considered Hamiltonian, namely J_{xy} and J_z, may therefore be derived from the values γ_j provided by the proposed adaptation procedure for given but unknown source states (5). For BQPT, k and m only yield sign indeterminacies in the exponentials of the process (6)–(10). Moreover, they can be set to zero if all other terms of (27) are known to be small enough.

This yields our first reported method for performing complete BQPT (with the above indeterminacies) with classical processing means. Related BQPT capabilities could however be derived from our previous BQSS methods. First, our method in [6] here also yields (27), but that BQSS method requires quantum processing means and it is much more difficult to implement them than classical ones. Second, our previous classical-processing BQSS methods only estimate the single parameter of their mixing model, which is different from here because they only use measurements along one direction. This parameter only yields J_{xy}. These BQSS methods then achieve complete BQPT for the isotropic Heisenberg model ($J_{xy} = J_z$) of [3], but only partial BQPT for the general cylindrical-symmetry Heisenberg model (arbitrary J_{xy} and J_z) used in [4,5].

6   Conclusion

Our contributions in this paper are twofold. We first proposed a new class of BQSS methods, by introducing an extended set of spin component measurements, which allowed us to restore an estimate of the entangled state $|\psi(t)\rangle$ and to develop corresponding classical-processing inverting and adapting blocks of the separating system. We then explicitly introduced a new research field, namely *Blind* Quantum Process Tomography (BQPT), as the extension of its existing non-blind version. Although not detailed in our previous papers, BQPT could be obtained as a spin-off of our corresponding BQSS methods, but with limitations (only partial identification or need for quantum processing means), which are here avoided. We plan to further develop this new class of BQSS and BQPT methods and to test their performance with simulated data.

References

1. Comon, P.: Blind identification and source separation in 2x3 under-determined mixtures. IEEE Trans. Signal Process. **52**, 11–22 (2004)
2. Comon, P., Jutten, C. (eds.): Handbook of Blind Source Separation. Independent Component Analysis and Applications. Academic Press, Oxford (2010)
3. Deville, Y., Deville, A.: Blind separation of quantum states: estimating two qubits from an isotropic Heisenberg spin coupling model. In: Davies, M.E., James, C.J., Abdallah, S.A., Plumbley, M.D. (eds.) ICA 2007. LNCS, vol. 4666, pp. 706–713. Springer, Heidelberg (2007). *Erratum: replace two terms $E\{r_i\}E\{q_i\}$ in (33) of [3] by $E\{r_i q_i\}$, since q_i depends on r_i*
4. Deville, Y., Deville, A.: Classical-processing and quantum-processing signal separation methods for qubit uncoupling. Quantum Inf. Proc. **11**, 1311–1347 (2012)

5. Deville, Y., Deville, A.: Quantum-source independent component analysis and related statistical blind qubit uncoupling methods. In: Naik, G.R., Wang, W. (eds.) Blind Source Separation: Advances in Theory, Algorithms and Applications. Springer, Berlin (2014)
6. Deville, Y., Deville, A.: Blind qubit state disentanglement with quantum processing: principle, criterion and algorithm using measurements along two directions. In: Proceedings of ICASSP 2014, Florence, Italy, pp. 6262–6266, 4–9 May 2014
7. Nielsen, M.A., Chuang, I.L.: Quantum Computation and Quantum Information. Cambridge University Press, Cambridge (2000)

Blind Source Separation in Nonlinear Mixture for Colored Sources Using Signal Derivatives

Bahram Ehsandoust[1,2(✉)], Masoud Babaie-Zadeh[2], and Christian Jutten[1]

[1] Univ. Grenoble Alpes, CNRS, GIPSA-lab, Grenoble, France
[2] Electrical Engineering Department, Sharif University of Technology, Tehran, Iran
{bahram.ehsandoust,christian.jutten}@gipsa-lab.grenoble-inp.fr,
mbzadeh@sharif.edu

Abstract. While Blind Source Separation (BSS) for linear mixtures has been well studied, the problem for nonlinear mixtures is still thought not to have a general solution. Each of the techniques proposed for solving BSS in nonlinear mixtures works mainly on specific models and cannot be generalized for many other realistic applications. Our approach in this paper is quite different and targets the general form of the problem. In this advance, we transform the nonlinear problem to a time-variant linear mixtures of the source derivatives.

The proposed algorithm is based on separating the derivatives of the sources by a modified novel technique that has been developed and specialized for the problem, which is followed by an integral operator for reconstructing the sources. Our simulations show that this method separates the nonlinearly mixed sources with outstanding performance; however, there are still a few more steps to be taken to get to a comprehensive solution which are mentioned in the discussion.

Keywords: Blind Source Separation · Nonlinear mixtures · Independent Component Analysis

1 Introduction

Blind Source Separation (BSS) is the problem of extracting the source signals that have been mixed together in a number of observations without any information about the mixture model or the sources [7]. In the simplest form of the problem, the number of sources and the observations are the same, and the sources are assumed to be statistically independent. The problem is formulated as

$$\mathbf{x}(t) = \mathbf{f}(\mathbf{s}(t)), \tag{1}$$

where $\mathbf{x}(t) = [x_1(t), ..., x_n(t)]^T$ and $\mathbf{s}(t) = [s_1(t), ..., s_n(t)]^T$ are the observation and source vectors, respectively (and n is the number of sources which is considered to be equal to the number of observations). The problem has been intensively studied since 1985. The first idea for performing the separation was trying to make independent signals combining the observations; hence named

© Springer International Publishing Switzerland 2015
E. Vincent et al. (Eds.): LVA/ICA 2015, LNCS 9237, pp. 193–200, 2015.
DOI: 10.1007/978-3-319-22482-4_22

Independent Component Analysis (ICA) [6,10]. It leaded to many noticeable achievements when the mixing model is linear, and several algorithms have been proposed such as INFOMAX, [2], JADE [5], Normalized EASI. [4], SOBI. [3], and FastICA [16].

Many practical applications for BSS have been considered by now, which is still being diversified. Separating the signals in a Multiple Input Multiple Output (MIMO) situation, separating EEG signals from different parts of the brain, discovering the hidden parameters affecting economical indexes, discriminating different layers of the earth from the reflections of the emitted electromagnetic wave, etc. are a few samples of the applications of BSS in real world [12].

However, extending theses accomplishments to the nonlinear case is not straightforward. As indicated in [11], ICA (i.e. source independence) is not sufficient for separating sources which are nonlinearly mixed.

But there are many realistic applications for which the linear BSS model could not be applied; e.g. smart chemical sensor arrays [8], hyperspectral imaging [9], and removing show-through in scanned documents [17]. Therefore, studies on this issue were focused on specific applications with restricted mixing models. Post Nonlinear composition (PNL) [18], Convolutive Post Nonlinear mixture [1], Bi-Linear model [17], conformal mappings [15], and mappings that can be transformed to linear mixtures via a nonlinear function [13] are some of the objectives that have been attained in this regard. Moreover, the problem has been addressed in [14] using the velocity state-space of the observations for continuous-time sources more recently.

Our approach to this problem is quite distinct. In this work, we add a few assumptions that should be met in addition to the conventional ones which are discussed in the following section. However, the conditions are not restrictive in practical applications and, so, the approach is more general.

2   The Main Idea

Our approach in this paper is based on the assumption that if the mixed sources meet a few conditions, they can be blindly separated utilizing signal derivatives. The assumptions consist of:

1. The derivatives of the sources need to meet the separability conditions of the conventional BSS problem for linear mixtures. Especially, they are statistically independent.
2. The sources are supposed to be colored; they need to have temporal correlation.
3. The nonlinear mixing model is required to be an invertible mapping.
4. The nonlinear mixing model is time-invariant.

Apparently, these conditions are met when the sources are not originally identical. In other words, if the sources are mutually independent in terms of stochastic processes, their derivatives are mutually independent as well (which guarantees meeting the required conditions).

The proposed method is based on the fact that the derivatives of the sources are mixed linearly even though the mixture model is nonlinear in general. Let $s_1(t)$ to $s_n(t)$ and $x_1(t)$ to $x_n(t)$ represent the source signals and the observations respectively. The above assumption can be written as

$$x(t) = f(s(t)) \Rightarrow \frac{\partial x}{\partial t} = J(s(t))\frac{\partial s}{\partial t} \qquad (2)$$

where f and J correspond the nonlinear mixing function and its Jacobian matrix respectively. The key point here is that, as the above equation shows, *although the signals are mixed nonlinearly, their derivatives are mixed linearly.* Accordingly, the main steps of the separating approach are summarized below:

Algorithm 1: Adaptive Algorithm for Time-Variant mixtures (AATV)

1. Compute the derivatives of the observations;
2. Consider them as the inputs for a linear BSS algorithm for linear mixtures;
3. Separate the derivatives of the observations to get the derivatives of the sources;
4. The sources are supposed to be the integral of the results.

However, except the sensitivity of the proposed method to the noise due to the derivative computation which is not addressed in the current work, there are two main challenges that highly affect the performance of the mentioned algorithm. The first one is that the mixing matrix (Jacobian) is time-variant and changes over the time. Therefore the existing methods for BSS in linear mixtures, which usually assume that the mixing matrix is constant, must be modified before being used for this problem. It is worth noting that the alterations of the mixing matrix depend on both the nonlinear mixing model, and the dynamics of the sources. However if the nonlinear mixing function changes slowly with respect to the sources (at the state of saturation, as an extreme example), the variations of the sources may lead to relatively small changes in the mixing matrix.

The second issue is the cumulative error in the integral of the separated signals because of the slow convergence of the BSS procedure. It should be noted that because the mixing matrix is time variant, it is not possible to run a batch algorithm for separating the derivatives that would prevent the convergence error in the signals: an adaptive (and fast) algorithm is mandatory.

To overcome the first challenge, we use an adaptive and iterative algorithm, which tracks the separating matrix in each iteration. In this context, the traditional convergence never happens, since the mixing model may vary over the time even though the output signals become completely separated at a moment.

Utilizing an adaptive algorithm causes the slow convergence and hence the error accumulation (the second mentioned concern). As it is stated before, even converging to the exact separating matrix at a point does not mean that it will properly work for the remained samples of the signals; for the reason that the mixing matrix (Jacobian) may alter rapidly, which should be followed by the adaptive algorithm again.

But the Jacobian matrix is not inherently time-variant. Actually, due to the assumption of mixing system to be time-invariant, its Jacobian is a time-invariant function of the *source* as well. In fact, considering the linear BSS model resulted by derivative computation, it is seen a time-variant mixing matrix; because the Jacobian matrix is calculated at different points (of the varying sources). Thus, trying to find the nonlinear nature of the separating matrix in the BSS procedure, we face a time-invariant problem, which may be solved by a batch algorithm.

$$\dot{\mathbf{x}} = \mathbf{J}(\mathbf{s})\dot{\mathbf{s}} \longleftrightarrow \frac{\partial \mathbf{x}}{\partial t} = \mathbf{J}(\mathbf{s}(t))\frac{\partial \mathbf{s}}{\partial t} \tag{3}$$

In other words, we propose to estimate the nonlinear mixing model which exists behind the linear time-variant matrix. This idea is similar to the one used in batch algorithms for linear mixtures, where the estimated mixing model is adjusted after data convergence, and is then applied to the whole data. The only difference here is that the system is nonlinear and, so, nonlinear modeling is used for the function to be extracted.

As a consequence, the BSS procedure is split into two phases. At the first phase, an adaptive algorithm is run on the observation derivatives and the separated signals are constructed. Then we extract the nonlinear time-invariant model from the result. To do this, the data after convergence (of the algorithm) is used for nonlinear modeling of the mapping. The obtained model is then applied to the data as the second run to compensate the slow convergence error.

Finally, the separating algorithm for the blind separation of nonlinearly mixed sources can be summarized as follows:

Algorithm 2: Adaptive Algorithm and Nonlinear Estimation (AANE)

1. Compute the derivatives of the observations;
2. Consider them as the inputs for an adaptive and iterative linear BSS algorithm;
3. Estimate (e.g. by spline interpolation) the time-invariant nonlinear model (nonlinear modeling) of the time-variant linear separating system derived from the previous step (by utilizing only the data after convergence);
4. Apply the obtained model to the whole data (observations) again in order to separate the signals correctly and get the derivatives of the sources;
5. The sources are the integral of the results.

3   Simulation Results

Simulating the proposed method, we have chosen a simple two-input two-output system with the integrals of a saw-tooth (named $s_1(t)$) and a sinusoid (named $s_2(t)$) signal as the input. The mixing model (4) is the same as the counter example [1], which was thought not to be separable at all.

$$\begin{bmatrix} x_1 \\ x_2 \end{bmatrix} = \begin{bmatrix} \cos\alpha & -\sin\alpha \\ \sin\alpha & \cos\alpha \end{bmatrix} \begin{bmatrix} s_1 \\ s_2 \end{bmatrix} \qquad \text{where } \alpha = 0.1 \times \sqrt{s_1^2 + s_2^2}. \tag{4}$$

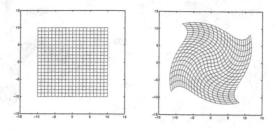

Fig. 1. Nonlinear mapping graph

Figure 1 shows the nonlinear mapping of the mixing model (it can be shown that the model is invertible).

From the given model, calculating the Jacobian matrix is straightforward.

$$J = \begin{bmatrix} \frac{\partial x_1}{\partial s_1} & \frac{\partial x_1}{\partial s_2} \\ \frac{\partial x_2}{\partial s_1} & \frac{\partial x_2}{\partial s_2} \end{bmatrix} = \begin{bmatrix} \cos\alpha & -\sin\alpha \\ \sin\alpha & \cos\alpha \end{bmatrix} \begin{bmatrix} 1 - s_2\frac{\partial\alpha}{\partial s_1} & -s_2\frac{\partial\alpha}{\partial s_2} \\ s_1\frac{\partial\alpha}{\partial s_1} & 1 + s_1\frac{\partial\alpha}{\partial s_2} \end{bmatrix} \tag{5}$$

In our simulation, Normalized-EASI [4] has been used as the linear BSS algorithm. This method is adaptive, iterative, and hence, suitable for the conditions of the problem. In addition, its equivariancy causes that its performance remains the same with respect to how much the sources are mixed together.

Performing the nonlinear modeling of the time-variant mixing model, we have used the down-sampled version (for increasing the speed of calculations) of the second half of the result (to make sure that we are using the outputs after convergence) after running Normalized-EASI on the data, to be modeled through a smoothing spline manner.

Obviously, it is not necessary to find the nonlinear *mixing* system for performing the proposed algorithm; it is sufficient to model the *separating* function instead. The models for the two outputs of the considered simulation are depicted in Fig. 2. In this figure, the blue circles are the converged samples of the elements of the separating matrix (which are resulted from the Normalized-EASI algorithm), and the plotted surfaces are the nonlinear model fitted to the data using smoothing spline. It should be also noted that we may directly model the nonlinear mapping of the two output signals as a function of the inputs (the derivatives of the observations) as well.

Finally, the obtained model is then applied to the whole data to adjust the result and is plotted for both sources in Fig. 3. This figure compares the original sources with the reconstructed ones by performing both the algorithm 1: AATV (signal named "AATV Result") and the algorithm 2: AANE (signal named "AANE Result"). In addition, Table 1 shows the normalized mean squared error for both results of the both algorithms.

It is also shown in Fig. 3 that Normalized-EASI is a fast adaptive algorithm and overcomes the first challenge (the mixing matrix changes over the time) pointed out in the previous section. Nevertheless, the second issue (slow convergence and the cumulative error) dramatically damages the result if not amended by the nonlinear modeling.

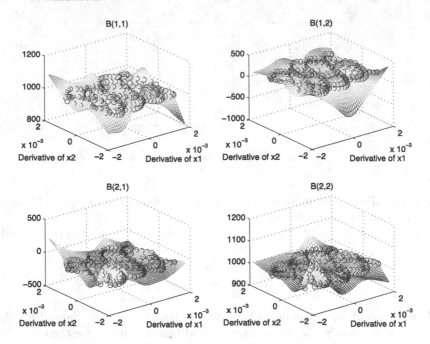

Fig. 2. Nonlinear modeling result of the elements of the separating matrix

Table 1. NMSE for AATV and AANE

	AATV	AANE
NMSE for the Source 1	1.6083	0.0032
NMSE for the Source 2	0.4875	0.0244

4   Discussion

In this paper, a new method is proposed for BSS in nonlinear mixtures based on separating the derivatives of the signals. Our simulations show that the proposed algorithm works well and can be considered as a simple approach to a new class of techniques in this field.

This approach targets a more general class of practical BSS problems in nonlinear mixtures than the existing ones and is not limited to a specific mixing model. However, there are a few concerns that should be thought about in this regard.

Firstly, the sources are supposed to be colored in the BSS problem (see the assumptions of the algorithm). This supposition is to make the derivatives suited for being separated by the adaptive BSS method for linear mixtures. It should be such that the adaptive algorithm can follow the variations. Solving the BSS for nonlinear mixtures through the proposed framework, we need that the BSS for the time-varying linear mixtures works properly.

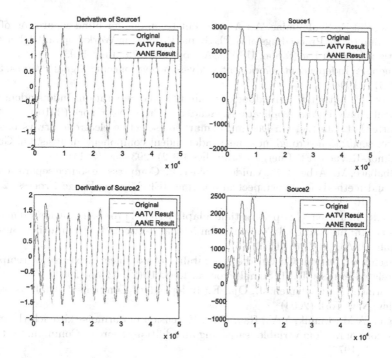

Fig. 3. Final extracted signals in comparison with the sources

Secondly, we have not considered the noise in the proposed algorithm. As we know, taking the derivatives amplifies the noise, and hence, makes the method more sensitive to the noise. At last, we should note that there exist the traditional ambiguities of BSS in linear mixtures (scaling and permutation) in the proposed method as well. However, these ambiguities do not differ for different samples of the signals (because in the proposed method, the *unique* extracted nonlinear model is applied on the whole data).

Acknowledgement. This work is partly funded by the European project 2012-ERC-Adv-320684 CHESS.

References

1. Babaie-Zadeh, M.: On blind source separation in convolutive and nonlinear mixtures. Ph.D. thesis, INP Grenoble (2002)
2. Bell, T., Sejnowski, T.: An information-maximization approach to blind separation and blind deconvolution. Neural Comput. **7**(6), 1004–1034 (1995)
3. Belouchrani, A., Abed-Meraim, K., Cardoso, J., Moulines, E.: A blind source separation technique using second-order statistics. IEEE Trans. Signal Process. **45**(2), 434–444 (1997)
4. Cardoso, J.F., Laheld, B.: Equivariant adaptive source separation. IEEE Trans. Signal Process. **44**(12), 3017–3030 (1996)

5. Cardoso, J.F., Souloumiac, A.: An efficient technique for blind separation of complex sources. In: Proceedings of IEEE Signal Processing Workshop on Higher-Order Statistics, South Lac Tahoe, USA (CA), pp. 275–279, June 1993
6. Comon, P.: Independent component analysis, a new concept? Signal Process. **36**(3), 287–314 (1994)
7. Comon, P., Jutten, C.: Handbook of Blind Source Separation: Independent Component Analysis And Applications. Academic Press, New York (2010)
8. Duarte, L.T., Jutten, C.: Design of smart ion-selective electrode arrays based on source separation through nonlinear independent component analysis. Oil Gas Sci. Technol.-Revue dIFP Energies Nouvelles **69**(2), 293–306 (2014)
9. Golbabaee, M., Arberet, S., Vandergheynst, P.: Compressive source separation: theory and methods for hyperspectral imaging. IEEE Trans. Image Process. **22**(12), 5096–5110 (2013)
10. Hérault, J., Jutten, C.: Space or time adaptive signal processing by neural networks models. In: International Conference on Neural Networks for Computing, Snowbird (Utah, USA), pp. 206–211 (1986)
11. Hosseini, S., Jutten, C.: On the separability of nonlinear mixtures of temporally correlated sources. IEEE Signal Process. Lett. **10**(2), 43–46 (2003)
12. Hyvärinen, A., Karhunen, J., Oja, E.: Independent Component Analysis, vol. 46. Wiley, New York (2004)
13. Kagan, A.M., Linnik, Y.V., Rao, C.R.: Extension of darmois-skitovich theorem to functions of random variables satisfying an addition theorem. Commun. Stat. **1**(5), 471–474 (1973)
14. Levin, D.N.: Performing nonlinear blind source separation with signal invariants. IEEE Trans. Signal Process. **58**(4), 2131–2140 (2010)
15. Hyvärinen, A., Pajunen, P.: Nonlinear independent component analysis: existence and uniqueness results. Neural Networks **12**, 429–439 (1999)
16. Hyvärinen, A.: Fast and robust fixed-point algorithms for independent component analysis. IEEE Trans. Neural Networks **10**(3), 626–634 (1999)
17. Merrikh-Bayat, F., Babaie-Zadeh, M., Jutten, C.: Linear-quadratic blind source separating structure for removing show-through in scanned documents. IJDAR **14**(4), 319–333 (2011)
18. Taleb, A., Jutten, C.: Source separation in post nonlinear mixtures. IEEE Trans. Signal Process. **47**(10), 2807–2820 (1999)

Sparse and Low-Rank Modeling for Acoustic Signal Processing

A Study on Manifolds of Acoustic Responses

Bracha Laufer-Goldshtein[1], Ronen Talmon[2], and Sharon Gannot[1](✉)

[1] Bar-Ilan University, 5290002 Ramat-gan, Israel
bracha.gold@walla.com, Sharon.Gannot@biu.ac.il
[2] Technion – Israel Institute of Technology, Technion City, 3200003 Haifa, Israel
ronen@ee.technion.ac.il

Abstract. The construction of a meaningful metric between acoustic responses which respects the source locations, is addressed. By comparing three alternative distance measures, we verify the existence of the acoustic manifold and give an insight into its nonlinear structure. From such a geometric view point, we demonstrate the limitations of linear approaches to infer physical adjacencies. Instead, we introduce the diffusion framework, which combines local and global processing in order to find an intrinsic nonlinear embedding of the data on a low-dimensional manifold. We present the diffusion distance which is related to the geodesic distance on the manifold. In particular, simulation results demonstrate the ability of the diffusion distance to organize the samples according to the source direction of arrival (DOA).

1 Introduction

Speech processing in reverberant environments facilitates a very complex relation between the emitted speech and the signal received by the microphones. Many algorithms, such as beamformers and localizers, try to distinguish between signals based on their propagation vector. In some scenarios, e.g. meeting rooms or cars, it can be assumed that the source position is confined to a predefined area. It can be reasonable to assume that representative samples from the region of the interest can be measured in advance. Due to reverberation, common practice is to represent the acoustic responses using a large number of variables, corresponding to the vast amount of reflections from the different surfaces characterizing the enclosure. In fact, the acoustic responses are only influenced by a small set of parameters related to the physical characteristics of the environment, such as: the enclosure dimensions and shape, the surfaces' materials and the positions of the microphones and the source. As a result, the high-dimensional acoustic responses are not uniformly scattered in their original space, but are rather concentrated on a manifold of much lower dimension. We therefore investigate the manifold of the acoustic responses and examine the proper distance between them.

In the context of multi-channel echo cancellation, Fozunbal et al. [6] presented a system identification algorithm by learning a low dimensional *linear* model of the room. Talmon and Gannot [10] proposed a different approach for supervised system identification, utilized for generalized sidelobe canceller (GSC) beamformer, based on the diffusion maps concept [2]. A similar approach is discussed in this paper.

© Springer International Publishing Switzerland 2015
E. Vincent et al. (Eds.): LVA/ICA 2015, LNCS 9237, pp. 203–210, 2015.
DOI: 10.1007/978-3-319-22482-4_23

The manifold perspective was also examined in light of the source localization problem. The existence of a binaural manifold was discussed by Deleforge et al. in [3–5] and a localization algorithm was presented. Another approach for supervised source localization based on the diffusion framework was introduced in [8].

In the current study, we show how to construct an informative metric between acoustic responses which respects the position of the source in the enclosure. For simplicity, the demonstration will focus only on the DOA of the source. We are interested in a static configuration, in which the properties of the enclosure and the position of the microphones remain fixed. In such an acoustic environment, the only varying degree of freedom is the source location. This is the latent variable which distinguishes between different acoustic responses. Accordingly, we will embed the acoustic responses in an intrinsic low-dimensional space representing the manifold and show that this embedding corresponds with the position of the source.

2   Problem Formulation

We consider a single source generating an unknown speech signal $s(n)$, which is received by a pair of microphones. Both the speaker and the microphones are located in an enclosure, e.g., a conference room or a car interior, with moderate reverberation time. The received signals, denoted by $x(n)$ and $y(n)$, are contaminated by additive stationary noise sources and are given by:

$$x(n) = a_1(n) * s(n) + u_1(n)$$
$$y(n) = a_2(n) * s(n) + u_2(n) \tag{1}$$

where n is the time index, $a_i(n)$, $i = \{1,2\}$ are the corresponding acoustic impulse response (AIRs) relating the source and each of the microphones, and $u_i(n)$, $i = \{1,2\}$ are uncorrelated white Gaussian noise (WGN) signals. Each of the AIRs is composed of the direct path between the source and the microphone, as well as reflections from the surfaces characterizing the enclosure. Consequently, even in moderate reverberation, the AIR is typically modelled as a long finite impulse response (FIR) filter.

Common practice is to define an appropriate feature vector that faithfully represents the characteristics of the acoustic path and is invariant to the other factors, i.e., the stationary noise and the varying speech signals. An equivalent representation of (1) is given by:

$$y(n) = h(n) * x(n) + v(n)$$
$$v(n) = u_2(n) - h(n) * u_1(n) \tag{2}$$

where $h(n)$ is the relative impulse response between the microphones with respect to the source, satisfying $a_2(n) = h(n)*a_1(n)$. In (2), the relative impulse response represents the system relating the measured signal $x(n)$ as an input and the measured signal $y(n)$ as an output.

For convenience, we represent (2) in the frequency domain. Assuming high signal to noise ratio (SNR) conditions, the Fourier transform of the relative impulse response, termed the relative transfer function (RTF), is obtained by:

$$H(k) = \frac{S_{yx}(k)}{S_{xx}(k)} = \frac{S_{ss}(k)A_2(k)A_1^*(k)}{S_{ss}(k)|A_1(k)|^2} = \frac{A_2(k)}{A_1(k)} \quad k = 0, \dots, D-1 \qquad (3)$$

where $H(k)$ is the RTF, $S_{yx}(k)$ is the cross power spectral density (CPSD) between $y(n)$ and $x(n)$, $S_{xx}(k)$ is the power spectral density (PSD) of $x(n)$ and $S_{ss}(k)$ is the PSD of the source. $A_1(k)$ and $A_2(k)$ are the acoustic transfer functions (ATFs) of the respective AIRs, and k denotes a discrete frequency index. Since $A_1(k)$ and $A_2(k)$ are unavailable, we use the estimated RTF $\hat{H}(k) \equiv \frac{\hat{S}_{yx}(k)}{\hat{S}_{xx}(k)}$, based on the estimated PSD and CPSD. The choice of the value of D should balance the tradeoff between correspondence to the relative impulse response length (large value) and latency considerations (small value). Accordingly, we define the feature vector $\mathbf{h} = [\hat{H}(0), \dots, \hat{H}(D-1)]^T$ as the concatenation of estimated RTF values in all frequency bins. In practice, we discard high frequencies in which the ratio in (3) is meaningless due to the lack of speech components. When noise influence cannot be neglected, we use, instead, an RTF estimator based on the non-stationarity of the speech signal [7].

3   Manifold-Based Distance Measures

Three alternative distance measures for quantifying the affinity between different RTFs, are addressed. We start with linear distance measures, namely, the Euclidean distance and the distance derived by principal component analysis (PCA) mapping. Next, we describe the concept of diffusion maps and present the diffusion distance [2]. For deriving the PCA-based distance and the diffusion distance, we assume the availability of a considerable amount of representative RTFs from various locations in the region of interest in the enclosure. The geometric interpretation of each of the distance measures is examined, and their hidden assumptions are highlighted and discussed.

3.1   Linear Distance Measures

The Euclidean distance between RTFs is denoted by:

$$D_{\mathrm{Euc}}(\mathbf{h}_i, \mathbf{h}_j) = \|\mathbf{h}_i - \mathbf{h}_j\|. \qquad (4)$$

The Euclidean distance does not assume an existence of a manifold and compares two RTFs in their original space. In particular, the Euclidean distance is equal to the geodesic distance when the manifold is flat, thus, inexplicitly, the Euclidean distance respects flat manifolds. Therefore, it is a good affinity measure only when the RTFs are uniformly scattered all over the space, or when they lie on a flat manifold.

The second distance we consider is based on PCA, which is the most common method to date for *linear* dimensionality reduction. First, the empirical mean and covariance matrix of the data are computed by:

$$\boldsymbol{\mu} = \frac{1}{N} \sum_{i=1}^{N} \mathbf{h}_i, \quad \mathbf{R} = \frac{1}{N} \sum_{i=1}^{N} (\mathbf{h}_i - \boldsymbol{\mu})(\mathbf{h}_i - \boldsymbol{\mu})^T \tag{5}$$

where N is the number of available representative RTFs in the region of interest. Then, by applying eigenvalue decomposition to the covariance matrix $\mathbf{R}$, we obtain a set of D eigenvectors and eigenvalues, denoted by $\{\mathbf{v}_i, \lambda_i\}_{i=0}^{D-1}$. The d eigenvectors, corresponding to the d largest eigenvalues, are viewed as the principal components of the data and form a new low-dimensional coordinate system. Finally, the RTFs are linearly projected onto the new coordinates/principal components:

$$\boldsymbol{\nu}(\mathbf{h}_i) = [\mathbf{v}_1, \ldots \mathbf{v}_d]^T (\mathbf{h}_i - \boldsymbol{\mu}). \tag{6}$$

The corresponding distance is given by the Euclidean distance between the projections:

$$D_{\mathrm{PCA}}(\mathbf{h}_i, \mathbf{h}_j) = \|\boldsymbol{\nu}(\mathbf{h}_i) - \boldsymbol{\nu}(\mathbf{h}_j)\|. \tag{7}$$

PCA is essentially a global approach; the principal directions of the *entire* set of RTFs are extracted from the covariance matrix. Then, the RTFs are *linearly* projected onto these directions, assuming that the manifold is linear/flat. As a result, the algorithm filters undesired samples' perturbations with respect to the manifold, which are caused by artifacts, such as: noise, estimation error and non-uniform sampling. Assuming that the manifold is indeed flat, PCA performs better than the Euclidean distance, due to this element of filtering.

3.2 Diffusion Distance

The concept of diffusion maps was introduced by Coifman and Lafon [2] as a general method for data-driven nonlinear dimensionality reduction. The diffusion framework consists of the following steps [9].

First, the affinity between RTFs is measured based on a pairwise weight function $k(\cdot, \cdot)$. Typically, the affinity is defined by a Gaussian function:

$$k(\mathbf{h}_i, \mathbf{h}_j) = \exp\left\{ -\frac{\|\mathbf{h}_i - \mathbf{h}_j\|^2}{\varepsilon} \right\}. \tag{8}$$

Such an affinity preserves locality since it defines local neighbourhoods according to the value of the scale parameter ϵ: for $\|\mathbf{h}_i - \mathbf{h}_j\| \ll \epsilon$, $k(\mathbf{h}_i, \mathbf{h}_j) \to 1$, and for $\|\mathbf{h}_i - \mathbf{h}_j\| \gg \epsilon$, $k(\mathbf{h}_i, \mathbf{h}_j) \to 0$.

Second, $\{\mathbf{h}_i\}$ are interpreted as nodes in a Graph. A Markov process can be defined on the graph via a construction of the transition matrix $\mathbf{P} = \mathbf{D}^{-1}\mathbf{W}$, where $\mathbf{W}$ is the Gram matrix defined by $W_{ij} = k(\mathbf{h}_i, \mathbf{h}_j)$, and $\mathbf{D}$ is a diagonal matrix whose elements are the row sums of $\mathbf{W}$. Accordingly, $p(\mathbf{h}_i, \mathbf{h}_j) \equiv P_{ij}$

represents the probability of transition in a single Markov step from node $\mathbf{h}_i$ to node $\mathbf{h}_j$.

In the next step, a nonlinear mapping of the RTFs into a new low-dimensional Euclidean space is built according to:

$$\boldsymbol{\Phi}_d : \mathbf{h}_i \mapsto \left[\varphi_1^{(i)}, \ldots, \varphi_d^{(i)}\right]^T \tag{9}$$

where $\{\varphi_j\}_{j=1}^d$ are the d-principal right-singular vectors of the transition matrix $\mathbf{P}$, and $\varphi_k^{(i)}$ denotes the ith entry of the vector φ_k. Note that φ_0 is ignored since it is an all-ones column vector.

The diffusion distance, which describes the relationships between pairs of samples in terms of their graph connectivity, is defined by:

$$D_{\text{Diff}}(\mathbf{h}_i, \mathbf{h}_j) = \|p(\mathbf{h}_i, \cdot) - p(\mathbf{h}_j, \cdot)\|_{\phi_0} = \sum_{r=1}^N \left(p(\mathbf{h}_i, \mathbf{h}_r) - p(\mathbf{h}_j, \mathbf{h}_r)\right)^2 / \phi_0^{(r)}$$

$$\tag{10}$$

where ϕ_0 is the most dominant left-singular vector of $\mathbf{P}$. The diffusion distance reflects the flow between two RTFs on the manifold, which is related to the geodesic distance on the manifold. It can be shown that the diffusion distance is equal to the Euclidean distance in the diffusion maps space when using all N eigenvectors, and can be well approximated by only the first few d eigenvectors [2], i.e.,

$$D_{\text{Diff}}(\mathbf{h}_i, \mathbf{h}_j) \cong \|\boldsymbol{\Phi}_d(\mathbf{h}_i) - \boldsymbol{\Phi}_d(\mathbf{h}_j)\|. \tag{11}$$

Though both diffusion maps and PCA construct a low-dimensional representation of the data, the two algorithms differ by the following fundamental distinctions. First, in PCA the data is globally viewed as of one piece drawn from some probability distribution and only the second order statistics is regarded. In contrast, diffusion maps combines local connections via the kernel construction and global processing via the spectral decomposition. Second, in PCA the hidden assumption is that the manifold is flat, thus linear projections are appropriate. On the other hand, diffusion maps is nonlinear and the data is embedded in new coordinates rather than linearly projected.

4   Analysis of the Manifold

In this section we examine the ability of each of the distance measurements, discussed in this paper, to organize the RTFs according to the corresponding DOA. For this purpose, we used the following setup. A source located in a $6 \times 6.2 \times 3$ m room, is picked up by two microphones located in $(3, 3, 1)$ m and $(3.2, 3, 1)$ m, respectively. The position of the source is confined to an arc of $10° \div 60°$ at 2 m distance with respect to the first microphone. The manifold analysis is carried out using a set of $N = 400$ samples, generated uniformly in the specified range. For each location, we simulate a unique 3 s speech signal, sampled at 16 kHz. The received signals are obtained by convolving the clean speech signal with the corresponding AIR, simulated based on the image method [1], and contaminated by a WGN with 20 dB SNR. For each source location, the CPSD and the PSD are estimated with Welch's method with 0.128 s windows and 75 % overlap and are utilized for estimating the RTF in (3) for $D = 2048$ frequency bins.

Both PCA and the diffusion map procedures were applied to the data. For computing the PCA-based distance, we used the projections on the $d = 10$ eigenvectors associated with the 10 largest eigenvalues (d was chosen empirically to obtain maximal range of monotonic behaviour). For the diffusion distance, only the first element in the mapping ($d = 1$) was considered. This choice will be justified in the sequel. All the distance measures were averaged over 50 rotations of the constellation described above with respect to the first microphone.

Figure 1(a) depicts the average Euclidean distance between each of the RTFs and a reference RTF corresponding to 10°, as a function of the angle, for three different reverberation times: 150 ms, 300 ms and 500 ms. We observe a monotonic behavior of the Euclidean distance with respect to the angle, which is confined to a certain region that becomes smaller as the reverberation time increases. Consequently, we conclude that the Euclidean distance is meaningful for small arcs, whose size is determined by the amount of reverberation. This implies that, in general, the Euclidean distance is not a good distance measure between RTFs, however, it can be utilized in the Gaussian kernel for diffusion maps, which takes into account only nearby RTFs through ϵ.

Figure 1(b) depicts the same illustration as Fig. 1(a) for the PCA-based distance. We observe similar trends with respect to the reverberation time compared to that inspected by the Euclidean distance. Here as well, the monotonicity of the distance with respect to the angle is maintained only in a limited region. However, this region is larger than the one exhibited by the Euclidean distance.

It follows that both the Euclidean distance and the PCA-based distance are not appropriate for measuring angles' proximity. The reason is that they both

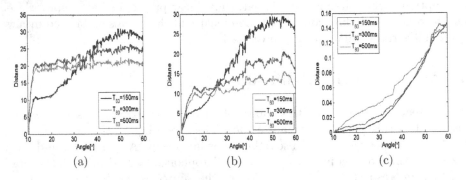

(a)   (b)   (c)

Fig. 1. Averages of the Euclidean distance (a), the PCA-based distance (b) and the diffusion distance ($\epsilon \approx 50$) (c) between each of the RTFs and the RTF corresponding to 10°, as functions of the angle.

rely on the assumption that the manifold is flat. However, monotonicity is preserved only for local environments indicating that the manifold is only locally linear, capturing its tangent plane, but generally, has a nonlinear structure. From the fact that locality is preserved for smaller regions when reverberation increases, we conclude that the complexity and nonlinearity of the manifold goes hand in hand with the amount of reverberation.

We now turn to the diffusion distance. The kernel scale ϵ should be adjusted to the distance at which monotonicity is maintained by the Euclidean distance, and should ignore longer distances. In Fig. 1(c) we examine the diffusion distance. It can be seen that for almost the entire range, the diffusion distance is monotonic with respect to the angle, indicating that it is an appropriate distance measure in terms of the source DOA. Moreover, by comparing the distance measures in Fig. 1(a)–(c), we observe that the diffusion distance is almost invariant with respect to the reverberation time, whereas the other distance measures significantly vary with the amount of reverberation.

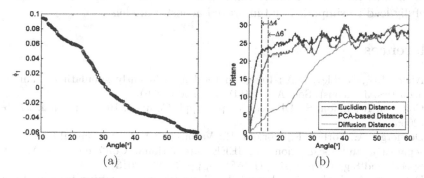

(a)   (b)

Fig. 2. (a) Single-element diffusion mapping $\Phi_1(\cdot)$ for $T_{60} = 300\,\text{ms}$. (b) The three distance measures (normalized to the same scale) between each of the RTFs and the RTF at $10°$, as a function of the angle for $T_{60} = 300\,\text{ms}$. The dashed lines show the boundary angles until which monotonicity is preserved.

Further insight into the mapping itself is gained by plotting the single-element mapping $\Phi_1(\cdot)$, as depicted in Fig. 2(a). We observe that the mapping corresponds well with the angle up to a monotonic distortion. Thus, the diffusion mapping successfully reveals the latent variable, namely, the position of the source. The almost perfect matching between the first element of the mapping and the corresponding position justifies the use of $d = 1$ for estimating the diffusion distance.

In practice we will be interested in a single scenario rather than in the average behaviour. Figure 2(b) compares between the range of monotonicity for each of the distance measures, for a single arbitrary scenario with moderate reverberation time of 300 ms. We observe that the monotonic behaviour is approximately maintained along $\Delta = 4°$ for the Euclidean distance, and along $\Delta = 6°$ for the PCA-based distance. The diffusion distance is monotonic for almost the entire range of $\Delta = 50°$. This confirms our previous conjecture that the diffusion distance is advantageous in measuring the physical position of the source.

5 Conclusions

In this paper we strengthen the claim on the existence of a nonlinear acoustic manifold, whose complexity is influenced by the amount of reverberation. We demonstrate the shortcomings of both the Euclidean distance and the PCA-based distance and their inability to measure the real physical distance. Instead, we propose to use the diffusion distance derived under the diffusion framework, which measures the distance between samples with respect to the manifold. Simulation results show that the diffusion distance properly arranges the RTFs according to the corresponding DOA.

This research lays the foundations for robust source localization algorithms based on a data-driven manifold. Moreover, the existence of an acoustic manifold paves the way for a better understanding of the acoustic environment, and will hopefully lead to simplified and improved acoustic models.

References

1. Allen, J.B., Berkley, D.A.: Image method for efficiently simulating small room acoustics. J. Acoust. Soc. Am. **65**(4), 943–950 (1979)
2. Coifman, R., Lafon, S.: Diffusion maps. Appl. Comput. Harmon. Anal. **21**, 5–30 (2006)
3. Deleforge, A., Forbes, F., Horaud, R.: Variational EM for binaural sound-source separation and localization. In: IEEE International Conference on Acoustics, Speech and Signal Processing (ICASSP), pp. 76–80 (2013)
4. Deleforge, A., Forbes, F., Horaud, R.: Acoustic space learning for sound-source separation and localization on binaural manifolds. Int. J. Neural Syst. **25**(1), 19 (2015)
5. Deleforge, A., Horaud, R.: 2D sound-source localization on the binaural manifold. In: IEEE International Workshop on Machine Learning for Signal Processing (MLSP). Santander, Spain, September 2012
6. Fozunbal, M., Kalker, T., Schafer, R.W.: Multi-channel echo control by model learning. In: The International Workshop on Acoustic Echo and Noise Control (IWAENC). Seattle, Washington, September 2008
7. Gannot, S., Burshtein, D., Weinstein, E.: Signal enhancement using beamforming and nonstationarity with applications to speech. IEEE Trans. Signal Process. **49**(8), 1614–1626 (2001)
8. Laufer, B., Talmon, R., Gannot, S.: Relative transfer function modeling for supervised source localization. In: IEEE Workshop on Applications of Signal Processing to Audio and Acoustics (WASPAA). New Paltz, NY, October 2013
9. Talmon, R., Cohen, I., Gannot, S., Coifman, R.: Diffusion maps for signal processing: a deeper look at manifold-learning techniques based on kernels and graphs. IEEE Signal Process. Mag. **30**(4), 75–86 (2013)
10. Talmon, R., Gannot, S.: Relative transfer function identification on manifolds for supervised GSC beamformers. In: 21st European Signal Processing Conference (EUSIPCO). Marrakech, Morocco, September 2013

A Polynomial Dictionary Learning Method for Acoustic Impulse Response Modeling

Jian Guan[1], Jing Dong[2], Xuan Wang[1]([✉]), and Wenwu Wang[2]

[1] Computer Application Research Center, Harbin Institute of Technology Shenzhen
Graduate School, Shenzhen 518055, China
{j.guan,wangxuan}@cs.hitsz.edu.cn
[2] University of Surrey, Guildford GU2 7XH, UK
{j.dong,w.wang}@surrey.ac.uk

Abstract. Dictionary design is an important issue in sparse representations. As compared with pre-defined dictionaries, dictionaries learned from training signals may provide a better fit to the signals of interest. Existing dictionary learning algorithms have focussed overwhelmingly on standard matrix (i.e. with scalar elements), and little attention has been paid to polynomial matrix, despite its widespread use for describing convolutive signals and for modelling acoustic channels in both room and underwater acoustics. In this paper, we present a method for polynomial matrix based dictionary learning by extending the widely used K-SVD algorithm to the polynomial matrix case. The atoms in the learned dictionary form the basic building components for the impulse responses. Through the control of the sparsity in the coding stage, the proposed method can be used for denoising of acoustic impulse responses, as demonstrated by simulations for both noiseless and noisy data.

Keywords: Dictionary learning · Polynomial matrix · Impulse responses

1 Introduction

Sparse representation has drawn intensive research interest for more than a decade. It aims to represent a signal with a linear combination of a small number of atoms chosen from an overcomplete dictionary in which the total number of atoms is greater than the dimension of the signal atoms [5]. The dictionary can be pre-defined using a mathematical equation. Alternatively, it can also be adapted from training data with a machine learning algorithm, leading to a category of algorithms called dictionary learning.

Several algorithms have been proposed for dictionary learning, such as the MOD, K-SVD and SimCO algorithms [1,3]. These algorithms have shown promising results in a number of tasks (e.g. denoising, super-resolution, and source separation) for a variety of natural signals, including acoustic and image data. In some practical applications, however, these algorithms cannot be directly applied since the signals that need to be dealt with may contain time delays. For example, in

© Springer International Publishing Switzerland 2015
E. Vincent et al. (Eds.): LVA/ICA 2015, LNCS 9237, pp. 211–218, 2015.
DOI: 10.1007/978-3-319-22482-4_24

acoustic modelling, the propagation channels between sources and microphones (or hydrophones) are often represented as acoustic impulse responses, which are usually described by polynomial matrix (with time delays) rather than a standard matrix with scalar elements. Polynomial matrices have been widely used in acoustic and communication channel modelling [7], e.g. for convolutive mixing and unmixing. Each element in a polynomial matrix can be represented as a finite impulse response (FIR) filter, e.g. for describing the input-output relationship [6].

In this paper, we develop a polynomial dictionary learning technique based on K-SVD algorithm. More specifically, we extend the K-SVD algorithm [1] for polynomial matrix based sparse representation model. Each atom in the learned dictionary is a polynomial represented as a FIR filter. All the atoms in the dictionary provides an overall description of the acoustic environment from which the acoustic impulse responses are used to train the dictionary. Such a dictionary has potential applications in denoising, dereverberation/deconvolution, and channel shortening of acoustic impulse responses, as partly demonstrated by simulations.

The remainder of the paper is organized as follows: Sect. 2 gives a brief review of the background of conventional dictionary learning and polynomial matrix decomposition. Section 3 presents the proposed polynomial dictionary learning method in detail. Section 4 shows the simulations and results. Section 5 concludes this paper.

2 Background

2.1 Conventional Dictionary Learning

Given a signal $\mathbf{y} \in \mathbb{R}^n$, the sparse representation of $\mathbf{y}$ can be expressed as

$$\mathbf{y} = \mathbf{D}\mathbf{x} \tag{1}$$

where $\mathbf{D} \in \mathbb{R}^{n \times K}$ ($n \ll K$) is an overcomplete dictionary containing K atoms, $\{\mathbf{d}_j\}_{j=1}^{K} \in \mathbb{R}^n$, and $\mathbf{x} \in \mathbb{R}^K$ is the sparse coefficient vector for representing $\mathbf{y}$. Two problems are often studied, namely, sparse coding and dictionary learning. Sparse coding aims to estimate $\mathbf{x}$, given $\mathbf{y}$ and $\mathbf{D}$, subject to the constraint that $\mathbf{x}$ is sparse, i.e. the number of non-zero elements measured by l_0 norm is below a pre-defined threshold (or relaxed by the l_1 norm of $\mathbf{x}$). In dictionary learning, the aim is to train a dictionary $\mathbf{D}$ based on a set of signals $\{\mathbf{y}_i\}_{i=1}^{N}$ which form a matrix $\mathbf{Y} \in \mathbb{R}^{n \times N}$, subject to the constraint that sparse coding coefficient matrix $\mathbf{X}$ is sparse. Here we focus on the dictionary learning problem.

2.2 Polynomial Matrix

A polynomial matrix is a matrix whose elements are polynomials. A $p \times q$ polynomial matrix $\mathbf{A}(z)$ can be expressed as

$$\mathbf{A}(z) = \sum_{\ell=0}^{L-1} \mathbf{A}(\ell)z^{-\ell} = \begin{bmatrix} a_{11}(z) & a_{12}(z) & \cdots & a_{1q}(z) \\ a_{21}(z) & \ddots & & \vdots \\ \vdots & & \ddots & \vdots \\ a_{p1}(z) & \cdots & \cdots & a_{pq}(z) \end{bmatrix} \tag{2}$$

where $\mathbf{A}(\ell) \in \mathbb{C}^{p \times q}$ is the coefficient matrix of $z^{-\ell}$, which denotes the impulse response at lag ℓ, and L is the length of impulse response. We use the coefficient of the polynomial to express the magnitude of the impulse responses. The F-norm of the polynomial matrix is defined as

$$\|\mathbf{A}(z)\|_F = \sqrt{\sum_{i=1}^{p}\sum_{j=1}^{q}\sum_{\ell=0}^{L-1} |a_{ij}(\ell)|^2} \tag{3}$$

Polynomial matrices have been widely used for acoustic impulse response modelling to describe the multi-path channel impulses propagating from the sources to the sensors. There are other forms of polynomial matrix decomposition techniques such as polynomial eigen-value or singular-value decompositions [4]. In this paper, we develop a polynomial dictionary learning algorithm by extending the conventional dictionary learning algorithm to the polynomial matrix case, as detailed next.

3   Polynomial Dictionary Learning

The conventional dictionary learning model (1) can be extended to the polynomial case as

$$\mathbf{Y}(z) = \mathbf{D}(z)\mathbf{X} \tag{4}$$

where $\mathbf{D}(z) \in \mathbb{C}^{n \times K}$ is an overcomplete polynomial dictionary matrix which contains polynomial atoms, $\mathbf{Y}(z) \in \mathbb{C}^{n \times N}$ is the "signals" to be represented, which can be an impulse response matrix, and $\mathbf{X} \in \mathbb{R}^{K \times N}$ is the sparse matrix which contains the representation coefficients of $\mathbf{Y}(z)$.

Suppose the length of the impulse response in $\mathbf{Y}(z)$ is L, according to Eq. (2), model (4) can be represented as

$$\sum_{\ell=0}^{L-1} \mathbf{Y}(\ell)z^{-\ell} = \sum_{\ell=0}^{L-1} \mathbf{D}(\ell)z^{-\ell}\mathbf{X} \tag{5}$$

where $\mathbf{Y}(\ell) \in \mathbb{R}^{n \times N}$ and $\mathbf{D}(\ell) \in \mathbb{R}^{n \times K}$ are the coefficient matrices of polynomial matrix $\mathbf{Y}(z)$ and $\mathbf{D}(z)$ at lag ℓ, respectively. As can be seen from Eq. (5), for any $\ell \in (0, L-1)$, $\mathbf{Y}(\ell)$ is represented as the linear combination of atoms in $\mathbf{D}(\ell)$, and $\mathbf{X}$ is the representation coefficients. This means that each coefficient matrix of the polynomial matrix $\mathbf{Y}(z)$ at different lags can also be represented by a coefficient matrix of $\mathbf{D}(z)$ at the corresponding lag weighted by the same representation matrix $\mathbf{X}$. Therefore, the coefficient matrices of polynomial matrices $\mathbf{Y}(z)$ and $\mathbf{D}(z)$ satisfy the form

$$\mathbf{Y}_i = \mathbf{D}_i\mathbf{X} \tag{6}$$

where $\mathbf{Y_i} \in \mathbb{R}^{n \times N}$ and $\mathbf{D_i} \in \mathbb{R}^{n \times K}$ are the coefficient matrices of $\mathbf{Y}(z)$ and $\mathbf{D}(z)$ at lag $i = 0, 1, ..., L-1$, respectively. Note that, for notational convenience, we have used $\mathbf{Y}_i$ to denote the coefficient matrix $\mathbf{Y}(\ell)$. We then define the following matrices by concatenating the coefficient matrices at all the time lags vertically

$$\underline{\mathbf{Y}} = [\mathbf{Y}_0; \mathbf{Y}_1; \mathbf{Y}_2; ...; \mathbf{Y}_{L-1}] \tag{7}$$

$$\underline{\mathbf{D}} = [\mathbf{D}_0; \mathbf{D}_1; \mathbf{D}_2; ...; \mathbf{D}_{L-1}] \tag{8}$$

As a result, Eq. (4) can be rewritten as

$$\underline{\mathbf{Y}} = \underline{\mathbf{D}}\mathbf{X} \tag{9}$$

where $\underline{\mathbf{Y}} \in \mathbb{R}^{nL \times N}$ and $\underline{\mathbf{D}} \in \mathbb{R}^{nL \times K}$. We can see from Eq. (9) that the polynomial dictionary learning problem (4) is now converted to a conventional dictionary learning model. Similarly, $\underline{\mathbf{D}}$ is overcomplete, hence, nL should be much smaller than K in Eq. (9).

The new dictionary $\underline{\mathbf{D}}$ can now be learned with a conventional dictionary learning algorithm such as the K-SVD algorithm [1], in which each atom and its corresponding sparse coefficient are updated simultaneously one by one with an iterative process. The learned $\underline{\mathbf{D}}$ can be used to reconstruct $\underline{\mathbf{Y}}$, and the reconstructed matrix is denoted as $\underline{\hat{\mathbf{Y}}}$. With a reverse operation to Eq. (7), we can obtain the coefficient matrix of the polynomial matrix at each time lag, as follows

$$\underline{\hat{\mathbf{Y}}} = \left[\hat{\mathbf{Y}}_0; \hat{\mathbf{Y}}_1; \hat{\mathbf{Y}}_2; ...; \hat{\mathbf{Y}}_{L-1}\right] \tag{10}$$

where $\underline{\hat{\mathbf{Y}}}$ is the restored matrix of $\underline{\mathbf{Y}}$, $\hat{\mathbf{Y}}_i$ is the coefficient matrix of polynomial channel matrix $\underline{\hat{\mathbf{Y}}}(z)$ at lag i, $i = 0, 1, 2..., L - 1$. With the coefficient matrices obtained above we can then construct the polynomial matrix $\underline{\hat{\mathbf{Y}}}(z)$ by using Eq. (2).

4   Simulations and Results

In this section, we evaluate the performance of the proposed method for learning a polynomial dictionary, and use it to recover a polynomial matrix. Two types of polynomial matrices were used, with one (i.e. the elements of its coefficient matrix) generated randomly, and the other as acoustic impulse responses (using a room image model). In both cases, noise is added to evaluate the capability of the proposed method for the recovery of noisy acoustic impulse responses.

4.1   Data Generation and Performance Measure

Polynomial Matrices Synthesis. The polynomial matrices were generated synthetically as follows. First, we generated a random scalar matrix $\underline{\mathbf{D}}$ with uniformly distributed entries, which was then used as the coefficient matrix for the polynomial matrix $\mathbf{D}(z)$ where each column of $\underline{\mathbf{D}}$ was normalized. Then, $\underline{\mathbf{Y}}$ was generated by the linear combination of different columns in $\underline{\mathbf{D}}$. At last, the polynomial matrices $\mathbf{Y}(z)$ and $\mathbf{D}(z)$ were generated by splitting their coefficient matrices according to Eqs. (7) and (8). The dimensions of the signals and dictionaries are designed according to the different experiments described later in this section.

Acoustic Impulse Response Generation and Modeling. The acoustic impulse responses were generated in a $20 \times 20 \times 3$ m³ room (to simulate a large hall) by the image model [2], where the reverberation time is 900 ms, sampling frequency is 16 KHz. The number of sampling points is set to be 14400 which means the number of time lags for each impulse response is 14400, and 1600 acoustic impulse responses were generated as the training set. We use a polynomial matrix to model acoustic signals by splitting each acoustic impulse response signal into 80 sections, where each section was modeled as a polynomial with 20 lags, so all these 1600 room acoustic impulse responses can be modeled as a 10×115200 polynomial matrix with 20 time lags for each element.

Performance Index. We define two performance indices to measure how well our proposed method performs. The *error rate* of the recovered polynomial matrix is defined as

$$E_r = \frac{\|\mathbf{Y}(z) - \hat{\mathbf{Y}}(z)\|_F^2}{\|\mathbf{Y}(z)\|_F^2} \tag{11}$$

where $\mathbf{Y}(z)$ denotes the original polynomial matrix without noise, and $\hat{\mathbf{Y}}(z)$ is the recovered polynomial matrix of $\mathbf{Y}(z)$. E_r means how similar the recovered $\hat{\mathbf{Y}}(z)$ to $\mathbf{Y}(z)$, the smaller the better. We also defined an *error rate* of noisy signal

$$E_n = \frac{\|\mathbf{Y}(z) - \mathbf{Y}_n(z)\|_F^2}{\|\mathbf{Y}(z)\|_F^2} \tag{12}$$

where $\mathbf{Y}_n(z)$ is the noisy polynomial matrix obtained by adding noise to $\mathbf{Y}(z)$, and E_n reflects the difference between $\mathbf{Y}(z)$ and $\mathbf{Y}_n(z)$ (i.e. the relative noise level).

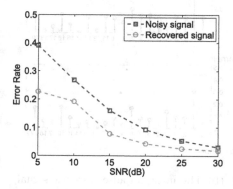

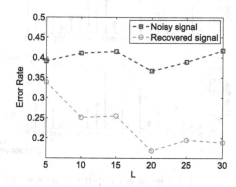

Fig. 1. The error rate comparison between E_n and E_r for the test signals at different SNR levels.

Fig. 2. The error rate comparison of E_n and E_r at different time lags for the test signal with the same noise level.

4.2   Experiments

We carry out several experiments on both synthetic data and simulated room acoustic impulse responses. The proposed method is tested on different noise levels, different impulse response lags, and the recovery of acoustic impulse response from noisy data.

Experiments on Synthetic Impulse Response Data. We synthesized a 10×1500 polynomial matrix $\mathbf{Y}(z)$ with 3 lags as training data. The dictionary $\mathbf{D}(z)$ was 10×50 polynomial matrix. The sparsity was set to be 3. In order to test whether our method can recover a signal (i.e. polynomial matrix) corrupted by noise of different levels, white Gaussian noise at different signal-to-noise (SNR) ratios was added to the test signal. Note that noise was added to the coefficient matrix of the polynomial matrix. Figure 1 shows how the error rate changes at different noise levels. We can see that the error rate of the recovered signal is smaller than that of the noisy signal at all tested noise levels. This means that the recovered signal is much more similar to the clean signal as compared with the noisy signal.

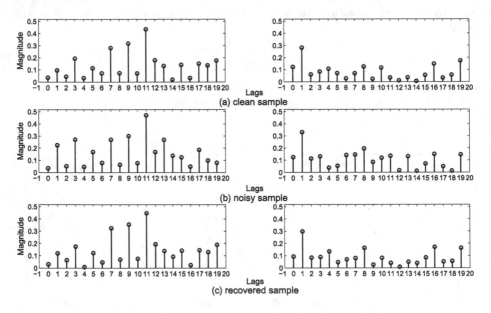

Fig. 3. (a) The clean impulse response signal. (b) The noisy impulse response signal which was obtained by adding 5 dB noise to the clean impulse response signal. (c) The impulse response recovered from the noisy version by the proposed method.

The impulse response at lag ℓ can be expressed as the coefficient matrix of $\mathbf{A}(\ell)z^{-\ell}$. As we use polynomial matrix to simulate impulse responses, we carried out another experiment to evaluate the recovery accuracy of the proposed

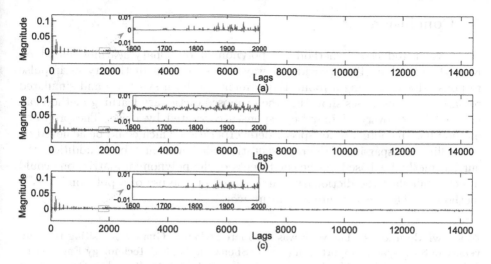

Fig. 4. (a) The clean acoustic impulse response signal. (b) The noisy acoustic impulse signal which was obtained by adding 5 dB white Gaussian noise to clean acoustic impulse signal. (c) The recovered acoustic impulse response signal.

method at different time lags with the same noise level, SNR = 5 dB. In this experiment, the training data was constructed as a 5×10000 polynomial matrix. As the polynomial dictionary has to be overcomplete, the dictionary was designed to be a 5×200 polynomial matrix. The test data was a 5×100 polynomial matrix, and the sparsity was set to be 3. Figure 2 shows the error rates at different time lags of the simulated impulse response. By comparing the error rates E_n and E_r, it can be seen that the error rate of recovered signal is also smaller than that of the noisy signal. Therefore, our proposed method can recover noisy signal at different time lags.

An illustration of the impulse responses is provided in Fig. 3, which shows two elements of the polynomial matrix (without noise), its corresponding noise added matrix and recovered matrix, where the lags of these polynomial matrices are 20. Compared with the clean signal, we can see from Fig. 3 that our proposed method can reconstruct the polynomial matrix very well from the noisy samples. This figure shows the denoising ability of the proposed method for the reconstruction of the impulse responses from noisy measurements.

Experiments on Impulse Responses Generated by Image Room. We conducted another experiment with a noisy impulse response signal generated by a room image model [2]. The clean impulse responses generated by the room image model were used as training data to train a polynomial dictionary. A noisy test signal was generated by adding noise at SNR = 5 dB. The proposed algorithm was used to recover the clean impulse responses. It can be seen from Fig. 4 that the recovered acoustic impulse is very similar to the clean one.

5   Conclusion

We have introduced a method for polynomial dictionary learning based on K-SVD algorithm. This provides a way for learning a dictionary of impulse responses for describing a room. Experiments on both synthetic and simulated room impulse responses show that the proposed dictionary learning method can be used for recovery of impulse responses corrupted by noise. The proposed method has the potential for speech dereverberation which could be achieved by controlling the sparsity of the representation coefficient matrix. In addition, the current method is based on the coefficients of the polynomial matrix, one could directly calculate the dictionary matrix based on the use of a polynomial SVD method [4]. These constitute our future work.

Acknowledgements. The work was conducted when J. Guan was visiting the University of Surrey, and supported in part by Shenzhen Applied Technology Engineering Laboratory for Internet Multimedia Application under Grants Shenzhen Development and Reform Commission, China (Grant Number 2012720).

References

1. Aharon, M., Elad, M., Bruckstein, A.: K-SVD: an algorithm for designing overcomplete dictionaries for sparse representation. IEEE Trans. Signal Process. **54**(11), 4311–4322 (2006)
2. Allen, J.B., Berkley, D.A.: Image method for efficiently simulating small-room acoustics. J. Acoust. Soc. Am. **65**(4), 943–950 (1979)
3. Dai, W., Xu, T., Wang, W.: Simultaneous codeword optimization (simco) for dictionary update and learning. IEEE Trans. Signal Process. **60**(12), 6340–6353 (2012)
4. Foster, J.A., McWhirter, J.G., Davies, M.R., Chambers, J.A.: An algorithm for calculating the QR and singular value decompositions of polynomial matrices. IEEE Trans. Signal Process. **58**(3), 1263–1274 (2010)
5. Kreutz-Delgado, K., Murray, J.F., Rao, B.D., Engan, K., Lee, T.W., Sejnowski, T.J.: Dictionary learning algorithms for sparse representation. Neural Comput. **15**(2), 349–396 (2003)
6. Rota, L., Comon, P., Icart, S.: Blind MIMO paraunitary equalizer. In: Proceedings of IEEE International Conference on Acoustics, Speech, and Signal Processing, ICASSP 2003, vol. 4, pp. 285–288. IEEE (2003)
7. Saramäki, T., Bregovic, R.: Multirate systems and filter banks. Multirate Syst. Des. Appl. **2**, 27–85 (2001)

A Local Model of Relative Transfer Functions Involving Sparsity

Zbyněk Koldovský$^{1(\boxtimes)}$, Jakub Janský^1, and Francesco Nesta^2

^{1} Faculty of Mechatronics, Informatics and Interdisciplinary Studies,
Technical University of Liberec, Studentská 2, 461 17 Liberec, Czech Republic
{jakub.jansky,zbynek.koldovsky}@tul.cz
^{2} Conexant System, 1901 Main Street, Irvine, CA, USA
Francesco.Nesta@conexant.com

Abstract. We propose a model of Relative Transfer Functions between two microphones which correspond to closed target positions within a certain spatially constrained area. Each RTF is modeled as the product of two transfer functions. One corresponds to a linear-phase filter and is the common factor of all the RTFs. The second transfer function is an individual factor that should be as sparse as possible in the time domain. A learning algorithm to identify the decomposition given a set of RTFs is proposed. The common factor is the main output, which we then apply to reconstruct an unknown RTF corresponding to a position within the assumed area, when only an incomplete measurement of it is available.

Keywords: Relative transfer function · Generalized sidelobe canceler · Convex programming · Sparse modeling

1 Introduction

A noisy stereo recording of a target signal can be, in the short-term Discrete Fourier Transform (DFT) domain, described as

$$X_{\mathrm{L}}(k,\ell) = S_{\mathrm{L}}(k,\ell) + Y_{\mathrm{L}}(k,\ell), \tag{1}$$
$$X_{\mathrm{R}}(k,\ell) = H(k)S_{\mathrm{L}}(k,\ell) + Y_{\mathrm{R}}(k,\ell),$$

where k and ℓ denote, respectively, the frequency and the frame index; let the DFT window length be M. Further let X_{L} and X_{R} denote signals observed by microphones; S_{L} is the response (image) of the target signal on the left microphone; Y_{L} and Y_{R} are the remaining signals (noise and interferences) commonly referred to as noise. H is the Relative Transfer Function (RTF) between the microphones related to the target signal.

The RTF is an important component of multichannel audio signal processing systems [1]. Using the RTF, a multichannel filter can be designed such that it performs spatial null towards a target source, thereby yielding a noise signal reference on its output [3]. Specifically, let $\hat{H}$ be an estimate of H and the

This work was supported by California Community Foundation through Project No. DA-15-114599.

E. Vincent et al. (Eds.): LVA/ICA 2015, LNCS 9237, pp. 219–226, 2015.
DOI: 10.1007/978-3-319-22482-4_25

multichannel filter with inputs X_L and X_R be such that its output is $Z(k, \ell) = \widehat{H}(k)X_\text{L}(k, \ell) - X_\text{R}(k, \ell)$. By (1), it holds that

$$Z(k, \ell) = \underbrace{\left(\widehat{H}(k) - H(k)\right)S_\text{L}(k, \ell)}_{\text{target signal leakage}} + \underbrace{\widehat{H}(k)Y_\text{L}(k, \ell) - Y_\text{R}(k, \ell)}_{\text{noise reference}}. \tag{2}$$

For $\widehat{H} = H$, the target signal leakage vanishes, and $Z(k, \ell) = H(k)Y_\text{L}(k, \ell) - Y_\text{R}(k, \ell)$ provides the key noise reference signal.

However, real-world RTFs have many coefficients that can change quickly in a short period of time. The RTF estimation thus poses a difficult problem. Several estimators have been proposed to estimate the RTF directly from noisy recordings. A frequency-domain estimator assuming nonstationary target signal and stationary noise was proposed in [4]. Methods based on Blind Source Separation (BSS) and Independent Component Analysis (ICA) can cope with general situations (e.g., with nonstationary directional interfering sources) [5]. However, the accuracy of such estimation is limited.

During intervals where only the target source is active, conventional least-square approaches can be used to obtain highly accurate RTF estimates. These estimates can be used later when noise is active. However, the acoustic conditions must remain the same: in particular, the position of the target source must be preserved. The fact that the target source position is often limited to a confined area can be used as a priori knowledge. For example, it is possible to collect a bank of RTFs during noise-free intervals such that the area is covered by the bank [2]. The problem of estimating the RTF can then be simplified to one of choosing an appropriate RTF from the bank.

Recent methods aim to find suitable low-rank models for such banks of acoustic transfer functions, which are instrumental in computing highly accurate RTF estimates from noisy recordings; see, e.g., [6,7]. In this paper, we propose a novel sparsity-based model, which is applied when reconstructing incomplete RTF (iRTF), that is, an RTF estimate whose values are known only for certain frequencies. In [8], the iRTF is completed through finding its sparsest representation in the time domain. This is justified by the fact that relative impulse responses (ReIRs), i.e., the time-domain counterparts of RTFs, are compressible (approximately sparse) sequences. However, since exact sparsity is invoked in [8], there are performance limitations; see also [9]. The goal of our paper is to exploit the proposed model in order to lower this performance loss.

Throughout the paper, upper-case letters will denote transfer functions (DFT domain) while their time-domain counterparts will be denoted by lower-case letters. Bold letters will denote vectors comprising coefficients of the corresponding quantities. For example, $(\mathbf{H})_k = H(k)$. The time domain counterpart of H, called the relative impulse response (ReIR), is h, and $(\mathbf{h})_k = h(k)$.

2   Model Proposal

Let H_p be the RTF for the pth position of the target source within a confined area, and $p = 1, \ldots, P$; h_p be the corresponding ReIR. The positions $1, \ldots, P$

can form either a regular or an irregular grid [2,12]. Our goal is to model H_p as

$$H_p(k) \approx B(k)G_p(k), \quad k = 0, \ldots, M-1, \quad p = 1, \ldots, P, \tag{3}$$

where g_p is sparse. B is independent of the position index p, so it is a common factor of $H_1, \ldots, H_P$.

As a motivator, note that each h_p can be approximated by a sparse g_p such that neither $\|\mathbf{h}_p - \mathbf{g}_p\|_2$ nor the energy of the target signal leakage in (2) for $\hat{H} = G_p$ is higher than a chosen limit; see [10]. The conjecture behind the proposed model is that the residual $B = H_p/G_p$ might be independent of p since the RTFs come from the same area. Assuming that B is known, an iRTF for a target position within the area can be reconstructed as $B \cdot G$ by invoking the sparsity of g.

Linear-Phase Unit-Norm Constraint. Definition (3) is not unique. The scale of B can be arbitrarily changed, which can be compensated for by the reciprocal scaling of all G_p, $p = 1, \ldots, P$. To solve the scaling uncertainty, B will be constrained to have a unit norm. Similarly, B can be arbitrarily delayed. Therefore, we constrain B to have a constant group delay of $\lfloor M/2 \rfloor$, hence, linear phase. It means that b is constrained to be symmetric[1] along its $(\lfloor M/2 \rfloor + 1)$th coefficient; we will assume, without loss on generality, that M is odd. The set of the unit-norm symmetric filters (vectors) of length M will be denoted by $\mathcal{A}_M$.

A Method to Learn the Common Factor B. Let $H_1, \ldots, H_P$ be given. Before learning the common factor B, the causality of the filters in (3) must be ensured by delaying each h_p by $\lfloor M/2 \rfloor + D$ samples. The delay by $\lfloor M/2 \rfloor$ is due to the linear-phase constraint imposed on B. The value of $D \geq 0$ must be sufficiently high to ensure the causality of g_ps; we typically use $D = 10$.

To find B, we formulate an optimization problem with the above constraints on B as

$$\mathbf{B} = \arg \min_{\mathbf{B}, \mathbf{G}_1, \ldots, \mathbf{G}_P} \sum_{p=1}^{P} \|\mathbf{g}_p\|_0 \quad \text{s.t.} \quad \sum_{p=1}^{P} \|\mathbf{H}_p - \text{diag}(\mathbf{B})\mathbf{G}_p\|_2^2 \leq \epsilon, \quad \mathbf{b} \in \mathcal{A}_M. \tag{4}$$

where $\text{diag}(\cdot)$ denotes a diagonal matrix with the argument on its main diagonal; $\|\cdot\|_0$ denotes the ℓ_0 pseudo-norm that is equal to the number of nonzero elements of the argument; ϵ is a free positive constant. Note that $\mathbf{B} = \mathbf{Fb}$, $\mathbf{H}_p = \mathbf{Fh}_p$, and $\mathbf{G}_p = \mathbf{Fg}_p$ where $\mathbf{F}$ is the $M \times M$ unitary matrix of DFT.

However, the problem (4) is NP hard and *not* convex in general even if the ℓ_0 pseudo-norm is replaced by the ℓ_1 norm. Without a guarantee to find the global minimum of (4), we propose an alternating algorithm that can be used to find a satisfactory solution.

---

[1] This constraint excludes the set of anti-symmetric linear-phase filters. However, b tends to be close to the pure delay filter in practice, hence it is always symmetric. In experiments with real-world RTFs, we did not observe any cases where b was anti-symmetric.

In the beginning, it is assumed that feasible $B, G_1, \ldots, G_P$ are given as an initial guess. This can be, for example, $B(k) = e^{\frac{i2\pi\lfloor M/2\rfloor k}{M}}$ and $G_p(k) = \overline{B(k)}H_p(k)$, $\forall p, k$, where $\overline{\cdot}$ denotes the complex conjugate. Then, one iteration of the algorithm consists of two optimization steps. First, all $\mathbf{G}_p$ are fixed while $\mathbf{B}$ is computed such that it minimizes the constraint in (4); that is, the solution of

$$\min_{\mathbf{B}} \sum_{p=1}^{P} \|\mathbf{H}_p - \text{diag}(\mathbf{B})\mathbf{G}_p\|_2^2 \quad \text{s.t.} \quad \mathbf{b} \in \mathcal{A}_M. \tag{5}$$

This is a constrained least-squares problem that can be solved analytically. Let $\mathbf{c} = [c_1, \ldots, c_{\lfloor M/2\rfloor}]^T$ denote the symmetric part of $\mathbf{b}$, i.e., $\mathbf{b} = [\mathbf{c}; b_{\lfloor M/2\rfloor+1}; \underline{\mathbf{c}}]$ where $\underline{\cdot}$ denotes the upside down operator. For $\mathbf{b}$ being the solution of (5) it holds that

$$[\mathbf{c}; b_{\lfloor M/2\rfloor+1}] \propto \left(\mathbf{A}^H \text{diag}\Big(\sum_{p=1}^{P} \text{diag}(\mathbf{G}_p)\overline{\mathbf{G}_p}\Big)\mathbf{A}\right)^{-1} \mathbf{A}^H \left(\sum_{p=1}^{P} \text{diag}(\overline{\mathbf{G}}_p)\mathbf{H}_p\right), \tag{6}$$

where $\mathbf{A}$ is an $M \times (\lfloor M/2\rfloor + 1)$ matrix whose ith row is $[\mathbf{F}_{i,1} + \mathbf{F}_{i,M}, \mathbf{F}_{i,2} + \mathbf{F}_{i,M-1}, \ldots, \mathbf{F}_{i,\lfloor M/2\rfloor} + \mathbf{F}_{i,\lfloor M/2\rfloor+2}, \mathbf{F}_{i,\lfloor M/2\rfloor+1}]$. After computing $\mathbf{b}$ using (6), the vector is normalized to satisfy the constraint $\|\mathbf{b}\|_2 = 1$, so finally $\mathbf{b} \in \mathcal{A}_M$. Note that $B, G_1, \ldots, G_P$ remain feasible for (4) after the first step.

The goal of the second step is to improve the sparsity of $G_1, \ldots, G_P$ while preserving their feasibility. B is fixed while $G_1, \ldots, G_P$ are optimized to approach the solution of (4). To this end, P independent convex programs are solved, one for each G_p,

$$\min_{\mathbf{G_P}} \|\mathbf{g}_p\|_1 \quad \text{w.r.t.} \quad \|\mathbf{H}_p - \text{diag}(\mathbf{B})\mathbf{G}_p\|_2^2 \leq \epsilon/P. \tag{7}$$

This optimization problem is well-known under the name of *basis pursuit denoising* (BPDN) and can be efficiently solved using, e.g., SPGL1[2]; see [11]. Using the fact that $\mathbf{F}$ is a unitary matrix, (7) can be written in its equivalent form

$$\min_{\mathbf{g_P}} \|\mathbf{g}_p\|_1 \quad \text{w.r.t.} \quad \|\mathbf{h}_p - \mathbf{F}^H\text{diag}(\mathbf{B})\mathbf{F} \cdot \mathbf{g}_p\|_2^2 \leq \epsilon/P. \tag{8}$$

Since $\mathbf{b}$ is real-valued, the matrix $\mathbf{F}^H\text{diag}(\mathbf{B})\mathbf{F}$ as well as the whole program (8) are real-valued. The proposed optimization algorithm is summarized in Algorithm 1.

3   Application: Sparse Recovery of an Incomplete RTF

An iRTF is represented by an $|\mathcal{S}| \times 1$ vector $\mathbf{Y}$ whose kth element is

$$(\mathbf{Y})_k = \widehat{H}(i_k), \qquad k \in \mathcal{S}, \tag{9}$$

---

[2] http://www.cs.ubc.ca/~mpf/spgl1.

Algorithm 1. Learning algorithm to find B.

Input: $\mathbf{H}_1, \ldots, \mathbf{H}_P$ and ϵ

Output: $\mathbf{B}$

Initialization: $(\mathbf{B})_k = e^{\frac{i2\pi \lfloor M/2 \rfloor k}{M}}$, $(\mathbf{G}_p)_k = \overline{(\mathbf{B})_k}(\mathbf{H}_p)_k$, $\mathbf{B}_{old} = \mathbf{0}_{M \times 1}$

while $\|\mathbf{B}_{old} - \mathbf{B}\|_2 > \texttt{tol}$ **do**

    $\mathbf{B}_{old} = \mathbf{B}$;

    Compute $\mathbf{b}$ using (6);

    $\mathbf{b} \leftarrow \mathbf{b}/\|\mathbf{b}\|_2$;

    $\mathbf{B} = \texttt{fft}(\mathbf{b})$;

    $\mathbf{Q} = \mathbf{F}^H \texttt{diag}(\mathbf{B})\mathbf{F}$;

    for $p \in \{1, \ldots, P\}$ **do**

        $\mathbf{g}_p = \arg\min_{\mathbf{g}} \|\mathbf{g}\|_1$ w.r.t. $\|\mathbf{h}_p - \mathbf{Q}\mathbf{g}\|_2 \le \epsilon/P$;

        $\mathbf{G}_p = \texttt{fft}(\mathbf{g}_p)$;

    end

end

---

where $\mathcal{S} = \{i_1, \ldots, i_{|\mathcal{S}|}\} \subset \{1, \ldots, M\}$ is the set of indices of known values of $\widehat{H}$. In [8], it is proposed to retrieve the complete RTF estimate from $\mathbf{Y}$ through

$$\widehat{\mathbf{h}} = \arg\min_{\mathbf{h}} \|\mathbf{h}\|_1 \quad \text{w.r.t.} \quad \|\mathbf{Y} - \mathbf{F}_{\mathcal{S}}\mathbf{h}\|_2 \le \epsilon, \tag{10}$$

where the subscript $(\cdot)_{\mathcal{S}}$ denotes a vector/matrix with selected rows whose indices are in $\mathcal{S}$; ϵ is a free positive parameter[3]. This is the BPDN optimization program, which, in other words, seeks for the sparsest representation of $\mathbf{y}$ in the time domain. As mentioned in the Introduction, the performance of this method is limited due to the fact that the original h is typically not exactly sparse.

Now, assume that $\widehat{H}$ is a noisy estimate of RTF for the current (unknown) position of the target within an assumed area. Let $\mathbf{Y}$ be its incomplete version where $\mathcal{S}$ is selected on the basis of a certain hypothesis (e.g., we select only those elements of $\widehat{H}$ that are not affected by the noise [8]). Next, assume that a bank of RTFs $H_1, \ldots, H_P$ is given. These RTFs are valid for some positions within the area but can be different from the current ones estimated by $\widehat{H}$. Using the bank of RTFs, B from (3) can be identified using Algorithm 1 with some ϵ. The goal is now to exploit B as a priori knowledge when retrieving H from $\mathbf{Y}$.

We propose computing the complete RTF estimate as

$$\widehat{\mathbf{H}} = \texttt{diag}(\mathbf{B})\mathbf{G} \tag{11}$$

where $\mathbf{G} = \mathbf{F}\mathbf{g}$, and

$$\mathbf{g} = \arg\min_{\mathbf{g}} \|\mathbf{g}\|_1 \quad \text{w.r.t.} \quad \|\mathbf{Y} - \texttt{diag}(\mathbf{B}_{\mathcal{S}})(\mathbf{F}_{\mathcal{S}}\mathbf{g})\| \le \epsilon. \tag{12}$$

---

[3] In fact, $\widehat{\mathbf{h}}$ is in [8] sought through solving $\min_{\mathbf{h}} \|\mathbf{F}_{\mathcal{S}}\mathbf{h} - \mathbf{Y}\|_2^2 + \tau\|\mathbf{h}\|_1$ where $\tau > 0$. This is, nevertheless, an equivalent problem to (10) in the sense that there exists τ such that the solutions are the same.

Discussion. In view of the Compressed Sensing theory, (10) as well as (12) can be interpreted as sparse reconstructions of compressed measurements $\mathbf{Y}$ in the time domain when the sensing matrix is, respectively, $\mathbf{F}_S$ and $\mathtt{diag}(\mathbf{B}_S)\mathbf{F}_S$. In other words, the sensing domain is, respectively, the Fourier domain and the Fourier domain transformed through $\mathtt{diag}(\mathbf{B})$. The principal difference depends on the distance of $\mathbf{B}$ from the pure delay by $\lfloor M/2 \rfloor$ samples.

4   Experimental Verification

The first experiment was based on simulated data. An artificial RTF of length $M = 1025$ was generated as $H = B \cdot G$ where $(\mathbf{g})_k = a_k e^{-0.008(|k|-20)}$, $k = 1, \ldots, M$, with only q random nonzero a_k from $[-0.5; 0.5]$, and $a_{20} = 1$. Hence, $\mathbf{g}$ is q-sparse and has an exponential decay with the highest peak at $k = 20$. B was generated as $(\mathbf{B})_k = d_k e^{\frac{i2\pi k \lfloor M/2 \rfloor}{M}}$ where d_k were taken at random from $[0.5; 1.5]$. Such B has a linear phase and is close to the pure delay by $\lfloor M/2 \rfloor$ samples.

A 10 s female voice signal was taken to simulate a noise-free recording according to (1). Then, H was estimated from each 1 s interval of the recording using the least-squares estimator. Then only p percent of the most active frequencies in the original signal were put in S, and the iRTF was created. This procedure simulates situations when the signal by the target source does not excite the full frequency range or when some frequency bins are contaminated by noise. Therefore, only iRTF is available. The reconstructed estimates of H were computed through (10) and (12) using known B, respectively, both with $\epsilon = 0.1$; the sampling frequency was 16 kHz.

The resulting RTFs were evaluated in terms of attenuation rate (ATR). The ATR is defined as the ratio between the power of the original signal on the left microphone and the power of the target signal leakage term in (2). The more negative the ATR (in dB), the better the target signal blocking. The resulting ATRs averaged over the respective intervals are shown in Fig. 1. The "model-based solution" through (12) using known B achieves significantly better ATR than the sparse solution by (10) for all levels of the incompleteness p. For $p \rightarrow 100$, both solutions approach the ATR by the true RTF.

The second experiment was conducted with real-world recordings. In an office room with $T_{60} \approx 300$ ms, a 12 s sequence of audio signals (4 s white noise + 4 s male speech + 4 s female speech) played by a loudspeaker was recorded by two microphones. The loudspeaker was placed in front of the microphones at a distance of 1.5 m and rotated at nine different angles from -90 through $90°$ ($0°$ is the direction towards the microphones). The microphone spacing was 5 cm.

The RTFs for the loudspeaker's rotations were computed using the first second of the recorded white noise and the least-square estimator (from this point forward referred to as "true RTF"). Algorithm 1 was applied with $\epsilon = 0.01$ and $\mathtt{tol} = 10^{-3}$ to learn the common factor B of true RTFs corresponding to the rotations by $-30°$, $0°$, and $30°$. The method converged after 46 iterations.

The recordings were divided into twelve 1 s intervals. On each interval, the least-square RTF estimate was computed and 10 percent of its values

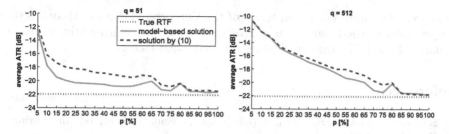

Fig. 1. Results of the first experiment for two levels of sparsity of the factor G in $H = B \cdot G$, namely, $q = 51$ and $q = 512$.

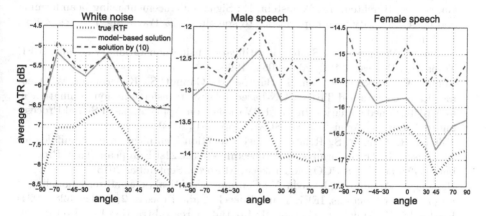

Fig. 2. Average ATR achieved by RTFs reconstructed using (10) and the model-based solution (12). The ATR achieved by the true RTF is shown and corresponds to an optimum achievable performance.

corresponding to the most active frequencies on the left microphone was selected to build up the iRTF.

The relative ATRs averaged over 4-s intervals are shown in Fig. 2. These results show that the knowledge of B helps to reconstruct the RTF with the average relative improvement of ATR by up to 2 dB. It should be noted that although B was derived only using known RTFs for rotations by $-30°$, $0°$, and $30°$, no overlearning is observed in Fig. 2 for these rotations: The ATR improvement is rather uniform over all rotations and mainly depends on the original signal. Unlike white noise, the speech signals are easier to attenuate by the reconstructed RTF as they do not span the whole frequency range.

5   Conclusions

Algorithm 1 was shown to be able to learn the common factor of RTFs that are known for positions of a target source within a confined area. This factor can be used when reconstructing an incomplete measurement of an RTF from the same area. An interpretation in terms of the Compressed Sensing theory is that the proposed method learns a new sensing domain (or, equivalently, a sparsity domain) of the RTFs through finding their common factor.

References

1. Benesty, J., Makino, S., Chen, J. (eds.): Speech Enhancement, 1st edn. Springer, Heidelberg (2005)
2. Koldovský, Z., Málek, J., Tichavský, P., Nesta, F.: Semi-blind noise extraction using partially known position of the target source. IEEE Trans. Speech Audio Lang. Process. **21**(10), 2029–2041 (2013)
3. Gannot, S., Burshtein, D., Weinstein, E.: Signal enhancement using beamforming and nonstationarity with applications to speech. IEEE Trans. Sig. Process. **49**(8), 1614–1626 (2001)
4. Shalvi, O., Weinstein, E.: System identification using nonstationary signals. IEEE Trans. Sig. Process. **44**(8), 2055–2063 (1996)
5. Reindl, K., Markovich-Golan, S., Barfuss, H., Gannot, S., Kellermann, W.: Geometrically constrained TRINICON-based relative transfer function estimation in underdetermined scenarios. In: IEEE Workshop on Applications of Signal Processing to Audio and Acoustics (WASPAA), pp. 1–4 (2013)
6. Talmon, R., Gannot, S.: Relative transfer function identification on manifolds for supervised GSC beamformers. In: Proceedings of the 21st European Signal Processing Conference (EUSIPCO), Marrakech, Morocco, September 2013
7. Haneda, Y., Makino, S., Kaneda, Y.: Common acoustical pole and zero modeling of room transfer functions. IEEE Trans. Speech Audio Process. **2**(2), 320–328 (1994)
8. Koldovský, Z., Málek, J., Gannot, S.: Spatial source subtraction based on incomplete measurements of relative transfer function. Accepted in IEEE Trans. Speech Audio Lang. Process. April 2015
9. Koldovský, Z., Tichavský, P.: Sparse reconstruction of incomplete relative transfer function: discrete and continuous time domain. Accepted for a Special Session at EUSIPCO 2015, Nice, France, September 2015
10. Málek, J., Koldovský, Z.: Sparse target cancellation filters with application to semi-blind noise extraction. In: Proceedings of the 41st IEEE International Conference on Audio, Speech, and Signal Processing (ICASSP 2014), Florence, Italy, May 2014, pp. 2109–2113 (2014)
11. van den Berg, E., Friedlander, M.P.: Probing the Pareto frontier for basis pursuit solutions. SIAM J. Sci. Comput. **31**(2), 890–912 (2008)
12. Málek, J., Botka, D., Koldovský, Z., Gannot, S.: Methods to learn bank of filters steering nulls toward potential positions of a target source. In: Proceedings of the 4th Joint Workshop on Hands-free Speech Communication and Microphone Arrays (HSCMA 2014), Nancy, France, 12–14 May 2014

Improving Relative Transfer Function Estimates Using Second-Order Cone Programming

Zbyněk Koldovský[1,2]([✉]), Jiří Málek[1], and Petr Tichavský[2]

[1] Faculty of Mechatronics, Informatics and Interdisciplinary Studies, Technical University of Liberec, Studentská 2, 461 17 Liberec, Czech Republic
zbynek.koldovsky@tul.cz
[2] Institute of Information Theory and Automation, Pod vodárenskou věží 4, P.O. Box 18, 182 08 Praha 8, Czech Republic

Abstract. This paper addresses the estimation of Relative Transfer Function (RTF) between microphones from noisy recordings. We utilize an incomplete initial measurement of the RTF, which is known for only several frequency bins. The measurement is completed by finding its sparsest representation in the time domain. We propose to perform this reconstruction by solving a Second-Order Cone Program (SOCP). Free parameters of this formulation represent distance of the completed RTF from the initial estimate. We select these parameters based on the theoretical performance of the initial estimate. In experiments with real-world data, this approach achieves a significant refinement of the RTF, especially in scenarios with low signal-to-noise ratios.

Keywords: Audio signal processing · Relative transfer function · Compressed sensing · Second-order cone programming · Incomplete measurement

1 Introduction

A noisy recording of a target signal observed through two microphones can be, in the short-term Discrete Fourier Transform (DFT) domain, described as

$$X_{\mathrm{L}}(k,\ell) = H_{\mathrm{L}}(k)S(k,\ell) + Y_{\mathrm{L}}(k,\ell)$$
$$X_{\mathrm{R}}(k,\ell) = H_{\mathrm{R}}(k)S(k,\ell) + Y_{\mathrm{R}}(k,\ell) \tag{1}$$

where k and ℓ denote, respectively, the frequency and the frame index; let the DFT length be M; S denotes the target signal; X_{L} and X_{R} correspond, respectively, to the signals observed on the left and right microphones; Y_{L} and Y_{R} are the remaining signals (noise and interferences) commonly referred to as noise. H_{L} and H_{R} denote the acoustic transfer functions between the microphones and the target, which are assumed to be approximately constant (independent of ℓ) during short intervals.

This work was supported by The Czech Sciences Foundation through Project No. 14-11898S.

E. Vincent et al. (Eds.): LVA/ICA 2015, LNCS 9237, pp. 227–234, 2015.
DOI: 10.1007/978-3-319-22482-4_26

Define the relative transfer function (RTF) as $H_{\mathrm{RTF}}(k) = H_{\mathrm{R}}(k)H_{\mathrm{L}}(k)^{-1}$. Then, (1) can be re-written as

$$
\begin{aligned}
X_{\mathrm{L}}(k,\ell) &= S_{\mathrm{L}}(k,\ell) + Y_{\mathrm{L}}(k,\ell) \\
X_{\mathrm{R}}(k,\ell) &= H_{\mathrm{RTF}}(k)S_{\mathrm{L}}(k,\ell) + Y_{\mathrm{R}}(k,\ell)
\end{aligned}
\tag{2}
$$

where $S_{\mathrm{L}}(k,\ell) = H_{\mathrm{L}}(k)S(k,\ell)$. The time domain counterpart of H_{RTF}, called the relative impulse response (ReIR), will be denoted as h_{rel}.

Knowing H_{RTF} (or h_{rel}) enables to design an efficient spatial filter with two inputs X_{L} and X_{R} such that it cancels the target source and only pass through the noise signals. The output of the spatial filter[1] is $Z(k,\ell) = H(k)X_{\mathrm{L}}(k,\ell) - X_{\mathrm{R}}(k,\ell)$, which is determined by the transfer function H. By (2), it holds that

$$
Z(k,\ell) = \underbrace{\big(H(k) - H_{\mathrm{RTF}}(k)\big)S_{\mathrm{L}}(k,\ell)}_{\text{target signal leakage}} + \underbrace{H(k)Y_{\mathrm{L}}(k,\ell) - Y_{\mathrm{R}}(k,\ell)}_{\text{noise reference}}.
\tag{3}
$$

For $H = H_{\mathrm{RTF}}$ the target signal leakage vanishes, and $Z(k,\ell) = H_{\mathrm{RTF}}(k)Y_{\mathrm{L}}(k,\ell) - Y_{\mathrm{R}}(k,\ell)$. Hence, $Z(k,\ell)$ provides the key noise reference signal, which is important for audio applications such as source separation or speech enhancement.

The signal-to-noise ratio (SNR) in $Z(k,\ell)$ can be used as a practical evaluator of $H(k)$. We will therefore use *attenuation ratio* (ATR), which is the ratio between the initial SNR in (1) and the SNR in $Z(k,\ell)$.

The RTF estimation when noise is active is a challenging problem. During noise-free intervals, conventional time-domain or frequency-domain estimators can be used. The obtained RTF can be used later when noise is active, however, the position of the target must remain the same. To estimate the RTF from noisy data, Shalvi and Weinstein proposed a method assuming model where nonstationary target signal is interfered by a stationary noise [2]. Methods based on Blind Source Separation (BSS) can cope also with directional nonstationary interfering sources [3]. There are also methods based on low-rank models of the RTF that can be learned in noise-free conditions. This class embodies, e.g., an approach utilizing bank of pre-learned RTFs [1] or a model based on diffusive maps [4].

Recently, a possibility to estimate the RTF using its *incomplete measurement* was studied in [5,6]. The incomplete RTF is an RTF estimate whose values are known only for some frequencies. The estimate is completed (reconstructed) through finding its sparsest representation in the time domain. The motivation for the latter step is that typical ReIRs are fast decaying sequences, thus, appear to be compressible (approximately sparse).

In [5], the reconstruction is done through solving a weighted LASSO optimization problem. However, the optimum choice of weights is highly nontrivial, so only a heuristic choice is proposed. In this paper, we propose to use a different formulation based on second-order cone programming. Parameters of this

---

[1] The right channel X_{R} as well as H are typically delayed by a few samples due to possible acausality of H_{RTF}. We omit this detail here for the sake of simplicity of the notation.

formulation have clear meaning: Each parameter limits the distance of the reconstructed RTF value from its initial estimate.

2 Relative Transfer Function Estimators

Conventional Frequency-Domain Estimator. It follows from (2) that during intervals where noise signals are not active ($Y_L = Y_R = 0$), it is possible to estimate the RTF as

$$\widehat{H}_{FD}(k) = \frac{\hat{\Phi}_{X_R X_L}(k)}{\hat{\Phi}_{X_L X_L}(k)}. \qquad (4)$$

$\hat{\Phi}_{AB}$ denotes the sample-based estimate of (cross-)power spectral density between A and B. When signals are contaminated by noise, the estimator becomes biased where the bias (as well as its variance) depends on noise characteristics. This estimator will be abbreviated by FD (Frequency Domain estimator).

The authors of [2] considered the model where the target signal is wide-sense stationary (WSS) during short intervals (subintervals) but nonstationary over longer segments (piecewise stationary). The assumption about the noise is such that $V(k, \ell) = Y_R(k, \ell) - H_{RTF}(k)Y_L(k, \ell)$ is WSS. Under this model, it was derived that the bias[2] of FD is

$$\mathrm{E}[\widehat{H}_{FD}(k)] - H_{RTF}(k) = \frac{\Phi_{V X_L}(k)}{\langle \Phi^p_{X_L X_L}(k) \rangle} \qquad (5)$$

where $\mathrm{E}[\cdot]$ stands for the expectation operator, and $\langle \cdot \rangle$ denotes the average of the argument over the subintervals indexed by the superscript p, $p = 1, \ldots, P$. Note that the model assumes that $\Phi_{V X_L}(k)$ is independent of p. To estimate the bias, the cross-spectral densities on the right-hand side of (5) can be replaced by their sample-based estimates; $V(k, \ell)$ can be estimated as $-Z(k, \ell)$ from (3).

Estimator Admitting Presence of Stationary Noise. An estimator that is unbiased under the validity of the above model can be computed as the least-square solution of the following overdetermined set of equations [2]

$$\begin{bmatrix} \hat{\Phi}^1_{X_R X_L}(k) \\ \vdots \\ \hat{\Phi}^P_{X_R X_L}(k) \end{bmatrix} = \begin{bmatrix} \hat{\Phi}^1_{X_L X_L}(k) & 1 \\ \vdots & \vdots \\ \hat{\Phi}^P_{X_L X_L}(k) & 1 \end{bmatrix} \begin{bmatrix} \widehat{H}_{NSFD}(k) \\ \hat{\Phi}_{V X_L}(k) \end{bmatrix}. \qquad (6)$$

The variance of this estimator, from here referred to as NSFD (Non-Stationarity based Frequency Domain estimator), is equal to

$$\mathrm{var}[\widehat{H}_{NSFD}(k)] = \frac{1}{N} \frac{\Phi_{VV}(k)\langle 1/\Phi^p_{X_L X_L}(k)\rangle}{\langle \Phi^p_{X_L X_L}(k)\rangle \langle 1/\Phi^p_{X_L X_L}(k)\rangle - 1}. \qquad (7)$$

---

[2] The variance of FD under the model is also derived in [2] and could be taken into account. The bias, however, seems to have a larger influence on the entire accuracy of FD; we therefore focus on the bias.

Note that $\langle\Phi^p_{X_L X_L}(k)\rangle\langle 1/\Phi^p_{X_L X_L}(k)\rangle$ is close to 1 when $\Phi^p_{X_L X_L}(k)$ does not depend much on p, which happens when the target signal is almost stationary (as well as the noise). By contrast, $\langle\Phi^p_{X_L X_L}(k)\rangle\langle 1/\Phi^p_{X_L X_L}(k)\rangle \gg 1$ when the signal is sufficiently dynamical. Therefore, NSFD is suitable in situations when the target signal is speech and the noise is (approximately) stationary.

It is worth to point here to a problem that is unintentionally hidden in the analysis. Speech signals are sparse in the time-frequency domain. It thus often happens that $S_L(k,\ell) = 0$ for some k, which means that $H_{RTF}(k)$ vanishes from the model (2). The behavior of NSFD then depends on the character of the (stationary) noise source. If the noise is diffused, the variance (7) approaches infinity, so we are aware of the inaccuracy of the estimate for the given frequency. However, if the noise comes from a spatial source, NSFD yields an estimate of the RTF which is related to the noise source, not to the target source.

It is important to avoid the latter case. Otherwise a large error is introduced into the estimator although (7) need not signalize it. If this case is detected through some additional hypothesis (e.g., by means of a voice-activity detector), the RTF estimate for the given k can be dropped and replaced using the method proposed in this paper. In experiments, we will focus on the described situation by considering speech as the target signal interfered by a directional quasi-stationary noise.

There are many other RTF estimators that can be taken into account in the following considerations; see, e.g., [7,8]. Nevertheless, we will constrain our focus on the estimators FD a NSFD in this paper.

3   Sparse Reconstruction of Incomplete RTF

As already mentioned, an incomplete RTF is obtained by taking values of an RTF estimate but only for those frequencies where the estimate appears to be accurate enough. Let the set of the accepted frequencies $\{i_1, \ldots, i_{|S|}\}$ be denoted by S; we can constrain $|S| \leq \lceil M/2 + 1 \rceil$ due to the symmetry of the DFT and due to the fact that the ReIR is real-valued.

The method in [5] aims to find the sparsest representation of the incomplete RTF in the time domain using weighted LASSO. The reconstructed ReIR is sought as the solution of

$$\mathbf{h}_{WLASSO} = \arg\min_{\mathbf{h}} \|\mathbf{F}_S\mathbf{h} - \mathbf{f}\|^2_2 + \|\mathbf{w} \odot \mathbf{h}\|_1, \tag{8}$$

where $\mathbf{f}$ is a $|S| \times 1$ vector with elements $f_k = \widehat{H}(i_k)$, $i_k \in S$, $k = 1, \ldots, |S|$; $\mathbf{F}$ is the $M \times M$ matrix of the DFT and $\mathbf{F}_S$ denotes its submatrix comprised of rows whose indices are in S; $\mathbf{h}$ denotes an $M \times 1$ vector of coefficients of the estimate of h_{rel}; $\mathbf{w}$ is an $M \times 1$ vector of nonnegative weights; $\odot$ denotes the element-wise product.

The weights control the sparsity level of the solution. They can incorporate a priori knowledge, because elements of $\mathbf{h}_{WLASSO}$ with higher weights tend to be closer to or equal to zero and vice versa. A heuristic selection respecting the expected shape of h_{rel} was proposed in [5]; similar idea is used in [10].

A drawback of (8) is that the influence of the weights on the quality of the reconstructed RTF (ReIR) is not clear. In this paper, we therefore consider a different formulation where the reconstructed ReIR is defined as the solution of

$$\mathbf{h}_{\text{SOCP}} = \arg\min_{\mathbf{h}} \|\mathbf{h}\|_1 \quad \text{w.r.t.} \quad |(\mathbf{F}_{\mathcal{S}}\mathbf{h} - \mathbf{f})_{i_k}| \leq \epsilon_{i_k}, \quad \forall i_k \in \mathcal{S}, \quad (9)$$

which is a second-order cone program (SOCP). In this formulation, the distance of the i_kth element of the reconstructed RTF from $\widehat{H}(i_k)$ is constrained to be less or equal to ϵ_{i_k} (in absolute value). For example, it is reasonable to choose ϵ_{i_k} proportional to a theoretical bias or variance of the estimate $\widehat{H}(i_k)$.

Practical Implementation. Assume a stereo noisy recording obeying (2) is given. Let M be the length of DFT, which corresponds to the length of the to be estimated ReIR; for simplicity, let M be even. The proposed estimation procedure consists of four steps.

1. Compute the initial RTF estimate $\widehat{H}(k)$, $k = 0, \ldots, M/2 + 1$, using some known method. In this paper, we will consider FD given by (4) and NSFD computed through (6).
2. Compute a theoretical estimation error of $\widehat{H}(k)$, $k = 0, \ldots, M - 1$, denoted as δ_k. Here, we compute the theoretical bias (5) in case of FD and the theoretical variance (7) in case of NSFD.
3. Select $\mathcal{S}$. In this paper, we select p percents of frequency bins that yield the highest SNR (oracle selection) or the highest normalized kurtosis (kurtosis-based selection)[3]. The parameter p will be referred to as *percentage*.
4. Solve the SOCP given by (9) where $\epsilon_{i_k} = \alpha\delta_{i_k}$, $i_k \in \mathcal{S}$, using the ECOS package [9] to obtain the reconstructed ReIR; its DFT gives the reconstructed RTF; α is a free positive constant (we select $\alpha = 1$ in case of NSFD and $\alpha = 0.2$ in case of FD).

4   Experiments

In experiments, the above proposed procedures to estimate the RTF from noisy recordings are verified on real-world audio signal mixtures. A female utterance from SiSEC 2013[4] from the task "Two-channel mixtures of speech and real-world background noise" is used as the target signal. The signal has 10 s in length; the sampling frequency is 16 kHz.

The noise signal is a fan hum "FanRear.wav" by user Otakua taken from the repository of free audio samples[5]. Note that this signal is approximately

---

[3] The kurtosis-based selection appears to be efficient when the frequency components of the target signal have non-Gaussian distribution while those of the noise are Gaussian; see Sect. 5 in [5]. In real-world situations, this is often satisfied when the target signal is speech and the noise is quasi-stationary.

[4] http://sisec.wiki.irisa.fr/.

[5] http://www.freesound.org/.

stationary (as assumed by NSFD) as well as directional (spatial source). The densities of its spectral components are close to Gaussian, hence their kurtosis is close to zero. By contrast, the kurtosis of active spectral components of speech is often positive. This enables to utilize the kurtosis as a contrast to select S.

To simulate spatial sources, the target and noise signal are convolved with room impulse responses from [11][6]. The reverberation time T_{60} is 160 ms; the distance of the microphones is 3 cm; the source-microphone distance is 2 m. The target and noise source is located, respectively, in the direction of 0° and 75° on the left-hand side.

The spatial images of the signals are mixed together at a specified input SNR (averaged over both microphones). The mixed signal is divided into 1s-intervals with 75 % overlap, and the RTF estimation is conducted independently on each of total 37 intervals. The results are then given in the form of ATR averaged over all intervals.

Results and Discussions. Figures 1 and 2 summarize results of the experiments with NSFD and FD, respectively. The figures show the average ATR of the estimated RTFs as functions of percentage p and of input SNR.

For both initial estimates, the RTFs reconstructed by SOCP yield ATR that is comparable or higher to that obtained by weighted LASSO. This holds for both the oracle and kurtosis-based selection of S. The improvements compared to LASSO are achieved for low values of p (below $10-20\%$) and for scenarios with the lower input SNR (see Figs. 1(a,c) and 2(a)). When input SNR is -10 dB, only few frequency components of the initial RTF estimate are accurate "enough", so small p should be selected. Then, SOCP appears to be more robust than LASSO. For higher values of p and higher input SNR, both approaches achieve comparable ATRs (see Figs. 1(b,d) and 2(b)).

The overall ATRs with FD are significantly lower than those with NSFD; cf. Figs. 1(a) and 2(a). This confirms the assumption that NSFD yields more accurate RTF estimate, when target speech is interfered by a stationary noise.

Figures 1(c,d) and 2(b) show that the reconstructed RTFs from incomplete measurements yield significant improvement in terms of ATR, especially, when input SNR is low. When input SNR is sufficiently high, the ATR by the reconstructed RTFs can be lower than that of the initial estimates, depending on p. When p is too low, the loss of the ATR signalizes that too much information was lost from the incomplete RTF (see Fig. 1(c) where input SNR> 0). By contrast, when the RTF measurement is almost complete (p close to 100 %), the ATRs by the reconstructed RTFs are getting closer to those of the initial estimators.

The oracle selection yields higher ATR compared to kurtosis-based selection of $\widehat{H}(k)$ for all values of p and all considered input SNR levels (by up to 3 dB). This is due to the strong prior knowledge utilized by the oracle selection (the true input SNR within the frequency bins).

---

[6] http://www.eng.biu.ac.il/gannot/downloads/.

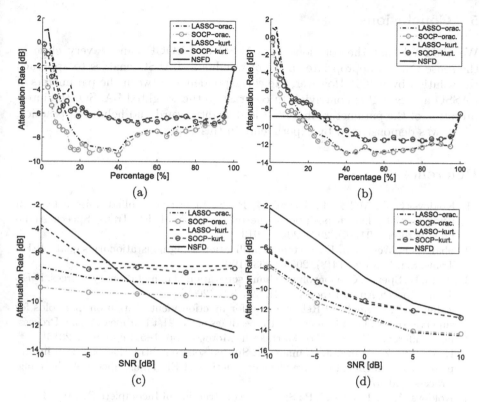

Fig. 1. Attenuation of female speech in the presence of directional fan hum. The initial estimate $\widehat{H}(k)$ is NSFD (6). (a,b) Dependence on the percentage of frequency bins included in $\mathcal{S}$ (input SNR = -10 dB or 0 dB, respectively), (c,d) dependence on input SNR ($p = 15\%$ or $p = 65\%$, respectively). The more negative the value (in dBs) of ATR is, the better the target signal blocking.

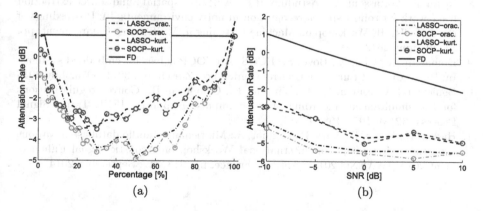

Fig. 2. Attenuation of female speech in the presence of directional fan hum. The initial estimate $\widehat{H}(k)$ is FD (4). (a) Dependence on the percentage of frequency bins included in $\mathcal{S}$ (input SNR = -10 dB), (b) the dependance on input SNR ($p = 45\%$).

5   Conclusions

We observed that the solutions by LASSO and SOCP can be very close in the sense that an appropriate choice of weights in LASSO enables to approach the solution by SOCP. However, the correspondence between the parameters of LASSO and SOCP is nontrivial. In contrast to the weighted LASSO, the interpretation of the parameters in SOCP is straightforward and helpful in practice, which was demonstrated by experiments in this paper.

References

1. Koldovský, Z., Málek, J., Tichavský, P., Nesta, F.: Semi-blind noise extraction using partially known position of the target source. IEEE Trans. Speech Audio Lang. Process. **21**(10), 2029–2041 (2013)
2. Shalvi, O., Weinstein, E.: System identification using nonstationary signals. IEEE Trans. Sig. Process. **44**(8), 2055–2063 (1996)
3. Parra, L., Spence, C.: Convolutive blind separation of non-stationary sources. IEEE Trans. Speech Audio Process. **8**(3), 320–327 (2000)
4. Talmon, R., Gannot, S.: Relative transfer function identification on manifolds for supervised GSC beamformers. In: Proceedings of the 21st European Signal Processing Conference (EUSIPCO), Marrakech, Morocco, pp. 1–5, September 2013
5. Koldovský, Z., Málek, J., Gannot, S.: Spatial source subtraction based on incomplete measurements of relative transfer function. IEEE Trans. Speech Audio Lang. Process. (2015)
6. Koldovský, Z., Tichavský, P.: Sparse Reconstruction of Incomplete Relative Transfer Function: Discrete and Continuous Time Domain. In: Accepted for a Special Session at EUSIPCO 2015, Nice, France, September 2015
7. Takahashi, Y., Takatani, T., Osako, K., Saruwatari, H., Shikano, K.: Blind spatial subtraction array for speech enhancement in noisy environment. IEEE Trans. Audio Speech Lang. Process. **17**(4), 650–664 (2009)
8. Nesta, F., Matassoni, M., Astudillo, R.F.: A flexible spatial blind source extraction framework for robust speech recognition in noisy environments. In: Proceedings of the 2nd CHiME Workshop on Machine Listening in Multisource Environment, pp. 33–40, June 2013
9. Domahidi, A., Chu, E., Boyd, S.: ECOS: An SOCP solver for embedded systems. In: Proceedings of European Control Conference, Zurich, pp. 3071–3076, July 2013
10. Benichoux, A., Simon, L.S.R., Vincent, E., Gribonval, R.: Convex regularizations for the simultaneous recording of room impulse responses. IEEE Trans. Signal Process. **62**(8), 1976–1986 (2014)
11. Hadad, E., Heese, F., Vary, P., Gannot, S.: Multichannel audio database in various acoustic environments. In: International Workshop on Acoustic Signal Enhancement 2014 (IWAENC 2014), Antibes, France, pp. 313–317, September 2014

System Identification in the Behavioral Setting
A Structured Low-Rank Approximation Approach

Ivan Markovsky[✉]

Department ELEC, Vrije Universiteit Brussel (VUB),
Pleinlaan 2, Building K, 1050 Brussels, Belgium
imarkovs@vub.ac.be

Abstract. System identification is a fast growing research area that encompasses a broad range of problems and solution methods. It is desirable to have a unifying setting and a few common principles that are sufficient to understand the currently existing identification methods. The behavioral approach to system and control, put forward in the mid 80's, is such a unifying setting. Till recently, however, the behavioral approach lacked supporting numerical solution methods. In the last 10 years, the structured low-rank approximation setting was used to fulfill this gap. In this paper, we summarize recent progress on methods for system identification in the behavioral setting and pose some open problems. First, we show that errors-in-variables and output error system identification problems are equivalent to Hankel structured low-rank approximation. Then, we outline three generic solution approaches: (1) methods based on local optimization, (2) methods based on convex relaxations, and (3) subspace methods. A specific example of a subspace identification method—data-driven impulse response computation—is presented in full details. In order to achieve the desired unification, the classical ARMAX identification problem should also be formulated as a structured low-rank approximation problem. This is an outstanding open problem.

Keywords: System identification · Errors-in-variables modeling · Behavioral approach · Hankel matrix · Low-rank approximation · Impulse response estimation · ARMAX identification

1 Introduction

System identification aims at deriving a dynamical model $\widehat{\mathscr{B}}$ (*i.e.*, a mathematical description) from observed data $\mathscr{D}$ of a to-be-modeled physical plant. The data is typically obtained by sampling and quantization in time-domain from one or more independent measurement experiments. Each measurement point is a real-valued vector of the observed variables from the system and the model postulates a relation among the variables.

Prior knowledge and/or assumptions about the plant are incorporated in the identification problem by restricting the model to belong to a set of models $\mathscr{M}$, called the *model class*. Therefore, a system identification problem is a mapping:

© Springer International Publishing Switzerland 2015
E. Vincent et al. (Eds.): LVA/ICA 2015, LNCS 9237, pp. 235–242, 2015.
DOI: 10.1007/978-3-319-22482-4_27

$$\text{data} \xrightarrow[]{\substack{\text{system} \\ \text{identification}}} \text{model} \tag{ID}$$
$$\mathscr{D} \qquad\qquad\qquad \widehat{\mathscr{B}} \in \mathscr{M}$$

The mapping (ID) is defined implicitly as a solution to an optimization problem, *i.e.*, the model $\widehat{\mathscr{B}}$ minimizes (among all feasible models) a specified cost function. Different identification problems correspond to different choices of a model class and the cost function.

Two contradictory objectives in system identification are:

1. "simple" model,
2. "good" fit of the data by the model.

Typically the model class is used to impose a hard bound on the model complexity and the cost function is used to measure the model-data misfit (lack of fit). It is possible, however, to minimize the model complexity subject to a hard bound on the misfit or, more generally, consider the bi-objective minimization of the complexity and the misfit.

In exact identification [11, Chap. 7], the model complexity is minimized subject to the constraint that the model fits the data exactly (zero misfit). If such a model exists in the model class, it is called the *most powerful unfalsified model* (of $\mathscr{D}$ in $\mathscr{M}$) [14]. Exact identification is a theoretical tool which is a generalization of the realization problem in system theory and appears in approximate and stochastic identification problems [7].

The data collected in a real-life experiment is "inexact" due to disturbances (unobserved variables), measurement noises, discretization, and quantization errors. Methods for computing the most powerful unfalsified model, however, lead through simple modifications to a class of practical identification methods known as subspace methods.

In this paper, we consider the model class of linear time-invariant systems of bounded complexity $\mathscr{L}_{\mathtt{m},\ell}$ — number of input variables at most $\mathtt{m}$ and lag at most ℓ. In the behavioral setting [15], no a priori separation of the variables into inputs and outputs is made, however any model allows a nonunique input/output partition. Although the choice of the input variables is in general not unique, the number of inputs is a model invariant, *i.e.*, it does not depend on the partitioning.

In Sect. 2, we define the model class $\mathscr{L}_{\mathtt{m},\ell}$ and the approximation criterion, which specify the identification problem (ID). The misfit has the geometric interpretation of the Euclidean distance between the data and the model. In the stochastic setting, misfit minimization corresponds to errors-invariables system identification [13], *i.e.*, (ID) is a maximum-likelihood estimator in the errors-invariables setting. Section 3 related the identification problem (ID) to the weighted Hankel structured low-rank approximation. Three generic classes of solution methods are outlined: local optimization based methods, convex relaxation based methods, and subspace methods. As a specific example of a subspace method, in Appendix A, we present a data-driven algorithm for impulse response estimation. Section 4 draws conclusions and states some open problems. One of them is integration of the classical ARMAX setting in the behavioral setting.

2   Problem Formulation

A dynamical system $\mathscr{B}$ is a set of trajectories. In discrete-time, a trajectory is a q-variable time-series $w : \mathbb{Z} \to \mathbb{R}^q$. The class of finite dimensional linear time-invariant systems with at most $\mathtt{m}$ inputs is denoted by $\mathscr{L}_\mathtt{m}$. This class admits a representation

$$\mathscr{B} = \mathscr{B}(R) := \{\, w \mid R_0 w + R_1 \sigma w + \cdots + R_\ell \sigma^\ell w = 0 \,\}, \qquad \text{(DE)}$$

where σ is the shift operator $(\sigma w)(t) = w(t+1)$. The smallest number ℓ, for which there is ℓth order representation $\mathscr{B} = \mathscr{B}(R)$ is called the *lag* of the system. The pair $(\mathtt{m}, \ell)$ specify the model complexity. The model class of bounded complexity is denoted by $\mathscr{L}_{\mathtt{m},\ell}$.

The model variables w can be partitioned into inputs u and outputs y, *i.e.*, there is a permutation matrix $\varPi$, such that $w = \varPi \left[\begin{smallmatrix} u \\ y \end{smallmatrix}\right]$. The system can then be represented in the classical form

$$\mathscr{B} = \mathscr{B}(A, B, C, D, \varPi) := \{\, w = \varPi \left[\begin{smallmatrix} u \\ y \end{smallmatrix}\right] \mid \text{there is } x,$$
$$\text{such that } \sigma x = Ax + Bu, \; y = Cx + Du \,\}.$$
$$\text{(I/S/O)}$$

We will assume that $\varPi$ can be chosen equal to I and the block $P_\ell \in \mathbb{R}^{\mathtt{p} \times \mathtt{p}}$ of $R_\ell = \begin{bmatrix} Q_\ell & -P_\ell \end{bmatrix}$ in a difference equation representation is nonsingular.

Let the identification data $\mathscr{D}$ be an observed trajectory

$$w_\mathrm{d} = \big(w_\mathrm{d}(1), \ldots, w_\mathrm{d}(T)\big), \quad w_\mathrm{d}(t) \in \mathbb{R}^q$$

of the to-be-identified system. The approximation criterion, called data-model misfit, is defined as follows:

$$M(\mathscr{D}, \mathscr{B}) := \min_{\widehat{w} \in \mathscr{B}} \|w_\mathrm{d} - \widehat{w}\|_2,$$

where $\widehat{w}$ is the optimal approximation of w_d by $\mathscr{B}$. Note that $\widehat{w}$ is the projection of w_d on $\mathscr{B}$.

The identification problem considered is misfit minimization over all system $\widehat{\mathscr{B}}$ in the model class $\mathscr{L}_{\mathtt{m},\ell}$:

$$\text{minimize} \quad \text{over } \mathscr{B} \in \mathscr{L}_{\mathtt{m},\ell} \quad M(w_\mathrm{d}, \mathscr{B}). \qquad \text{(SYSID)}$$

Generalizations of problem (SYSID) (see [8]) are weighted 2-norm approximation criteria, specification of exact and missing variables, and data consisting on multiple trajectories.

3   Hankel Low-Rank Approximation

In what follows, we will use the block-Hankel matrix

$$\mathscr{H}_{\ell+1}(w) := \begin{bmatrix} w(1) & w(2) & \cdots & w(T-\ell) \\ w(2) & w(3) & \cdots & w(T-\ell+1) \\ \vdots & \vdots & & \vdots \\ w(\ell+1) & w(\ell+2) & \cdots & w(T) \end{bmatrix}.$$

The fundamental link between the system identification problem (SYSID) and structured low-rank approximation is the following equivalence

$$w \in \mathscr{B} \in \mathscr{L}_{\mathrm{m},\ell} \quad \Longleftrightarrow \quad \operatorname{rank}\left(\mathscr{H}_{\ell+1}(w)\right) \leq (\ell+1)\mathrm{m} + \mathrm{p}\ell. \qquad (*)$$

In words, w is an (exact) trajectory of the linear time-invariant system $\mathscr{B}$ if and only if the Hankel structured matrix $\mathscr{H}_{\ell+1}(w_{\mathrm{d}})$ is rank deficient. Note that the complexity of the model $\mathscr{B}$ (number of inputs m and lag ℓ) is directly related to the rank constraint of the Hankel matrix.

Using (*), we can rewrite the identification problem (SYSID) as an equivalent Hankel low-rank approximation problem

$$\begin{aligned} \text{minimize} \quad &\text{over } \widehat{w} \quad \|w_{\mathrm{d}} - \widehat{w}\|^2 \\ \text{subject to} \quad &\operatorname{rank}\left(\mathscr{H}_{\ell+1}(\widehat{w})\right) \leq r, \end{aligned} \qquad \text{(SLRA)}$$

where $r := (\ell+1)\mathrm{m} + \mathrm{p}\ell$. The main issue in system identification is that Problem (SLRA) is nonconvex. Therefore, various heuristics, reviewed later, are used for its solution.

3.1 Variable Projections Approach

One approach for dealing with the rank constraint in (SLRA) is to use the kernel representation

$$\operatorname{rank}\left(\mathscr{H}_{\ell+1}(\widehat{w})\right) \leq r \iff R\mathscr{H}_{\ell+1}(\widehat{w}) = 0$$
$$\text{and} \quad R \in \mathbb{R}^{\mathrm{p} \times (\ell+1)q} \text{ is full row rank. (KER)}$$

Using (KER), (SLRA) becomes a classical parameter optimization problem,

$$\begin{aligned} \text{minimize} \quad &\text{over } \widehat{w} \text{ and } R \quad \|w_{\mathrm{d}} - \widehat{w}\|^2 \\ \text{subject to} \quad &R\mathscr{H}_{\ell+1}(\widehat{w}) = 0 \quad \text{and} \quad R \text{ is f.r.r.} \end{aligned} \qquad (\text{SLRA}_R)$$

(SLRA_R) is furthermore equivalent to

$$\text{minimize} \quad \text{over f.r.r. } R \in \mathbb{R}^{(m-r) \times m} \quad M(R), \qquad \text{(OUTER)}$$

where

$$M(R) := \min_{\widehat{w}} \|w_{\mathrm{d}} - \widehat{w}\|^2 \qquad \text{(INNER)}$$
$$\text{subject to} \quad R\mathscr{H}_{\ell+1}(\widehat{w}) = 0.$$

Note that (INNER) is a classical linear least squares problem.

The approach for solving (SLRA_R) by minimizing (OUTER) is closely related to the variable projection method in numerical linear algebra [4]. In [4], however, an explicit function $\widehat{b} = A(\theta)x$, where x is unconstrained, is considered, while in the context of the structured low-rank approximation problem, an implicit relation $R\mathscr{H}_{\ell+1}(\widehat{w}) = 0$ is considered, where the variable R is constrained to have full row rank. This fact requires new type of algorithms where the nonlinear least squares problem is an optimization problem on a Grassmann manifold, see [1,2].

In (OUTER), the cost function M is minimized over the set of full row rank matrices R. Indeed, M depends only on the space spanned by the rows of R. In order to find a minimum of M, the search space in (OUTER) can be replaced by the matrices satisfying the constraint

$$RR^\top = I_\mathrm{p}.$$

A software package for Hankel structured low-rank approximation is presented in [9]. The Levenberg-Marquardt algorithm [12] implemented the GNU Scientific Library [3], are used for the solution of the nonlinear least squares problem. This structured low-rank approximation package is used in [8] for system identification. The software as well as simulation examples demonstrating its usage and comparing it with alternative methods are available from:

http://slra.github.io/software-ident.html.

3.2 Alternating Projections Approach

The second approach is based on the image representation of the rank constraint

$$\mathrm{rank}\left(\mathscr{H}_{\ell+1}(\widehat{w})\right) \le r \quad \Longleftrightarrow \quad \mathscr{H}_{\ell+1}(\widehat{w}) = PL \quad \text{where}$$
$$P \in \mathbb{R}^{\bullet \times r} \text{ and } L \in \mathbb{R}^{r \times \bullet}$$

and a representation of a structured matrix by an linear equality constraint

$$\Pi(PL) - PL = 0,$$

where Π is a projection of a matrix to the nearest one with Hankel structure.

3.3 Nuclear Norm Heuristic

A convex relaxation of (SLRA) is obtained by replacing the rank constraint with a constraint $\|\mathscr{H}_{\ell+1}(\widehat{w})\|_* \le \gamma$, where $\|\cdot\|_*$ is the nuclear norm, *i.e.*, the sum of the singular values. The parameter γ is selected in [6] by bisection aiming to achieve the desired rank of the approximation $\mathscr{H}_{\ell+1}(\widehat{w})$. Other authors, see, *e.g.*, [5] select a value of γ that does not necessarily lead to rank deficient matrix. In this case, the nuclear norm minimization is used as a preprocessing operation. The sequence $\widehat{w}$ completed by the nuclear norm minimization is then given as an input to a subspace method, which does the rank reduction.

3.4 Subspace Methods

The subspace methods for approximate system identification originate from corresponding methods for exact system identification, by replacing exact operations such as rank revealing factorization and solution of a system of linear equations by approximate methods — unstructured low-rank approximation (achieved via the singular value decomposition) and approximation solution of a system of linear equations in the least squares sense. Since two or more steps of the algorithm are adapted in this way, the result heuristic methods can be called multi-stage methods. They are suboptimal, however, they are fast and effective methods for approximate system identification. A detailed specific example of a subspace method is shown in Appendix A.

4   Conclusions and Future Perspectives

In this paper, we described a unifying setting for system identification as a biobjective optimization problem. The identified model is defined in the behavioral sense as a set of trajectories. The two objectives are (1) minimization of the fitting error and (2) minimization of the model complexity. As a specific example of a fitting error, we gave the misfit, *i.e.*, the projection of the data on the model. This error criterion corresponds to the class of the errors-in-variables problems in the system identification literature. Another error criterion is the latency, which corresponds to the class of the ARMAX identification problems.

The main computation tool in the behavioral setting is Hankel structured low-rank approximation. The link to low-rank approximation follows from the fact that a time series is a trajectory of a linear time invariant system if and only if a Hankel structured matrix composed of the data is rank deficient. Once the identification problem is re-formulated as a structured low-rank approximation problem, it can be solved by various methods. The methods however are classified into three groups: local optimization based methods, convex relaxations, and subspace methods. In general, the subspace methods are faster but less efficient than the optimization based methods.

The class of methods based on convex relaxations are currently actively developed. The main challenges in this area of research are finding theoretical bounds for the distance to global optimality and development of efficient computational methods.

Another research challenge is formulation of the latency minimization (ARMAX system identification) as a structured low-rank approximation and solution of the resulting problem by existing methods for low-rank approximation.

Acknowledgements. The research leading to these results has received funding from the European Research Council under the European Union's Seventh Framework Programme (FP7/2007-2013)/ERC Grant agreement number 258581 "Structured low-rank approximation: Theory, algorithms, and applications".

A   Subspace Method for Impulse Response Estimation

Let $\mathscr{B}$ be a linear time-invariant system of order n with lag ℓ and let $w = (u, y)$ be an input-output partitioning of the variables. In [16], it is shown that, under the following conditions,

- the data w_d is exact, *i.e.*, $w_d \in \mathscr{B}$,
- $\mathscr{B}$ is controllable,
- u_d is persistently exciting, *i.e.*, $\mathscr{H}_{n+\ell+1}(u_d)$ is full rank,

the Hankel matrix $\mathscr{H}_t(w_d)$ with t block-rows, composed from w_d, spans the space $\mathscr{B}|_{[1,t]}$ of all t-samples long trajectories of the system $\mathscr{B}$, *i.e.*,

$$\text{image}\left(\mathscr{H}_t(w_d)\right) = \mathscr{B}|_{[1,t]}.$$

This implies that there exists a matrix G, such that

$$\mathscr{H}_t(y_d)G = H,$$

where H is the vector of the first t samples of the impulse response of $\mathscr{B}$. The problem of computing the impulse response H from the data w_d reduces to the one of finding a particular G.

Define U_p, U_f, Y_p, Y_f as follows

$$\mathscr{H}_{\ell+t}(u_d) =: \begin{bmatrix} U_p \\ U_f \end{bmatrix}, \qquad \mathscr{H}_{\ell+t}(y_d) =: \begin{bmatrix} Y_p \\ Y_f \end{bmatrix},$$

where

$$\text{row dim}(U_p) = \text{row dim}(Y_p) = \ell$$

and

$$\text{row dim}(U_f) = \text{row dim}(Y_f) = t.$$

Then if $w_d = (u_d, y_d)$ is a trajectory of a controllable linear time-invariant system $\mathscr{B}$ of order $\mathbf{n}$ and lag ℓ and if u_d is persistently exciting of order $t + \ell + \mathbf{n}$, the system of equations

$$\begin{bmatrix} U_p \\ U_f \\ Y_p \end{bmatrix} G = \begin{bmatrix} \begin{bmatrix} 0_{m\ell \times m} \\ I_m \\ 0_{m(t-1) \times m} \end{bmatrix} \\ 0_{p\ell \times m} \end{bmatrix}, \qquad (**)$$

is solvable for $G \in \mathbb{R}^{\bullet \times m}$, and for any particular solution G, the matrix $Y_f G$ contains the first t samples of the impulse response of $\mathscr{B}$, *i.e.*,

$$Y_f G = H.$$

This gives Algorithm 1 for the computation of H.

Algorithm 1 computes the first t samples of the impulse response; however, the persistency of excitation condition imposes a limitation on how big t can be. This limitation can be avoided by a modification of the algorithm. L consecutive samples, where L is a user specified parameter that is small enough to allow the application of Algorithm 1, are computed iteratively. Then, provided the system is stable, by monitoring the decay of H in the course of the computations, gives a way to determine how many samples are needed to capture the transient behavior of the system.

Algorithm 1. Block computation of the impulse response from data.

Require: u_d, y_d, ℓ, and t.
1: Solve the system of equations (**) and let G be the computed solution.
2: Compute $H = Y_f G$.
Ensure: H.

In case of noisy data, the system of Eq. (**) on step 1 in Algorithm 1 has no exact solution. Using a least squares approximate solution instead, turns Algorithm 1 in a heuristic for approximate system identification. The algorithm is heuristic because the maximum likelihood estimator requires structured total least squares solution of (**). The structured total least squares problem, however, is nonlinear optimization problem [10].

References

1. Absil, P.A., Mahony, R., Sepulchre, R.: Optimization Algorithms on Matrix Manifolds. Princeton University Press, Princeton (2008)
2. Absil, P.A., Mahony, R., Sepulchre, R., Dooren, P.V.: A Grassmann-Rayleigh quotient iteration for computing invariant subspaces. SIAM Rev. **44**(1), 57–73 (2002)
3. Galassi, M., et al.: GNU scientific library reference manual. http://www.gnu.org/software/gsl/
4. Golub, G., Pereyra, V.: Separable nonlinear least squares: the variable projection method and its applications. Inst. Phys. Inverse Prob. **19**, 1–26 (2003)
5. Liu, Z., Hansson, A., Vandenberghe, L.: Nuclear norm system identification with missing inputs and outputs. Control Lett. **62**, 605–612 (2013)
6. Markovsky, I.: How effective is the nuclear norm heuristic in solving data approximation problems? In: Proceedings of the 16th IFAC Symposium on System Identification, Brussels, pp. 316–321 (2012)
7. Markovsky, I.: Low Rank Approximation: Algorithms, Implementation, Applications. Springer, London (2012)
8. Markovsky, I.: A software package for system identification in the behavioral setting. Control Eng. Prac. **21**, 1422–1436 (2013)
9. Markovsky, I., Usevich, K.: Software for weighted structured low-rank approximation. J. Comput. Appl. Math. **256**, 278–292 (2014)
10. Markovsky, I., Van Huffel, S., Pintelon, R.: Block-Toeplitz/Hankel structured total least squares. SIAM J. Matrix Anal. Appl. **26**(4), 1083–1099 (2005)
11. Markovsky, I., Willems, J.C., Van Huffel, S., De Moor, B.: Exact and Approximate Modeling of Linear Systems: A Behavioral Approach. Monographs on Mathematical Modeling and Computation, vol. 11. SIAM, Bangkok (2006)
12. Marquardt, D.: An algorithm for least-squares estimation of nonlinear parameters. SIAM J. Appl. Math. **11**, 431–441 (1963)
13. Söderström, T.: Errors-in-variables methods in system identification. Automatica **43**, 939–958 (2007)
14. Willems, J.C.: From time series to linear system–part II. Exact modelling. Automatica **22**(6), 675–694 (1986)
15. Willems, J.C.: From time series to linear system–part I. Finite dimensional linear time invariant systems, part II. Exact modelling, part III. Approximate modelling. Automatica **22**, **23**, 561–580, 675–694, 87–115 (1986, 1987)
16. Willems, J.C., Rapisarda, P., Markovsky, I., Moor, B.D.: A note on persistency of excitation. Control Lett. **54**(4), 325–329 (2005)

Sparsity and Cosparsity for Audio Declipping: A Flexible Non-convex Approach

Srđan Kitić[✉], Nancy Bertin, and Rémi Gribonval

Inria/IRISA, Panama Team, Rennes, France
{srdjan.kitic,remi.gribonval}@inria.fr, nancy.bertin@irisa.fr
https://team.inria.fr/panama

Abstract. This work investigates the empirical performance of the sparse synthesis versus sparse analysis regularization for the ill-posed inverse problem of audio declipping. We develop a versatile non-convex heuristics which can be readily used with both data models. Based on this algorithm, we report that, in most cases, the two models perform almost similarly in terms of signal enhancement. However, the analysis version is shown to be amenable for real time audio processing, when certain analysis operators are considered. Both versions outperform state-of-the-art methods in the field, especially for the severely saturated signals.

Keywords: Clipping · Audio · Sparse · Cosparse · Non-convex · Real-time

1 Introduction

Clipping, or magnitude saturation, is a well-known problem in signal processing, from audio [1,13] to image processing [2,18] and digital communications [17]. The focus of this work is audio declipping, to restore clipped audio signals. Audio signals become saturated usually during acquisition, reproduction or A/D conversion. The perceptual manifestation of clipped audio depends on the level of clipping degradation and the audio content. In case of mild to moderate clipping, the listener may notice occasional "clicks and pops" during playback. When clipping becomes severe, the audio content is usually perceived as if it was contaminated with a high level of additive noise, which may be explained by the introduction of a large number of harmonics caused by the discontinuities in the degraded signal. In addition to audible artifacts, some recent studies have shown that clipping has a negative impact on Automatic Speech Recognition (ASR) performance [11,22].

In the following text, a sampled audio signal is represented by the vector $x \in \mathbb{R}^n$ and its clipped version is denoted by $y \in \mathbb{R}^n$. The latter can be easily

R. Gribonval—This work was supported in part by the European Research Council, PLEASE project (ERC-StG-2011-277906).

© Springer International Publishing Switzerland 2015
E. Vincent et al. (Eds.): LVA/ICA 2015, LNCS 9237, pp. 243–250, 2015.
DOI: 10.1007/978-3-319-22482-4_28

deduced from x through the following nonlinear observation model, called *hard clipping*:

$$y_i = \begin{cases} x_i & \text{for } |x_i| \le \tau, \\ \text{sign}(x_i)\tau & \text{otherwise.} \end{cases} \tag{1}$$

While idealized, this clipping model is a convenient approximation allowing to clearly distinguish the clipped parts of a signal by identifying the samples having the highest absolute magnitude. Indices corresponding to "reliable" samples of y (not affected by clipping) are indexed by Ω_r, while Ω_c^+ and Ω_c^- index the clipped samples with positive and negative magnitude, respectively.

Our goal is to estimate the original signal x from its clipped version y, *i.e.* to "declip" the signal y. Ideally, the estimated signal $\hat{x}$ should satisfy natural magnitude constraints in order to be consistent with the clipped observations. Thus, we seek an estimate $\hat{x}$ which fulfills the following criteria:

$$M_r\hat{x} = M_r y \qquad M_c^+\hat{x} \ge M_c^+ y \qquad M_c^-\hat{x} \le M_c^- y, \tag{2}$$

where the matrices M_r, M_c^- and M_c^+ are *restriction operators*. These are simply row-reduced identity matrices used to extract the vector elements indexed by the sets Ω_r, Ω_c^+ and Ω_c^-, respectively. We write the constraints (2) as $\hat{x} \in \Gamma(y)$.

Obviously, consistency alone is not sufficient to ensure uniqueness of $\hat{x}$, thus one needs to further regularize the inverse problem. The declipping inverse problem is amenable to several regularization approaches proposed in the literature, such as based on linear prediction [12], minimization of the energy of high order derivatives [11], psychoacoustics [6], sparsity [1,6,15,21,23] and cosparsity [14] (where we introduced a simplified version of the analysis-based algorithm presented in this paper). The last two *priors*, briefly explained in the next section, enable some state-of-the-art methods in clipping restoration.

In this paper we empirically compare the performance of the two priors, by means of a declipping algorithm which is easily adaptable to both cases. Our findings are that the sparsity-based version of the algorithm marginally outperforms the cosparsity-based one, but this fact may be attributed to the choice of the stopping criterion. On the other hand, for a class of analysis operators, the cosparsity-based algorithm has very low complexity per iteration, which makes it suitable for real-time audio processing.

2   The Sparse Synthesis and Sparse Analysis Data Models

It is well-known that the energy of audio signals is often concentrated either in a small number of frequency components, or in short temporal bursts [20], *i.e.* they are (approximately) time-frequency sparse. The traditional sparse synthesis viewpoint [8,9] on this property is that audio signals are well approximated by linearly combining few columns of a *dictionary* matrix $D \in \mathbb{C}^{n \times d}$, $d \ge n$ such as a Gabor dictionary, *i.e.* $x \approx Dz$, where $z \in \mathbb{C}^d$ is sparse. A less explored alternative is the cosparse analysis perspective [19] asserting that Ax is approximately sparse, with $A \in \mathbb{C}^{p \times n}$, $p \ge n$ and *analysis operator*. The two data models are different [7,19], unless $p = n$ and $\boldsymbol{A} = \boldsymbol{D}^{-1}$. Finding the sparsest (in the sense of synthesis or analysis) vector $\boldsymbol{x}$ satisfying constraints such as (2) is in general intractable, but convex or greedy heuristics provide efficient algorithms with certain performance guarantees [8,9,19].

3   Algorithms

Some empirical evidence [6,23] suggests that standard ℓ_1 convex relaxation does not perform well for sparse synthesis regularization of the declipping inverse problem. Therefore, we developed an algorithmic framework based on non-convex heuristics, that can be straightforwardly parametrized for use in both the synthesis and the analysis setting. To allow for possible real-time implementation, the algorithms operate on individual blocks (chunks) of audio data, which is subsequently resynthesized by means of the overlap-add scheme.

The heuristics should approximate the solution of the following synthesis- and analysis-regularized inverse problems[1]:

$$\underset{\boldsymbol{x},\boldsymbol{z}}{\text{minimize}} \ \|\boldsymbol{z}\|_0 + \boldsymbol{1}_{\Gamma(\boldsymbol{y})}(\boldsymbol{x}) + \boldsymbol{1}_{\ell_2 \le \varepsilon}(\boldsymbol{x} - \boldsymbol{D}\boldsymbol{z}) \tag{3}$$

$$\underset{\boldsymbol{x},\boldsymbol{z}}{\text{minimize}} \ \|\boldsymbol{z}\|_0 + \boldsymbol{1}_{\Gamma(\boldsymbol{y})}(\boldsymbol{x}) + \boldsymbol{1}_{\ell_2 \le \varepsilon}(\boldsymbol{A}\boldsymbol{x} - \boldsymbol{z}). \tag{4}$$

The indicator function $\boldsymbol{1}_{\Gamma(\boldsymbol{y})}$ of the constraint set $\Gamma(\boldsymbol{y})$ forces the estimate $\boldsymbol{x}$ to satisfy (2). The additional penalty $\boldsymbol{1}_{\ell_2 \le \varepsilon}$ is a *coupling* functional. Its role is to enable the end-user to explicitly bound the distance between the estimate and its sparse approximation. These are difficult optimization problems: besides inherited NP-hardness, the two problems are also non-convex and non-smooth.

We can represent (3) and (4) in an equivalent form, using the indicator function on the cardinality of $\boldsymbol{z}$ and an integer-valued unknown k:

$$\underset{\boldsymbol{x},\boldsymbol{z},\text{k}}{\text{minimize}} \ \boldsymbol{1}_{\ell_0 \le \text{k}}(\boldsymbol{z}) + \boldsymbol{1}_{\Gamma(\boldsymbol{y})}(\boldsymbol{x}) + F_c(\boldsymbol{x},\boldsymbol{z}) \tag{5}$$

where $F_c(\boldsymbol{x},\boldsymbol{z})$ is the appropriate coupling functional. For a fixed k, problem (5) can be seen as a variant of the *regressor selection* problem, which is (locally) solvable by the Alternating Direction Method of Multipliers (ADMM) [3,5]:

Synthesis version	Analysis version
$\bar{\boldsymbol{z}}^{(i+1)} = \mathcal{H}_\text{k}(\hat{\boldsymbol{z}}^{(i)} + \boldsymbol{u}^{(i)})$	$\bar{\boldsymbol{z}}^{(i+1)} = \mathcal{H}_\text{k}(\boldsymbol{A}\hat{\boldsymbol{x}}^{(i)} + \boldsymbol{u}^{(i)})$
$\hat{\boldsymbol{z}}^{(i+1)} = \arg\min_{\boldsymbol{z}} \|\boldsymbol{z} - \bar{\boldsymbol{z}}^{(i+1)} + \boldsymbol{u}^{(i)}\|_2^2$	$\hat{\boldsymbol{x}}^{(i+1)} = \arg\min_{\boldsymbol{x}} \|\boldsymbol{A}\boldsymbol{x} - \bar{\boldsymbol{z}}^{(i+1)} + \boldsymbol{u}^{(i)}\|_2^2$
subject to $\boldsymbol{D}\boldsymbol{z} \in \Gamma(\boldsymbol{y})$	subject to $\boldsymbol{x} \in \Gamma(\boldsymbol{y})$
$\boldsymbol{u}^{(i+1)} = \boldsymbol{u}^{(i)} + \hat{\boldsymbol{z}}^{(i+1)} - \bar{\boldsymbol{z}}^{(i+1)}$	$\boldsymbol{u}^{(i+1)} = \boldsymbol{u}^{(i)} + \boldsymbol{A}\hat{\boldsymbol{x}}^{(i+1)} - \bar{\boldsymbol{z}}^{(i+1)}.$

$$\tag{6}$$

---

[1] Observe that if $\boldsymbol{D}$ and $\boldsymbol{A}$ are unitary matrices, the two problems become identical.

The operator $\mathcal{H}_k(\boldsymbol{v})$ performs hard thresholding, *i.e.* sets all but k highest in magnitude components of $\boldsymbol{v}$ to zero. Unlike the standard regressor selection algorithm, for which the ADMM multiplier [5] needs to be appropriately chosen to avoid divergence, the above formulation is independent of its value.

In practice, it is difficult to guess the optimal value of k beforehand. An adaptive estimation strategy is to periodically increase k (starting from some small value), perform several runs of (6) for a given k and repeat the procedure until the constraint embodied by F_c is satisfied. This corresponds to *sparsity relaxation*: as k gets larger, the estimated $\boldsymbol{z}$ becomes less sparse.

The proposed algorithm, dubbed *SParse Audio DEclipper (SPADE)*, comes in two flavors. The pseudocodes for the synthesis version ("*S-SPADE*") and for the analysis version ("*A-SPADE*") are given in Algorithms 1 and 2.

Algorithm 1. S-SPADE	**Algorithm 2.** A-SPADE
Require: $\boldsymbol{D}, \boldsymbol{y}, \boldsymbol{M}_r, \boldsymbol{M}_c^+, \boldsymbol{M}_c^-, \mathrm{s}, \mathrm{r}, \varepsilon$	**Require:** $\boldsymbol{A}, \boldsymbol{y}, \boldsymbol{M}_r, \boldsymbol{M}_c^+, \boldsymbol{M}_c^-, \mathrm{s}, \mathrm{r}, \varepsilon$
1: $\hat{\boldsymbol{z}}^{(0)} = \boldsymbol{D}^{\mathsf{H}}\boldsymbol{y}, \boldsymbol{u}^{(0)} = \boldsymbol{0}, \mathrm{i} = 1, \mathrm{k} = \mathrm{s}$	1: $\hat{\boldsymbol{x}}^{(0)} = \boldsymbol{y}, \boldsymbol{u}^{(0)} = \boldsymbol{0}, \mathrm{i} = 1, \mathrm{k} = \mathrm{s}$
2: $\bar{\boldsymbol{z}}^{(i)} = \mathcal{H}_k\left(\hat{\boldsymbol{z}}^{(i-1)} + \boldsymbol{u}^{(i-1)}\right)$	2: $\bar{\boldsymbol{z}}^{(i)} = \mathcal{H}_k\left(\boldsymbol{A}\hat{\boldsymbol{x}}^{(i-1)} + \boldsymbol{u}^{(i-1)}\right)$
3: $\hat{\boldsymbol{z}}^{(i)} = \arg\min_{\boldsymbol{z}} \|\boldsymbol{z} - \bar{\boldsymbol{z}}^{(i)} + \boldsymbol{u}^{(i-1)}\|_2^2$ s.t. $\boldsymbol{x} = \boldsymbol{D}\boldsymbol{z} \in \Gamma$	3: $\hat{\boldsymbol{x}}^{(i)} = \arg\min_{\boldsymbol{x}} \|\boldsymbol{A}\boldsymbol{x} - \bar{\boldsymbol{z}}^{(i)} + \boldsymbol{u}^{(i-1)}\|_2^2$ s.t. $\boldsymbol{x} \in \Gamma$
4: **if** $\|\hat{\boldsymbol{z}}^{(i)} - \bar{\boldsymbol{z}}^{(i)}\|_2 \leq \varepsilon$ **then**	4: **if** $\|\boldsymbol{A}\hat{\boldsymbol{x}}^{(i)} - \bar{\boldsymbol{z}}^{(i)}\|_2 \leq \varepsilon$ **then**
5: terminate	5: terminate
6: **else**	6: **else**
7: $\boldsymbol{u}^{(i)} = \boldsymbol{u}^{(i-1)} + \hat{\boldsymbol{z}}^{(i)} - \bar{\boldsymbol{z}}^{(i)}$	7: $\boldsymbol{u}^{(i)} = \boldsymbol{u}^{(i-1)} + \boldsymbol{A}\hat{\boldsymbol{x}}^{(i)} - \bar{\boldsymbol{z}}^{(i)}$
8: $\mathrm{i} \leftarrow \mathrm{i} + 1$	8: $\mathrm{i} \leftarrow \mathrm{i} + 1$
9: **if** $i \bmod \mathrm{r} = 0$ **then**	9: **if** $i \bmod \mathrm{r} = 0$ **then**
10: $\mathrm{k} \leftarrow \mathrm{k} + \mathrm{s}$	10: $\mathrm{k} \leftarrow \mathrm{k} + \mathrm{s}$
11: **end if**	11: **end if**
12: go to 2	12: go to 2
13: **end if**	13: **end if**
14: **return** $\hat{\boldsymbol{x}} = \boldsymbol{D}\hat{\boldsymbol{z}}^{(i)}$	14: **return** $\hat{\boldsymbol{x}} = \hat{\boldsymbol{x}}^{(i)}$

The relaxation rate and the relaxation stepsize are controlled by the integer-valued parameters $\mathrm{r} > 0$ and $\mathrm{s} > 0$, while the parameter $\varepsilon > 0$ is the stopping threshold.

Lemma 1. *The SPADE algorithms terminate in no more than* $\mathrm{i} = \lceil \mathrm{dr}/\mathrm{s} + 1 \rceil$ *iterations.*

Proof. Once $\mathrm{k} \geq \mathrm{d}$, the hard thresholding operation $\mathcal{H}_k$ becomes an identity mapping. Then, the minimizer of the constrained least squares step 3 is $\hat{\boldsymbol{z}}^{(i-1)}$ (respectively, $\hat{\boldsymbol{x}}^{(i-1)}$) and the distance measure in the step 4 is equal to $\|\boldsymbol{u}^{(i-1)}\|_2$. But, in the subsequent iteration, $\boldsymbol{u}^{(i-1)} = \boldsymbol{0}$ and the algorithm terminates.

This bound is quite pessimistic: in practice, we observed that the algorithm terminates much sooner, which suggest that there might be a sharper upper bound on the iteration count.

4   Computational Aspects

The general form of the *SPADE* algorithms does not impose restrictions on the choice of the dictionary nor the analysis operator. From a practical perspective, however, it is important that the complexity per iteration is kept low. The dominant cost of *SPADE* is in the evaluation of the linearly constrained least squares minimizer step, whose computational complexity can be generally high. Fortunately, for some choices of D and A this cost is dramatically reduced.

Namely, if the matrix A^H forms a *tight frame* ($A^H A = \zeta I$), it is easy to show that the step 3 of A-SPADE reduces to[2]:

$$x^{(i)} = \mathcal{P}_\Xi \left(\frac{1}{\zeta} A^H (\bar{z}^{(i)} - u^{(i-1)}) \right), \text{ where:}$$

$$\Xi = \{ x \mid \begin{bmatrix} -M_c^+ \\ M_c^- \end{bmatrix} x \leq \begin{bmatrix} -M_c^+ \\ M_c^- \end{bmatrix} y \text{ and } M_r x = M_r y \}.$$

The projection $\mathcal{P}_\Xi(\cdot)$ is straightforward and corresponds to component-wise mappings, thus the per iteration cost of the algorithm is reduced to the cost of evaluating matrix-vector products.

On the other hand, for S-SPADE this simplification is not possible and the constrained minimization in step 3 needs to be computed iteratively. However, by exploiting the tight frame property of $D = A^H$ and the Woodbury matrix identity, one can build an efficient algorithm that solves this optimization problem with low complexity.

Finally, the computational cost can be further reduced if the matrix-vector products with D and A can be computed with less than quadratic cost. Some transforms that support both tight frame property and fast product computation are also favorable in our audio (co)sparse context. Such well-known transforms are Discrete Fourier Transform, (Modified) Discrete Cosine Transform, (Modified) Discrete Sine Transform and Discrete Wavelet Transform, for instance.

5   Experiments

The experiments are aimed to highlight differences in signal enhancement performance between S-SPADE and A-SPADE, and implicitly, the sparse and cosparse data models. It is noteworthy that in the formally equivalent setting ($A = D^{-1}$), the two algorithms become identical. As a sanity-check, we include this setting in the experiments. The relaxation parameters are set to r = 1 and s = 1, and the stopping threshold is $\varepsilon = 0.1$.

In addition to *SPADE* algorithms, we also include Consistent IHT [15] and social sparsity declipping algorithm [21] as representatives of state-of-the-art. The latter two algorithms use the sparse synthesis data model for regularizing the declipping inverse problem. Consistent IHT is a low-complexity algorithm based on famous Iterative Hard Thresholding for Compressed Sensing [4], while the social sparsity declipper is based on a structured sparsity prior [16].

---

[2] Recall that the matrices M_r, M_c^+ and M_c^- are tight frames by design.

As mentioned before, this work is not aimed towards investigating the appropriateness of various time-frequency transforms in the context of audio recovery, which is why we choose traditional Short Time Fourier Transform (STFT) for all experiments. We use sliding square-rooted Hamming window of size 1024 samples with 75 % overlap. The redundancy level of the involved frames (corresponding to *per-chunk* inverse DFT for the dictionary and forward DFT for the analysis operator) is 1 (no redundancy), 2 and 4. The social sparsity declipper, based on Gabor dictionary, requires batch processing of the whole signal. We adjusted the temporal shift, the window and the number of frequency bins in accordance with previously mentioned STFT settings[3].

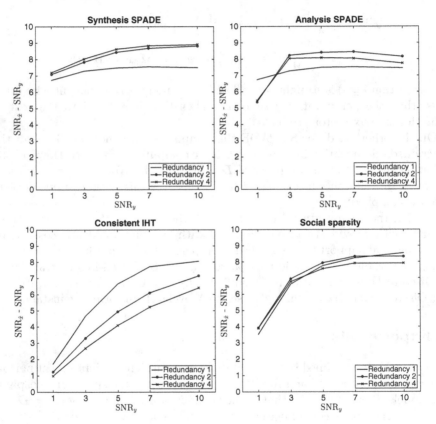

Fig. 1. Declipping performance in terms of the SDR improvement.

For a measure of performance, we use a simple difference between signal-to-distortion ratios of clipped ($\mathrm{SDR}_{\boldsymbol{y}}$) and processed ($\mathrm{SDR}_{\hat{\boldsymbol{x}}}$) signals:

$$\mathrm{SDR}_{\boldsymbol{y}} = 20\log_{10}\frac{\left\|\begin{bmatrix} \boldsymbol{M}_c^+ \\ \boldsymbol{M}_c^- \end{bmatrix}\boldsymbol{x}\right\|_2}{\left\|\begin{bmatrix} \boldsymbol{M}_c^+ \\ \boldsymbol{M}_c^- \end{bmatrix}\boldsymbol{x} - \begin{bmatrix} \boldsymbol{M}_c^+ \\ \boldsymbol{M}_c^- \end{bmatrix}\boldsymbol{y}\right\|_2}, \mathrm{SDR}_{\hat{\boldsymbol{x}}} = 20\log_{10}\frac{\left\|\begin{bmatrix} \boldsymbol{M}_c^+ \\ \boldsymbol{M}_c^- \end{bmatrix}\boldsymbol{x}\right\|_2}{\left\|\begin{bmatrix} \boldsymbol{M}_c^+ \\ \boldsymbol{M}_c^- \end{bmatrix}\boldsymbol{x} - \begin{bmatrix} \boldsymbol{M}_c^+ \\ \boldsymbol{M}_c^- \end{bmatrix}\hat{\boldsymbol{x}}\right\|_2}$$

[3] We use the implementation kindly provided by the authors.

Hence, only the samples corresponding to clipped indices are taken into account. Concerning *SPADE*, this choice makes no difference, since the remainder of the estimate $\hat{x}$ perfectly fits the observations y. However, it may favor the other two algorithms that do not share this feature.

Audio examples consist of 10 music excerpts taken from RWC database [10], which significantly differ in tonal and vocal content. The excerpts are of approximately similar duration ($\sim 10\,s$), and are sampled at 16 kHz with 16 bit encoding. The inputs are generated by artificially clipping the audio excerpts at five levels, ranging from severe ($\text{SDR}_y = 1\,\text{dB}$) towards mild ($\text{SDR}_y = 10\,\text{dB}$).

According to the results presented in Fig. 1, the *SPADE* algorithms yield highest improvement in SDR among the four considered approaches. As assumed, S-SPADE and A-SPADE achieve similar results in a non-redundant setting, but when the overcomplete frames are considered, the synthesis version performs somewhat better. Interestingly, the overall best results for the analysis version are obtained for the twice-redundant frame, while the performance slightly drops for the redundancy four. This is probably due to the absolute choice of the parameter ε, and suggests that in the analysis setting, this value should be replaced by a relative threshold instead. In the non-redundant case, declipping by A-SPADE and Consistent IHT took (on the average) 3 min and 7 min, respectively, while the other two algorithms were much slower[4] (on the order of hours).

6   Conclusion

We presented a novel algorithm for non-convex regularization of the declipping inverse problem. The algorithm is flexible in terms that it can easily accommodate sparse (S-SPADE) or cosparse (A-SPADE) prior, and as such has been used to compare the recovery performance of the two data models. The empirical results are slightly in favor of the sparse synthesis data model. However, the analysis version does not fall far behind, which makes it attractive for practical applications. Indeed, due to the natural way of imposing clipping consistency constraints, it can be implemented in an extremely efficient way, even allowing for a real-time signal processing. Benchmark on real audio data demonstrates that both versions outperform considered state-of-the-art algorithms in the field.

Future work will be dedicated to theoretical analysis of the algorithm, with emphasis on convergence. A possible extension is envisioned by introducing structured (co)sparsity priors in the presented algorithmic framework.

References

1. Adler, A., Emiya, V., Jafari, M.G., Elad, M., Gribonval, R., Plumbley, M.D.: Audio inpainting. IEEE Trans. Audio Speech Lang. Process. **20**(3), 922–932 (2012)
2. Aydin, T.O., Mantiuk, R., Myszkowski, K., Seidel, H.: Dynamic range independent image quality assessment. In: ACM Transactions on Graphics (TOG), vol. 27, p. 69. ACM (2008)

---

[4] All algorithms were implemented in Matlab®, and run in single-thread mode.

3. Bertsekas, D.P.: Nonlinear Programming. Athena Scientific, Belmont (1999)
4. Blumensath, T., Davies, M.E.: Iterative hard thresholding for compressed sensing. Appl. Computat. Harmonic Anal. **27**(3), 265–274 (2009)
5. Boyd, S., Parikh, N., Chu, E., Peleato, B., Eckstein, J.: Distributed optimization and statistical learning via the alternating direction method of multipliers. Found. Trends Mach. Learn. **3**(1), 1–122 (2011)
6. Defraene, B., Mansour, N., De Hertogh, S., van Waterschoot, T., Diehl, M., Moonen, M.: Declipping of audio signals using perceptual compressed sensing. IEEE Trans. Audio Speech Lang. Process. **21**(12), 2627–2637 (2013)
7. Elad, M., Milanfar, P., Rubinstein, R.: Analysis versus synthesis in signal priors. Inverse Probl. **23**(3), 947 (2007)
8. Eldar, Y.C., Kutyniok, G.: Compressed Sensing: Theory and Applications. Cambridge University Press, Cambridge (2012)
9. Foucart, S., Rauhut, H.: A Mathematical Introduction to Compressive Sensing. Springer, New York (2013)
10. Goto, M., Hashiguchi, H., Nishimura, T., Oka, R.: RWC music database: popular, classical and jazz music databases. ISMIR **2**, 287–288 (2002)
11. Harvilla, M.J., Stern, R.M.: Least squares signal declipping for robust speech recognition. In: INTERSPEECH (2014)
12. Janssen, A., Veldhuis, R., Vries, L.: Adaptive interpolation of discrete-time signals that can be modeled as autoregressive processes. IEEE Trans. Acoust. Speech Sig. Process. **34**(2), 317–330 (1986)
13. Kahrs, M., Brandenburg, K.: Applications of Digital Signal Processing to Audio and Acoustics, vol. 437. Springer Science and Business Media, New York (1998)
14. Kitić, S., Bertin, N., Gribonval, R.: Audio declipping by cosparse hard thresholding. In: iTwist-2nd International-Traveling Workshop on Interactions Between Sparse Models and Technology (2014)
15. Kitić, S., Jacques, L., Madhu, N., Hopwood, M.P., Spriet, A., De Vleeschouwer, C.: Consistent iterative hard thresholding for signal declipping. In: IEEE ICASSP, pp. 5939–5943. IEEE (2013)
16. Kowalski, M., Siedenburg, K., Dorfler, M.: Social sparsity! neighborhood systems enrich structured shrinkage operators. IEEE Trans. Sig. Process. **61**(10), 2498–2511 (2013)
17. Li, X., Cimini, L.J.: Effects of clipping and filtering on the performance of OFDM. In: 47th IEEE Vehicular Technology Conference, vol. 3, pp. 1634–1638. IEEE (1997)
18. Naik, S.K., Murthy, C.A.: Hue-preserving color image enhancement without gamut problem. IEEE Trans. Image Process. **12**(12), 1591–1598 (2003)
19. Nam, S., Davies, M.E., Elad, M., Gribonval, R.: The cosparse analysis model and algorithms. Appl. Comput. Harmonic Anal. **34**(1), 30–56 (2013)
20. Plumbley, M.D., Blumensath, T., Daudet, L., Gribonval, R., Davies, M.E.: Sparse representations in audio and music: from coding to source separation. Proc. IEEE **98**(6), 995–1005 (2010)
21. Siedenburg, K., Kowalski, M., Dorfler, M.: Audio declipping with social sparsity. In: IEEE ICASSP, pp. 1577–1581. IEEE (2014)
22. Tachioka, Y., Narita, T., Ishii, J.: Speech recognition performance estimation for clipped speech based on objective measures. Acoust. Sci. Technol. **35**(6), 324–326 (2014)
23. Weinstein, A.J., Wakin, M.B.: Recovering a clipped signal in sparseland (2011). arXiv preprint arXiv:1110.5063

Joint Audio Inpainting and Source Separation

Çağdaş Bilen$^{(\boxtimes)}$, Alexey Ozerov, and Patrick Pérez

Technicolor, 975 avenue des Champs Blancs, CS 17616,
35576 Cesson Sévigné, France
{cagdas.bilen,alexey.ozerov,patrick.perez}@technicolor.com

Abstract. Despite being two important problems in audio signal processing that are interconnected in practice, audio inpainting and audio source separation have not been considered jointly. It is not uncommon in practice to have the mixtures to be separated which also suffer from artifacts due to clipping or other losses. In present work, we consider this problem of source separation using partially observed mixtures. We introduce a flexible framework based on non-negative tensor factorisation (NTF) to attack this new task, and we apply it to source separation with clipped mixtures. It allows us to perform declipping and source separation either in turn or jointly. We investigate experimentally these two regimes and report large performance gains compared to source separation with clipping artefacts being ignored, which is the common approach in practice.

1 Introduction

Audio inpainting and audio source separation are two important problems in audio signal processing. The former is defined as the one of reconstructing the missing parts in an audio signal [1]. It's been coined "audio inpainting" to draw an analogy with visual inpainting, a widely studied problem where the goal is to reconstruct regions in images, for restoration or editing purposes [2]. We consider here the problem of audio inpainting in which some temporal audio samples are lost (as opposed to earlier works where losses are in time frequency domain), such as with saturation of amplitude (clipping) or interfering high amplitude impulsive noise (clicking), and need to be recovered (called declipping and declicking for these two specific cases respectively).

The problem of audio source separation is the one of separating an audio signal into meaningful, distinctive sources which add up to a known mixture, such as separating a music signal into signals from different instruments. Even though audio source separation and audio inpainting have been studied extensively, these two problems have not yet been considered jointly. There are common situations, however, where one task should benefit from the other and vice versa: many audio signals to be de-clipped are in fact composed of multiple sources and, conversely, the audio mixtures in various source separation tasks might be clipped due to the

---

Ç. Bilen and A. Ozerov—This work was partially supported by ANR JCJC program MAD (ANR-14-CE27-0002).

E. Vincent et al. (Eds.): LVA/ICA 2015, LNCS 9237, pp. 251–258, 2015.
DOI: 10.1007/978-3-319-22482-4_29

nature of recording equipment. Hence considering these two tasks simultaneously could help improve the performance of both.

In this paper we propose a first approach toward this goal. To this end, we extend our recent work on audio inpainting with application to declipping [3]. This approach, based on non-negative matrix factorization (NMF) performs as well or better than the state of the art group sparsity based methods such as [10]. It builds on the recent successes of NMF [7] and non-negative tensor factorization (NTF) in audio inpainting [3,9,11][1]. Since, NMF/NTF framework is also very powerful in source separation [5,8,13], it lends itself to addressing the joint problem of audio inpainting and audio source separation.

Extending [3], we estimate individual sources using a low rank NTF model, with the help of some temporal source activity information as in [8]. The proposed algorithm not only can perform audio inpainting and source separation sequentially (i.e., first inpaint the mixture, then separate the sources), but also can perform these two tasks jointly (i.e., simultaneously inpaint the mixture and separate the sources). It is shown that joint inpainting and separation benefits both tasks greatly, especially when the loss due to clipping is significant. The performance of both the sequential and the joint approaches are shown to be much better than the performance of source separation when the degradation due to clipping is ignored as it is usually the case in practice. Section 2 is devoted to problem formulation and modeling description. The main algorithm is outlined in Sect. 3. The experiments are presented in Sect. 4, and some conclusions are drawn in Sect. 5.

2   Signal Model and Problem Formulation

Let us consider the following single-channel[2] mixing equation in time domain:

$$x_t'' = \sum_{j=1}^{J} s_{jt}'' + a_t'', \quad t \in [\![1,T]\!], j \in [\![1,J]\!] \tag{1}$$

where t is the discrete time index, j is the source index, and x_t'', s_{jt}'', and a_t'' denote respectively mixture, source and quantization noise samples.[3] It is assumed that the mixture is only observed on a subset of time indices $\Xi'' \subset [\![1,T]\!]$ called *mixture observation support* (MOS). For clipped signals this support indicates the indices with signal magnitude smaller than the clipping threshold. For the

---

[1] As opposed to [3], earlier works on audio inpainting with NMF/NTF models [9,11] cannot optimally address arbitrary losses in time domain, since the missing data are formulated in time frequency domain.

[2] This work would be readily extended to the multi-channel case. For sake of simplicity, we only consider the single-channel case here.

[3] Throughout, letters with two primes, e.g., x'', denote time domain signals, letters with one prime, e.g., x' denote framed and windowed-time domain signals and letters with no primes, e.g., x, denote complex-valued short-time Fourier transform (STFT) coefficients.

rest of this paper, we assume for sake of simplicity that there is no mixture quantization ($a_t'' = 0$).

The sources are unknown. We assume, however, that it is known which sources are *active* at which time periods. For a multi-instrument music for instance, this information corresponds to knowing which instruments are playing at any instant. Furthermore it is also assumed that if the mixture is clipped, the clipping threshold is known.

In order to compute the STFT coefficents, the mixture and the sources are first converted to windowed-time domain with a window length M and a total of N windows with resulting coefficients denoted by s_{jmn}' and x_{mn}' representing the original sources and the mixture in windowed-time domain respectively for $m = [\![1, M]\!], n = [\![1, N]\!], j = [\![1, J]\!]$. We also introduce the set $\Xi' \subset [\![1, N]\!] \times [\![1, M]\!]$ that is the MOS within the framed representation corresponding to Ξ'' in time domain, and its frame-level restriction $\Xi_n' = \{m | (m, n) \in \Xi'\}$. We will denote the observed clipped mixture in windowed-time domain as $\mathbf{x}_c' = \{\mathbf{x}_{c,n}'\}_{n=1}^N$ and its restriction to un-clipped instants as $\bar{\mathbf{x}}' = \{\bar{\mathbf{x}}_n'\}_{n=1}^N$, where $\bar{\mathbf{x}}_n' = [x_{mn}']_{m \in \Xi_n'}$. The STFT coefficients of the sources, s_{jfn}, and the mixture, x_{fn}, are computed via applying a unitary fourier transform, $\mathbf{U} \in \mathbb{C}^{F \times M} (F = M)$, to each window of the windowed-time domain counterparts. For example, $[x_{1n}, \cdots, x_{Fn}]^T = \mathbf{U}[x_{1n}', \cdots, x_{Mn}']^T$ [4].

The sources are modelled in the STFT domain with a normal distribution $(s_{jfn} \sim \mathcal{N}_c(0, v_{jfn}))$ where the variance tensor $\mathbf{V} = [v_{jfn}]_{j,f,n}$ has the low-rank NTF structure (with a small K) [8] such that $v_{jfn} = \sum_{k=1}^K q_{jk} w_{fk} h_{nk}$ with $q_{jk}, w_{fk}, h_{nk} \in \mathbb{R}_+$. This model is parametrized by $\theta = \{\mathbf{Q}, \mathbf{W}, \mathbf{H}\}$, with $\mathbf{Q} = [q_{jk}]_{j,k} \in \mathbb{R}_+^{J \times K}$, $\mathbf{W} = [w_{fk}]_{f,k} \in \mathbb{R}_+^{F \times K}$ and $\mathbf{H} = [h_{nk}]_{n,k} \in \mathbb{R}_+^{N \times K}$.

The assumed information on which sources are active at which time periods is captured by constraining certain entries of $\mathbf{Q}$ and $\mathbf{H}$ to be zero as in [8]. Each of the K components being assigned to a single source through $\mathbf{Q}(\Psi_Q) \equiv 0$ for some appropriate set Ψ_Q of indices, the components of each source are marked as silent through $\mathbf{H}(\Psi_H) \equiv 0$ with an appropriate set Ψ_H of indices.

3   Separation and Declipping

Similar to the algorithm introduced in [3], we propose to estimate model parameters using a generalized expectation-maximization (GEM) algorithm [4] and to estimate the signals using the Wiener filtering [6]. The proposed algorithm is briefly described in Algorithm 1, and its steps described below. Note that it can be used not only for joint audio inpainting and source separation, but also for audio inpainting only, setting the number of sources to 1, and for source separation only, when the observed indices of the mixture cover the entire time axis.

---

[4] $\mathbf{x}^T$ and $\mathbf{x}^H$ represent the non-conjugate transpose and the conjugate transpose of the vector (or matrix) $\mathbf{x}$ respectively.

Algorithm 1. GEM algorithm for NTF model estimation

1: **procedure** JOINT-INPAINTING-SSEPARATION-NTF($\mathbf{x}'_c, \Xi', \bar{\mathbf{x}}', \Psi_H, \Psi_Q$)
2:       Initialize non-negative $\mathbf{Q}, \mathbf{W}, \mathbf{H}$ randomly, set $\mathbf{H}(\Psi_H)$ and $\mathbf{Q}(\Psi_Q)$ to 0
3:    **repeat**
4:          Estimate $\hat{\mathbf{s}}$ (sources), given $\mathbf{Q}, \mathbf{W}, \mathbf{H}, \bar{\mathbf{x}}', \Xi'$ ▷ E-step, see Sect. 3.1
5:          Estimate $\tilde{\mathbf{s}}$ (sources obeying clipping constraint) and $\tilde{\mathbf{P}}$ (posterior power
              spectra), given $\hat{\mathbf{s}}, \mathbf{Q}, \mathbf{W}, \mathbf{H}, \bar{\mathbf{x}}', \Xi'$ and $\mathbf{x}'_c$
                                                    ▷ Applying clipping constraint, see Sect. 3.2
6:          Update $\mathbf{Q}, \mathbf{W}, \mathbf{H}$ given $\tilde{\mathbf{P}}$ ▷ M-step, see Sect. 3.3
7:    **until** convergence criteria met
8: **end procedure**

3.1   Estimation of Sources

All the underlying distributions are assumed to be Gaussian and all the relations between the source signal and the observations are linear, except the clipping constraint that will be addressed specifically in Sect. 3.2. Hence, Thus, without taking into account the clipping constraint, the source can be estimated in the minimum mean square error (MMSE) sense via Wiener filtering [6] given the covariance tensor $\mathbf{V}$ defined in Sect. 2 by the model parameters $\mathbf{Q}, \mathbf{W}, \mathbf{H}$.

We can write the posterior distribution of each source frame $\mathbf{s}_{jn}$ given the corresponding observed mixture frame $\bar{\mathbf{x}}'_n$ and NTF model θ as $\mathbf{s}_{jn}|\bar{\mathbf{x}}'_n; \theta \sim \mathcal{N}_c(\hat{\mathbf{s}}_{jn}, \widehat{\Sigma}_{\mathbf{s}_{jn}\mathbf{s}_{jn}})$ with $\hat{\mathbf{s}}_{jn}$ and $\widehat{\Sigma}_{\mathbf{s}_{jn}\mathbf{s}_{jn}}$ being, respectively, posterior mean and posterior covariance tensor, each of which can be computed by Wiener filtering as

$$\hat{\mathbf{s}}_{jn} = \Sigma^H_{\bar{\mathbf{x}}'_n\mathbf{s}_{jn}} \Sigma^{-1}_{\bar{\mathbf{x}}'_n\bar{\mathbf{x}}'_n} \bar{\mathbf{x}}'_n, \qquad \widehat{\Sigma}_{\mathbf{s}_{jn}\mathbf{s}_{jn}} = \Sigma_{\mathbf{s}_{jn}\mathbf{s}_{jn}} - \Sigma^H_{\bar{\mathbf{x}}'_n\mathbf{s}_{jn}} \Sigma^{-1}_{\bar{\mathbf{x}}'_n\bar{\mathbf{x}}'_n} \Sigma_{\bar{\mathbf{x}}'_n\mathbf{s}_{jn}}, \quad (2)$$

with the definitions $\Sigma_{\mathbf{s}_{jn}\mathbf{s}_{jn}} = \mathrm{diag}([v_{jfn}]_f)$, $\Sigma_{\bar{\mathbf{x}}'_n\mathbf{s}_{jn}} = \mathbf{U}(\Xi'_n)^H \mathrm{diag}([v_{jfn}]_f)$ and $\Sigma_{\bar{\mathbf{x}}'_n\bar{\mathbf{x}}'_n} = \mathbf{U}(\Xi'_n)^H \mathrm{diag}([\sum_j v_{jfn}]_f)\mathbf{U}(\Xi'_n)$ where $\mathbf{U}(\Xi'_n)$ is the $M = F \times |\Xi'_n|$ matrix of columns from $\mathbf{U}$ with index in Ξ'_n.

Note that when there is no noise in the mixture, the resulting estimates for the sources with the wiener filtering will always add up exactly to the observed mixture at the non-clipped support.

3.2   Clipping Constraint

For a declipping application, the estimated mixture must have amplitude larger than clipping threshold outside OS in windowed time domain, that is:

$$\mathbf{U}(\{m\})^H \sum_j \hat{\mathbf{s}}_{jn} \times \mathrm{sign}(x'_{mn}) \geq |x'_{mn}|, \; \forall n, \; \forall m \notin \Xi'_n. \qquad (3)$$

In order to update the NTF model as described in the following section, the posterior power spectra, $\tilde{\mathbf{P}} = \left[\tilde{p}_{jfn} = \mathbb{E}\left[|s_{jfn}|^2 \big| \bar{\mathbf{x}}'_n; \theta\right]\right]_{j,f,n}$, must be computed. However under the clipping constraint (3), the distribution is no longer Gaussian and the computation of posterior power spectra is no longer computationally simple. Instead we use the *Covariance Projection* method introduced in [3], in which the samples not obeying the constraint (3) after the wiener filtering stage are assumed to be known and equal to the clipping threshold and the wiener filtering step is repeated with rest of the unknowns still assumed to be gaussian distributed. As a result, final estimates of the sources $\tilde{\mathbf{s}}$, which satisfy (3) and the corresponding posterior covariance matrix, $\tilde{\boldsymbol{\Sigma}}_{\mathbf{s}_{jn}\mathbf{s}_{jn}}$, are obtained. Therefore the posterior power spectra can be computed as

$$\tilde{p}_{jfn} = \mathbb{E}\left[\left|s_{jfn}\right|^2 \middle| \bar{\mathbf{x}}'_n; \boldsymbol{\theta}\right] \cong \left|\tilde{s}_{jfn}\right|^2 + \tilde{\boldsymbol{\Sigma}}_{\mathbf{s}_{jn}\mathbf{s}_{jn}}(f,f). \tag{4}$$

3.3 Updating the Model

NTF model parameters can be re-estimated using the multiplicative update (MU) rules minimizing the IS divergence [5] between the the 3-valence tensor of estimated source power spectra $\tilde{\mathbf{P}}$ and the 3-valence tensor of the NTF model approximation $\mathbf{V}$ defined as $D_{IS}(\tilde{\mathbf{P}}\|\mathbf{V}) = \sum_{j,f,n} d_{IS}(\tilde{p}_{jfn}\|v_{jfn})$, where $d_{IS}(x\|y) = x/y - \log(x/y) - 1$ is the IS divergence; $\tilde{p}_{jfn}$ is specified by (4) and v_{jfn} is as defined in Sect. 2. As a result, $\mathbf{Q}, \mathbf{W}, \mathbf{H}$ can be updated with the multiplicative update (MU) rules presented in [8]. These MU rules can be repeated several times to improve the model estimate.

4 Experimental Results

In order to observe the performance of declipping and source separation using the proposed algorithm, 5 different music mixtures[5], each composed of 3 sources (bass, drums and vocals), are considered under 3 different clipping conditions. For each mixture with a maximum magnitude of 1 in time domain, 3 clipping levels at the thresholds of 0.2 (heavy clipping), 0.5 (moderate clipping) and 0.8 (light clipping) are considered, resulting in a total of 15 mixtures with different clipping levels. Each mixture is reconstructed by joint declipping and source separation, sequential declipping and source separation and only source separation ignoring the clipping artefacts. The proposed algorithm has been used for all the reconstructions[6] with 15 components ($K = 15$ with 5 components assigned to each source). The STFT is computed using a half-overlapping sine window of 1024 samples (64 ms) and the proposed GEM algorithm is run for 100 iterations. The sources in the mixtures are artificially silenced during a percentage of the total time. An example of the activation periods of the sources and corresponding indices set to zero in $\mathbf{Q}$ and $\mathbf{H}$ during reconstruction are shown in Fig. 1.

The results of the optimizations can be seen in Table 1. Signal to noise ratio on the clipped support (SNR_m) for the declipped mixture is used to measure the declipping performance and signal to distortion ratio (SDR) is used to measure the source separation performance which are computed as described in [12].

---

[5] The mixtures are taken from the professionally produced music recordings of SiSEC 2015 (https://sisec.inria.fr/).

[6] For only declipping, the algorithm is used with a single source, and for only the source separation, the algorithm is used with the observed support set as the entire time axis.

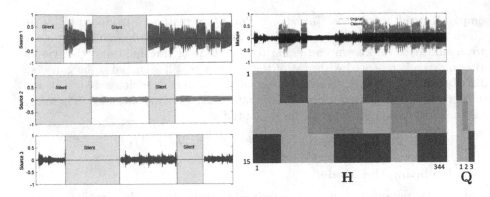

Fig. 1. Typical experimental set-up: a mixture (purple time-domain signal) of 3 sources (red, green, blue) is clipped at 0.2 (black). It will be un-clipped and separated using the low-rank NTF model with each source assigned to 5 out of the $K = 15$ components, as specified by $\mathbf{Q}(\varPsi_Q) \equiv 0$ (grey entries), and each source being silent at known instants as specified at the component level by $\mathbf{H}(\varPsi_H) \equiv 0$ (grey entries) (Color figure online).

The results in Table 1 show that when the clipping is severe, joint optimization is almost always preferable since it provides improvement on both the quality of the mixture and the quality of the separated sources with respect to source separation without declipping. This is as opposed to sequential approach which provides comparable quality improvement in the mixture at the expense of the performance in source separation. In fact, for heavy clipping the declipping in sequential approach often reduced the performance of source separation noticably with respect to separation without declipping. As the clipping gets lighter, the performance of sequential method approaches to that of joint method, and finally performs slightly better for light clipping. The joint optimization, however, still has few drawbacks which could be improved upon. The declipping in the sequential approach is performed with 15 components without any restrictions whereas the joint optimization is performed with the additional limitation that each source uses 5 components *independently*. Hence it is not possible that two sources share a common component in the joint optimization. This can be overcome by devising better methods to inject the prior information regarding the sources. It should be also noted that the sequential optimization is approximately twice as fast as joint optimization due to handling much less complicated problems in either phases of the sequential processing. The fact that the wiener filtering stage is independent for each window and can be parallelized to provide significant speed improvements, can be helpful to overcome this problem in the future.

5   Conclusion

Leveraging low-rank NTF techniques, we have proposed a novel framework to attack simultaneously audio inpainting and audio source separation. Focusing

Table 1. Performance of joint declipping and source separation ("Joint"), of sequential declipping and source separation ("Sequential") and of source separation only using clipped signal ("S. Separation"), on 5 mixtures of 3 sources and for three levels of clipping from light to heavy. The energy loss percentage due to declipping is also shown in the third column. The declipping performance is measured with SNR_m while the source separation performance is measured with SDR.

	Clipping loss	Energy loss	Joint		Sequential		S. Separation	
			SNR_m	SDR	SNR_m	SDR	SNR_m	SDR
Mixture 1	Heavy (th. 0.2)	42.56 %	14.64	9.22	14.14	7.08	7.22	6.01
	Mod. (th. 0.5)	2.60 %	18.78	8.09	19.30	8.10	15.84	8.13
	Light (th. 0.8)	0.04 %	24.49	8.08	25.61	8.07	20.75	8.09
Mixture 2	Heavy (th. 0.2)	50.86 %	9.72	5.13	9.72	5.58	6.62	4.27
	Mod. (th. 0.5)	4.43 %	17.97	6.98	18.57	6.57	14.53	7.11
	Light (th. 0.8)	0.08 %	24.25	6.81	25.15	6.73	21.85	6.76
Mixture 3	Heavy (th. 0.2)	49.28 %	16.64	11.82	17.21	−0.03	7.31	7.79
	Mod. (th. 0.5)	2.52 %	22.41	8.97	22.18	7.07	14.74	9.08
	Light (th. 0.8)	0.09 %	25.45	7.15	24.69	9.12	20.78	9.25
Mixture 4	Heavy (th. 0.2)	50.78 %	7.89	6.11	6.25	4.84	7.44	5.86
	Mod. (th. 0.5)	2.31 %	19.42	9.47	17.88	8.95	15.05	9.14
	Light (th. 0.8)	0.02 %	27.67	9.80	29.45	10.10	19.12	10.14
Mixture 5	Heavy (th. 0.2)	37.11 %	13.60	6.61	13.34	2.76	8.26	5.15
	Mod. (th. 0.5)	1.19 %	18.58	7.85	20.22	8.23	15.20	8.23
	Light (th. 0.8)	0.04 %	17.82	8.10	17.97	8.65	17.73	8.54
Average	Heavy (th. 0.2)	46.12 %	12.50	7.78	12.13	4.05	7.37	5.82
	Mod. (th. 0.5)	2.61 %	19.43	8.27	19.63	7.78	15.07	8.34
	Light (th. 0.8)	0.05 %	23.93	7.99	24.57	8.54	20.05	8.56

on the case where signal loss is due to clipping, we investigated audio declipping and source separation, either jointly or sequentially, with comparison to source separation ignoring the clipping. The results have shown that the clipping arte-facts must not be ignored in order to have a good source separation performance, especially in the case of severe clipping. We also showed that, the source separation also improves the performance of declipping and the joint optimization provides better source separation performance in almost all the cases when there is considerable clipping.

The proposed algorithm still has some limitations, such as the reduced flexibility in utilizing the components in the low rank NTF structure. Hence improved methods to provide prior information on the sources without such limitations is future work. It is also observed that the joint algorithm is slower than the other approaches, and increasing the speed of optimization through better parallel processing is also a promising direction for future research.

References

1. Adler, A., Emiya, V., Jafari, M., Elad, M., Gribonval, R., Plumbley, M.D.: Audio inpainting. IEEE Trans. Audio Speech Lang. Process. **20**(3), 922–932 (2012)
2. Bertalmio, M., Sapiro, G., Caselles, V., Ballester, C.: Image inpainting. In: SIGGRAPH 2000, pp. 417–424 (2000)
3. Bilen, Ç., Ozerov, A., Pérez, P.: Audio declipping via nonnegative matrix factorization. In: IEEE Workshop on Applications of Signal Processing to Audio and Acoustics, October 2015 (2015, submitted)
4. Dempster, A., Laird, N., Rubin, D.: Maximum likelihood from incomplete data via the EM algorithm. J. R. Stat. Soc. Ser. B (Methodol.) **39**, 1–38 (1977)
5. Févotte, C., Bertin, N., Durrieu, J.: Nonnegative matrix factorization with the Itakura-Saito divergence. With application to music analysis. Neural Comput. **21**(3), 793–830 (2009)
6. Kay, S.M.: Fundamentals of Statistical Signal Processing: Estimation Theory. Prentice Hall, Englewood Cliffs (1993)
7. Lee, D., Seung, H.: Learning the parts of objects with nonnegative matrix factorization. Nature **401**, 788–791 (1999)
8. Ozerov, A., Févotte, C., Blouet, R., Durrieu, J.L.: Multichannel nonnegative tensor factorization with structured constraints for user-guided audio source separation. In: IEEE International Conference on Acoustics, Speech, and Signal Processing (ICASSP 2011), Prague, May 2011, pp. 257–260 (2011)
9. Roux, J.L., Kameoka, H., Ono, N., de Cheveigné, A., Sagayama, S.: Computational auditory induction as a missing-data model-fitting problem with Bregman divergence. Speech Commun. **53**(5), 658–676 (2011)
10. Siedenburg, K., Kowalski, M., Dörfler, M.: Audio declipping with social sparsity. In: 2014 IEEE International Conference on Acoustics, Speech and Signal Processing (ICASSP), May 2014, pp. 1577–1581 (2014)
11. Simsekli, U., Cemgil, A.T., Yilmaz, Y.K.: Score guided audio restoration via generalised coupled tensor factorisation. In: International Conference on Acoustics Speech and Signal Processing (ICASSP 2012), pp. 5369–5372 (2012)
12. Vincent, E., Gribonval, R., Févotte, C.: Performance measurement in blind audio source separation. IEEE Trans. Audio Speech Lang. Process. **14**(4), 1462–1469 (2006)
13. Virtanen, T.: Monaural sound source separation by nonnegative matrix factorization with temporal continuity and sparseness criteria. IEEE Trans. Audio Speech Lang. Process. **15**(3), 1066–1074 (2007)

Audio Source Separation with Discriminative Scattering Networks

Pablo Sprechmann[1]([⊠]), Joan Bruna[2], and Yann LeCun[1,3]

[1] Courant Institute of Mathematical Sciences, New York University,
New York, USA
{pablo,yann}@cims.nyu.edu
[2] Department of Statistics, University of California, Berkeley, USA
joan.bruna@berkeley.edu
[3] Facebook AI Research, New York, USA

Abstract. Many monaural signal decomposition techniques proposed in the literature operate on a feature space consisting of a time-frequency representation of the input data. A challenge faced by these approaches is to effectively exploit the temporal dependencies of the signals at scales larger than the duration of a time-frame. In this work we propose to tackle this problem by modeling the signals using a time-frequency representation with multiple temporal resolutions. For this reason we use a signal representation that consists of a pyramid of wavelet scattering operators, which generalizes Constant Q Transforms (CQT) with extra layers of convolution and complex modulus. We first show that learning standard models with this multi-resolution setting improves source separation results over fixed-resolution methods. As study case, we use Non-Negative Matrix Factorizations (NMF) that has been widely considered in many audio application. Then, we investigate the inclusion of the proposed multi-resolution setting into a discriminative training regime. We discuss several alternatives using different deep neural network architectures, and our preliminary experiments suggest that in this task, finite impulse, multi-resolution Convolutional Networks are a competitive baseline compared to recurrent alternatives.

Keywords: Source separation · Scattering · Non-negative matrix factorization · Deep learning

1 Introduction

Monaural Source Separation is a fundamental inverse problem in audio and speech processing. In recent years, non-negative matrix factorization (NMF) [1] has been widely used for solving this and other challenging tasks, see [2] for a recent review. The basic idea is to decompose a time-frequency representation of the signal in terms of elementary atoms of dictionaries representing the different sources present in the mixture.

© Springer International Publishing Switzerland 2015
E. Vincent et al. (Eds.): LVA/ICA 2015, LNCS 9237, pp. 259–267, 2015.
DOI: 10.1007/978-3-319-22482-4_30

More recently, several works have observed that the efficiency of these methods can be improved with discriminative training. Discriminatively trained dictionary learning techniques [3–5] show the importance of adapting the modeling task to become discriminative at the inverse problem at hand. Going further in this direction, a number of works completely bypass the modeling aspect and approach inverse problems as non-linear regression problems using Deep Neural Networks (DNN) with differet levels of structure ranging from simple frame-by-frame regressors to more sophisticated Recurrent Neural Networks (RNN). These ideas have been applied to speech separation [6] and speech enhancement [7,8].

Although NMF applied on spectral features is highly efficient, it fails to model long range geometrical features that characterize speech signals. Increasing the temporal window is not the solution, since it increases significantly the dimensionality of the problem and reduces the discriminative power of the model. In order to overcome this limitation, many works have proposed regularized extensions of NMF to promote learned structure in the codes. Examples of these approaches are, temporal smoothness of the activation coefficients [9], including co-occurrence statistics of the basis functions [10], and learned temporal dynamics with Kalman filtering like techniques [11–13] or integrating RNN into the NMF framework [14].

Our main contribution is to show that using a stable and robust multi-resolution representation of the data can benefit the source separation algorithms in both discriminative and non-discriminative settings. Previous works have shown that the choice of the input features plays a key role on source separation [8,15] and speech recognition [16]. We take this observation a step further to the multi-resolution setting. We consider a deep representation based on the wavelet scattering pyramid, which produces information at different temporal resolutions and defines a metric which is increasingly contracting and can be thought of as a generalization of the CQT. Discriminative features having longer temporal context can be constructed with the scattering transform [17] and have been sucessfully applied to audio signals by [18]. While these features have shown excellent performance in various classification tasks, in the context of source separation we require a representation that not only captures long-range temporal structures, but also preserves as much temporal discriminability as possible.

As a non-discriminative setting, we extend the NMF framework to the proposed pyramid representation. We model the signals using non-negative dictionaries at every level of the hierarchy. While NMF dictionaries at the first level are very selective to temporally localized energy patterns, deeper layers provide additional modeling of the longer temporal dynamics [15]. Further, we also consider the discriminative training regime using several neural architectures. We evaluate both settings on a multi-speaker speech separation task, and we observe that in both training regimes the multi-resolution setting leads to better performance with respect to the baselines.

2   Single-Channel Source Separation

We consider the setting in which we observe a temporal signal $y(t)$ that is the sum of two sources $y(t) = x_1(t) + x_2(t)$, and we aim at finding estimates $\hat{x}_i(t)$, with $i = 1, 2$. We consider the supervised monoaural source separation problem, in which we have representative training data for each of the sources.

Most recent techniques typically operate on a non-negative time-frequency representation. Let us denote as $\Phi(y) \in \mathbb{R}^{m \times n}$ the transformed version of $y(t)$, comprising m frequency bins and n temporal frames. This transform can be thought as a non-linear analysis operator and is typically defined as the magnitude (or power) of a time-frequency representation such as the Short-Time Fourier Transform (STFT). Other robust alternatives have also been explored [6,8]. In all cases, the temporal resolution of the features is fixed and given by the frame duration.

Performing the separation in the non-linear representation is key to the success of these algorithms. The transformed domain is in general invariant to some irrelevant variability of the signals (such as local shifts), thus relieving the algorithms from learning it. This comes at the expense of inverting the unmixed estimates in the feature space.

The most common choice is to use the magintude STFT as the feature space. In this case, the phase recovery problem can be approximately solved using soft masks to filter the mixture signal [19]. The strategy resembles Wiener filtering and has demonstrated very good results in practice. Specifically, $\Phi(y) = |\mathcal{S}\{y\}|$, where $\mathcal{S}\{y\} \in \mathbb{C}^{m \times n}$ denotes the STFT of y. The estimated signals are obtained by filtering the mixture,

$$\hat{x}_i = \mathcal{S}^{-1}\{M_i \circ \mathcal{S}\{y\}\}, \quad \text{with} \quad M_i = \frac{\widehat{\Phi(x_i)^p}}{\sum_{l=1,2} \widehat{\Phi(x_l)^p}}, \tag{1}$$

where multiplication denoted $\circ$, division, and exponentials are element-wise operations. The parameter p defines the smoothness of the mask, we use $p = 2$ in our experiments.

2.1   Non-negative Matrix Factorization

Source separation methods based on matrix factorization approaches have received a lot of attention in the literature in recent years. NMF-based source separation techniques attempt to find the non-negative activations $Z_i \in \mathbb{R}^{q \times n}$, $i = 1, 2$ best representing the different speech components in non-negative dictionaries $D_i \in \mathbb{R}^{m \times q}$. In practice, first a separation is obtained in the feature spaces by solving a clasic NMF problem,

$$\min_{Z_i \geq 0} \mathcal{D}(\Phi(x) | \sum_{i=1,2} D_i Z_i) + \lambda \sum_{i=1,2} \mathcal{R}(Z_i), \tag{2}$$

where the first term in the optimization objective measures the dissimilarity between the input data and the estimated channels in the feature space. Common choices of $\mathcal{D}$ are the squared Euclidean distance, the Kullback-Leibler divergence, and the Itakura-Saito divergence. The second term in the minimization objective is included to promote some desired structure of the activations. In this work we use $\mathcal{D}$ reweighted squared Euclidean distance and the ℓ_1 norm as the regularization function $\mathcal{R}$, for which there exist standard optimization algorithms, see for example [20]. Once the optimal activations are solved for, the spectral envelopes of the speech are estimated as $\widehat{\Phi(x_i)} = D_i Z_i$, and the phase recovery is solved using (1).

In this supervised setting, the dictionaries are obtained from training data. The classic approach is to build a model for each source independently and later use them together at testing time. Many works have observed that sparse coding inference algorithms can be improved by directly optimizing the parameters of the model on the evaluation cost function. Task-aware (or discriminative) sparse modeling from [3] shows how to back-propagate through the Lasso. These ideas have been used in the context of source separation and enhancement [4,5]. The goal is to obtain dictionaries such that the solution of (2) also minimizes the reconstruction given the ground truth separation. It is important to note that the level of supervision is very mild, as in the training of autoencoders. We are artificially generating the mixtures, and consequently obtaining the ground truth.

2.2 Purely Discriminative Settings

With the mindset of the discriminative learning, one is tempted to simply replace the inference step by a generic neural network architecture, having enough capacity to perform non-linear regression. The systems are trained so as to optimize the fit between the ground truth separation and the output, the mean squared error (MSE) being the most common. Note that this can be performed in the feature space or in the time domain [8] (when the phase recovery is simple). Other alternatives studied in the literature consist of predicting the masks given in (1) as described by [6].

3 Pyramid Wavelet Scattering

In this section we present briefly the proposed wavelet scattering pyramid, which is conceptually similar to standard scattering networks introduced by [21], but creates features at different temporal resolutions at every layer.

Wavelet Filter Bank: We consider a complex wavelet with a quadrature phase. We assume that the center frequency of $\mathcal{F}\psi$ is 1 and that its bandwidth is of the order of Q^{-1}. Wavelet filters centered at the frequencies $\lambda = 2^{j/Q}$ are computed by dilating ψ: $\psi_\lambda(t) = \lambda\psi(\lambda t)$, and hence $\mathcal{F}\psi_\lambda(\omega) = \widehat{\psi}(\lambda^{-1}\omega)$. We denote by Λ the index set of $\lambda = 2^{j/Q}$ over the signal frequency support, with $j \leq J_1$.

The resulting filter bank has a constant number Q of bands per octave and J_1 octaves. Let us define $\phi_1(t)$ as a low-pass filter with bandwidth 2^{-J_1}. The wavelet transform of a signal $x(t)$ is $Wx = \{x * \phi_1(t), x * \psi_\lambda(t)\}_{\lambda \in \Lambda}$.

Pyramid Scattering Transform: Instead of using a fixed bandwidth smoothing kernel that is applied at all layers, we sample at critical rate in order to preserve temporal locality as much as possible. We start by removing the complex phase of wavelet coefficients in Wx with a complex modulus nonlinearity. Then, we arrange these first-layer coefficients as nodes in the first level of a tree. Each node of this tree is down sampled at the critical sampling rate of the layer Δ_1, given by the reciprocal of the largest bandwidth present in the filter bank:

$$|W^1|x = \{x_i^1\}_{i=1\ldots1+|\Lambda|} = \{x * \phi_1(\Delta_1 n), \ |x * \psi_\lambda(\Delta_1 n)|\}_{\lambda \in \Lambda}.$$

These first-layer coefficients give localized information both in time and frequency, with a trade-off dictated by the Q factor. They are however sensitive to local time-frequency warps, which are often uninformative. In order to increase the robustness of the representation, we transform each of the down sampled signals with a new wavelet filter bank and take the complex modulus of the oscillatory component. For simplicity, we assume a dyadic transformation, which reduces the filter bank to a pair of conjugate mirror filters $\{\phi_2, \psi_2\}$ [22], carrying respectively the low-frequencies and high-frequencies of the discrete signal from above the tree:

$$|W^2|x = \{x_i^1 * \phi_2(2n), \ |x_i^1 * \psi_2(2n)|\}_{i=1\ldots|W^1|}.$$

Every layer thus produces new feature maps at a lower temporal resolution. As shown in [17], only coefficients having gone through $m \leq m_{max}$ non-linearities are in practice computed, since their energy quickly decays. We fix $m_{max} = 2$ in our experiments. We can reapply the same operator as many times k as desired until reaching a temporal context $T = 2^k \Delta_1$. If the wavelet filters are chosen such that they define a non-expansive mapping [17], it results that every layer defines a metric which is increasingly contracting. Every layer thus produces new feature maps at a lower temporal resolution. In the end we obtain a tree of different representations, $\Phi^j(x) = |W^j|x$, $j = 1, \ldots, k$.

4   Source Separation Algorithms

This section shows how the pyramid scattering features could be used for performing source separation. Let us suppose two different sources X_1 and X_2, and let us consider for simplicity the features $\Phi^j(x_i)$, $j = 1, 2$, $i = 1, 2$, $x_i \in X_i$, obtained by localizing the scattering features of two different resolutions at their corresponding sampling rates. Therefore, Φ^1 carries more discriminative and localized information than Φ_2.

Non-discriminative Training: In the non-discriminative training, we train independent models for each source. Given training examples X_i from each source, we consider a NMF of each of the features:

$$\min_{D_i^j, Z_i^j \geq 0} \sum_{x_i \in X_i} \frac{1}{2} \|\Phi^j(x_i) - D_i^j Z_i^j\|^2 + \lambda_i^j \|Z_i^j\|_1,$$

where here the parameters λ_i^j control the sparsity-reconstruction trade-off in the sparse coding at level j. In our experiments we used a fixed value for all of them $\lambda_i^j = \lambda$. At test time, given $y = x_1 + x_2$, we estimate $\hat{x}_1$ and $\hat{x}_2$ as the solution of

$$\min_{\hat{x}_1 + \hat{x}_2 = y, Z_i^j \geq 0} \sum_{i=1,2} \frac{1}{2} \|\Phi^1(\hat{x}_i) - D_i^1 Z_i^1\|_2^2 + \lambda_i^1 \|Z_i\|_1 + \frac{1}{2} \|\Phi^2(\hat{x}_i) - D_i^2 Z_i^2\|_2^2 + \lambda_i^2 \|Z_i^2\|_1.$$

This problem is a coupled phase recovery problem under linear constraints. It can be solved using gradient descent as in [23], but in our setting we use a greedy algorithm, which approximates the unknown complex phases using the phase of $W_1 y$ and $W_2 |W_1 y|$ respectively. Similarly as in [8], we simplify the inference by using a stronger version of the linear constraint $y = x_1 + x_2$.

Discriminative Training: The pyramid scattering features can also be used to train end-to-end models. The most simple alternative is to train a DNN directly from features having the same temporal context as second layer scattering features. For simplicity, we replace the second layer of complex wavelets and modulus with a simple Haar transform: $\Phi_2(x) = \{|x * \psi_\lambda| * h_k(\Delta_1 n)\}_{\lambda \in \Lambda, k=0,\ldots J_2}$, where h_k is the Haar wavelet at scale 2^k, and we feed this feature into a DNN with the same number of hidden units as before. Unlike the non-discriminative case, we do not take the absolute value as in standard scattering to leave the chance to the DNN to recombine coefficients before the first non-linearity. We report results for $J_2 = 5$ which corresponds to a temporal context of 130 ms. We will refer to this alternative as *DNN-multi*. As a second example, we also consider a multi-resolution Convolutional Neural Network (CNN), constructed by creating contexts of three temporal frames at resolutions 2^j, $j = 0 \ldots, J_2 = 5$. We will refer to this alternative as *CNN-multi*. This setting has the same temporal context as the *DNN-multi* but rather than imposing separable filters we leave extra freedom. This architecture can access relatively large temporal context with a small number of learnable parameters. Since the phase recovery problem cannot be approximated with softmax as in (1), we use as the cost function the MSE of the reconstructed feature at all resolutions and then solve a phase recovery.

5   Experiments

As a proof of concept, we evaluated the different alternatives in a multi-speaker setting in which we aim at separating male and female speech. In each case, we trained two gender-specific modeles. The training data consists of recordings of a generic group of speakers per gender, none of which were included in the test set. The experiments were carried out on the TIMIT corpus. We adopted the standard test-train division, using all the training recordings (containing 462

Table 1. Source separation results on a multi-speaker settings. Average SDR, SIR and SAR (in *dB*) for different methods. Standard deviation of each result shown between brackets.

	SDR	SIR	SAR
NMF	6.1 [2.9]	14.1 [3.8]	7.4 [2.1]
scatt-NMF₁	6.2 [2.8]	13.5 [3.5]	7.8 [2.2]
scatt-NMF₂	6.9 [2.7]	16.0 [3.5]	7.9 [2.2]
CQT-DNN	9.4 [3.0]	17.7 [4.2]	10.4 [2.6]
CQT-DNN-5	9.2 [2.8]	17.4 [4.0]	10.3 [2.4]
CQT-DNN-multi	9.7 [3.0]	19.6 [4.4]	10.4 [2.7]
CQT-CNN-multi	**9.9** [3.1]	**19.8** [4.2]	**10.6** [2.8]

different speakers) for building the models and a subset of 12 different speakers (6 males and 6 females) for testing. For each speaker we randomly chose two clips and compared all female-male combinations (144 mixtures). All signals where mixed at 0 dB and resampled to 16 kHz. We used the *source-to-distortion ratio* (SDR), *source-to-interference ratio* (SIR), and *source-to-artifact ratio* (SAR) from the BSS-EVAL metrics [24]. We report the average over the both speakers.

Non-discriminative Settings: As a baseline for the non-discriminative setting we used standard NMF with STFT of frame lengths of 1024 samples and 50 % overlap, leading to 513 feature vectors. The dictionaries were chosen with 200 and 400 atoms. We evaluated the proposed scattering features in combination with NMF with one and two layers, referred as *scatt-NMF₁* and *scatt-NMF₂* respectively. We use complex Morlet wavelets with $Q_1 = 32$ voices per octave in the first level, and dyadic Morlet wavelets ($Q_2 = 1$) for the second level, for a review on Morlet wavelets refer to [22]. The resulting representation had 175 coefficients for the first level and around 2000 for the second layer. We used 400 atoms for *scatt-NMF₁* and 1000 atoms for *scatt-NMF₂*. Features were frame-wise normalized and we used $\lambda = 0.1$. All parameters were obtained using cross-validation on a few clips separated from the training as a validation set.

Discriminative Settings: We use a single and multi-frame *DNN*s as a baseline for this training setting. The network architectures consist of two hidden layers using the outputs of the first layer of scattering, that is, the CQT coefficients at a given temporal position. It uses rectified linear units (ReLU's) as in the rest of the architectures and the output is normalize so that it corresponds to the spectral mask discussed in (1). The multi-frame version considers the concatenation of 5 frames as inputs matching the temporal context of the tested multi-resolution versions. We used 512 and 150 units for the single-frame DNN (referred as *CQT-DNN*) and 1024 and 512 for the multi-frame one (referred as *CQT-DNN-5*), increasing the number of parameters did not improve the results. We optimize the network to optimize the MSE to each of the sources. We also include the architectures *DNN-multi* and *CNN-multi* described in Sect. 4. In all cases the weights are randomly initialized and training is performed using stochastic gradient descent with momentum. We used the GPU-enabled package Matconvnet [25].

Table 1 shows the results obtained for the speaker-specific and multi-speaker settings. In all cases we observe that the one layer scattering transform outperforms the STFT in terms of SDR. Furthermore, there is a tangible gain in including a deeper representation; $scatt$-NMF_2 performs always better than $scatt$-NMF_1. While the gain in the SDR and SAR are relatively small the SIR is 3 dB higher. It is thus benefitial to consider a longer temporal context in order to perform the separation sucessfully. On the other hand, as expected, the discriminative training yields very significant improvements. The same reasons that produced the improvements in the non-discriminative setting also have an impact in the discriminative case. Adding enough temporal contexts to the neural regressors improves their performance. The multi-temporal representation plays a key role as simply augmenting the number of frames does not lead to better performance (at least using baseline DNNs).

Discussion: We have observed that the performance of baseline source separation algorithms can be improved by using a temporal multi-resolution representation. The representation is able to integrate information across longer temporal contexts while removing uninformative variability with a relatively low parameter budget. In line with recent findings in the literature, we have observed that including discriminative criteria in the training leads to significant improvements in the source separation performance. While this report presents some promising initial results, several interesting comparisons need to be made and are subject of current research. Recent studies have evaluate the use of deep RNN's for solving the source separation problem [6, 8]. We are currently addressing the question of comparing different neural network architectures that exploit temporal dependancies and assessing whether the use of multi-resolution representation can play a role as in this initial study.

References

1. Lee, D.D., Seung, H.S.: Learning parts of objects by non-negative matrix factorization. Nature **401**(6755), 788–791 (1999)
2. Smaragdis, P., Fevotte, C., Mysore, G., Mohammadiha, N., Hoffman, M.: Static and dynamic source separation using nonnegative factorizations: a unified view. IEEE Sig. Process. Mag. **31**(3), 66–75 (2014)
3. Mairal, J., Bach, F., Ponce, J.: Task-driven dictionary learning. IEEE Trans. Pattern Anal. Mach. Intel. **34**(4), 791–804 (2012)
4. Sprechmann, P., Bronstein, A.M., Sapiro, G.: Supervised non-euclidean sparse NMF via bilevel optimization with applications to speech enhancement. In: HSCMA, pp. 11–15. IEEE (2014)
5. Weninger, F., Le Roux, J., Hershey, J.R., Watanabe, S.: Discriminative NMF and its application to single-channel source separation. In: Proceedings of ISCA Interspeech (2014)

6. Huang, P.-S., Kim, M., Hasegawa-Johnson, M., Smaragdis, P.: Deep learning for monaural speech separation. In: ICASSP, pp. 1562–1566 (2014)
7. Sprechmann, P., Bronstein, A., Bronstein, M., Sapiro, G.: Learnable low rank sparse models for speech denoising. In: ICASSP, pp. 136–140 (2013)
8. Weninger, F., Le Roux, J., Hershey, J.R., Schuller, B.: Discriminatively trained recurrent neural networks for single-channel speech separation. In: Proceedings IEEE GlobalSIP 2014 Symposium on Machine Learning Applications in Speech Processing (2014)
9. Févotte, C.: Majorization-minimization algorithm for smooth itakura-saito non-negative matrix factorization. In: ICASSP, pp. 1980–1983. IEEE (2011)
10. Wilson, K.W., Raj, B., Smaragdis, P., Divakaran, A.: Speech denoising using non-negative matrix factorization with priors. In: ICASSP, pp. 4029–4032 (2008)
11. Mysore, G.J., Smaragdis, P.: A non-negative approach to semi-supervised separation of speech from noise with the use of temporal dynamics. In: ICASSP, pp. 17–20 (2011)
12. Han, J., Mysore, G.J., Pardo, B.: Audio imputation using the non-negative hidden markov model. In: Theis, F., Cichocki, A., Yeredor, A., Zibulevsky, M. (eds.) LVA/ICA 2012. LNCS, vol. 7191, pp. 347–355. Springer, Heidelberg (2012)
13. Févotte, C., Le Roux, J., Hershey, J.R.: Non-negative dynamical system with application to speech and audio. In: ICASSP (2013)
14. Boulanger-Lewandowski, N., Mysore, G.J., Hoffman, M.: Exploiting long-term temporal dependencies in NMF using recurrent neural networks with application to source separation. In: ICASSP, May 2014, pp. 6969–6973 (2014)
15. Bruna, J., Sprechmann, P., LeCun, Y.: Source separation with scattering non-negative matrix factorization (2014, submitted)
16. Mohamed, A., Hinton, G., Penn, G.: Understanding how deep belief networks perform acoustic modelling. In: 2012 IEEE International Conference on Acoustics, Speech and Signal Processing (ICASSP), pp. 4273–4276. IEEE (2012)
17. Bruna, J., Mallat, S.: Invariant scattering convolution networks. IEEE Trans. Pattern Anal. Mach. Intel. **35**(8), 1872–1886 (2013)
18. Andén, J., Mallat, S.: Deep scattering spectrum (2013). arXiv preprint arXiv:1304.6763
19. Schmidt, M.N., Larsen, J., Hsiao, F.-T.: Wind noise reduction using non-negative sparse coding. In: MLSP, August 2007, pp. 431–436 (2007)
20. Févotte, C., Idier, J.: Algorithms for nonnegative matrix factorization with the β-divergence. Neural Comput. **23**(9), 2421–2456 (2011)
21. Mallat, S.: Recursive interferometric representation. In: Proceedings of EUSICO Conference, Denmark (2010)
22. Mallat, S.: A Wavelet Tour of Signal Processing. Academic Press, New York (1999)
23. Bruna, J., Mallat, S.: Audio texture synthesis with scattering moments (2013). arXiv preprint arXiv:1311.0407
24. Vincent, E., Gribonval, R., Févotte, C.: Performance measurement in blind audio source separation. IEEE Trans. Audio Speech Lang. Proc. **14**(4), 1462–1469 (2006)
25. Simonyan, K., Zisserman, A.: Very deep convolutional networks for large-scale image recognition (2014). arXiv preprint arXiv:1409.1556

Theory

A Dictionary Learning Method for Sparse Representation Using a Homotopy Approach

Milad Niknejad[1](✉), Mostafa Sadeghi[2], Massoud Babaie-Zadeh[2],
Hossein Rabbani[3], and Christian Jutten[4]

[1] Islamic Azad University, Majlesi Branch, Isfahan, Iran
milad3n@gmail.com
[2] Department of Electrical Engineering, Sharif University of Technology, Tehran, Iran
[3] Biomedical Engineering Department, Medical Image and Signal Processing
Research Center, Isfahan University of Medical Sciences, Isfahan, Iran
[4] GIPSA-lab, Institut Universitaire de France, Grenoble, France

Abstract. In this paper, we address the problem of dictionary learning for sparse representation. Considering the regularized form of the dictionary learning problem, we propose a method based on a homotopy approach, in which the regularization parameter is overall decreased along iterations. We estimate the value of the regularization parameter adaptively at each iteration based on the current value of the dictionary and the sparse coefficients, such that it preserves both sparse coefficients and dictionary optimality conditions. This value is, then, gradually decreased for the next iteration to follow a homotopy method. The results show that our method has faster implementation compared to recent dictionary learning methods, while overall it outperforms the other methods in recovering the dictionaries.

Keywords: Dictionary learning · Sparse representation · Homotopy · Adaptive · Warm-start method

1 Introduction

In recent years, it has been shown that sparse representation leads to promising results in many applications of signal processing [3]. Sparse representation deals with approximating a signal as a linear combination of a few known signals, called atoms, chosen from a signal collection, called dictionary. The performance of sparse coding for a particular class of signals is highly related to a dictionary having the ability to represent all signals in the class by linear combinations of a few atoms. Learning sparsifying dictionaries has also been shown to outperform known and predetermined dictionaries in some applications for classes of signals such as images [4] and audio [6].

A common approach to obtain the dictionary is to use alternating minimization in an iterative procedure [5,13]. In the sparse coding stage, sparse coefficients

---

This work was partially funded by European project 2012-ERC-AdG-320684 CHESS.

E. Vincent et al. (Eds.): LVA/ICA 2015, LNCS 9237, pp. 271–278, 2015.
DOI: 10.1007/978-3-319-22482-4_31

are obtained while the previously found dictionary is fixed, and in the dictionary update stage, the dictionary is found based on the obtained coefficients. In the sparse coding stage, Orthogonal Matching pursuit (OMP) [10] and Iterative Shrinkage Thresholding (IST) algorithm [12] have been used in Method of Optimal Directions (MOD) [5] and Majorization Method (MM) [13] dictionary learning, respectively. Among some examples of dictionary update stages, the MOD used the observation matrix multiplied by pseudo inverse of representation matrix, and a Maximum A Posteriori (MAP)-based dictionary learning in [7] used a gradient descent method, both followed by normalization of dictionary columns. However, all methods proposed so far have not considered the adaptivity of uncertain parameters of the cost function, such as the regularization parameter, to the data.

In this paper, we propose a dictionary learning method for sparse representations, which benefits from a homotopy (continuation) method. Generally, the homotopy is a heuristic that, first, computes the solution of an initial simpler problem, in which the global minimum can be easily found, and then, gradually deforms the initial problem to the desired one. The homotopy has also been used in solving nonlinear equations [8], and in the optimization relating to sparse representation with fixed dictionary [2,12]. Inspired by a homotopy approach, we propose a method which starts solving the dictionary learning cost function from a higher value of the regularization parameter, and adaptively decreases this parameter along iterations. Although our method uses an alternating minimization approach, as we explain through this paper, being capable of changing the value of the regularization parameter enables us to choose a regularization parameter such that it keeps the sparse representation solutions near the optimal after updating the dictionary. Our method can also be seen as a method that uses a homotopy approach with an adaptive regularization parameter selection.

In the following sections, the dictionary learning problem is first discussed in Sect. 2. Then, Sect. 3 is devoted to the description of our proposed method. In Sect. 4, we evaluate the performance and speed of our method in comparison to other dictionary learning algorithms.

2   The Dictionary Learning Problem

Let $\{\mathbf{y}_l \in \mathbb{R}^p\}_{l=1}^L$ be the set of training signals, and $\{\mathbf{x}_l \in \mathbb{R}^q\}_{l=1}^L$ be the set of corresponding representation coefficients over the dictionary $\mathbf{D} \in \mathbb{R}^{p \times q}$. Forming a training data matrix by $\mathbf{Y} \triangleq [\mathbf{y}_1 \ldots \mathbf{y}_L]$, and the representation matrix by $\mathbf{X} \triangleq [\mathbf{x}_1 \ldots \mathbf{x}_L]$, the dictionary learning problem for sparse representations, as used in [13], can be mathematically modeled by the joint optimization problem of the form

$$\underset{\mathbf{D} \in \mathscr{D}, \mathbf{X}}{\operatorname{argmin}} \{\|\mathbf{Y} - \mathbf{DX}\|_F^2 + \lambda_d \|\mathbf{X}\|_{1,1}\} \tag{1}$$

where $\|.\|_F$ indicates the Frobenius norm, and $\|\mathbf{X}\|_{r,s} \triangleq \sum_i (\sum_j |x_{i,j}|^r)^{s/r}$. Although other matrix norms (generally $0 < r < 1$ and $0 < s < 1$), promotes sparsity in the representations in (1). $\|\mathbf{X}\|_{1,1}$ is used for this purpose due to its convexity and also its separability into the absolute sum of the individual entries of the matrix i.e. $\|\mathbf{X}\|_{1,1} = \sum_i \sum_j |x_{i,j}|$. In (1), λ_d is the desired value for the regularization parameter, and is set to achieve a suitable tradeoff between the accuracy of the representations and the sparsity level in $\mathbf{X}$. The desired value of the regularization parameter depends on the application in which the dictionary learning is employed. As an example in [4], this value is set proportional to the variance of Gaussian noise for an image denoising application. Since solving the optimization problem in (1) tends to increase the norms of the atoms, which unfavorably affects some sparse representation algorithms, it is desirable to constrain the norms of the dictionary atoms by defining the admissible set of

$$\mathscr{D} = \{\mathbf{D} \in \mathbb{R}^{p \times q} \quad \text{s.t} \quad \forall j \, \|\mathbf{d}_j\|_2 \leq 1\}. \tag{2}$$

3 Our Proposed Method

Using a homotopy approach, we start to solve the optimization problem in (1) with a high value of the regularization parameter, and then decrease it along the iterations adaptively until reaching the desired value of λ_d. The starting value for the regularization parameter and the procedure of choosing its values along iterations is discussed in Sect. 3.3. So our proposed method at the nth iteration, instead of a fixed value of λ_d, solves the optimization problem of the form

$$\underset{\mathbf{D} \in \mathscr{D}, \mathbf{X}}{\operatorname{argmin}} \{\|\mathbf{Y} - \mathbf{DX}\|_F^2 + \lambda^{(n)} \|\mathbf{X}\|_{1,1}\} \tag{3}$$

where as n grows, $\lambda^{(n)}$ decreases adaptively.

In order to solve the minimization problem in (3), our algorithm alternates among the sparse coding stage, the dictionary update stage and the update of $\lambda^{(n)}$. Our method also uses $\mathbf{X}^{(n)}$ and $\mathbf{D}^{(n)}$ found in the optimization problem with $\lambda^{(n)}$ as a warm-start for solving the optimization problem with the nearby value of $\lambda^{(n+1)}$. Using a warm-start strategy has been previously shown to be effective in improving the speed of dictionary learning algorithms [11,13].

3.1 Sparse Coding Stage

In our method, a sparse coding algorithm which belongs to the class of IST methods is used. These methods benefit from a proper initialization enabling us to use the warm-start strategy. At the kth iteration of the sparse coding algorithm in the nth dictionary learning iteration, the sparse coding solves

$$\mathbf{X}^{(k+1)} = \underset{\mathbf{X}}{\operatorname{argmin}} \{\|\mathbf{X} - \mathbf{U}^{(k)}\|_F^2 + \frac{\lambda^{(n)}}{c} \|\mathbf{X}\|_{1,1}\} \tag{4}$$

in which c should satisfy $c > \|\mathbf{D}^T\mathbf{D}\|$ where $\|.\|$ stands for the spectral norm, and $\mathbf{U}^{(k)} \triangleq \mathbf{X}^{(k)} + \frac{1}{c}(\mathbf{D}^T(\mathbf{Y} - \mathbf{DX}^{(k)}))$ [12]. The global optimum of the convex and non-smooth problem in (4), is the point with zero subgradient i.e.

$$2c(\mathbf{X} - \mathbf{U}^{(k)}) - \lambda^{(n)} \mathscr{P} \in 0 \tag{5}$$

where $\mathscr{P} = \nabla\|\mathbf{X}\|_{1,1}$ is a set of matrices whose entries satisfy

$$\begin{cases} p_{i,j} = 1 & \text{if } \mathbf{x}_{i,j} > 0 \\ p_{i,j} \in [-1\ 1] & \text{if } \mathbf{x}_{i,j} = 0 \\ p_{i,j} = -1 & \text{if } \mathbf{x}_{i,j} < 0. \end{cases}$$

The point that satisfies the optimality condition (5), is obtained by $\mathbf{X}^{(k)} = \mathcal{S}_{\frac{\lambda^{(n)}}{2c}}(u_{i,j})$ which is a soft thresholding operator on entries of $\mathbf{U}$ with the threshold value of $\frac{\lambda^{(n)}}{2c}$. Using the soft thresholding operator there is one single matrix $\mathbf{P}^{(k)}$ from the set $\mathscr{P}$ which makes the condition in (5) turn into

$$2c(\mathbf{X}^{(k)} - \mathbf{U}^{(k)}) - \lambda^{(n)}\mathbf{P}^{(k)} = 0. \tag{6}$$

3.2 Dictionary Update Stage

The dictionary update stage is to find the minimization problem in (1), while $\mathbf{X}$ is fixed with the value found in the previous sparse coding stage. Similar to [7], we use the gradient descent algorithm. So in the kth iteration of the gradient descent of the nth iteration of dictionary learning algorithm, our method updates the dictionary by

$$\mathbf{D}^{(k+1)} = \mathbf{D}^{(k)} + \rho(\mathbf{Y} - \mathbf{D}^{(k)}\mathbf{X}^{(n)})\mathbf{X}^{(n)T} \tag{7}$$

where ρ is an appropriate constant and is set to .001 in our implementations. Using the gradient descent has a fast implementation, and due to using a proper initialization, enables us to employ a warm-start strategy, similar to sparse coding algorithm in our method. Then, our algorithm normalizes the atoms whose norms are more than one to the unit norm and keep the other atoms intact.

3.3 Determining the Regularization Parameter

In many homotopy methods, decreasing the regularization parameter is done heuristically by a linear or exponential decay [9]. However in this section, we propose a more sophisticated choice for this value.

One of the disadvantages of the alternating minimization between the two stages of dictionary learning algorithms is that each stage may not preserve the optimality of the other one. So the solutions in an alternating minimization approach might oscillate around an optimal point. To understand this, assume without loss of generality that dictionary update is performed after sparse coding stage at each iteration. Updating the dictionary may not preserve the optimality condition derived for the sparse coding in (5) or equivalently in (6), since this condition is not considered in the dictionary update stage. So updating $\mathbf{D}$ might lead to a deviation from the optimality condition of sparse coding at end of each iteration of dictionary learning algorithm. Being capable of changing the regularization parameter in our method, in order to alleviate this, after the dictionary update stage we choose the regularization parameter in such a way that it best preserves the optimality condition in (6) for sparse coding. The criterion for optimality of sparse coding stage could be the Frobenius norm of the term that is set to zero in (6). After the alteration of $\mathbf{D}$, the value of $\mathbf{U}$ changes, and the mentioned term might not be equal to zero. So we find the value of the regularization parameter which minimizes the Frobenius norm of the term set to zero in (6), after updating the dictionary, based on the current estimate of $\mathbf{D}$ and $\mathbf{X}$ i.e.

$$\lambda_{opt} = \underset{\lambda}{\arg\min} \|\mathbf{R}^{(n)} - \lambda \mathbf{P}^{(n)}\|_F^2$$

$$= \underset{\lambda}{\arg\min}\{Tr(\mathbf{R}^{(n)^T}\mathbf{R}^{(n)}) + \lambda^2 Tr(\mathbf{P}^{(n)^T}\mathbf{P}^{(n)}) - 2\lambda Tr(\mathbf{P}^{(n)^T}\mathbf{R}^{(n)})\} \quad (8)$$

where $\mathbf{R}^{(n)} = 2c(\mathbf{X}^{(n)} - \mathbf{U}^{(n)})$. The global optimum of the above least square minimization problem can be found by setting its derivative to zero which leads

$$\lambda_{opt} = \frac{Tr(\mathbf{P}^{(n)^T}\mathbf{R}^{(n)})}{Tr(\mathbf{P}^{(n)^T}\mathbf{P}^{(n)})}. \quad (9)$$

Having found the optimal value for $\lambda^{(n)}$ based on the current estimation of $\mathbf{X}$ and $\mathbf{D}$, in order to follow a homotopy, we gradually decrease this value by a constant factor which leads to $\lambda^{(n+1)} = (1 - \epsilon)\lambda_{opt}$ where ϵ is a small constant. However, implementing some iterations of our algorithm without applying $\lambda^{(n+1)} = (1 - \epsilon)\lambda_{opt}$ is desirable, since it leads to equilibrium for a joint point of $(\mathbf{X}, \mathbf{D}, \lambda)$ (Note that the value of λ here is higher than the desired value). The value of regularization parameter is also forced to be bounded to the desired value of λ_d which makes final iterations be implemented with this value of regularization parameter. It is worth mentioning that the procedure of finding the optimal value for regularization parameter and decreasing it by a constant factor has been used in a homotopy based sparse coding (with fixed dictionary) in [12]. However the procedure of obtaining the optimal value is completely different and novel in our method, and is adapted to the dictionary learning application.

The initial optimal value of the regularization parameter is set to $\|D^T Y\|_\infty$, where $\|.\|_\infty$ returns the maximum absolute value of the matrix entries, since for $\lambda^{(1)} > \|D^T Y\|_\infty$, the solution of zero is optimal in (5) [12], and consequently, no update of initial dictionary is occurred in the dictionary update stage (Fig. 1).

It is worth mentioning that the performance of homotopy methods depends on the tracing the optimal solutions while the value of the regularization parameter changes. As we discussed in this subsection, by the proposed optimal choice for the regularization parameter, our algorithm tends to keep the optimal solutions along iterations for both dictionary and sparse coefficients.

4   Simulations Results

In this section, we compare our method with other methods using synthetic signals to evaluate the performance of algorithms in recovering the dictionary that produces the data.

- Initialization: Choose an initial dictionary $\mathbf{D} \in \mathbf{R}^{p \times q}$
- For $n = 1, \ldots, N$ (main loop)

  Sparse coding stage:
  1. Initialize with $\mathbf{D} = \mathbf{D}^{(n-1)}$, $\mathbf{X}^{(k=0)} = \mathbf{X}^{(n-1)}$
  2. For $k = 1, \ldots, K_s$
  $$\mathbf{U}^{(k)} = \mathbf{X}^{(k-1)} + \frac{1}{c}(\mathbf{D}^T(\mathbf{Y} - \mathbf{D}\mathbf{X}^{(k-1)})), \mathbf{X}^{(k)} = \mathcal{S}_{\frac{\lambda(n)}{2c}}(\mathbf{U}^{(k)})$$
  End For
  3. Set $\mathbf{X}^{(n)} = \mathbf{X}^{(K_s)}$

  Dictionary update stage:
  1. Initialize with $\mathbf{X} = \mathbf{X}^{(n)}$, $\mathbf{D}^{(k=0)} = \mathbf{D}^{(n-1)}$
  2. For $k = 1, \ldots, K_d$
  $$\mathbf{D}^{(k)} = \mathbf{D}^{(k-1)} + \rho(\mathbf{Y} - \mathbf{D}^{(k-1)}\mathbf{X})\mathbf{X}^T$$
  Normalize columns of dictionary whose norms are more than 1.
  3. End For
  4. Set $\mathbf{D}^{(n)} = \mathbf{D}^{(K_d)}$

  Regularization parameter selection:
  1. Obtain the optimum regularization parameter λ_{opt} by (9)
  2. decrease the regularization parameter by $\lambda^{(n+1)} = \max((1 - \epsilon)\lambda_{opt}, \lambda_d)$
- End For (main loop)
- Final answer is $\mathbf{D} = \mathbf{D}^{(N)}$

Fig. 1. Our proposed dictionary learning algorithm

A dictionary of size 30×60 is randomly generated with independent identically distributed (i.i.d.) Gaussian entries, and its columns are normalized to have unit norms. 4000 sample signals $\{\mathbf{y}_l\}_{l=1}^{4000}$ are produced by linear combination of a few (precisely determined by Q in each experiment) number of atoms with the coefficients which are i.i.d. Gaussian in random and independent locations. We compare our method with MOD [5] and K-SVD [1] as two well-known methods, and also with the Majorization dictionary learning algorithm [13] which has improved those methods and its sparse coding algorithm is similar to our method. For other methods, the MATLAB codes published online by the authors were used. All the experiments were done with core i5 CPU with 4 GB of memory using Matlab 2011a under Microsoft Windows 7 operating system.

The percentages of recovered atoms are compared for different methods during the execution time with data generated by $Q = 4$ number of atoms in Fig. 2. The CPU time is considered in this experiment to roughly compare the computational complexity of the algorithms. The value of ϵ for homotopy decreasing factor is set to 0.05 and applied every 4 iterations (to obtain an equilibrium point for a higher value of the regularization parameter, as described in the previous section). We found that this implementation leads to an appropriate tradeoff between the speed and preserving the performance in our algorithm. The desired value of λ_d in our algorithm and the value of the regularization parameter in MM method are both set to 0.18. Figure 2 shows that our method converges faster and more accurate in this case. In order to better compare the speed of the algorithms, the CPU times are reported for different algorithms while the sparsity level Q varies from 3 to 6. Also, the percentages of the recovered atoms for this experiment are shown in Fig. 3(a) to compare the performance of algorithms in recovering dictionary atoms. The corresponding implementation times are shown in Fig. 3(b). The values are averaged over three independent implementations of algorithms. Based on this figure, our algorithm is more successful in recovering the dictionary except for $Q = 3$, and is faster in all the cases, compared to other methods.

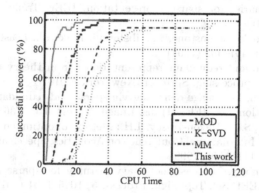

Fig. 2. Comparison of the percentage of recovered atoms vs. the computational time for different methods in $Q = 4$.

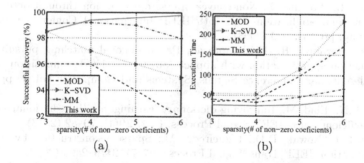

Fig. 3. Comparison of performance of dictionary learning algorithms for datasets with different values of sparsity level Q: (a) Percentage of recovered atoms, (b) implementation time.

5 Conclusion

In this paper, we proposed a homotopy-based method for dictionary learning for sparse representation in which the value of the regularization parameter decreases along iterations. We proposed an adaptive selection for the regularization parameter which best preserves the optimality of sparse coefficients after the dictionary update at each iteration. The results showed that our method is more successful in recovering the dictionaries compared to other methods, and it has faster implementation time.

References

1. Aharon, M., Elad, M., Bruckstein, A.: K-SVD: an algorithm for designing of overcomplete dictionaries for sparse representation. IEEE Trans. Sig. Process. **54**, 4311–4322 (2006)
2. Efron, B., Hastie, T., Johnstone, I., Tibshirani, R., et al.: Least angle regression. Ann. Stat. **32**(2), 407–499 (2004)
3. Elad, M.: Sparse and redundant representations: from theory to applications in signal and image processing. Springer, New York (2010)
4. Elad, M., Aharon, M.: Image denoising via sparse and redundant representations over learned dictionaries. IEEE Trans. Image Process. **15**, 3736–3745 (2006)
5. Engan, K., Aase, S.O., Hakon-Husoy, J.H.: Method of optimal directions for frame design. In: IEEE International Conference on Acoustics, Speech and Signal Processing, vol. 5, pp. 2443–2446 (1999)
6. Jafari, M., Plumbley, M.: Fast dictionary learning for sparse representations of speech signals. IEEE Sel. Top. Sign. Process. **5**, 1025–1031 (2011)
7. Kreutz-Delgado, K., Murray, J.F., Rao, B.D., Engan, K., Lee, T., Sejnowski, T.: Dictionary learning algorithms for sparse representation. Neural Comput. **15**(2), 349–396 (2003)
8. Liao, S.: Homotopy analysis method in nonlinear differential equations. Springer, Heidelberg (2012)
9. Mancera, L., Portilla, J.: Non-convex sparse optimization through deterministic annealing and applications. In: 15th IEEE International Conference on Image Processing, pp. 917–920 (2008)
10. Pati, Y., Rezaiifar, R., Krishnaprasad, P.: Orthogonal matching pursuit: recursive function approximation with applications to wavelet decomposition. In: 27th Annual Asilomar Conference Signals, Systems and Computers vol. 1, pp. 40–44 (1993)
11. Smith, L.N., Elad, M.: Improving dictionary learning: multiple dictionary updates and coefficient reuse. IEEE Signal Process. Lett. **20**(1), 79–82 (2013)
12. Wright, S.J., Nowak, R.D., Figueiredo, M.: Sparse reconstruction by separable approximation. IEEE Trans. Signal Process. **57**(7), 2479–2493 (2009)
13. Yaghoobi, M., Blumensath, T., Davies, M.E.: Dictionary learning for sparse approximations with the majorization method. IEEE Trans. Signal Process. **57**(6), 2178–2191 (2009)

Invertible Nonlinear Dimensionality Reduction via Joint Dictionary Learning

Xian Wei$^{(\boxtimes)}$, Martin Kleinsteuber, and Hao Shen

Department of Electrical and Computer Engineering,
Technische Universität München, Munich, Germany
{xian.wei,kleinsteuber,hao.shen}@tum.de

Abstract. This paper proposes an invertible nonlinear dimensionality reduction method via jointly learning dictionaries in both the original high dimensional data space and its low dimensional representation space. We construct an appropriate cost function, which preserves inner products of data representations in the low dimensional space. We employ a conjugate gradient algorithm on smooth manifold to minimize the cost function. By numerical experiments in image processing, our proposed method provides competitive and robust performance in image compression and recovery, even on heavily corrupted data. In other words, it can also be considered as an alternative approach to compressed sensing. While our approach can outperform compressed sensing in task-driven learning problems, such as data visualization.

Keywords: Invertible nonlinear dimensionality reduction · Joint dictionary learning · Inner products preservation · Compressed sensing

1 Introduction

Dimensionality reduction (DR) is a powerful instrument to tackle large scale signal processing problems. It often serves as a preprocessing step to transform original high dimensional data to a low dimensional space. Then specific tasks, such as filtering or 2D visualization, can be performed directly on the low dimensional representations, cf. [1]. Most classic DR algorithms focus on finding a low-dimensional embedding of original data, which are often not reversible. In other words, there is no reliable reconstruction from the low dimensional space back to the original high dimensional space. However, in many applications, such as communication transmission, image down-sampling and super-resolution, and modeling the time varying data (dynamic textures), it requires that such DR processes can be reversible. Hence, finding an invertible nonlinear DR mapping is a long standing problem in the community.

Recently, the technique of compressed sensing (CS) [2] has shown that high dimensional signals and images can be reconstructed from the measurements in far lower dimensional space than what is usually considered necessary. Formally, it assumes that a signal $x \in \mathbb{R}^m$ admits a factorization $x = D\alpha$ with respect

© Springer International Publishing Switzerland 2015
E. Vincent et al. (Eds.): LVA/ICA 2015, LNCS 9237, pp. 279–286, 2015.
DOI: 10.1007/978-3-319-22482-4_32

to a set of atoms D, also known as a dictionary, where $\alpha \in \mathbb{R}^k$ is sparse. Then the CS problem can be formulated as recovering x from its low dimensional representation $y = Ax$ or $y = AD\alpha$, with $y \in \mathbb{R}^{d \times m}$ for $d \ll m$. Here, $A \in \mathbb{R}^{d \times m}$ is called a projection matrix.

This paper considers an alternative process of DR associated with the dictionary learning (DL) models [3,4]. Namely, the dictionary D is not given as some orthonormal basis, but learned from training samples. This problem has been studied in the framework of CS, known as Blind CS in [5]. Moreover, D and A can also be simultaneously learned from data via some joint optimizations [6,7]. However, one challenge in the CS model is that it has to guarantee incoherence between the projection matrix A and the dictionary D, as well as mutual incoherence between pair atoms in D and A themselves [2]. It is commonly known to be difficult to achieve, when D is redundant. Additionally, specific learning tasks, such as 2D visualization, in compressed domain, are often difficult [8].

In this work, different from the methods of CS via optimizing the projection matrix [6,7], we propose an alternative approach to model the process of DR, using a couple dictionaries ($D \in \mathbb{R}^{m \times k}, P \in \mathbb{R}^{d \times k}$) with $d \ll m$, referred to as *DRCDL* in this work. It can successfully achieve the task of interest, while still avoid to learn the projection matrix directly.

2 Joint Dictionary Learning Under Inner Products Preservation

Let us denote by $X := [x_1, \ldots, x_n] \in \mathbb{R}^{m \times n}$ the data matrix containing n data samples $x_i \in \mathbb{R}^m$, and $Y := [y_1, \ldots, y_n] \in \mathbb{R}^{d \times n}$ with $d < m$ be its corresponding low dimensional representation via some DR mapping $g \colon x_i \mapsto y_i$ for all $i = 1, \ldots, n$. In this work, we assume that both the original data and its low dimensional representation share the same or quite similar sparse structure. Such an assumption is popularly adopted in the framework of coupled sparse representation [9].

We assume that all data points $x_i \in \mathbb{R}^m$ admit sparse representations with respect to a common dictionary $D := [d_1, \ldots, d_k] \in \mathbb{R}^{m \times k}$, i.e.

$$x_i = D\phi_i, \qquad \text{for all } i = 1, \ldots, n, \tag{1}$$

where $\phi_i \in \mathbb{R}^k$ is the corresponding sparse representation of x_i. In this work, we further assume that all columns of the dictionary D have unit norm. We then define the set

$$\mathcal{S}(m, k) := \{D \in \mathbb{R}^{m \times k} | \text{ ddiag}(D^\top D) = I_k\}, \tag{2}$$

where $\text{ddiag}(Z)$ is the diagonal matrix whose entries on the diagonal are those of Z, and I_k denotes the identity matrix. We assume that the low dimensional representations Y share the same sparse structure with respect to a low dimensional dictionary $P := [p_1, \ldots, p_k] \in \mathbb{R}^{d \times k}$, i.e. $y_i = P\phi_i$ with $P \in \mathcal{S}(d, k)$. By a slight abuse of notations, we denote by $\phi_D \colon x_i \mapsto \phi_i$ and $\phi_P \colon y_i \mapsto \phi_i$ the sparse coding in the original data space and the low dimensional representation space, respectively. Then, we denote a nonlinear DR mapping by

$$g\colon x_i \mapsto P\phi_D(x_i), \tag{3}$$

and reversely by

$$g^{-1}\colon y_i \mapsto D\phi_P(y_i). \tag{4}$$

The aim of DR is to find a DR mapping $g\colon x_i \mapsto y_i$, which is stable and preserves as much useful structure as possible. When the DR mapping is linear, according to the Johnson-Lindenstrauss (JL) lemma, cf. [10], every n-point subset of Euclidean space can be embedded in dimension $O(\epsilon^{-2}\log n)$ with $1+\epsilon$ distortion with $0 < \epsilon < 1/2$. Specifically, distance or inner product information of the high dimensional data is preserved in the low dimensional representation space, when ϵ is close to zero [11,12]. The loss introduced by the DR mapping g can be measured by the following function

$$\mathcal{G}(X;Y) := \sum_{i=1}^{n}(x_i^\top x_j - y_i^\top y_j)^2. \tag{5}$$

Recall the assumption that both the original data point x_i and its low dimensional representation $y_i := g(x_i)$ share the same sparse structure, i.e. $x_i = D\phi_i$ and $y_i = P\phi_i$. We adopt the loss function (5) directly to the current coupled sparse representation setting as

$$\mathcal{G}_{(D,P)}(X;Y) = \sum_{i=1}^{n}\left(\phi_i^\top(D^\top D - P^\top P)\phi_j\right)^2. \tag{6}$$

Roughly speaking, the loss $\mathcal{G}_{(D,P)}$ is small, if either the sparse representations are pair-wise conjugate with respect to $D^\top D - P^\top P$, or the difference $D^\top D - P^\top P$ is essentially small. In this work, we consider the second argument. Since both dictionaries D and P are often assumed to be full rank, P can be also considered as a low rank approximation of D.

In order to ensure stability of the proposed nonlinear DR mapping g, we need to guarantee moderate mutual incoherence in both the high and low dimensional dictionaries, i.e. $D \in \mathbb{R}^{m \times k}$ and $P \in \mathbb{R}^{d \times k}$, according to the theory in sparse representation, cf. [13]. However, when the difference $D^\top D - P^\top P$ is sufficiently small, the mutual coherence of D is ensured to be close to the mutual coherence of P. Hence, instead of penalizing on both D and P, we propose to apply a logarithmic barrier function to enforce the mutual coherence of P, i.e.

$$r(P) = - \sum_{1 \le i < j \le k} \log\left(1 - (p_i p_j^\top)^2\right). \tag{7}$$

Finally, let us denote $\Phi(Y,P) := [\phi_P(y_1), \dots, \phi_P(y_n)] \in \mathbb{R}^{k \times n}$. Then, by considering the reconstruction error in the original data space, we propose the following cost function

$$f\colon \mathcal{S}(m,k) \times \mathcal{S}(d,k) \times \mathbb{R}^{d\times n} \to \mathbb{R}$$

$$(D,P,Y) \mapsto \frac{1}{2n}\|X - D\Phi(Y,P)\|_F^2 + \frac{\mu_1}{2k^2}\left\|D^\top D - P^\top P\right\|_F^2 + \mu_2 r(P), \tag{8}$$

where $\mu_1 > 0$ weighs between the loss of distance preservation of DR and the reconstruction error of the training samples, and $\mu_2 > 0$ controls the mutual coherence of the dictionary to be learned.

As a special case, we can assume that the dictionary P is simply constructed from D via a linear mapping specified by $U \in \mathbb{R}^{m\times d}$, i.e. $P = U^\top D$. Then the low dimensional representations y_i can be directly obtained via $y_i = U^\top x_i = U^\top D\phi_i$. Minimizing the term $\left\|D^\top D - P^\top P\right\|_F^2$ leads to a simple solution that U is the eigenvectors of D corresponding to its first d largest eigenvalues. We call this model compressive coupled dictionaries learning (*CCDL*) in the rest of the paper.

3   A Conjugate Gradient DR Algorithm

Recall the fact that the set $\mathcal{S}(m,k)$ is the product of k unit spheres, i.e. a $k(m-1)$ dimensional smooth manifold. In what follows, we adopt the conjugate gradient algorithm on smooth manifolds, which has demonstrated its competitive performance in (co-)sparse dictionary learning, cf. [3,4], to minimize the cost function f on the product manifold $\mathcal{S}(m,k) \times \mathcal{S}(d,k) \times \mathbb{R}^{d\times n}$.

In this work, we employ the sparse solution given by solving an elastic-net problem, cf. [14], as

$$\phi^* := \operatorname*{argmin}_{\phi\in\mathbb{R}^k} \tfrac{1}{2}\|y - P\phi\|_2^2 + \lambda_1\|\phi\|_1 + \tfrac{\lambda_2}{2}\|\phi\|_2^2, \tag{9}$$

where $\lambda_1 > 0$ and $\lambda_2 > 0$ are regularization parameters, which ensures stability and uniqueness of solutions. Let us define the set of indices of the non-zero entries of the solution $\phi^* = [\varphi_1^*, \dots, \varphi_k^*]^\top \in \mathbb{R}^k$ as $\Lambda := \{i \in \{1,\dots,k\}|\varphi_i^* \neq 0\}$. Then the solution of the elastic net (9) has a closed-form expression as

$$\phi_D^*(y) := \left(D_\Lambda^\top D_\Lambda + \lambda_2 I_d\right)^{-1}\left(D_\Lambda^\top y - \lambda_1 s_\Lambda\right), \tag{10}$$

where $s_\Lambda \in \{\pm 1\}^{|\Lambda|}$ carries the signs of ϕ_Λ^*, $D_\Lambda \in \mathbb{R}^{m\times|\Lambda|}$ is the subset of D in which the index of atoms (rows) fall into support Λ. The solution $\phi_P^*(y)$ shares an algorithmically convenient property of being locally twice differentiable with respect to both P and y, cf. [15,16].

Recall the tangent space $T_D\mathcal{S}(m,k)$ of $\mathcal{S}(m,k)$ at $D \in \mathcal{S}(m,k)$ as

$$T_D\mathcal{S}(m,k) := \{\Xi \in \mathbb{R}^{m\times k}|\mathrm{ddiag}(\Xi^\top D) = 0\}, \tag{11}$$

and the orthogonal projection of a matrix $Z \in \mathbb{R}^{m\times k}$ onto the tangent space $T_D\mathcal{S}(m,k)$ with respect to the inner product $\langle \Xi, \Psi \rangle = \mathrm{tr}(\Xi^\top \Psi)$ as

$$\Pi_D(Z) := Z - D\mathrm{ddiag}(D^\top Z). \tag{12}$$

Then, by computing the first derivation of f at (D, P, Y) in tangent direction $(H_D, H_P, H_Y) \in T_{(D,P,Y)}\mathcal{S}(m,k) \times \mathcal{S}(d,k) \times \mathbb{R}^{d \times n}$, we get the Riemannian gradient of f at (D, P, Y) as

$$\operatorname{grad} f(D, P, Y) = \big(\Pi_D(\nabla_f(D)), \Pi_P(\nabla_f(P)), \nabla_f(Y)\big), \tag{13}$$

where $\nabla_f(D)$, $\nabla_f(P)$, and $\nabla_f(Y)$ are the Euclidean gradients of f with respect to the three arguments, respectively. Firstly, the Euclidean gradient $\nabla_f(D)$ of f with respect to D is computed as

$$\nabla_f(D) = \sum_{i=1}^{n} (D\phi_P(y_i) - 2x_i)\,\phi_P(y_i)^\top + \frac{2\mu_1}{k^2} D(D^\top D - P^\top P). \tag{14}$$

Using some shorthand notation, let Λ_i be the support of nonzero entries of $\phi_P(y_i)$, and denote $K_i := P_{\Lambda_i}^\top P_{\Lambda_i} - \lambda_2 I_k$, $r_i := P_{\Lambda_i}^\top y_i - \lambda_1 s_{\Lambda_i}$, $\Delta x_i := x_i - D\phi_P(y_i)$, and $q_i := r_i \Delta x_i^\top$. Then, the Euclidean gradient $\nabla_f(P)$ of f is computed as

$$\nabla_f(P) = \sum_{i=1}^{n} 2\mathcal{V}\{-y_i(\Delta x_i)^\top D_{\Lambda_i} K_i^{-1} + P_{\Lambda_i} K_i^{-1} D_{\Lambda_i}^\top q_i^\top K_i^{-1}$$

$$+ P_{\Lambda_i} K_i^{-1} q_i D_{\Lambda_i} K_i^{-1}\} + \frac{2\mu_1}{k^2} P(P^\top P - D^\top D) + \mu_1 \nabla_r(P), \tag{15}$$

with

$$\nabla_r(P) = P \sum_{1 \le i < j \le n} \frac{2p_i^\top p_j}{1 - (p_i^\top p_j)^2}(E_{ij} + E_{ji}) \tag{16}$$

being the gradient of the logarithmic barrier function (7). Here, $\mathcal{V}\{\cdot\}$ denotes the full vector of $\{\cdot\}$. By E_{ij}, we denote a matrix whose i^{th} entry in the jth column is equal to one, and all others are zero. Finally, the Euclidean gradient $\nabla_f(Y)$ is computed as

$$\nabla_f(Y) = \big[\mathcal{V}\{D_{\Lambda_1} K_1^{-1} \Delta_{x_1}\}, \dots, \mathcal{V}\{D_{\Lambda_n} K_n^{-1} \Delta_{x_n}\}\big]. \tag{17}$$

By assembling the Riemannian gradients, geodesics and parallel transports on the underlying manifolds, a conjugate gradient algorithm on $\mathcal{S}(m,k) \times \mathcal{S}(d,k) \times \mathbb{R}^{d \times n}$ is straightforward. Due to the page limit, we omit the presentation of the algorithm, and refer to [4] for more technical details.

4 Numerical Experiments

In this section, we investigate the performance of our proposed DR framework via couple dictionaries learning ($DRCDL$) and its linear version - compressive CDL ($CCDL$) for signal compression, reconstruction, and visualization. Before presenting our experiments, we briefly discuss the question of choosing the parameters in our formulation. Considering the high coherence among the images or

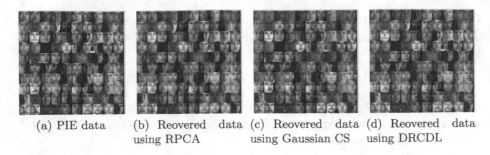

(a) PIE data   (b) Reovered data using RPCA   (c) Reovered data using Gaussian CS   (d) Reovered data using DRCDL

Fig. 1. From (b) to (d), recovering the reduced data from $d = 16$, using RPCA, Gaussian CS, and DRCDL, respectively. The PSNR is 23.01 dB, 25.28 dB and 31.12 dB.

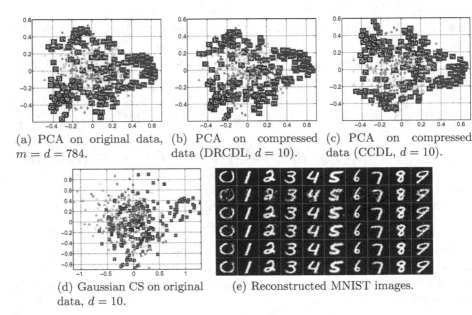

(a) PCA on original data, $m = d = 784$.   (b) PCA on compressed data (DRCDL, $d = 10$).   (c) PCA on compressed data (CCDL, $d = 10$).

(d) Gaussian CS on original data, $d = 10$.   (e) Reconstructed MNIST images.

Fig. 2. From (a) to (d), employing PCA on original data and reduced data, respectively. (e) shows that reconstructing MNIST images from $d = 16$ to $m = 784$, using CS and DRCDL, CCDL. From top to bottom (six rows): original data, recovered images using Gaussian CS, Gaussian CS based on K-SVD dictionary, DRCDL and CCDL.

imaginary patches, we prefer a dictionary with low redundancy, namely, $k \leq 2m$ for $D \in \mathbb{R}^{m \times k}$. For parameters (λ_1, λ_2) in Eq. (9), we put an emphasis on sparse solutions and choose $\lambda_2 \in (0, \frac{\lambda_1}{10})$, as proposed in [14]. The parameters for μ_1, μ_2 in (8) could be well tuned via performing cross validation. The CMU Multi-PIE [17] faces and MNIST handwritten digital databases[1] are used as the benchmark dataset for images compression, reconstruction and 2D visualization in our experiments. In order to evaluate our proposed method on DR and reconstruction,

---

[1] http://yann.lecun.com/exdb/mnist/.

(a) 33 Training images used for learning dictionary

(b) Lena   (c) DRCDL   (d) CCDL   (e) JPEG   (f) BCS

Fig. 3. Recovery performance on Lena with a compression rate $\eta = 32$. (b) is the corrupted image with PSNR = 16.12 dB; (c) to (f) are recovered images using DRCDL, CCDL, JPEG2000 and BCS. The PSNR (dB) is 26.52, 26.44, 16.60 and 22.28, respectively.

we compare it with the classic CS approach with the random Gaussian sensing matrix [2] (Gaussian CS), and robust principle components analysis (RPCA) [18]. In Figs. 1 and 2, 5000 images are randomly chosen for training D, and 500 images are randomly taken from remaining dataset for testing. We first reduce the dimensionality from $m = 1024, 784$ to $d = 16$ for PIE and MNIST, and then recover them using Gaussian CS, RPCA, and proposed DRCDL, CCDL, respectively. Figures 1 and 2(e) demonstrate that the proposed methods perform much better on signal reconstruction, in comparison with Gaussian CS and RPCA.

In Figure 2, we impose PCA on original data and reduced data respectively, to achieve a 2D visualization. Figure 2(b), (c) and (d) show that learning directly in the compressed domain, is feasible. Compared to Gaussian CS, our proposed methods (DRCDL and CCDL) exhibit more stable and competitive performance on the results of PCA, even in very low-dimensional compressed domain with $d = 10$. Figure 3 shows the results of image compression and recovery on single image - Lena. Compared to Bayesian CS (BCS) [19] and JPEG2000, our proposed methods *DRCDL* and *CCDL* exhibit a stronger performance, when the input data is heavy corrupted.

5   Conclusions

This paper proposes a coupled dictionary learning approach to achieve the task of invertible nonlinear DR. It aims to preserve distance information of the original high dimensional dataset in their low dimensional representations. Our experiments in image recovery and visualization demonstrate robust as well as competitive performance of the proposed framework, in comparison with the state of the art methods.

References

1. Van der Maaten, L.J., Postma, E.O., van den Herik, H.J.: Dimensionality reduction: a comparative review. J. Mach. Learn. Res. **10**(1–41), 66–71 (2009)

2. Donoho, D.L.: Compressed sensing. IEEE Trans. Inf. Theory **52**(4), 1289–1306 (2006)
3. Aharon, M., Elad, M., Bruckstein, A.: K-SVD: an algorithm for designing overcomplete dictionaries for sparse representation. IEEE Trans. Signal Process. **54**(11), 4311–4322 (2006)
4. Hawe, S., Seibert, M., Kleinsteuber, M.: Separable dictionary learning. In: IEEE Conference on Computer Vision and Pattern Recognition (CVPR), pp. 438–445. IEEE (2013)
5. Gleichman, S., Eldar, Y.C.: Blind compressed sensing. IEEE Trans. Inf. Theory **57**(10), 6958–6975 (2011)
6. Carvajalino, J.M.D., Sapiro, G.: Learning to sense sparse signals: simultaneous sensing matrix and sparsifying dictionary optimization. IEEE Trans. Image Process. **18**(7), 1395–1408 (2009)
7. Elad, M.: Optimized projections for compressed sensing. IEEE Trans. Signal Process. **55**(12), 5695–5702 (2007)
8. Calderbank, R., Jafarpour, S., Schapire, R.: Compressed learning: Universal sparse dimensionality reduction and learning in the measurement domain. Technical report, Computer Science, Princeton University (2009)
9. Zeyde, R., Elad, M., Protter, M.: On Single Image Scale-Up Using Sparse-Representations. In: Boissonnat, J.-D., Chenin, P., Cohen, A., Gout, C., Lyche, T., Mazure, M.-L., Schumaker, L. (eds.) Curves and Surfaces 2011. LNCS, vol. 6920, pp. 711–730. Springer, Heidelberg (2012)
10. Johnson, W.B., Lindenstrauss, J.: Extensions of Lipschitz mappings into a Hilbert space. Contemp. Math. **26**(189–206), 1 (1984)
11. Kim, H., Park, H., Zha, H.: Distance preserving dimension reduction for manifold learning. In: SDM, SIAM, pp. 527–532 (2007)
12. Baraniuk, R., Davenport, M., DeVore, R., Wakin, M.: A simple proof of the restricted isometry property for random matrices. Constructive Approximation **28**(3), 253–263 (2008)
13. Elad, M.: Sparse and Redundant Representations: From Theory to Applications in Signal and Image Processing. Springer, New York (2010)
14. Zou, H., Hastie, T.: Regularization and variable selection via the elastic net. J. Roy. Stat. Soc. Ser. B (Stat. Methodol.) **67**(2), 301–320 (2005)
15. Mairal, J., Bach, F., Ponce, J.: Task-driven dictionary learning. IEEE Trans. Pattern Anal. Mach. Intell. **34**(4), 791–804 (2012)
16. Wei, X., Shen, H., Kleinsteuber, M.: An adaptive dictionary learning approach for modeling dynamical textures. In: Proceedings of the 39th IEEE International Conference on Acoustics, Speech, and Signal Processing (ICASSP), pp. 3567–3571 (2014)
17. Sim, T., Baker, S., Bsat, M.: The CMU pose, illumination, and expression (PIE) database. In: Fifth IEEE International Conference on Automatic Face and Gesture Recognition (FG), pp. 46–51. IEEE (2002)
18. De la Torre, F., Black, M.J.: Robust principal component analysis for computer vision. In: Eighth IEEE International Conference on Computer Vision (ICCV), vol. 1, pp. 362–369. IEEE (2001)
19. Ji, S., Xue, Y., Carin, L.: Bayesian compressive sensing. IEEE Trans. Signal Process. **56**(6), 2346–2356 (2008)

Patchworking Multiple Pairwise Distances for Learning with Distance Matrices

Ken Takano[2], Hideitsu Hino[1]([✉]), Yuki Yoshikawa[2], and Noboru Murata[2]

[1] University of Tsukuba, 1–1–1 Tennoudai, Tsukuba, Ibaraki 305–8573, Japan
hinohide@cs.tsukuba.ac.jp
[2] Waseda University, 3–4–1 Ohkubo, Tokyo, Shinjuku-ku 169–8555, Japan

Abstract. A classification framework using only a set of distance matrices is proposed. The proposed algorithm can learn a classifier only from a set of distance matrices or similarity matrices, hence applicable to structured data, which do not have natural vector representation such as time series and graphs. Random forest is used to explore ideal feature representation based on the distance between points defined by a set of given distance matrices. The effectiveness of the proposed method is evaluated through experiments with point process data and graph structured data.

Keywords: Classification · Structured data · Decision trees · Random forest · Spike train · Graph kernel

1 Introduction

In recent years, the need for dealing with *structured* objects are increasing. Structured objects does not have natural vector representation, and the conventional Euclidean distance cannot be used for measuring distance or similarity of these objects. Examples of structured data include time series, graphs, character strings and genome sequences. For these structured data, if we could define an appropriate metric between each pair of objects, distance based classifier such as k-nearest neighbor (k-NN; [1]) or similarity based classifier such as the support vector machine (SVM [2]) are of the choice for the purpose of classification. For the objects represented as vectors, distance metric learning (DML [4–7]) is one of the standard approaches for optimizing the distance metric for the given task and given dataset. However, most of existing works on DML assume that objects are represented as vectors, and the problem of DML is formulated as a learning of the Mahalanobis distance matrix.

Different methods are developed in different research fields to define metrics for structured data. Designing a kernel function to capture the intrinsic similarity of structured objects has been extensively studied [8]. One example is found in graphs; graphs do not have fixed length vector representation in general, and there are various alternatives for similarity measures or kernels between graphs developed to analyse graph structured data such as Web graphs, power grid networks, and protein interaction networks [9–11]. Another example is analysis

© Springer International Publishing Switzerland 2015
E. Vincent et al. (Eds.): LVA/ICA 2015, LNCS 9237, pp. 287–294, 2015.
DOI: 10.1007/978-3-319-22482-4_33

of point processes, i.e., sequences of time points at which events occur [12]; since its discrete nature, the realization of point process cannot be identified with a function in L_2 space commonly done in continuous time series analysis, and there are considerable effort to define distance metrics for point process data in the literature of neuroscience [13–19]. As described above, structured data such as graphs and point processes have a number of different metrics, and given these kind of structured datasets, it is not obvious which metric is the most appropriate for the given task with the dataset at hand, and the problem of metric selection remains as an issue for the data analyst.

In this paper, we consider the situation that we are given a set of different distance matrices calculated for a set of structured objects, and solely based on them, we aim at learning a highly accurate classifier. The proposed approach is instantiated by using the random forest (RF [20]), hence the proposed method can handle both classification problems and regression problems.

2 Notation and Problem Setting

Let $O = \{o_i\}_{i=1}^n$ be a set of objects $o_i \in \mathcal{O}$ of size n, where $\mathcal{O}$ is some space to which objects belong to. We consider classification problems, and for an object $o_i \in \mathcal{O}$, there is an output $y_i \in \mathcal{Y}$, where $\mathcal{Y}$ is a discrete set of class labels. Given a set of observations $\mathcal{D} = \{(o_i, y_i)\}_{i=1}^n$, we consider a predictor $f : \mathcal{O} \to \mathcal{Y}$, which try to approximate the underlying correspondence of the input object and the output label.

We consider the case where the objects are not represented as vectors. Suppose there are m different distance measures for a pair of objects (o_i, o_j), namely, there are m distance functions $d^{(l)} : \mathcal{O} \times \mathcal{O} \to \mathbb{R}$. We assume that we observe m distance matrices $D^{(1)}, \ldots, D^{(m)}$, where $D^{(l)} \in \mathbb{R}^{n \times n}, [D^{(l)}]_{ij} = d^{(l)}(o_i, o_j)$. We assume that all we are given are distance matrices calculated based on those distance measures and a set of outputs $\{y_i\}_{i=1}^n$ for each objects. We propose to construct a predictor f based on whether a new input is close to or away from training samples. In the following section, we propose a concrete procedure for learning the predictor f by using a set of distance matrices.

We note that the problem setting in this paper is similar to that of the multiple kernel learning (MKL [22, 23]), in which the optimal combination of the given kernel matrices are explored. Distance matrix $D \in \mathbb{R}^{n \times n}$ and kernel matrix $K \in \mathbb{R}^{n \times n}$ are mutually convertible, though, the way of conversion is not unique. In this work, distance matrix D is transformed to a kernel matrix K by the double centering $2K_{ij} = -D_{ij} + \frac{1}{n} \sum_{l=1}^n D_{il} + \frac{1}{n} \sum_{l=1}^n D_{li} - \frac{1}{n^2} \sum_{k=1}^n \sum_{l=1}^n D_{kl}$ followed by truncation of negative eigenvalues to make the kernel matrix positive semi-definite. Given a kernel matrix K, the Euclidean distance matrix in the kernel-induced feature space is obtained by $D_{ij} = K_{ii} + K_{jj} - 2K_{ij}$. In experimental section, we interchangeably use distance and kernel (similarity) by using either of the above explained transformations, and the performance of some MKL algorithms and that of the proposed algorithm are compared. Also, by optimizing the convex combination of different distance matrices, a modified

LMNN (Large Margin Nearest Neighbor) method is proposed in [24]. This variant of LMNN for distance matrices is also included for the target of comparison in the experimental section.

3   Implementation Based on Trees

Given distance matrices

$$D = (D^{(1)}, \ldots, D^{(m)}) \in \mathbb{R}^{n \times mn}, \tag{1}$$

we consider their concatenation. We extract the feature vector corresponding to the i-th object o_i as

$$\xi_i = (d^{(1)}(o_i, o_1), d^{(1)}(o_i, o_2), \ldots, d^{(2)}(o_i, o_1), \ldots, d^{(m)}(o_i, o_n)) \in \mathbb{R}^{mn}, \tag{2}$$

which is illustrated in Fig. 1. When a new object o_{new} to be classified is given, we first calculate distances from all of the training samples with respect to m distance metrics[1], and concatenate the distances to obtain the mn-dimensional feature vector representation

$$\xi_{new} = (d^{(1)}(o_{new}, o_1), d^{(1)}(o_{new}, o_2), \ldots, d^{(m)}(o_{new}, o_n)) \in \mathbb{R}^{mn} \tag{3}$$

of the new object to be inputted to the learned classifier.

Elements of ξ_i are distances from o_i to other objects, measured by m different distance measures. We propose to use ξ_i as an input for the random forest (RF [20]). The RF is composed of a number of binary decision trees trained by using bootstrap samples of data with randomly sampled elements. We hereafter explain how each binary decision tree works. A binary decision tree is constructed by repeated splits of subsets of $\mathbb{R}^{nm}$ into two descendant subsets, beginning with $\mathbb{R}^{nm}$ itself. The terminal subsets form partitions of input space, in our case, $\mathbb{R}^{nm}$. Each terminal subset is designated by a class label.

In Fig. 2, we show an illustrative example showing the expected behaviour of the learnt decision tree with inputs ξ_i. For the purpose of illustration, we assume binary classification problem as shown in Fig. 2, where an object o_i is a point in $\mathbb{R}^2$. We introduce two distance measures $d^{(1)}$ and $d^{(2)}$, which are Euclidean distances calculated only using the first and the second dimension of the points, respectively. Two spirals correspond to two distinct classes (a point in class 1 is marked $\circ$ and a point in class 2 is marked $\triangle$), and each class has 50 samples. Since the number of samples $n = 100$ and the number of different distance metrics is $m = 2$, the dimensionality of the feature vector $\xi_i, i = 1, \ldots, n$ is $mn = 200$. Tree in Fig. 2 is a binary decision tree learned by using the dataset $\{\xi_i\}_{i=1}^n, \xi_i \in \mathbb{R}^{200}$. In the root node of the tree, the first decision is made based on the distance from the sample o_{98} with respect to the second distance metric $d^{(2)}$. When a new object o_{new} is inputted, if $d^{(2)}(o_{new}, o_{98}) > 3.65$, the object is immediately labelled as class 1. If the distance is ≤ 3.65 (shaded region in

---

[1] So, we need to access to the distance functions.

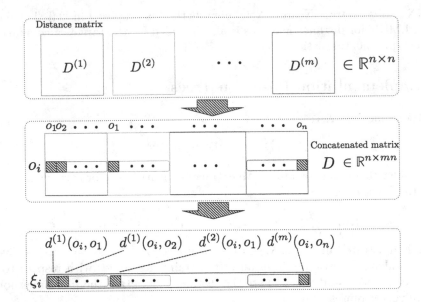

Fig. 1. Feature extraction procedure for an object o_i from a given set of distance matrices. m distance matrices $D^{(l)}, l = 1, \ldots, m$ are concatenated into a large matrix $D \in \mathbb{R}^{n \times mn}$. From the matrix D, a feature vector $\boldsymbol{\xi}_i$ for an object o_i is extracted as the i-th row of the matrix.

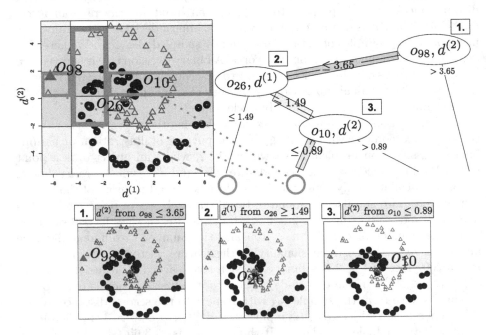

Fig. 2. Illustrative example of classification by tree with input $\boldsymbol{\xi} \in \mathbb{R}^{nm}$.

the bottom left panel), we proceed to the next step of the decision tree. In the next step, the distance from o_{26} with respect to $d^{(1)}$ is examined, and if it is less than 1.49, the sample is labeled as class 1. Otherwise, we proceed to the final step and $d^{(2)}(o_{new}, o_{10})$ is examined. The class assignment is done by averaging the outcomes of all decision trees. We note that the proposed method is similar to the *landmark method* for learning with similarity function [21]. The defining difference from the method used in [21] is that our method utilize the RF for adaptively select the elements of feature vectors used for classification.

4   Experiments

To show the effectiveness of the proposed method, two different classes of structured objects are considered. The one is the spike train, which is considered as examples of point process data and of importance in neuroscience, and the other is graphs, which is one of the most popular structured objects.

4.1   Experimental Protocol

Given a set of structured objects $O = \{o_i\}_{i=1}^n$, we make a set of distance or kernel matrices by using m different algorithms. Based on the set of matrices, we perform classification by using the following methods:

Single: Apply k-NN classification, the SVM, and the RF for each single matrix. Each matrix is (if necessary) transformed to a distance matrix for k-NNs and RF, and to a kernel matrix for SVMs. The RF is performed using the feature vector extracted from the distance matrix.

Averaging: Apply the three classifiers to a matrix D obtained by averaging the m matrices as $D = \frac{1}{m} \sum_{l=1}^m D_l$ for k-NNs and for RFs, and $K = \frac{1}{m} \sum_{l=1}^m K_l$ for SVMs.

MKL1,2: The m different matrices are transformed into kernel matrices, and then combined into one kernel matrix by MKL algorithms MKL1(called `align` in the original paper) and MKL2(called `alignf`), which are computationally efficient and shown to offer state-of-the-art classification accuracy [23]. The obtained matrices are used to train k-NNs, SVMs, and RFs.

LMNN: The m different matrices are transformed into distance matrices, and combined into one distance matrix by a variant of the LMNN algorithm [24] for convex combination of distance matrices. The obtained matrix is used to train the three classifiers.

Proposed: The m different distance matrices are concatenated by row, and the proposed method based on the RF is applied.

In summary, m different matrices plus four combined matrices (Averaging, MKL1,2, and LMNN) are prepared for three classifiers as targets for comparison to the proposed method, hence the total number of classifiers is $3 \times (m + 4) + 1$.

The given dataset is divided into 10 disjoint subsets, and the classification error rates are calculated by 10-fold cross validation. Tuning parameters of three

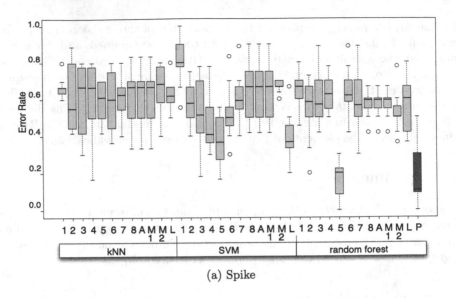

(a) Spike

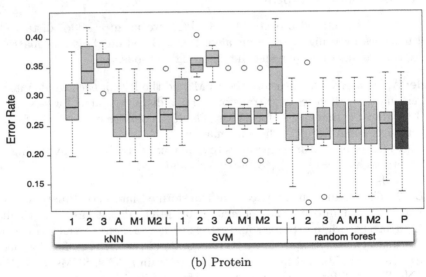

(b) Protein

Fig. 3. Classification errors of various methods for spike train and protein datasets. The x-axis corresponds to different methods divided into three categories, kNN, SVM, and RF, according to the base classifiers. The numbers 1 to 8 indicate single metrics. Other symbols indicate different strategies to combine matrices, where A, M1, M2, L, and P are Average, MKL1, MKL2, LMNN, and the Proposed method, respectively.

classifiers, the number k in k-NNs, soft margin parameter $C > 0$ in SVMs, and the number of trees and the depth of trees in the RF, are optimized by 10-fold cross validation using only training dataset.

4.2 Spike Train and Protein Graph Datasets

We evaluate the classification methods using two types of structured data, the spike train and graph.

As a spike train dataset, we use an artificial data created by Fellous et al. [25]. The original dataset with detailed description is available from http://cnl.salk.edu/fellous/data/JN2004data/data.html. We adopted a three classes dataset (Spike), which is a set of artificially generated spike patterns. Each class has 35 samples and the dataset Spike is composed of 105 samples. There are considerable number of metrics for spike trains. In this paper, we use three distance metrics ([13–15]) and five kernels ([16–19], where [19] proposed two kernels) for point process data. Since we have $m = 8$ different metrics, the number of the methods for comparison is $3 \times (m + 4) + 1 = 37$.

As a graph objects, we use a graph made on the Protein dataset, which comes with the task of classifying enzymes and non-enzymes. The number of samples in Protein is 1113. The GraphHopper kernel [11], the PROP kernel [10], and the Weisfeiler-Lehman kernel [9] are selected for the kernel functions for graph structured objects because of their computational efficiency and wide use in the literature. Implementation and datasets are downloaded from http://image.diku.dk/aasa/software.php. Since we have $m = 3$ different kernel matrices, the number of the methods for comparison is $3 \times (m + 4) + 1 = 22$. Boxplots of the classification errors are shown in Fig. 3, which shows that the proposed method performs comparable or superior to other classification methods.

5 Conclusion

In this paper, we constructed a simple classifier which can be trained using only a set of distance matrices. The aim of the proposed method is dealing with structured data, and frees us from the problem of metric selection for such structured data. Considering the increase of non-vectorial, complicated data, the need for a principled and automatic method for selecting or combining many candidate metrics for such data is increasing. In spite of its simplicity of the proposed method, the method offers favorable classification accuracy as shown by experiments using both point process datasets and graph datasets.

The proposed method can efficiently combine different distance metrics, and can find an appropriate distance measure using the supervised information $\{y_i\}$. Distance measure is an important latent structure for an abstract space of objects, and the analysis of the obtained distance structure by the proposed method is our important future work. As mentioned in Sect. 3, the proposed approach have similarity to the learning with similarity function. Theoretical analysis of our propose method will be investigated in the same line of [21].

Acknowledgements. Part of this work is supported by KAKENHI No.26120504, 25870811, and 25120009.

References

1. Cover, T., Hart, P.: Nearest neighbor pattern classification. IEEE Trans. Inf. Theor. **13**(1), 21–27 (1967)
2. Boser, B., et al.: A training algorithm for optimal margin classifiers. In: COLT (1992)
3. Yang, L., Jin, R.: Distance metric learning: A comprehensive survey. Technical report: Michigan State University (2006)
4. Kulis, B.: Metric learning: a survey. Found. Trends Mach. Learn. **5**(4), 287–364 (2013)
5. Goldberger, J., et al.: Neighborhood component analysis. In: NIPS (2004)
6. Weinberger, K., et al.: Distance metric learning for large margin nearest neighbor classification. In: NIPS (2006)
7. Davis, J.V., et al.: Information-theoretic metric learning. In: ICML (2007)
8. Shawe-Taylor, J., Cristianini, N.: Kernel Methods for Pattern Analysis. Cambridge University Press, New York (2004)
9. Shervashidze, N., Borgwardt, K.M.: Fast subtree kernels on graphs. In: NIPS (2009)
10. Neumann, M., Patricia, N., Garnett, R., Kersting, K.: Efficient graph kernels by randomization. In: Flach, P.A., De Bie, T., Cristianini, N. (eds.) ECML PKDD 2012, Part I. LNCS, vol. 7523, pp. 378–393. Springer, Heidelberg (2012)
11. Feragen, A., et al.: Scalable kernels for graphs with continuous attributes. In: NIPS (2013)
12. Reiss, R.-D.: A Course on Point Processes. Springer Series in Statistics. Springer, New York (1993)
13. Kreuz, T., et al.: Measuring spike train synchrony. J. Neurosci. Methods **165**(1), 151–161 (2007)
14. van Rossum, M.C.W.: A novel spike distance. Neural Compt. **13**(4), 751–763 (2001)
15. Houghton, C.: Studying spike trains using a van rossum metric with a synapse-like filter. J. Comput. Neurosci. **26**(1), 149–155 (2009)
16. Hunter, J.D., Milton, J.G.: Amplitude and frequency dependence of spike timing: implications for dynamic regulation. J. Neurophysiol. **90**(1), 387–394 (2003)
17. Quiroga, R.Q.: Event synchronization: a simple and fast method to measure synchronicity and time delay patterns. Phys. Rev. E **66**, 041904 (2002)
18. Schreiber, S., et al.: A new correlation-based measure of spike timing reliability. Neurocomputing **52**, 925–931 (2003)
19. Paiva, A.R.C., et al.: A reproducing kernel hilbert space framework for spike train signal processing. Neural Comput. **21**(2), 424–449 (2009)
20. Breiman, L.: Random forests. Mach. Learn. **45**(1), 5–32 (2001)
21. Balcan, M., Blum, A.: On a theory of learning with similarity functions. In: Proceedings of ICML (2006)
22. Lanckriet, G.R.G., et al.: Learning the kernel matrix with semidefinite programming. J. Mach. Learn. Res. **5**, 27–72 (2004)
23. Cortes, C., et al.: Algorithms for learning kernels based on centered alignment. J. Mach. Learn. Res. **13**, 795–828 (2012)
24. Suryanto, C.H., et al.: Combination of multiple distance measures for protein fold classification. In: ACPR (2013)
25. Fellous, J.M., et al.: Discovering spike patterns in neuronal responses. J. Neurosci. **24**(12), 2989–3001 (2004)

Robust Structured Low-Rank Approximation on the Grassmannian

Clemens Hage$^{(\boxtimes)}$ and Martin Kleinsteuber

Department of Electrical Engineering and Information Technology,
Technische Universität München, Arcisstr. 21, 80333 Munich, Germany
{hage,kleinsteuber}@tum.de
http://www.gol.ei.tum.de

Abstract. Over the past years Robust PCA has been established as a standard tool for reliable low-rank approximation of matrices in the presence of outliers. Recently, the Robust PCA approach via nuclear norm minimization has been extended to matrices with linear structures which appear in applications such as system identification and data series analysis. At the same time it has been shown how to control the rank of a structured approximation via matrix factorization approaches. The drawbacks of these methods either lie in the lack of robustness against outliers or in their static nature of repeated batch-processing. We present a Robust Structured Low-Rank Approximation method on the Grassmannian that on the one hand allows for fast re-initialization in an online setting due to subspace identification with manifolds, and that is robust against outliers due to a smooth approximation of the ℓ_p-norm cost function on the other hand. The method is evaluated in online time series forecasting tasks on simulated and real-world data.

1 Introduction

Many applications such as system identification and time series analysis motivate the problem of Structured Low-Rank Approximation (SLRA). While common low-rank approximations like PCA aim to find a low-dimensional subspace to represent high-dimensional data optimally with respect to some norm or divergence, in the structured case this problem is extended by the additional constraint that the low-rank approximation has to meet a certain linear structure (Hankel, Toeplitz, Sylvester).

For the prominent case of Hankel matrices a method dubbed Singular Spectrum Analysis (SSA) [4] has been presented, which performs the simplest way of SLRA in that it computes a low-rank approximation of a (Hankel-) structured matrix followed by a so-called diagonal averaging step, which is the projection onto the space of Hankel matrices. An obvious drawback of this method is that this projection can destroy the low-rank property established before. The method of Cadzow [6] alternates between these two steps until the algorithm converges to a solution that is indeed low-rank and structured. However, as Chu et al. [8] and Markovsky [12] state, this solution can be far away from the initialization with no guarantees of finding an actually meaningful approximation to the data.

E. Vincent et al. (Eds.): LVA/ICA 2015, LNCS 9237, pp. 295–303, 2015.
DOI: 10.1007/978-3-319-22482-4_34

Recently, Ishteva et al. [11] have proposed a factorization approach with a cost function that joints the structural and low-rank constraint. The dimension of the two matrix factors is an upper bound on the rank of the approximation and its structure is enforced with a side condition. The approximation is fitted to the data according to an ℓ_2-norm, although it is well known that low-rank approximations of this kind can be vulnerable against outliers. For the unstructured case this has been the major incentive to move from PCA to robustified PCA methods such as the Robust PCA method by Candes et al. [7] which recovers a subspace in the presence of sparse outliers of great magnitude. Ayazoglu et al. [1] have proposed to extend this concept to structured matrices by introducing additional Lagrangian multipliers. Their method is called Structured Robust PCA (SRPCA), and its performance is demonstrated in visual applications like Target Location Prediction, Tracklet Matching (both matrix completion problems) and Outlier Removal from trajectories, which can be interpreted as outlier identification through robust subspace estimation.

One of the drawbacks of many Robust PCA approaches and thus also the SRPCA approach is their batch-processing nature. In an online setting the algorithm needs to be re-run from scratch if the data set grows or changes over time. This can be alleviated by factorizing the low-rank approximation of data into an orthogonal matrix representing the subspace and a coefficient matrix containing the coordinates of the currently observed data in this subspace, cf. [2,9] for the unstructured case. Whenever new data comes in, it is possible to initialize the subspace optimization with the previous estimate and to update both the subspace and the coordinates to the new data. Obviously, whenever the subspace does not change significantly this saves computational effort compared to a random initialization. We will firstly derive a batch algorithm for Robust Structured Low-Rank Approximation on the Grassmannian and then outline how to process structured data online in an efficient way. We illustrate the performance of the proposed algorithm on several time series forecasting tasks.

2   Robust Structured Low-Rank Approximation on the Grassmannian Using a Smoothed ℓ_p-norm Cost Function

2.1   Low-Rank and Sparsity Constraints in the Unstructured Case

Low-rank approximation of data is a well-studied problem, cf. [13] for a recent overview. In the past years a trend can be seen towards Robust PCA methods that are tolerant against outliers in the data. An often considered data model is $\mathbf{X} = \mathbf{L} + \mathbf{S}$, where the input data $\mathbf{X} \in \mathbb{R}^{m \times n}$ is assumed to be composed of a low-rank part $\mathbf{L}$ with $\mathrm{rank}(\mathbf{L}) \leq k$ and a sparse matrix $\mathbf{S}$ that contains few non-zero entries. While in [7] the low-rank constraint is enforced via minimization of the nuclear norm, in this work we will consider the Robust PCA setting on the *Grassmannian*

$$\mathrm{Gr}_{k,m} := \{\mathbf{P} \in \mathbb{R}^{m \times m} | \mathbf{P} = \mathbf{U}\mathbf{U}^\top, \mathbf{U} \in \mathrm{St}_{k,m}\}. \tag{1}$$

The subspace is hereby represented by an element $\mathbf{U}$ of the *Stiefel* manifold $\mathrm{St}_{k,m} = \{\mathbf{U} \in \mathbb{R}^{m \times k} | \mathbf{U}^\top \mathbf{U} = \mathbf{I}_k\}$, where $\mathbf{I}_k$ is the $(k \times k)$-identity matrix. The low-rank approximation is then $\mathbf{L} = \mathbf{U}\mathbf{Y}$ with $\mathbf{Y} \in \mathbb{R}^{k \times n}$.

In contrast to the ℓ_2-norm in common PCA, robust approaches use relaxations of the ℓ_0-norm for fitting the low-rank approximation to the data, as gross outliers might otherwise distort the estimation. The smoothed ℓ_p-norm

$$h_\mu \colon \mathbb{R}^{m \times n} \to \mathbb{R}^+, \quad X \mapsto \sum_{j=1}^{n} \sum_{i=1}^{m} \left(x_{ij}^2 + \mu \right)^{\frac{p}{2}}, \quad 0 < p < 1 \tag{2}$$

presented in [9] behaves similarly to the ℓ_0-norm for sparse outliers, but treats small additive Gaussian noise like an ℓ_2 norm. A Robust Low-Rank Approximation problem using the sparsifying function (2) writes as

$$\left(\hat{\mathbf{U}}, \hat{\mathbf{Y}} \right) = \arg \min_{\mathbf{U} \in \mathrm{St}_{m,k}, \mathbf{Y} \in \mathbb{R}^{k \times n}} h_\mu (\mathbf{X} - \mathbf{U}\mathbf{Y}), \qquad \hat{\mathbf{L}} = \hat{\mathbf{U}}\hat{\mathbf{Y}}. \tag{3}$$

The sparse component can be recovered via $\hat{\mathbf{S}} = \mathbf{X} - \hat{\mathbf{L}}$, possibly followed by a thresholding operation to remove residual noise.

2.2 Extension to Linear Matrix Structures and Algorithmic Description

Besides the low-rank and sparsity decomposition no other assumptions are made on the data in (3). In many applications however, structured matrices like Hankel, Sylvester or Toeplitz matrices play an important role. Following the notation of [11] we denote by $\mathcal{S}$ a linear matrix structure, by $\mathcal{S}(\mathbf{d})$ a structured matrix obtained from a data series $\mathbf{d}$ and by $\mathcal{P}_\mathcal{S}(\mathbf{X})$ the orthogonal projection of any (possibly unstructured) matrix $\mathbf{X}$ onto the image of $\mathcal{S}$ w.r.t. the standard inner product. For example, if $\mathcal{H}$ is a Hankel structure the orthogonal projection $\mathcal{P}_\mathcal{H}$ is equivalent to the diagonal averaging step in SSA [4].

In order to include the structural constraint we extend the cost function (3) by the side condition $\mathbf{L} \in \mathcal{S}$. This motivates the Lagrangian Multiplier scheme

$$\min_{\mathbf{U} \in \mathrm{St}_{m,k}, \mathbf{Y} \in \mathbb{R}^{k \times n}, \Lambda \in \mathbb{R}^{m \times n}} h_\mu (\mathbf{X} - \mathbf{U}\mathbf{Y}) + \langle \Lambda, \mathbf{U}\mathbf{Y} - \mathcal{P}_\mathcal{S}(\mathbf{U}\mathbf{Y}) \rangle + \tfrac{\rho}{2} \|\mathbf{U}\mathbf{Y} - \mathcal{P}_\mathcal{S}(\mathbf{U}\mathbf{Y})\|_F^2.$$

$$\tag{4}$$

Algorithm 1 outlines the extension of the Robust PCA method from Hage and Kleinsteuber [9] to the case of structured matrices. The algorithm considers a partial observation $\hat{\mathbf{X}}$ of the data with $\mathcal{A}$ defining which entries are actually observed. In each iteration, three steps are performed. Firstly, the subspace estimate is updated. In this realization the subspace is uniquely identified with a Grassmannian projector $\mathbf{P} = \mathbf{U}\mathbf{U}^\top$ with $\mathbf{U} \in \mathrm{St}_{k,m}$. The optimization problem is solved via Conjugate Gradient (CG) descent with backtracking line-search using a QR-decomposition based retraction on the Grassmannian. Once $\mathbf{P}$ and thereby $\mathbf{U}$ have been found the coordinates $\mathbf{Y}$

are updated with a CG method in Euclidean space, such that a new optimum low-rank estimate $\mathbf{L} = \mathbf{U}\mathbf{Y}$ is found. In a third step the Lagrangian multiplier Λ is updated, then ρ is increased and μ is decreased. While μ controls the behavior of the sparsifying function $h_\mu(\cdot)$, the parameter ρ weighs between the data fitting term and the structural side condition. More precisely, as long as ρ is small the Robust PCA term is the leading power and a low-rank approximation is fitted to the data. With increasing ρ the structural condition is more and more enforced until it is the dominating term in the cost function. In a practical application the optimization can also be terminated if the residual Hankel penalty ϵ, i.e. the Frobenius distance to the next Hankel-structured matrix normalized by the number of matrix entries falls below a certain threshold τ.

Algorithm 1. Alternating minimization scheme for Grassmannian Robust Structured Low-Rank Approximation

---

Initialize:

Choose $\mathbf{X}_0 \in \mathbb{R}^{m \times n}$, s.t. $\mathcal{A}(\mathbf{X}_0) = \hat{\mathbf{X}}$.

Initialize $\mathbf{U}^{(0)}$ randomly or from k left singular vectors of $\mathbf{X}_0$.

$\mathbf{Y}^{(0)} = \mathbf{U}^{(0)\top}\mathbf{X}_0, \mathbf{L}^{(0)} = \mathbf{U}^{(0)}\mathbf{Y}^{(0)}$

$\mathbf{P}^{(0)} = \mathbf{U}^{(0)}\mathbf{U}^{(0)\top}$

Choose $\mu^{(0)}$ and $\mu^{(I)}$, compute $c_\mu = \left(\frac{\mu^{(I)}}{\mu^{(0)}}\right)^{1/(I-1)}$

Choose $\rho^{(0)}$ and $\rho^{(I)}$, compute $c_\rho = \left(\frac{\rho^{(I)}}{\rho^{(0)}}\right)^{1/(I-1)}$

for $i = 1 : I$ **do**

$$\mathbf{P}^{(i+1)} = \underset{\mathbf{P} \in \mathrm{Gr}_{k,m}}{\arg\min} \, h_{\mu^{(i)}}(\hat{\mathbf{X}} - \mathcal{A}(\mathbf{P}\mathbf{L}^{(i)}))$$
$$- \left\langle \Lambda^{(i)}, \mathbf{P}\mathbf{L}^{(i)} - \mathcal{P}_\mathcal{S}\left(\mathbf{P}\mathbf{L}^{(i)}\right)\right\rangle$$
$$+ \frac{\rho^{(i)}}{2}\|\mathbf{P}\mathbf{L}^{(i)} - \mathcal{P}_\mathcal{S}\left(\mathbf{P}\mathbf{L}^{(i)}\right)\|_F^2$$

find $\mathbf{U}^{(i+1)}$ s.t. $\mathbf{U}^{(i+1)}\mathbf{U}^{(i+1)\top} = \mathbf{P}^{(i+1)}$

Subspace Step

$$\mathbf{Y}^{(i+1)} = \underset{\mathbf{Y} \in \mathbb{R}^{k \times n}}{\arg\min} \, h_{\mu^{(i)}}(\hat{\mathbf{X}} - \mathcal{A}(\mathbf{U}^{(i+1)}\mathbf{Y}))$$
$$- \left\langle \Lambda^{(i)}, \mathbf{U}^{(i+1)}\mathbf{Y} - \mathcal{P}_\mathcal{S}\left(\mathbf{U}^{(i+1)}\mathbf{Y}\right)\right\rangle$$
$$+ \frac{\rho^{(i)}}{2}\|\mathbf{U}^{(i+1)}\mathbf{Y} - \mathcal{P}_\mathcal{S}\left(\mathbf{U}^{(i+1)}\mathbf{Y}\right)\|_F^2$$

$\mathbf{L}^{(i+1)} = \mathbf{U}^{(i+1)}\mathbf{Y}^{(i+1)}$

Coordinate Step

$$\Lambda^{(i+1)} = \Lambda^{(i)} - \rho^{(i)}\left(\mathbf{L}^{(i+1)} - \mathcal{P}_\mathcal{S}\left(\mathbf{L}^{(i+1)}\right)\right)$$

Multiplier Update

$$\mu^{(i+1)} = c_\mu \mu^{(i)}, \quad \rho^{(i+1)} = c_\rho \rho^{(i)}$$

$$\epsilon = \frac{1}{mn}\|\mathbf{L}^{(i+1)} - \mathcal{P}_\mathcal{S}\left(\mathbf{L}^{(i+1)}\right)\|_F$$

end for

$\hat{\mathbf{L}} = \mathcal{P}_\mathcal{S}\left(\mathbf{L}^{(I)}\right), \quad \hat{\mathbf{S}} = \mathbf{X} - \hat{\mathbf{L}}$

---

3   Efficient Online Time Series Forecasting via Robust SLRA

We have outlined how to use manifold optimization for Robust SLRA on the Grassmannian. In the important case of a Hankel structure SLRA corresponds to identifying an LTI system, cf. [13]. In practical applications, however, an observed system might be time-variant. Or the observed data is not related to a physical system at all but still exhibits repetitive or periodic behavior. The field of Time Series Analysis [3] deals with these signals and numerous auto-regressive methods for filtering and forecasting data series exist. In the SLRA context a low-rank Hankel matrix (and thus an LTI) is fitted to the observed data and the future development is extrapolated from this approximation. Thereby, the rank bounds the complexity of the approximation. If the behavior of the data changes over time a new model needs to be determined for each observation instance. For our proposed Grassmannian Robust SLRA method this means that both a new subspace and new coordinates need to be computed. However, when the signal characteristics vary moderately over time it is likely that the new subspace lies close to the previously found one. Therefore, the subspace should not be randomly initialized but rather updated with the new data point, in a similar way as Robust Subspace Tracking ([10,14]) in the unstructured case. As discussed earlier, however, we do not optimize directly on the space of structured low-rank matrices. Instead we relax the structural constraint, update the subspace and then tighten the structural side condition again by varying the parameter ρ in the cost function.

In Algorithm 2 we describe an Online Time Series Forecasting method based on Robust Structured Low-Rank Approximation on the Grassmannian. The algorithm receives as inputs the time series of data $\mathbf{d}$ to be analyzed as well as the desired order of the system and the forecasting range, i.e. the number of samples to be predicted. Since a Hankel matrix of size $m \times m$ contains $(2m-1)$ samples of data, the prediction starts at $\mathbf{d}(2m)$. A data vector $\mathbf{x}_{(j)}$ of length $(2m-1)$ is extracted from the data up to the present position, padded with zeros according to the forecasting range and structured to form a Hankel matrix. The first subspace estimate $\mathbf{U}^{(0)}$ is not initialized randomly but with $\mathbf{U}_{(j-1)}$, the final subspace estimate of the previous set of data samples. Note that the subscript (j) counts the position of the current set of data samples in the data stream while in Algorithm 1 the superscript (i) denoted the iter-

---

Algorithm 2. Online Time Series Forecasting via Grassmannian Robust SLRA

Input: data series $\mathbf{d} \in \mathbb{R}^N$, system order k, analysis dimension m, forecasting range r

for $j = 2m : N$ **do**

  Define $\mathbf{x}_{(j)} = \left[\mathbf{d}(j-(2m-1):j)^\top \mid \mathbf{0}_r^\top \right]^\top$
  and $\mathcal{A}$ according to forecasting range

  Obtain $\mathbf{X}_{(j)} = \mathcal{H}(\mathbf{x}_{(j)})$
  Initialize $\mathbf{U}^{(0)} = \mathbf{U}_{(j-1)}$
  $\mathbf{Y}^{(0)} = \mathbf{U}^{(0)\top}\mathbf{X}_{(j)}$, $\mathbf{L}^{(0)} = \mathbf{U}^{(0)}\mathbf{Y}^{(0)}$
  $\mathbf{P}^{(0)} = \mathbf{U}^{(0)}\mathbf{U}^{(0)\top}$
  Select number of iterations $I_{(j)}$
  Choose $\rho^{(0)}$ and $\rho^{(I)}$
  Compute $c_{\rho(j)} = \left(\frac{\rho^{(I)}}{\rho^{(0)}} \right)^{1/(I_{(j)}-1)}$
  for $i = 1 : I_{(j)}$ **do**

    Subspace Step from Alg. 1
    Coordinate Step from Alg. 1
    Multiplier Update from Alg. 1
    $\rho^{(i+1)} = c_{\rho(j)}\rho^{(i)}$
  end for

  $\mathbf{l}_{(j)} = \mathcal{P}_S \left(\mathbf{L}^{(I_{(j)})} \right)$
  $\hat{\mathbf{d}}_{(j)}(j+1 : j+r) = \mathbf{l}_{(j)}(2m : 2m+r-1)$

end for

---

ation count of the alternating minimization steps in the batch process. Accordingly, for each set of data samples at position j the number of alternating minimization steps $I_{(j)}$ for the current estimation must be set beforehand. This number varies between a predefined I_{min} and I_{max}, and the choice is based on the residual Hankel penalty $\epsilon_{(j-1)}$ of the previous iteration. This corresponds to the observation that significant changes in the system behavior lead to higher values for ϵ and require more iterations in the optimization process, whereas less update steps are required if the subspace changes slowly or does not change at all. We start with the previous iterate on the Stiefel manifold and execute the three steps from Algorithm 1 in an alternating manner until convergence. Notice that the cost function parameter μ is fixed here and only the Lagrangian parameter ρ is changed in each iteration to steer between data fit and structure. The forecasting is realized as a robust matrix completion problem, i.e. the respective entries in the lower right corner of $\mathbf{X}$ are considered unobserved entries. Once a structured low-rank estimate has been found, the predicted entries of $\mathbf{d}$ can easily be read from the last entries of $\mathbf{l}_{(j)}$.

4   Experimental Results

In a first experiment we evaluate our method on an impulse response prediction task for a noisy observation of a simulated SISO Linear Time Varying (LTV) system. The data is generated via

$$\dot{\mathbf{x}}(t+1) = \mathbf{A}(t)\mathbf{x}(t) + \mathbf{b}^{\top}\mathbf{u}(t), \quad \mathbf{A}(t) = e^{0.001t\,\mathbf{Z}}, \quad \mathbf{Z}^{\top} = -\mathbf{Z}$$
$$\mathbf{y}(t) = \mathbf{c}^{\top}\mathbf{x}(t) + \mathbf{n}(t)$$

with $\mathbf{b}, \mathbf{c}$ being uniform random vectors and $\mathbf{Z}$ being a random skew-symmetric matrix. The additive observation noise $\mathbf{n}(t)$ contains of two parts, Gaussian noise with $\sigma = 0.01$ and randomly appearing (rate 0.05) salt and pepper noise samples that take on values of ± 0.5. The degree of the system is chosen as $k = 5$ and we generate the impulse response of the system for 300 samples. Our method is compared to the SLRA method from [11] and both algorithms are implemented in *MATLAB* on a desktop computer. We predict three time steps into the future from an observation of $2m-1$ samples, and the parameters are empirically chosen as $m = 20$, $\rho \in [10^{-6}, 10]$ (both methods), $p = 0.5$, $\mu = 0.005$, $\tau = 5 \times 10^{-4}$. The SLRA method is randomly initialized in each step, converges within 30 iterations and requires about 0.4 s. The iteration number of our method varies between $I_{min} = 16$ and $I_{max} = 128$ iterations with an average of 22 iterations that add up to 0.7 s per forecasting step. The forecasting results in Fig. 1 indicate that both methods are able to cope with the Gaussian noise quite well and predict the system behavior quite reliably, but the spurious outliers introduce errors in the SLRA extrapolation due to the ℓ_2-error weighting. Our proposed method is much more robust at the price of a higher computational effort. However, due to the beneficial subspace initialization the computation time is still competitive.

In a second experiment we compare our method on real-world data with SLRA and the *forecast* routine in MATLAB with a 12-month-seasonal ARIMA(0,1,1) model[1]. The time series is the well-known *Airline Passenger* dataset from [3] normalized to the range [0 1]. The upper bound on the rank of the approximation is chosen as $k = 8$, and we forecast 6 samples from $2m - 1$ samples with $m = 18$, which corresponds to projecting the monthly amount of passengers half a year into the future from observing the past three years. Figure 2 shows that all three methods succeed in forecasting the data, with average absolute deviations of 0.060 for SLRA, 0.036 for ARIMA and 0.044 for

---

[1] http://mathworks.com/help/econ/forecast-airline-passenger-counts.html.

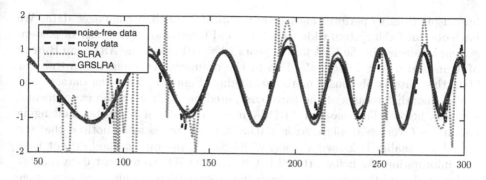

Fig. 1. Forecasting of 3 samples of a SISO-LTV impulse response with additive noise and ouliers

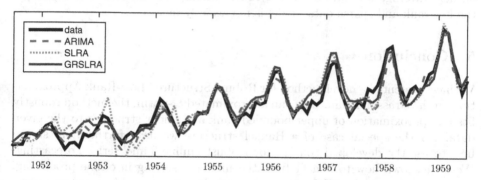

Fig. 2. Six month forecast of monthly airline passenger data from the years 1952–1960 based on 3 year observation period.

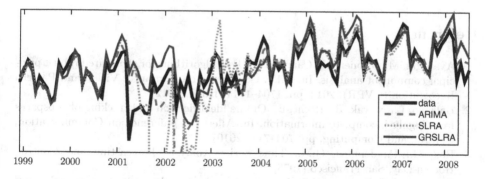

Fig. 3. Six month forecast of monthly airline passenger data from the years 1996–2009 based on 3 year observation period.

our method. On average, the ARIMA implementation requires 1.3s, SLRA 0.7s and our method ($I_{min} = 8, I_{max} = 64$, 12 iterations on average) is the fastest with 0.3s. The popularity of this well-known but also well-behaving dataset has inspired us to perform another experiment on airline passenger data. We have obtained the system-wide (domestic and international) number of passenger enplanements in the USA for the years 1996–2014 from the American Bureau of Transportation Statistics [5]. Due to the dramatic developments in the year 2001 this data is obviously more challenging. Figure 3 shows the dataset and the six month forecasts of the three compared methods with the experimental setup as before. The seasonal ARIMA model copes best with the challenging conditions (average absolute error of 0.086), but it needs to be noticed that the actual seasonality is known a priori while SLRA and our method do not have this information. As before, the SLRA method is able to forecast data reliably under good conditions but suffers from the gross outliers, resulting in an average absolute error of 0.182. Finally, our method shows more robustness against gross outliers (average absolute error of 0.102), although in this real-world example the low-rank and sparse data model is not exactly met.

5   Conclusion

We have presented a novel method for Robust Structured Low-Rank Approximation on the Grassmannian. Using an approximated ℓ_p-norm, the method robustly fits an approximation of upper-bounded rank and linear structure to the given data. For the special case of a Hankel structure we have furthermore shown how to use the developed concept for Robust Online Time Series Forecasting. We have shown how to benefit from the manifold setting in online processing, as we can increase the efficiency by re-using the previously identified subspace. Experimental results show that our method performs effectively and efficiently in simulated and real-world applications.

References

1. Ayazoglu, M., Sznaier, M., Camps, O.I.: Fast algorithms for structured robust principal component analysis. In: IEEE Conference on Computer Vision and Pattern Recognition (CVPR), 2012, pp. 1704–1711. IEEE (2012)
2. Balzano, L., Nowak, R., Recht, B.: Online identification and tracking of subspaces from highly incomplete information. In: Allerton Conference on Communication, Control, and Computing, pp. 704–711 (2010)
3. Box, G.E.P., Jenkins, G.M.: Time Series Analysis: Forecasting and Control. Holden-Day, San Francisco (1976)
4. Broomhead, D.S., King, G.P.: Extracting qualitative dynamics from experimental data. Phys. D 20(2), 217–236 (1986)
5. Bureau of Transportation Statistics, United States Department of Transportation: US Air Carrier travel statistics (2015). http://www.rita.dot.gov/bts
6. Cadzow, J.A.: Signal enhancement-a composite property mapping algorithm. IEEE Trans. Acoust. Speech Signal Process. 36(1), 49–62 (1988)
7. Candès, E., Li, X., Ma, Y., Wright, J.: Robust principal component analysis? J. ACM 58(3), 1–37 (2011)

8. Chu, M.T., Funderlic, R.E., Plemmons, R.J.: Structured low rank approximation. Linear Algebra Appl. **366**, 157–172 (2003)
9. Hage, C., Kleinsteuber, M.: Robust PCA and subspace tracking from incomplete observations using ℓ_0-surrogates. Comput. Stat. **29**, 1–21 (2014)
10. He, J., Balzano, L., Szlam, A.: Incremental gradient on the grassmannian for online foreground and background separation in subsampled video. In: Computer Vision and Pattern Recognition, pp. 1568–1575 (2012)
11. Ishteva, M., Usevich, K., Markovsky, I.: Factorization approach to structured low-rank approximation with applications. SIAM J. Matrix Anal. Appl. **35**(3), 1180–1204 (2014)
12. Markovsky, I.: Structured low-rank approximation and its applications. Automatica **44**(4), 891–909 (2008)
13. Markovsky, I.: Low-Rank Approximation - Algorithms, Implementation. Springer, Applications (2014)
14. Seidel, F., Hage, C., Kleinsteuber, M.: pROST : a smoothed ℓ_p-norm robust online subspace tracking method for realtime background subtraction in video. Mach. Vis. Appl. **25**(5), 1227–1240 (2014)

Blind Separation of Mixtures of Piecewise AR(1) Processes and Model Mismatch

Petr Tichavský[1]([✉]), Ondřej Šembera[1], and Zbyněk Koldovský[1,2]

[1] Institute of Information Theory and Automation of the CAS,
Prague, Czech Republic
`tichavsk@utia.cas.cz`, `sembeond@fjfi.cvut.cz`
[2] Faculty of Mechatronics, Informatics, and Interdisciplinary Studies,
Technical University of Liberec, Liberec, Czech Republic

Abstract. Modeling real-world acoustic signals and namely speech signals as piecewise stationary random processes is a possible approach to blind separation of linear mixtures of such signals. In this paper, the piecewise AR(1) modeling is studied and is compared to the more common piecewise AR(0) modeling, which is known under the names Block Gaussian SEParation (BGSEP) and Block Gaussian Likelihood (BGL). The separation based on the AR(0) modeling uses an approximate joint diagonalization (AJD) of covariance matrices of the mixture with lag 0, computed at epochs (intervals) of stationarity of the separated signals. The separation based on the AR(1) modeling uses the covariances of lag 0 and covariances of lag 1 jointly. For this model, we derive an approximate Cramér-Rao lower bound on the separation accuracy for estimation based on the full set of the statistics (covariance matrices of lag 0 and lag 1) and covariance matrices with lag 0 only. The bounds show the condition when AR(1) modeling leads to significantly improved separation accuracy.

Keywords: Autoregressive processes · Cramér-Rao bound · Blind source separation

1 Introduction

Blind source separation has found applications namely[1] in biomedical signal processing, for separating signals of interest from unwanted parasitic signals and noises, and in acoustical signal processing [6]. Modeling real-world acoustic signals and namely speech signals as piecewise stationary random processes is a possible approach to blind separation of linear mixtures of such signals. It appears that many times (depending on properties of the separated signals), methods utilizing nonstationarity of the separated signals outperform the more classical methods based on non-Gaussianity of the separated signals, or perform

---

[1] This work was supported by The Czech Science Foundation through Project No. 14-13713S.

ⓒ Springer International Publishing Switzerland 2015
E. Vincent et al. (Eds.): LVA/ICA 2015, LNCS 9237, pp. 304–311, 2015.
DOI: 10.1007/978-3-319-22482-4_35

equally well but with much lower computational complexity [5,10]. The methods
using signal nonstationarity divide the received signals (mixtures) to epochs, in
each epoch the signals are modeled as stationary, but properties of the signals
(namely their power) are assumed to change significantly in different epochs
[1,7–9].

The separation is called determined or overdetermined, if the number of the
available mixtures is higher or equal to the number of the sources, and is called
underdetermined otherwise. The latter case is studied in [4]. In this paper we
focus on the squared mixtures, where the number of the mixtures is equal to the
number of the sources.

The simplest nonstationarity-based separation methods use only covariance
matrices with lag 0. The mixing/demixing matrix can be found through an
approximate joint diagonalization of these matrices [1]. This method can be
statistically efficient (attaining a Cramér-Rao lower bound, CRB) [2,8], if the
separated data obey the assumed model, i.e. when the signals are i.i.d. in all
epochs. Real-world signals such as speech signals rarely obey the condition. Our
experiments with natural speech signals sampled at 16 kHz show that the correlation between two consecutive samples of the signals is typically 0.75 – 0.95.
This fact indicates that the separation methods using only the covariance matrices with lag 0 may not be optimal, and more accurate modeling of the separated
signals may increase accuracy of the separation.

A method called Block-AutoRegressive Blind Identification (BARBI) [11]
uses an autoregressive model of a general order n in each epoch of the source
signals. We refer to the method as BARBI(n). The number of the estimated
parameters grows with increasing model order n and the method seems to suffer
of overfitting, if $n > 2$. In this paper we provide a theoretical justification for
improved performance of BARBI(1) compared to BARBI(0) through the CRB
analysis.

2   Data Model

Consider linear instantaneous square mixing model

$$X_t = AZ_t, \tag{1}$$

where Z_t denotes a single time instance of the input signals, $A \in \mathbb{R}^{d \times d}$ is a mixing
matrix and $X_t \in \mathbb{R}$ is a time instance of the resulting mixtures. The input signals
are modeled by mutually independent piecewise stationary processes. We divide
the data into M epochs of the length T and assume that on each epoch the $i-$th
signal z_{it} takes a form of order one autoregressive process

$$z_{it} = -\rho_{im} z_{it-1} + \sigma_{im} w_{it}, \tag{2}$$

for $t = (m - 1)T + 1, \ldots mT$, where w_{it} is a Gaussian white noise with zero
mean and unit variance, ρ_{im} is an autoregressive coefficient corresponding to
the $i-$th input signal and the $m-$th epoch, and white noise sequences satisfy
independence relation

$$\mathbf{Cov}[w_{it}, w_{jt'}] = \delta_{ij}\delta_{tt'}.$$

The covariance of the input vector Z_t in the m−th epoch with lag 0 is then given by

$$D_m = \mathbf{Cov}\,[Z_t, Z_t] = \mathrm{diag}(d_{1m}, d_{2m}, \ldots d_{dm}), \tag{3}$$

where

$$d_{im} = \frac{\sigma_{im}^2}{1 - \rho_{im}^2}.$$

Covariance of the mixture X_t with lag 1 in the m−th epoch is

$$\mathbf{Cov}\,[Z_t, Z_{t+1}] = D_m Q_m \tag{4}$$

where $Q_m = -\,\mathrm{diag}(\rho_{1m}, \rho_{2m}, \ldots \rho_{dm})$ is a diagonal matrix of the m−th epoch autoregressive coefficients. The covariance matrices of the mixture X_t in the m−th epoch with lag 0 and with lag 1

$$R_m = A D_m A^T, \qquad S_m = A D_m Q_m A^T \tag{5}$$

are estimated from the data as

$$\hat{R}_m = \frac{1}{T} \sum_{t=(m-1)T+1}^{mT} X_t X_t^T, \qquad \hat{S}_m = \frac{1}{T-1} \sum_{t=(m-1)T+1}^{mT-1} X_{t+1} X_t^T. \tag{6}$$

The vector of the unknown parameters is

$$\theta = [\mathrm{vec}(A)^T; \mathrm{vec}(D)^T; \mathrm{vec}(Q)^T]^T \tag{7}$$

where D and Q are $d \times M$ matrices with elements d_{im} and ρ_{im}, $i = 1, \ldots, d$, $m = 1, \ldots, M$, respectively. Matrix A is the main parameter of interest and D, Q are nuisance parameters. Since each change in scale of the signals can be compensated by adequate change of the mixing matrix, the parameter D is constrained by the condition $\sum_m d_{im} = 1$ for all $i = 1, \ldots, d$. It means that the sum of the variances of each signal over all epochs is 1. Indeed, there are inequality constraints $0 \le d_{im}$ and $-1 < \rho_{im} < 1$ so that all signals in all epochs are *stable* AR processes.

3   Cramér-Rao Bound

The Cramér-Rao Bound is defined as an inverse of the Fisher information matrix. We shall assume, for simplicity, that the available data are Gaussian. It holds that for normally distributed data with a mean $\mu(\theta)$ and covariance matrix $C(\theta)$ the Fisher information matrix has elements

$$F_{\theta_i \theta_j} = \left(\frac{\partial \mu}{\partial \theta_i}\right)^T C^{-1} \left(\frac{\partial \mu}{\partial \theta_j}\right) + \frac{1}{2}\,\mathrm{tr}\left(C^{-1}\frac{\partial C}{\partial \theta_i} C^{-1}\frac{\partial C}{\partial \theta_j}\right). \tag{8}$$

In our case, the data have zero mean and only its covariance matrix C depends on the estimated parameter. In particular,

$$C = \text{Cov}([X_1, \ldots, X_{MT}]) = \text{blockdiag}(C_1, \ldots, C_M) \tag{9}$$

where

$$C_m = \text{btoeplitz}(AD_m A^T, AD_m Q_m A^T, \ldots, AD_m Q_m^{(T-1)} A^T). \tag{10}$$

C_m is the covariance matrix of the data in the m−th epoch, it is a symmetric block-Toeplitz matrix with the displayed first block-row.

Now, the CRB on $\text{vec}(A)$ is given as the left-upper corner submatrix of F^{-1} of the size $d^2 \times d^2$, obeys

$$\text{CRB}(\text{vec}(A)) = \text{CRB}_A = (A^{-1} \otimes I)\text{CRB}_I (A^{-T} \otimes I). \tag{11}$$

CRB_I is esentially (i.e. after a suitable re-ordering its columns and rows) block diagonal, with diagonal blocks of size 1×1 and 2×2,

$$\text{CRB}_I (A_{kk}) = \frac{1}{T} \tag{12}$$

for $k = 1, \ldots, d$, and

$$\text{CRB}_I ([A_{k\ell}, A_{\ell k}]) = \frac{1}{MT} \frac{1}{\phi_{k\ell}\phi_{\ell k} - 1} \begin{bmatrix} \phi_{k\ell} & -1 \\ -1 & \phi_{\ell k} \end{bmatrix} \tag{13}$$

for $k, \ell = 1, \ldots, d$, $k \neq \ell$, where [11]

$$\phi_{k\ell} = \frac{1}{M} \sum_{m=1}^{M} \frac{d_{km}}{d_{\ell m}} \frac{1 - 2\rho_{km}\rho_{\ell m} + \rho_{\ell m}^2}{1 - \rho_{\ell m}^2}. \tag{14}$$

Note that in the special case of all autoregressive parameters identical, $\rho_{km} = \rho$ for $k = 1, \ldots, d$, $m = 1, \ldots, M$, the resultant CRB expressions are independent of ρ.

3.1   CRB for Estimates Based on the Statistics

In this subsection, we investigate the maximum possible accuracy of the separation using only the statistics $\{\hat{R}_m\}$ and $\{\hat{R}_m, \hat{S}_m\}$, respectively. Thanks to the central limit theorem it holds that for $T \to \infty$ these statistics have asymptotically normal distribution with the asymptotic mean equal to the theoretical covariances $\{R_m\}$ and $\{R_m, S_m\}$, respectively, and have asymptotical covariance of errors proportional to $\frac{1}{T}$. The CRB for the estimates based on the statistics means computing the information content about the estimated parameter θ in the "concentrated" data $\{\hat{R}_m\}$ and $\{\hat{R}_m, \hat{S}_m\}$, assuming that the noise in the "concentrated" data is exactly zero mean and has exactly Gaussian distribution with the covariance structure that follows from analysis of the true statistics.

For computing the CRB as the inverse of the Fisher information matrix we will use the general formula (8) again. In this case, both the mean μ and the covariance matrix C depend on the estimated parameter. A detailed computation, not included here due to lack of space, shows that only the former term in (8) is dominant, has asymptotic order $O(T)$ for large T, and the latter term is negligible, having the order $O(1)$ only. The $O(1)$ terms will be neglected with respect to the leading term proportional to T. The asymptotic CRB is inversely proportional to T.

The concentrated data using the covariances with only lag 0 denoted $\hat{Y}_0$ are composed of $\mathcal{L}(\hat{R}_m)$ for $m = 1, \ldots, M$, where $\mathcal{L}(R)$ is the vector of elements of a lower triangular part of a matrix R,

$$\mathcal{L}(R) = [R_{11}, R_{21} \ldots R_{d1}, R_{22}, R_{32} \ldots R_{d2}, R_{33} \ldots R_{dd}]^T. \tag{15}$$

The concentrated data using the covariances with lag 0 and 1 denoted $\hat{Y}_{0+1}$ will be composed of $\mathcal{L}(\hat{R}_m)$ and $\mathcal{L}((\hat{S}_m + \hat{S}_m^T)/2)$ for $m = 1, \ldots, M$. Note that while S_m is symmetric, its sample estimate $\hat{S}_m$ may not be symmetric and thus we symmetrize it.

The covariance matrices of $\hat{Y}_0$ and $\hat{Y}_{0+1}$ can be computed as functions of parameter θ in (7). Again, they are block diagonal, having M blocks, because data in individual epochs and also the sample covariance matrices in them are mutually statistically independent. A straightforward but lengthy computation leads to the result that the asymptotic CRB for estimates based on the statistics $\hat{Y}_{0+1}$, denoted $\mathrm{CRB}^{(0+1)}(A)$ are identical to those in (13). It follows that $\hat{Y}_{0+1}$ is asymptotically sufficient. CRB for estimates based on $\hat{Y}_0$, denoted $\mathrm{CRB}^{(0)}(A)$, is higher, sometimes significantly. In particular,

$$\mathrm{CRB}^{(0)}(\mathrm{vec}(A)) = \mathrm{CRB}_A^{(0)} = (A^{-1} \otimes I)\mathrm{CRB}_I^{(0)}(A^{-T} \otimes I) \tag{16}$$

where $\mathrm{CRB}_I^{(0)}$ is block diagonal and independent of A again, and

$$\mathrm{CRB}_I^{(0)}([A_{k\ell}, A_{\ell k}]) = \frac{1}{MT} \frac{1}{\varphi_{k\ell}\varphi_{\ell k} - \omega_{k\ell}^2} \begin{bmatrix} \varphi_{k\ell} & -\omega_{k\ell} \\ -\omega_{k\ell} & \varphi_{\ell k} \end{bmatrix} \tag{17}$$

for $k, \ell = 1, \ldots, d$, $k \neq \ell$, with

$$\varphi_{k\ell} = \frac{1}{M} \sum_{m=1}^M \frac{d_{km}}{d_{\ell m}} \frac{1 - \rho_{km}\rho_{\ell m}}{1 + \rho_{\ell m}^2}, \qquad \omega_{k\ell} = \frac{1}{M} \sum_{m=1}^M \frac{1 - \rho_{km}\rho_{\ell m}}{1 + \rho_{\ell m}^2}. \tag{18}$$

In the special case $\rho_{in} = \rho$ for all $i = 1, \ldots, d$, $m = 1, \ldots, M$, it holds

$$\mathrm{CRB}^{(0)}(A) = \mathrm{CRB}(A) \frac{1 + \rho^2}{1 - \rho^2}. \tag{19}$$

If ρ is close to ± 1, the difference is significant.

4 Estimating A

From (10) it follows that A^{-1} can be sought as a matrix that jointly diagonalizes the matrices $\hat{R}_m$ and $\hat{S}_m$, $m = 1, \ldots, M$ [3]. The ordinary (unweighted) approximate joint diagonalization algorithms as UWEDGE [1] produce consistent but not optimal estimates of A. The asymptotically optimum estimate of θ can be found by minimizing the expression

$$\hat{\theta} = \operatorname{argmin}_\theta (\hat{Y}_{0+1} - Y_{0+1}(\theta))^T [\operatorname{Cov}(\hat{Y}_{0+1})]^{-1} (\hat{Y}_{0+1} - Y_{0+1}(\theta)). \tag{20}$$

The matrix $C_{0+1}(\theta) = \operatorname{Cov}(\hat{Y}_{0+1})$ is a function of the unknown parameter θ. In practice, $C_{0+1}(\theta)$ can be replaced by $C_{0+1}(\hat{\theta}_c)$, where $\hat{\theta}_c$ is a consistent estimate of θ, to achieve an asymptotically optimum estimate. Note that $C_{0+1}(\theta)$ is nearly block diagonal if its columns and rows are appropriately sorted and $A \approx I$. The weighted AJD algorithm WEDGE [1], and also BARBI [11] estimate a demixing matrix V. Let $V^{[i]}$ be an estimate of $V = A^{-1}$ from the i−th iteration. Then, the partially demixed covariance matrices are given as $\hat{R}_m^{[i]} = V^{[i]} \hat{R}_m V^{[i]T}$ and $\hat{S}_m^{[i]} = V^{[i]} \hat{S}_m V^{[i]T}$. These matrices are used to estimate parameters of the separated signals, i.e. $d_{jm}^{[i]} = (\hat{R}_m^{[i]})_{jj}$ and $\rho_{jm}^{[i]} = -(\hat{S}_m^{[i]})_{jj}/(\hat{R}_m^{[i]})_{jj}$ where $(X)_{jj}$ means the (j,j)−th element of matrix X.

The main iteration of WEDGE is

$$V^{[i+1]} = [A^{[i]}]^{-1} V^{[i]},$$

where the diagonal elements of $A^{[i]}$ are set to 1, and the off-diagonal elements of $A^{[i]}$ obey the 2×2 linear systems

$$\begin{bmatrix} A_{k\ell}^{[i]} \\ A_{\ell k}^{[i]} \end{bmatrix} = \left\{ \sum_{m=1}^{M} \begin{bmatrix} \hat{p}_{\ell\ell m}^T W_{k\ell m} \hat{p}_{\ell\ell m} & \hat{p}_{kkm}^T W_{k\ell m} \hat{p}_{\ell\ell m} \\ \hat{p}_{kkm}^T W_{k\ell m} \hat{p}_{\ell\ell m} & \hat{p}_{kkm}^T W_{k\ell m} \hat{p}_{kkm} \end{bmatrix} \right\}^{-1} \sum_{m=1}^{M} \begin{bmatrix} \hat{p}_{\ell\ell m}^T W_{k\ell m} \hat{p}_{k\ell m} \\ \hat{p}_{kkm}^T W_{k\ell m} \hat{p}_{k\ell m} \end{bmatrix}, \tag{21}$$

where $\hat{p}_{k\ell m} = [(\hat{R}_m^{[i]})_{k\ell}, (\hat{S}_m^{[i]})_{k\ell}]$ and $W_{k\ell m}$ should be proportional to the inverse of a 2×2 covariance matrix of $\hat{p}_{k\ell m}$ for $k, \ell = 1, \ldots, d$, $k \neq \ell$. We use the choice

$$W_{k\ell m}^{-1} = \frac{d_{km} d_{\ell m}}{1 - \rho_{km} \rho_{\ell m}} \begin{bmatrix} 1 + \rho_{km} \rho_{\ell m} & -\rho_{km} - \rho_{\ell m} \\ -\rho_{km} - \rho_{\ell m} & (1 + (\rho_{km} + \rho_{\ell m})^2 - \rho_{km}^2 \rho_{\ell m}^2)/2 \end{bmatrix}. \tag{22}$$

In BARBI, the relation (21) is replaced by

$$\begin{bmatrix} \hat{A}_{k\ell}^{[i]} \\ \hat{A}_{\ell k}^{[i]} \end{bmatrix} = \left\{ \sum_{m=1}^{M} \begin{bmatrix} \hat{p}_{\ell\ell m}^T q_{km} & \hat{p}_{kkm}^T q_{km} \\ \hat{p}_{kkm}^T q_{km} & \hat{p}_{kkm}^T q_{\ell m} \end{bmatrix} \right\}^{-1} \sum_{m=1}^{M} \begin{bmatrix} q_{km}^T \hat{p}_{k\ell m} \\ q_{\ell m}^T \hat{p}_{k\ell m} \end{bmatrix}, \tag{23}$$

where $q_{km} = W_{k\ell m} p_{\ell\ell m}$, and in the case of the AR order 1 it reads

$$q_{km} = \frac{1}{2 d_{km}(1 - \rho_{km}^2)} \begin{bmatrix} 1 + \rho_{km}^2 \\ -2\rho_{km} \end{bmatrix}. \tag{24}$$

5   Simulations

In the first simulation we consider a mixture of three piecewise AR(1) signals. The signals are composed of $M = 10$ epochs, each of the length $T = 100$. The signals have the same AR coefficient ρ in all epochs. The variances of the signals are increasing, $1, 2, ..., 10$, decreasing $10, 9, ..., 1$ and constant $5, ..., 5$, respectively, in the 10 epochs. We mix the signals using a random orthogonal (for simplicity) mixing matrix and demix them by BARBI(0) and BARBI(1) algorithms. The resultant average interference-to-signal ratios (ISR) obtained in 100 independent trials and corresponding CRB and $CRB^{(0)}$ are plotted as function of ρ in Fig. 1. We can see that BARBI(1) is nearly statistically efficient unless ρ is in a vicinity of -1. BARBI(0) does not achieve the $CRB^{(0)}$ except for ρ close to zero, but it follows the trend of $CRB^{(0)}$.

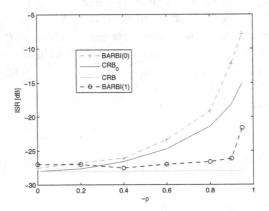

Fig. 1. Average ISR for separation of a mixture of artificial piecewise AR(1) signals achieved by BARBI(0) and BARBI(1) and corresponding CRBs versus the AR coefficient.

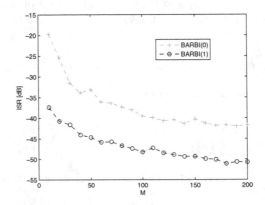

Fig. 2. Average ISR for separation of a mixture 16 natural speech signals achieved by BARBI(0) and BARBI(1) versus the number of epochs.

In the second simulation, we consider a mixture of 16 natural speech signals sampled at 16 kHz of the total length of 8.375 s, taken from the database in [4]. The average correlation between two consecutive samples in these signals is from 0.65 to 0.95, and the overall average is 0.81. Average ISR achieved by BARBI(0) and BARBI(1) versus the number of epochs is shown in Fig. 2.

6 Conclusions

We have proved that in blind separation of natural signals, piecewise AR(1) modeling represented by the algorithm BARBI(1) gives significantly improved separation accuracy if the sample lag-1 correlation of the original signals is close to 1. We plan to extend these results to underdetermined mixtures.

References

1. Tichavský, P., Yeredor, A.: Fast approximate joint diagonalization incorporating weight matrices. IEEE Trans. Signal Process. **57**, 878–891 (2009)
2. Doron, E., Yeredor, A., Tichavský, P.: Cramér-Rao-induced bound for blind separation of stationary parametric Gaussian sources. IEEE Sig. Process. Lett. **14**, 417–420 (2007)
3. Chabriel, G., Kleinsteuber, M., Moreau, E., Shen, H., Tichavský, P., Yeredor, A.: Joint matrices decompositions and blind source separation. IEEE Sig. Process. Mag. **31**, 34–43 (2014)
4. Tichavský, P., Koldovský, Z.: Weight adjusted tensor method for blind separation of underdetermined mixtures of nonstationary sources. IEEE Trans. Sig. Process. **59**, 1037–1047 (2011)
5. Tichavský, P., Koldovský, Z.: Fast and accurate methods of independent component analysis: a survey. Kybernetika **47**, 426–438 (2011)
6. Koldovský, Z., Tichavský, P.: Time-domain blind separation of audio sources on the basis of a complete ICA decomposition of an observation space. IEEE Trans. Audio Speech Lang. Process. **19**, 406–416 (2011)
7. Pham, D.-T., Garat, P.: Blind separation of mixtures of independent sources through a quasi maximum likelihood approach. IEEE Tr. Signal Process. **45**, 1712–1725 (1997)
8. Pham, D.-T.: Blind separation of instantaneous mixture of sources via the Gaussian mutual information criterion. Sig. Process. **81**, 855–870 (2001)
9. Dégerine, S., Zaïdi, A.: Separation of an instantaneous mixture of Gaussian autoregressive sources by the exact maximum likelihood approach. IEEE Trans. Sig. Process. **52**, 1492–1512 (2004)
10. Koldovský, Z., Tichavský, P.: A comparison of independent component and independent subspace analysis algorithms. In: Proceedings of the EUSIPCO 2009, Glasgow, Scotland, pp. 1447–1451 (2009)
11. Tichavský, P., Yeredor, A., Koldovský, Z.: A fast asymptotically efficient algorithm for blind separation of a linear mixture of block-wise stationary autoregressive processes. In: Proceedings of the ICASSP 2009, Taipei Taiwan, pp. 3133–3136 (2009)

Theoretical Studies and Algorithms Regarding the Solution of Non-invertible Nonlinear Source Separation

D.F.F. Baptista[1], R.A. Ando[1,2]([✉]), L.T. Duarte[3], C. Jutten[2], and R. Attux[1]

[1] School of Electrical and Computer Engineering, UNICAMP, Campinas, Brazil
[2] GIPSA-Lab, Université Joseph Fourier (UJF), Grenoble, France
`rafael.ando@gipsa-lab.grenoble-inp.fr`
[3] School of Applied Sciences, UNICAMP, Campinas, Brazil

Abstract. In this paper, we analyse and solve a source separation problem based on a mixing model that is nonlinear and non-invertible at the space of mixtures. The model is relevant considering it may represent the data obtained from ion-selective electrode arrays. We apply a new approach for solving the problems of local stability of the recurrent network previously used in the literature, which allows for a wider range of source concentration. In order to achieve this, we utilize a second-order recurrent network which can be shown to be locally stable for all solutions. Using this new network and the priors that chemical sources are continuous and smooth, our proposal performs better than the previous approach.

1 Problem Statement

The general blind source separation (BSS) problem consists in estimating sources, represented by the vector $\mathbf{s} = [s_1, s_2, ..., s_n]^T$, that have been mixed by an unknown function $\mathcal{F}(.)$, given only the mixtures $\mathbf{x} = [x_1, x_2, ..., x_n]^T$ and prior information on the model or the sources.

$$\mathbf{x} = \mathcal{F}(\mathbf{s}). \tag{1}$$

For a linear function $\mathcal{F}(.)$, the problem can be uniquely solved – up to scale and permutation ambiguities – by formulating a criterion of statistical independence, but this is no longer possible for a generic nonlinear mapping. [2,5,6]. It is well known that the nonlinear BSS problem is very difficult to solve, since the generic nonlinearities can cause multiple statistical independent solutions that are still mixtures of the sources.

In the literature, we can find several approaches for dealing with specific nonlinear mixtures, such as the Post-Nonlinear (PNL) [2,7] and the Linear-Quadratic (LQ) [4]. For the latter, a recurrent network has been proposed as

The authors would like to thank FAPESP, CNPq, CNRS and ERC project 2012-ERC-AdG-320684 CHESS for the funding.

© Springer International Publishing Switzerland 2015
E. Vincent et al. (Eds.): LVA/ICA 2015, LNCS 9237, pp. 312–319, 2015.
DOI: 10.1007/978-3-319-22482-4_36

part of the solving system, and, with some modifications, can also be used for a different mixing model which shall be analyzed in this paper, represented as follows:

$$x_1 = s_1 + a_1 s_2^2$$
$$x_2 = s_2 + a_2 \sqrt{s_1} \tag{2}$$

This model describes data obtained from ion-selective-electrodes (ISEs), where we have two sources (i.e. chemical species) and two sensors. More details about the suitability of model (2) to the problem can be found in [3].

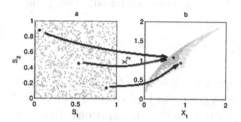

Fig. 1. Illustration of the folding of the source space

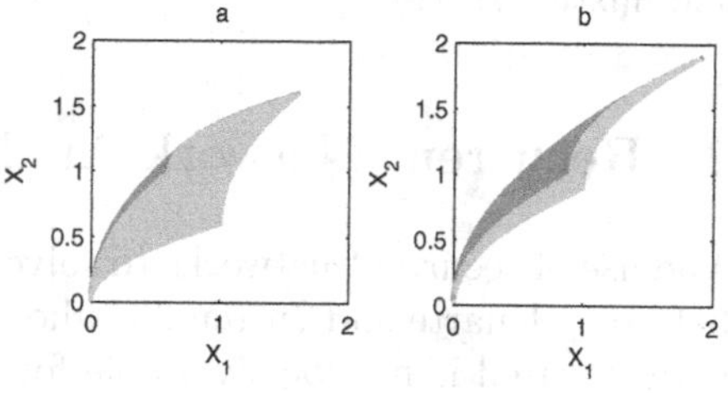

Fig. 2. Increase of the folded area with the selectivity coefficients

Our main goal in this paper is to obtain an estimate $\mathbf{y} = (y_1, y_2)$ of the sources $\mathbf{s} = (s_1, s_2)$, when the exact selectivity coefficients a_{ij} and the mixtures $\mathbf{x} = (x_1, x_2)$ are known. For the mixing model (2), the nonlinearity creates a difficulty for being non-invertible, since two distinct source points can be mapped onto the same mixture. Indeed, as can be seen in Fig. 1, the model effectively "folds" the source space, and this folding depends on the selectivity coefficients, as seen in Fig. 2. For a given point in the mixture space $\mathbf{x} = (x_1, x_2)$, it can be seen that the possible solutions are the sources $\mathbf{y} = (s_1, s_2)$ and a mixture of the sources, given by:

$$\mathbf{y}^* = \left(\frac{\left(\sqrt{s_1} \left(a_1 a_2^2 - 1 \right) + 2 a_1 a_2 s_2 \right)^2}{\left(a_1 a_2^2 + 1 \right)^2}, \frac{-a_1 a_2^2 s_2 + 2 a_2 \sqrt{s_1} + s_2}{a_1 a_2^2 + 1} \right) \tag{3}$$

The folding frontier (i.e., the *locus* of the points for which $\mathbf{y} = \mathbf{y}^*$) is given by:

$$a_1 a_2 s_2 = \sqrt{s_1}. \tag{4}$$

Since we are dealing with aqueous ionic solutions, we can restrict the sources concentrations to $s_i \in [0, 1]$. In this region of interest, we shall henceforth call the points below and above the folding frontier regions 1 and 2 respectively, as seen in Fig. 3. It can also be seen that while some of the points in region 1 are mapped to an invertible area (i.e., $\mathbf{y}^*$ is outside the region of interest, and therefore no ambiguity ensues), all points from region 2 are mapped to the non-invertible area.

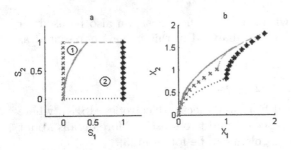

Fig. 3. Mapping of the frontiers in the source and mixture spaces

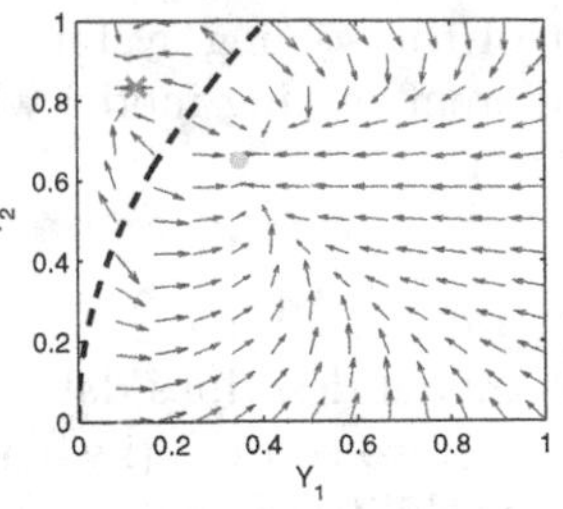

Fig. 4. Direction of network step based on the initialization point

2   Recurrent Network Analysis

The use of recurrent networks to solve this problem has already been considered before by Duarte and Jutten [3], who achieved good but limited results since the network used is not locally stable for all possible ionic concentrations. In order to solve this issue, a new approach that implements a second-order recurrent network based on the Newton-Raphson's method has been proposed [1]. The goal in using a recurrent network is to obtain the solutions of the model (2) numerically, since obtaining the analytical solution is not always straightforward.

2.1   Recurrent Network

The main idea to obtain a separating system is to represent (2) as

$$\mathbf{G}(\mathbf{s}) = \begin{bmatrix} s_1 + a_1 s_2^2 - x_1 \\ s_2 + a_2 \sqrt{s_1} - x_2 \end{bmatrix} = 0 \tag{5}$$

Then, the source separation problem can be interpreted as a homogeneous non-linear equation system and can be solved by root-finding algorithms such as the Newton-Raphson's method [1]. This leads to the following equation:

$$\mathbf{y}(m+1) = \mathbf{y}(m) - \mu \mathbf{J_G}^{-1} \mathbf{G}(\mathbf{y}(m)) \tag{6}$$

where $\mathbf{y}$ is the estimate of the sources, $\mu \in (0,1]$ is an adjustment scale factor, m is the iteration index of the network and $\mathbf{J_G}$ is the Jacobian matrix of $\mathbf{G}$. To avoid discontinuities in the derivatives when extending (2) to $\mathbb{R}^2$, we can use:

$$\begin{aligned} x_1 &= y_1 + a_1 y_2^2 \\ x_2 &= y_2 + a_2 \sqrt{|y_1|}\,\text{sign}(y_1) \end{aligned} \tag{7}$$

which is the same model as (2) in the region of interest $[0,1] \times [0,1]$.

When the network converges, we obtain $\mathbf{y}(m+1) = \mathbf{y}(m)$, which is defined as a fixed point of the network. From (6), we can show that this happens if and only if $\mathbf{G} = 0$, since $\mathbf{J_G}^{-1}$ is non-singular.

2.2 Stability of Solutions

We can calculate the stability condition of Eq. (6), for all fixed points. It is well known that for discrete dynamic systems, if all eigenvalues λ_i of the Jacobian matrix of the recurrence evaluated at the fixed point satisfy $|\lambda_i| < 1$, the system is locally stable. Computing the Jacobian matrix at the fixed points, leads to:

$$\mathbf{J}|_{\mathbf{G}=0} = \begin{bmatrix} \frac{\partial y_1(m+1)}{\partial y_1(m)} & \frac{\partial y_1(m+1)}{\partial y_2(m)} \\ \frac{\partial y_2(m+1)}{\partial y_1(m)} & \frac{\partial y_2(m+1)}{\partial y_2(m)} \end{bmatrix}_{\mathbf{G}=0} = \begin{bmatrix} 1-\mu & 0 \\ 0 & 1-\mu \end{bmatrix} \tag{8}$$

From (8) we conclude that for $\mu \in (0, 1]$, the fixed points are always locally stable. According to this result, the stability problem found in [3] no longer exists, since the solution (s_1, s_2) is stable for all source concentrations. However, we now have to deal with a non-separating solution (3) that is also stable. Since the two solutions are always in different regions, it is important to know how to control which solution we want the network to obtain. Given a sufficiently small step for the network in (6) – e.g., $\mu = 0.1$ –, if we initialize in $(0.9, 0.1)$ and $(0.1, 0.9)$ we converge to solutions in regions 1 and 2 respectively, as seen in Fig. 4. As a result, specifying the initial point can be used to reach the separating solution of the model.

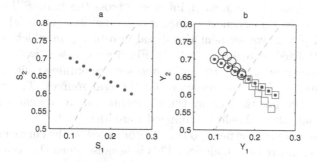

Fig. 5. Trajectories near the folding frontier.

2.3 Pivot Points

To have more knowledge about the solutions' trajectories near the folding frontier, we can look at Fig. 5. In Fig. 5a, we see the trajectory of the sources when it crosses the frontier. In Fig. 5b, we see two different trajectories in the (y_1, y_2) plane: the one represented by squares, corresponding to the estimate when we initialize in region 1, and one represented by circles when we initialize in region 2.

From the initial point of the trajectory, we can see that the solution corresponds to the square estimate before crossing the frontier, and the circle one afterwards. Let us define the point when the solution crosses the frontier as the *pivot*. Moreover, the *pivot* is also the point at which the initialization should be changed.

After observing the behaviour of the trajectory of the sources and the estimates in several simulations, it was suggested that the correct solution should combine the estimates in such a way that the resulting signal would be as smooth as possible near the folding frontier. As can also be seen in Fig. 5, we can verify that, near the border, the correct solution is smoother than the alternative in which no change of initialization occurs. Therefore, using as priors that the source are continuous and smooth, we propose an algorithm that determines which initialization should be applied to the points at the non-bijective region in order to obtain the correct solution.

3   Proposed Algorithm

In this section, we propose an algorithm that identifies the pivots and the initializations that should be used to recover the signal. In the following sections, we will present the technical aspects of the proposed algorithm.

3.1   Identification of Potential Pivots

In order to identify the pivots, we calculate the entire estimate with only the initialization in region 1 and make a system that predicts, at a given point of the calculated signal, if the next point would cross the border. The prediction does not have to be very accurate, since it only identifies potential pivots. After identifying all pivots, we segment the signal into blocks in which all the points are either initialized in region 1 or 2. In Fig. 6 we can see two signals in each graph, a thin one representing the estimated sources obtained with initialization at region 1, and a thick one representing the real sources. The vertical lines represent potential pivots given by the algorithm, and as we can see, all points where the initialization should be changed were identified.

For the prediction of the next point, a factor based on the curvature of the trajectory of the estimates was included. This is because when the sources cross the border, the trajectories of the estimates tend to rapidly change their direction, yielding high curvature. The algorithm therefore considers such abrupt variations as an indicative that the next point is a probable pivot. For more details about how the prediction is done, one can look at the pseudocode in Table 1.

3.2   Classification of Each Block

After segmenting the signal, we need to identify which initialization each block should be given. For that, we initially check for points outside the non-bijective area. As previously mentioned, we know that such points can only come from region 1, so we can safely initialize its corresponding block in it. For the remaining blocks, we select the initialization that maximizes the measure of smoothness around the pivot. We define smoothness as:

$$S(\mathbf{y}) = \mathrm{Var}(\mathbf{y}^2(n+1) - \mathbf{y}^2(n)) \tag{9}$$

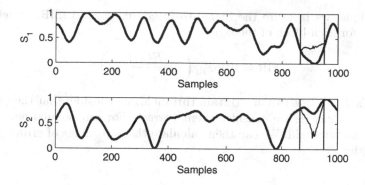

Fig. 6. Block segmentation of the estimates. Thick and thin curves represent the sources and the estimate, respectively.

where Var(.) is the empirical variance calculated in a N-sized window ($N = 11$ samples) centered around the pivot and n is the sample index of the signal. A pseudocode of the entire algorithm can be seen in Table 1:

Table 1. Pseudocode of the proposed algorithm

1. Using (6), obtain $\mathbf{y}$, the estimate from initialization point 1.
2. For each point of $\mathbf{y}$, predict the next point using:

$$\hat{\mathbf{y}}(n+1) = \mathbf{y}(n) + \lambda(n)\frac{\boldsymbol{\Delta}\mathbf{y}(n)}{||\boldsymbol{\Delta}\mathbf{y}(n)||} \tag{10}$$

$$\lambda(n) = \frac{\sum\limits_{i=n-4}^{n}||\boldsymbol{\Delta}\mathbf{y}(i)||}{5}\exp(1 - |cos\theta|) \tag{11}$$

where $\boldsymbol{\Delta}\mathbf{y}(n) = \mathbf{y}(n) - \mathbf{y}(n-1)$ is the last step, $\theta(n)$ the angle between $\boldsymbol{\Delta}\mathbf{y}(n+1)$ and $\boldsymbol{\Delta}\mathbf{y}(n+3)$, and $\lambda(n)$ a scalar factor based on the curvature of the trajectory.
3. If $\hat{\mathbf{y}}(n+1)$ is in region 2, store it as a potential pivot.
4. Segment signal into blocks separated by potential pivots.
5. Assign non-ambiguous blocks to region 1.
6. For the remaining blocks, test initializations in both regions and select the combination that minimizes the smoothness (9).

4   Simulation Results

In this section, we analyse the performance of the proposed algorithm and compare it to the previous methods for solving the problem [3]. We simulate sources by filtering a uniform random signal with a high order low-pass filter, with variable cut-off frequency. As one may notice, the smaller the cut-off frequency, the smoother the estimated sources obtained. To estimate the performance of the technique, we measure the Signal-to-Interference ratio (SIR) – defined in Eq. (12) – for each block of the signal.

$$\text{SIR} = 10 \log_{10} \left(\frac{E\{\mathbf{s}^2\}}{E\{(\mathbf{s} - \mathbf{y})^2\}} \right) \tag{12}$$

If a block's SIR is lower than a certain threshold, we consider that the estimates are wrong, otherwise we consider them correct. For our simulation, we used 15 dB as the threshold. We can then calculate the percentage of errors given by the algorithm.

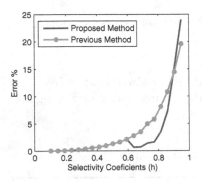

Fig. 7. Simulation results for fixed source smoothness and varying selectivity coefficient

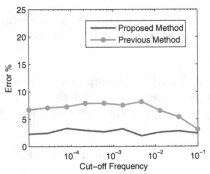

Fig. 8. Simulation results for fixed selectivity coefficient ($h = 0.8$) and varying source smoothness

4.1 First Scenario

In this simulation, we keep the sources' smoothness constant and vary the selectivity coefficients h, defined, for the sake of illustration, as $a_1 = a_2 = h$. Figure 7 shows the average percentage of errors, for 300 trials for each h and 20 different $h \in [0.1, 0.95]$. We can verify that for selectivity coefficients $h < 0.6$, results for both the proposed network and the previous one can be considered similar. However, for $h \in [0.6, 0.9]$ our approach yields considerably better results. For $h > 0.9$ (i.e. for ill-conditionned mixtures), the errors with the two methods are above 10 % and grow rapidly with h, and we can consider that both methods fail.

4.2 Second Scenario

In the second scenario, we keep the selectivity coefficients constant ($h = 0.8$) and vary the sources' smoothness by adjusting the filter's cutoff frequency. As we can see in Fig. 8, for all cutoff frequencies tested, the proposed algorithm performed better than the previous technique.

We can also see that regardless of the signal's smoothness, the percentage of errors in our technique remains roughly constant and very low (about 2 %), which suggests the smoothness of the sources do not interfere in the algorithm's performance.

5 Conclusion

In this paper we considered a problem of source separation in non-invertible nonlinear mixtures derived from an application in which chemical sensor arrays are used to measure ionic concentrations. For the simple case of two mixtures and two sources, we proposed a new method for solving the problem which has better stability properties. The nonlinear mapping studied presented difficulties caused by the existence of multiple solutions, which makes the model non-invertible. Nonetheless, our proposed method was capable of solving it using the prior source smoothness, and experimental results attested to the efficiency of the method even when the mixture is ill-conditioned.

Future works include extending the method to the blind case (i.e. when the mixing coefficients are unknown) and investigating other approaches, such as how the overdetermined scenario (i.e., with more sensors than sources), could provide additional information which would help solve the non-invertibility problem.

References

1. Ando, R.A.: Blind Source Separation in the Context of Polynomial Mixtures. Master's thesis, School of Electrical and Computer Engeneering, University of Campinas (2013)
2. Comon, P., Jutten, C.: Handbook of Blind Source Separation: Independent Component Analysis and Applications. Academic press, USA (2010)
3. Duarte, L.T., Jutten, C.: Design of smart ion-selective electrode arrays based on source separation through nonlinear independent component analysis. Oil Gas Sci. Technol.-Rev. d'IFP Energies Nouvelles 69(2), 293–306 (2014)
4. Hosseini, S., Deville, Y.: Blind separation of linear-quadratic mixtures of real sources using a recurrent structure. In: Mira, J., Álvarez, J.R. (eds.) IWANN 2003. LNCS, vol. 2687, pp. 241–248. Springer, Heidelberg (2003)
5. Hyvärinen, A., Pajunen, P.: Nonlinear independent component analysis: existence and uniqueness results. Neural Netw. 12(3), 429–439 (1999)
6. Jutten, C., Karhunen, J.: Advances in blind source separation (BSS) and independent component analysis (ICA) for nonlinear mixtures. Int. J. Neural Syst. 14(05), 267–292 (2004)
7. Taleb, A., Jutten, C.: Source separation in post-nonlinear mixtures. IEEE Trans. Sig. Process. 47(10), 2807–2820 (1999)

A Block-Jacobi Algorithm for Non-Symmetric Joint Diagonalization of Matrices

Hao Shen[(✉)] and Martin Kleinsteuber

Department of Electrical and Computer Engineering,
Technische Universität München, Munich, Germany
{hao.shen,kleinsteuber}@tum.de

Abstract. This paper studies the problem of Non-symmetric Joint Diagonalization (NsJD) of matrices, namely, jointly diagonalizing a set of complex matrices by one matrix multiplication from the left and one multiplication with possibly another matrix from the right. In order to avoid ambiguities in the solutions, these two matrices are restricted to lie in the complex oblique manifold. We derive a necessary and sufficient condition for the uniqueness of solutions, and characterize the Hessian of global minimizers of the off-norm cost function, which is typically used to measure joint diagonality. By exploiting existing results on Jacobi algorithms, we develop a block-Jacobi method that guarantees local convergence to a pair of joint diagonalizers at a quadratic rate. The performance of the proposed algorithm is investigated by numerical experiments.

Keywords: Non-symmetric joint diagonalization of matrices · Complex oblique manifold · Uniqueness conditions · Block Jacobi algorithm · Local quadratic convergence

1 Introduction

Joint Diagonalization (JD) of a set of matrices has attracted considerable attentions in the areas of statistical signal processing and multivariate statistics. Its applications include linear blind source separation (BSS), beamforming, and direction of arrival (DoA) estimation, cf. [1]. Classic literature focuses on the problem of symmetric joint diagonalization (SJD) of matrices. Namely, a set of matrices are to be diagonalized via matrix congruence transforms, i.e. multiplication from the left and the right with a matrix and its (Hermitian) transposed, respectively.

In this work, we consider the problem of *Non-symmetric Joint Diagonalization (NsJD)* of matrices, where the two matrices multiplied from the left and the right are different. Such a general form has been studied in the scheme of multiple-access multiple-input multiple-output (MIMO) wireless transmission, cf. [2]. In the work of [3], NsJD approaches have demonstrated its application in solving the problem of independent vector analysis. Moreover, the problem of NsJD is closely related to the problem of the canonical polyadic decomposition of tensors. We refer to [4,5] for further discussions on the latter subject.

© Springer International Publishing Switzerland 2015
E. Vincent et al. (Eds.): LVA/ICA 2015, LNCS 9237, pp. 320–327, 2015.
DOI: 10.1007/978-3-319-22482-4_37

One contribution of this work is the development of a block Jacobi method for solving the problem of NsJD at a super-linear rate of convergence. Jacobi-type methods have a long history in solving problems of matrix joint diagonalization. Early work in [6] employs a Jacobi algorithm for solving the unitary joint diagonalization problem based on the common off-norm cost function. In the non-unitary setting, Jacobi-type methods have been developed for both the log-likelihood formulation, cf. [7], and the common off-norm case, cf. [8,9]. The current work provides an extension of [10], which only considers the problem in the real symmetric setting.

2   Uniqueness of Non-symmetric JD

In this work, we denote by $(\cdot)^\top$ the matrix transpose, $(\cdot)^\mathsf{H}$ the Hermitian transpose, $\overline{(\cdot)}$ the complex conjugate of entries of a matrix, and by $Gl(m)$ the set of all $m \times m$ invertible complex matrices. Let $\{C_i\}_{i=1}^n$ be a set of $m \times m$ complex matrices, constructed by

$$C_i = A_L \Omega_i A_R^\mathsf{H}, \quad i = 1, \ldots n, \tag{1}$$

where $A_L, A_R \in Gl(m)$ and $\Omega_i = \mathrm{diag}\left(\omega_{i1}, \ldots, \omega_{im}\right) \in \mathbb{C}^{m \times m}$ with $\Omega_i \neq 0$. It is worth noticing that there is no relation between the matrices A_L and A_R. The task of NsJD is to find a pair of matrices $X_L, X_R \in Gl(m)$ such that the set of matrices

$$\left\{ X_L^\mathsf{H} C_i X_R \,\middle|\, i = 1, \ldots, n \right\} \tag{2}$$

are jointly diagonalized. In order to investigate properties of any algorithm that aims at finding such a joint diagonalizer, it is fundamental to understand under what conditions there exists a unique solution. In the remainder of this section, we therefore elaborate the uniqueness properties of the non-symmetric joint diagonalization problem (2).

As it is known from the SJD case, there is also an inherent permutation and scaling ambiguity here. Let $D_L, D_R \in Gl(m)$ be diagonal and $P \in Gl(m)$ be a permutation matrix. If $X_L^* \in Gl(m)$ and $X_R^* \in Gl(m)$ are the joint diagonalizers of problem (2), then so is the pair of $(X_L^* D_L P, X_R^* D_R P)$. In other words, the joint diagonalizers can only be identified up to individual scaling and a joint permutation. We define the set of two jointly column-wise permuted diagonal $(m \times m)$-matrices by

$$\mathcal{G}(m) := \left\{ (D_1 P, D_2 P) \middle| D_1, D_2 \in Gl(m) \text{ are diagonal and} \right.$$

$$\left. P \in Gl(m) \text{ is a permutation matrix} \right\}. \tag{3}$$

As the set $\mathcal{G}(m)$ admits a group structure, we can define an equivalence class on $Gl(m) \times Gl(m)$ as follows.

Definition 1 (Essential Equivalence). *Let $(X_L, X_R) \in Gl(m) \times Gl(m)$, then (X_L, X_R) is said to be essentially equivalent to $(Y_L, Y_R) \in Gl(m) \times Gl(m)$, and vice versa, if there exists $(E_L, E_R) \in \mathcal{G}(m)$ such that*

$$X_L = Y_L E_L \quad and \quad X_R = Y_R E_R. \tag{4}$$

Moreover, we say that the solution of problem (2) *is* essentially unique, *if it admits a unique solution on the set of equivalence classes.*

Due to the fact that

$$X_L^H C_i X_R = (X_L^H A_L) \Omega_i (A_R^H X_R), \tag{5}$$

we assume without loss of generality that the $C_i = \Omega_i$, $i = 1, \ldots, n$, are already diagonal. In other words, we investigate the question of under what conditions the set $\mathcal{G}(m)$ admits the only solutions to the joint diagonalization problem (2), when the C_i's are already diagonal.

In order to characterize the uniqueness conditions, we need to define a measure of collinearity for diagonal matrices. Recall $\Omega_i = \text{diag}(\omega_{i1}, \ldots, \omega_{im}) \in \mathbb{C}^{m \times m}$ for $i = 1, \ldots, n$. For a fixed diagonal position k, we denote by $z_k := [\omega_{1k}, \ldots, \omega_{nk}]^\top \in \mathbb{C}^n$ the vector consisting of the k-th diagonal element of each matrix, respectively. Then, the collinearity measure for the set of Ω_i's is defined by

$$\rho(\Omega_1, \ldots, \Omega_n) := \max_{1 \le k < l \le n} |c(z_k, z_l)|, \tag{6}$$

where $c(z_k, z_l)$ is the cosine of the complex angle between two vectors $v, w \in \mathbb{C}^n$, computed as

$$c(v, w) := \begin{cases} \frac{v^H w}{\|v\| \|w\|} & \text{if } v \ne 0 \wedge w \ne 0, \\ 1 & \text{otherwise.} \end{cases} \tag{7}$$

Here, $\|v\|$ denotes the Euclidean norm of a vector v. Note, that $0 \le \rho \le 1$ and that $\rho = 1$ if and only if there exists a complex scalar $\omega \in \mathbb{C}$ and a pair z_k, z_l, $k \ne l$ so that $z_k = \omega z_l$. In other words, $\rho = 1$ if and only if there exist two positions (k, k) and (l, l) such that the corresponding entries in the matrices Ω_i only differ by multiplication with a complex scalar ω. We adopt the methods for uniqueness analysis of symmetric joint diagonalization cases, developed in [11], to the NsJD setting.

Lemma 1. *Let* $\Omega_i \in \mathbb{C}^{m \times m}$, *for* $i = 1, \ldots, n$, *be diagonal, and let* $X_L, X_R \in Gl(m)$ *so that* $X_L^H \Omega_i X_R$ *is diagonal as well. Then the pair* (X_L, X_R) *is essentially unique if and only if* $\rho(\Omega_1, \ldots, \Omega_n) < 1$.

Proof. First, consider the case $m = 2$ and let

$$X_L = \begin{bmatrix} l_1 & l_2 \\ l_3 & l_4 \end{bmatrix} \in Gl(2), \quad \text{and} \quad X_R = \begin{bmatrix} r_1 & r_2 \\ r_3 & r_4 \end{bmatrix} \in Gl(2). \tag{8}$$

Then $X_L^H \Omega_i X_R$ is diagonal for $i = 1, \ldots, n$, if and only if

$$\begin{cases} \omega_{i1} \overline{l_1} r_2 + \omega_{i2} \overline{l_3} r_4 = 0 \\ \omega_{i1} \overline{l_2} r_1 + \omega_{i2} \overline{l_4} r_3 = 0 \end{cases} \tag{9}$$

for $i = 1, \ldots, n$. The corresponding system of linear equations reads as

$$\begin{bmatrix} \omega_{11} & \omega_{21} & \cdots & \omega_{n1} \\ \omega_{12} & \omega_{22} & \cdots & \omega_{n2} \end{bmatrix}^H \begin{bmatrix} l_1 \overline{r_2} & l_2 \overline{r_1} \\ l_3 \overline{r_4} & l_4 \overline{r_3} \end{bmatrix} = 0, \tag{10}$$

which only has a unique trivial solution if and only if the coefficient matrix in ω's has rank 2, which in turn is equivalent to $\rho(\Omega_1, \ldots, \Omega_n) < 1$. Specifically, the trivial solution, i.e.

$$l_1 \overline{r_2} = l_2 \overline{r_1} = l_3 \overline{r_4} = l_4 \overline{r_3} = 0, \tag{11}$$

together with the invertibility of X_L and X_R yields that either

$$l_1 = r_1 = l_4 = r_4 = 0, \quad \text{or} \quad l_2 = r_2 = l_3 = r_3 = 0. \tag{12}$$

Therefore, one can conclude that $(X_L, X_R) \in \mathcal{G}(2)$. For the case $m > 2$, if $\rho = 1$ then there exists a pair (k, l) such that $|c(z_k, z_l)| = 1$ and the same argument as above shows that $\rho = 1$ implies the non-uniqueness of the joint diagonalizer.

For the reverse direction of the statement, we assume that the joint diagonalizer is not in $\mathcal{G}(m)$. We further assume that one of the Ω_i's, say Ω_1, is invertible. Then

$$X_L^H \Omega_i X_R (X_L^H \Omega_1 X_R)^{-1} = X_L^H \Omega_i \Omega_1^{-1} (X_L^H)^{-1}, \tag{13}$$

for $i = 1, \ldots, n$, gives the simultaneous eigendecomposition of the diagonal matrices $\Omega_i \Omega_1^{-1}$. Since we assume $(X_L, X_R) \notin \mathcal{G}(m)$, this eigendecomposition is not unique and thus each matrix $\Omega_i \Omega_1^{-1}$ must have at least two identical eigenvalues and these eigenvalues must be located at the same positions (k, k) and (l, l) for all the matrices $\Omega_i \Omega_1^{-1}$. In other words, there exists a pair (k, l) with $k \neq l$ such that

$$\frac{\omega_{ik}}{\omega_{1k}} = \frac{\omega_{il}}{\omega_{1l}}, \tag{14}$$

which is equivalent to $|c(z_k, z_l)| = 1$ and hence $\rho(\Omega_1, \ldots, \Omega_n) = 1$. If all the Ω_i's are singular, we distinguish between two cases. Firstly, assume that there is a position on the diagonals, say k, where all $\omega_{ik} = 0$. Then $|c(z_k, z_l)| = 1$ holds true for any $k \neq l$ and thus $\rho = 1$. Secondly, if there is no common position where all the Ω_i's have a zero entry, there exists an invertible linear combination, say Ω_0, which can also be diagonalized via the same transformations. Then by considering a new set $\{\Omega_i\}_{i=0}^n$, the same argument as from (13) to (14) for the invertible case applies by replacing Ω_1 with Ω_0. ∎

3   Analysis of the Joint Diagonality Measure

To deal with the scaling ambiguity, we restrict the search space for the diagonalizing matrices to the quotient space

$$Op(m) = Gl(m)/\{D \in Gl(m) \mid D \text{ is diagonal.}\}, \tag{15}$$

cf. [12] for further details. As representatives for one equivalent class, we choose elements from the set of complex oblique matrices, i.e.

$$\mathcal{OB}(m) := \{X \in \mathbb{C}^{m \times m} \mid \text{ddiag}(X^H X) = I_m, \text{rk } X = m\}, \tag{16}$$

where $\mathrm{ddiag}(Z)$ forms a diagonal matrix, whose diagonal entries are just those of Z, and I_m is the $m \times m$ identity matrix. As a measure for the joint diagonality, we choose

$$f: Op(m) \times Op(m) \to \mathbb{R}, \qquad (X_L, X_R) \mapsto \frac{1}{4} \sum_{i=1}^{n} \left\| \mathrm{off}(X_L^{\mathsf{H}} C_i X_R) \right\|_F^2, \qquad (17)$$

where $\mathrm{off}(Z) = Z - \mathrm{ddiag}(Z)$ and $\| \cdot \|_F$ is the Frobenius norm of matrices.

The convergence rate of the Jacobi method depends on a non-degenerated Hessian form of the cost function at the optimal solution. In what follows, we therefore characterize the critical points of (17) and specify their Hessian form. Let us denote the set of all $m \times m$ matrices with zero diagonal entries by

$$\mathrm{off}(m) = \left\{ Z \in \mathbb{C}^{m \times m} | z_{ii} = 0, \text{ for } i = 1, \ldots, m \right\}. \qquad (18)$$

Then, for any $X \in Op(m)$, the following map

$$\mu_X : \mathrm{off}(m) \to Op(m), \ Z \mapsto X(I_m + Z) \, \mathrm{diag} \left\{ \tfrac{1}{\|X(e_1 + z_1)\|}, \ldots, \tfrac{1}{\|X(e_m + z_m)\|} \right\}, \qquad (19)$$

where $Z = [z_1, \ldots, z_m] \in \mathrm{off}(m)$ and e_i is the i-th standard basis vector, is a local and smooth parameterization around X. Let $\mathcal{X} := (X_L, X_R) \in Op^2(m) := Op(m) \times Op(m)$ and denote by $\mathcal{H} := (H_L, H_R) \in T_{\mathcal{X}} Op^2(m)$ the tangent vector at $\mathcal{X}$. The first derivative of f evaluated at $\mathcal{X}$ in tangent direction $\mathcal{H}$ yields

$$\mathrm{D}f(\mathcal{X})(\mathcal{H}) = \sum_{i=1}^{n} \mathrm{tr}\Big(\mathrm{off}\left(X_L^{\mathsf{H}} C_i X_R \right) X_R^{\mathsf{H}} C_i^{\mathsf{H}} H_L + \mathrm{off}\left(H_L^{\mathsf{H}} C_i X_R \right) X_R^{\mathsf{H}} C_i^{\mathsf{H}} X_L +$$

$$\mathrm{off}\left(X_L^{\mathsf{H}} C_i X_R \right) H_R^{\mathsf{H}} C_i^{\mathsf{H}} X_L + \mathrm{off}\left(X_L^{\mathsf{H}} C_i H_R \right) X_R^{\mathsf{H}} C_i^{\mathsf{H}} X_L \Big). \qquad (20)$$

Since any joint diagonalizer $\mathcal{X}^*$ is a global minimum, it satisfies $\mathrm{D}f(\mathcal{X}^*)\mathcal{H} = 0$. We use $\mu_{\mathcal{X}} := (\mu_{X_L}, \mu_{X_R})$ as the local parameterization of $Op^2(m)$ and compute the Hessian form $\mathsf{H}_f(\mathcal{X}^*) : T_{\mathcal{X}^*} Op^2(m) \times T_{\mathcal{X}^*} Op^2(m) \to \mathbb{R}$. Let $\mathcal{X}^* = (X_L^*, X_R^*)$ be a pair of joint diagonalizers. Without loss of generality assume that $X_L^{*\mathsf{H}} A_L$ is diagonal and denoted by $\Lambda_L := \mathrm{diag}(\lambda_{L1}, \ldots, \lambda_{Lm})$. Similarly, we denote $\Lambda_R := \mathrm{diag}(\lambda_{R1}, \ldots, \lambda_{Rm}) = A_R^{\mathsf{H}} X_R^*$. A tedious but direct calculation leads to

$$\mathsf{H}_f(\mathcal{X}^*)(\mathcal{H}, \mathcal{H}) = \frac{\mathrm{d}^2}{\mathrm{d}t^2}(f \circ \mu_{\mathcal{X}^*})(t\Theta, t\Xi)\Big|_{t=0} = \sum_{j \neq k}^{m} \begin{bmatrix} \theta_{jk} \\ \xi_{kj} \end{bmatrix}^{\top} \underbrace{\begin{bmatrix} \sum_{i=1}^{n} |\delta_{ij}|^2 & \sum_{i=1}^{n} \delta_{ij}\overline{\delta_{ik}} \\ \sum_{i=1}^{n} \overline{\delta_{ij}}\delta_{ik} & \sum_{i=1}^{n} |\delta_{ik}|^2 \end{bmatrix}}_{=:B_{jk}} \begin{bmatrix} \theta_{jk} \\ \xi_{kj} \end{bmatrix}, \quad (21)$$

where δ_{ij} is the j-th diagonal entry of the diagonal matrix $\Delta_i := X_L^* C_i X_R^* = \Lambda_L \Omega_i \Lambda_R$.

Clearly, the Hessian of f at $\mathcal{X}^*$ is at least positive semi-definite, and diagonal in terms of 2×2 blocks, with respect to the standard basis of the parameter space $\mathfrak{off}(m) \times \mathfrak{off}(m)$. Then, the definiteness of the Hessian simply depends on the determinant of B_{jk}'s, which is computed by

$$
\det(B_{jk}) = \left(\sum_{i=1}^{n} |\delta_{ij}|^2 \right) \left(\sum_{i=1}^{n} |\delta_{ik}|^2 \right) - \left| \sum_{i=1}^{n} \delta_{ij} \overline{\delta_{ik}} \right|^2
$$

$$
= |\lambda_{Lj}|^2 |\lambda_{Rj}|^2 |\lambda_{Lk}|^2 |\lambda_{Rk}|^2 \left(\left(\sum_{i=1}^{n} |\omega_{ij}|^2 \right) \left(\sum_{i=1}^{n} |\omega_{ik}|^2 \right) - \left| \sum_{i=1}^{n} \omega_{ij} \overline{\omega_{ik}} \right|^2 \right). \quad (22)
$$

By the Cauchy-Schwarz inequality, $\det(B_{jk})$ is non-negative, and is equal to zero if and only if there is a pair of positions (j, k), so that $z_j \in \mathbb{R}^n$ and z_k are linearly dependent, i.e. $\rho(\Omega_1, \ldots, \Omega_n) = 1$. Thus, we conclude.

Lemma 2. *Let the NsJD problem (2) have a unique joint diagonalizer. Then the Hessian of the off-norm function (17) at the joint diagonalizer is positive definite.*

4   Block-Jacobi for Non-symmetric Joint Diagonalization

In this section, we develop a block Jacobi algorithm to minimize the cost function (17). Firstly, let us define the complex one dimensional subspace

$$
V_{ij} := \{ Z = (z_{kl}) \in \mathbb{C}^{m \times m} | z_{kl} = 0 \text{ for } (k, l) \neq (i, j) \}. \quad (23)
$$

It is clear that $\oplus_{i \neq j}(V_{ij} \times V_{ji}) = \mathfrak{off}(m) \times \mathfrak{off}(m)$. We then define

$$
\mathcal{V}_{ij}(\mathcal{X}) := \{ \tfrac{\mathrm{d}}{\mathrm{d}t} \mu_{\mathcal{X}}(t \cdot Z)|_{t=0} | Z \in (V_{ij} \times V_{ji}) \}, \quad (24)
$$

yielding a vector space decomposition of the tangent space $T_{\mathcal{X}} Op^2(m)$.

   The block Jacobi-type method iteratively employs the search along the subspaces $\mathcal{V}_{ij}(X)$ in a cyclic manner. More precisely, let $(\Theta_{jk}, \Theta_{kj}) \in V_{ij} \times V_{ji}$ and denote $\theta = [\theta_{jk} \ \theta_{kj}]^\top \in \mathbb{C}^2$. For any point $\mathcal{X} \in Op^2(m)$, we construct a family of maps $\{ \nu_{jk}^{(\mathcal{X})} \}_{j \neq k}^m$ by

$$
\nu_{jk}^{(\mathcal{X})} : \mathbb{C}^2 \to Op^2(m), \qquad \theta \mapsto \mu_{\mathcal{X}}(\Theta_{jk}, \Theta_{kj}). \quad (25)
$$

The algorithm is presented in Algorithm 1. It is readily seen with (21) that the $\mathcal{V}_{ij}$ are orthogonal with respect to $H_f(\mathcal{X}^*)$. Therefore, the following result guarantees the super linear convergence rate of the block Jacobi method to an exact joint diagonalizer in case that this diagonalizer is essentially unique.

Theorem 1 ([13]). *Let M be an n-dimensional manifold and let x^* be a local minimum of the smooth cost function $f : M \to \mathbb{R}$ with nondegenerate Hessian $H_f(x^*)$. Let μ_x be a family of local parameterizations of M and let $\oplus_i V_i$ be a decomposition of $\mathbb{R}^n$. If the subspaces $V_i := T_0 \mu_{x^*}(V_i) \subset T_{x^*} M$ are orthogonal with respect to $H_f(x^*)$, then the Block-Jacobi method is locally quadratic convergent to x^*.*

Algorithm 1. *Jacobi Algorithm for Non-Symmetric Joint Diagonalization*

Step 1: Given an initial guess $\mathcal{X}^{(0)} = (X_L^{(0)}, X_R^{(0)}) \in Op^2(m)$ and set $s = 0$.

Step 2: Set $s = s + 1$ and let $\mathcal{X}_s = \mathcal{X}_{s-1}$.

For $1 \leq j < k \leq m$, update
$$\mathcal{X}_s \leftarrow \nu_{jk}^{(\mathcal{X}_s)}(\theta^*),$$
with $\theta^* = -\mathsf{H}_\varphi(0)^{-1}\nabla_\varphi(0)$. Here, H and ∇ denote the usual Hessian and the gradient of the function
$$\varphi \colon \mathbb{C}^2 \to \mathbb{R}, \quad \theta \mapsto f \circ \nu_{jk}^{(\mathcal{X}_s)}(\theta)$$

Step 3: If $\|\mathcal{X}_s - \mathcal{X}_{s-1}\|$ is small enough, stop

Otherwise, go to Step 2.

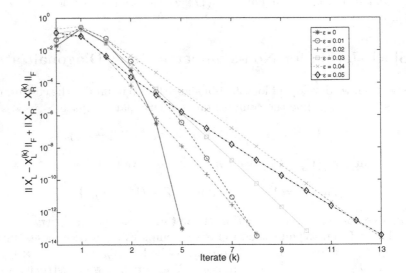

Fig. 1. Convergence properties of the proposed block Jacobi algorithm.

For an experimantal evaluation of this result, we consider the task of jointly diagonalizing a set of non-symmetric matrices $\{\widetilde{C}_i\}_{i=1}^n$, constructed by

$$\widetilde{C}_i = A_L \Lambda_i A_R^{\mathsf{H}} + \varepsilon E_i, \qquad i = 1, \ldots, n, \tag{26}$$

where $A \in \mathbb{C}^{m \times m}$ is a randomly picked matrix in $\mathcal{OB}(m)$, the modulus of diagonal entries of Λ_i are drawn from a uniform distribution on the interval $(9, 11)$, $E_i \in \mathbb{C}^{m \times m}$ represents the additive noise, whose entries are generated from a uniform distribution on the unit interval $(-0.5, 0.5)$, and $\varepsilon \in \mathbb{R}$ is the noise level. We set $m = 5$, $n = 20$, and run six tests in accordance with increasing noise, by using $\varepsilon = d \times 10^{-2}$ where $d = 0, \ldots, 5$. Each experiment was initialized with the same point, which is randomly drawn within an appropriate neighbourhood of the true joint diagonalizer.

The convergence of algorithms is measured by the distance of the accumulation point $\mathcal{X}^* := (X_L^*, X_R^*)$ to the current iterate $\mathcal{X}^{(k)} := (X_L^{(k)}, X_R^{(k)})$, i.e., by $\|X_L^{(k)} - X_L^*\|_F + \|X_R^{(k)} - X_R^*\|_F$. According to Fig. 1, our proposed algorithm converges locally quadratically fast to a pair of joint diagonalizers under the NsJD setting, i.e., when $\varepsilon = 0$, whereas with an increasing level of noise, the convergence rate slows down accordingly with a tendency of more gradual slopes.

References

1. Chabriel, G., Kleinsteuber, M., Moreau, E., Shen, H., Tichavsky, P., Yeredor, A.: Joint matrices decompositions and blind source separation: a survey of methods, identification, and applications. IEEE Sig. Process. Mag. **31**(3), 34–43 (2014)
2. Chabriel, G., Barrere, J.: Non-symmetrical joint zero-diagonalization and mimo zero-division multiple access. IEEE Trans. Sig. Process. **59**(5), 2296–2307 (2011)
3. Shen, H., Kleinsteuber, M.: A matrix joint diagonalization approach for complex independent vector analysis. In: Theis, F., Cichocki, A., Yeredor, A., Zibulevsky, M. (eds.) LVA/ICA 2012. LNCS, vol. 7191, pp. 66–73. Springer, Heidelberg (2012)
4. Kolda, T.G., Bader, B.W.: Tensor decompositions and applications. SIAM Rev. **51**(3), 455–500 (2009)
5. Cichocki, A., Mandic, D.P., Phan, A.H., Caiafa, C.F., Zhou, G., Zhao, Q., de Lathauwer, L.: Tensor decompositions for signal processing applications: From two-way to multiway component analysis. IEEE Sig. Process. Mag. **32**(2), 145–163 (2015)
6. Cardoso, J.F., Souloumiac, A.: Jacobi angles for simultaneous diagonalisation. SIAM J. Matrix Anal. Appl. **17**(1), 161–164 (1996)
7. Pham, D.T.: Joint approximate diagonalization of positive definite Hermitian matrices. SIAM J. Matrix Anal. Appl. **22**(4), 1136–1152 (2001)
8. Wang, F., Liu, Z., Zhang, J.: Nonorthogonal joint diagonalization algorithm based on trigonometric parameterization. IEEE Trans. Sig. Process. **55**(11), 5299–5308 (2007)
9. Souloumiac, A.: Nonorthogonal joint diagonalization by combining givens and hyperbolic rotations. IEEE Trans. Sig. Process. **57**(6), 2222–2231 (2009)
10. Shen, H., Hüper, K.: Block Jacobi-type methods for non-orthogonal joint diagonalisation. In: Proceedings of the 34th IEEE ICASSP, pp. 3285–3288 (2009)
11. Kleinsteuber, M., Shen, H.: Uniqueness analysis of non-unitary matrix joint diagonalization. IEEE Trans. Sig. Process. **61**(7), 1786–1796 (2013)
12. Kleinsteuber, M., Shen, H.: A geometric framework for non-unitary joint diagonalization of complex symmetric matrices. In: Nielsen, F., Barbaresco, F. (eds.) GSI 2013. LNCS, vol. 8085, pp. 353–360. Springer, Heidelberg (2013)
13. Kleinsteuber, M., Shen, H.: Block-jacobi methods with newton-steps andnon-unitary joint matrix diagonalization. In: Accepted at the 2nd Conference on Geometric Science of Information (GSI) (2015)

An Affine Equivariant Robust Second-Order BSS Method

Pauliina Ilmonen[1]([✉]), Klaus Nordhausen[2,3], Hannu Oja[2], and Fabian Theis[4,5]

[1] Department of Mathematics and Systems Analysis,
Aalto University, P.O. Box 11100, 00076 Aalto, Finland
pauliina.ilmonen@gmail.com
[2] Department of Mathematics and Statistics,
University of Turku, 20014 Turku, Finland
[3] School of Health Sciences, University of Tampere, 33014 Tampere, Finland
[4] Institute of Computational Biology, German Research Center for Environmental
Health, Helmholtz Zentrum München, 85764 Neuherberg, Germany
[5] Department of Mathematics, Technische Universitat München, Munich, Germany

Abstract. The interest in robust methods for blind source separation
has increased recently. In this paper we shortly review what has been
suggested so far for robustifying ICA and second order blind source separation. Furthermore do we suggest a new algorithm, eSAM-SOBI, which
is an affine equivariant improvement of (already robust) SAM-SOBI. In
a simulation study we illustrate the benefits of using eSAM-SOBI when
compared to SOBI and SAM-SOBI. For uncontaminated time series
SOBI and eSAM-SOBI perform equally well. However, SOBI suffers a
lot when the data is contaminated by outliers, whereas robust eSAM-
SOBI does not. Due to the lack of affine equivariance of SAM-SOBI,
eSAM-SOBI performs clearly better than it for both, contaminated and
uncontaminated data.

Keywords: ICA · SOBI · Location and scatter functionals · Time series

1 Introduction

Statistical procedures whose behavior is not influenced much by atypical observations or slight deviations from the model assumptions are considered robust.
Many robust methods have been developed to make classical statistical procedures valid and efficient in a neighborhood of the normal model, valid in many
nonparametric models, and to cope with outliers. For an overview see [1].

The interest in robust methods has recently increased in blind source separation problems. In this paper, we will shortly review what has been done so
far (as well as pointing out some pitfalls too) and suggest an improvement to
SAM-SOBI [2], which is a non-affine equivariant robustification of SOBI.

© Springer International Publishing Switzerland 2015
E. Vincent et al. (Eds.): LVA/ICA 2015, LNCS 9237, pp. 328–335, 2015.
DOI: 10.1007/978-3-319-22482-4_38

2   Robust Statistics

The main tools for robust multivariate methods are location and scatter functionals. Let $\mathbf{x}$ be a p-variate random vector with cumulative distribution function (cdf) $F_{\mathbf{x}}$.

A p-variate vector valued functional $\mathbf{T}(F_{\mathbf{x}}) = \mathbf{T}(\mathbf{x})$ is called a location functional, if it is affine equivariant in the sense that $\mathbf{T}(F_{\mathbf{Ax+b}}) = \mathbf{AT}(F_{\mathbf{x}}) + \mathbf{b}$ for all full rank $p \times p$ matrices $\mathbf{A}$ and all p-vectors $\mathbf{b}$. A $p \times p$ matrix valued functional $\mathbf{S}(F_{\mathbf{x}}) = \mathbf{S}(\mathbf{x})$ is called a scatter functional, if it is positive definite and affine equivariant in the sense that $\mathbf{S}(F_{\mathbf{Ax+b}}) = \mathbf{AS}(F_{\mathbf{x}})\mathbf{A}'$ for all full rank $p \times p$ matrices $\mathbf{A}$ and all p-vectors $\mathbf{b}$. Sample versions of location functionals and scatter functionals are called location statistics and scatter statistics, respectively. They are obtained by replacing $F_{\mathbf{x}}$ by the empirical cdf F_n. In many applications the equality in the previous equation can be relaxed to $\mathbf{S}(F_{\mathbf{Ax+b}}) \propto \mathbf{AS}(F_{\mathbf{x}})\mathbf{A}'$, in which case $\mathbf{S}(F_{\mathbf{x}})$ is called a shape matrix.

Clearly the mean vector $\mathbf{E}(\mathbf{x})$ is a location functional and the covariance matrix $\mathbf{COV}(\mathbf{x}) = \mathbf{E}((\mathbf{x} - \mathbf{E}(\mathbf{x}))(\mathbf{x} - \mathbf{E}(\mathbf{x}))')$ is a scatter functional. There exist many other location and scatter functionals which all have different properties. For example, M-functionals are defined by the two implicit equations $\mathbf{T}(\mathbf{x}) = \mathbf{E}(w_1(r))^{-1}\mathbf{E}(w_1(r)\mathbf{x})$ and $\mathbf{S}(\mathbf{x}) = \mathbf{E}(w_2(r)(\mathbf{x} - \mathbf{T}(\mathbf{x}))(\mathbf{x} - \mathbf{T}(\mathbf{x}))')$, where $w_1(r)$ and $w_2(r)$ are nonnegative continuous functions of the Mahalanobis distance $r = \|\mathbf{S}(\mathbf{x})^{-1/2}(\mathbf{x} - \mathbf{T}(\mathbf{x}))\|_2$. Later in this paper we use the Hettmansperger-Randles estimator [3], which has the weight functions $w_1(r) = 1/r$ and $w_2(r) = p/r^2$. This estimator can be seen as the joint estimate of the spatial median and Tyler's shape matrix. Tyler's shape matrix is considered the most robust M-estimate and it is especially efficient in high-dimensions.

If $\mathbf{S}(\mathbf{x})$ is a diagonal matrix for all $\mathbf{x}$ having independent components, it is said to have the independence property. This is important as in many applications zero off-diagonal values are taken as indicators of independence. Most robust scatter functionals do posses that property only if all the components (or all the components except for one) are symmetric. However, every scatter matrix $\mathbf{S}(\mathbf{x})$ can be symmetrized by defining $\mathbf{S}_{sym}(\mathbf{x}) = \mathbf{S}(\mathbf{x}_1 - \mathbf{x}_2)$, where $\mathbf{x}_1$ and $\mathbf{x}_2$ are independent copies of $\mathbf{x}$. All symmetrized scatter matrices have the independence property.

Many robust statistical methods replace the mean vector and the covariance matrix with robust functionals - but this requires the assumption that the robust sample statistics estimate the same population quantity, or the corresponding functionals have similar properties (like independence property). Most of these functionals have been designed for the elliptical model and hence it is important to note that depending on the underlying model, different scatter functionals estimate different population quantities. However, in all symmetric models, $\mathbf{T}(\mathbf{x})$ estimates the center of symmetry and in the elliptical model all scatter matrices estimate the same population quantity up to multiplication by a constant. Note that this is not the case for example in the independent component model. For detailed discussion about this see [4].

3   Robust Blind Source Separation

For simplicity and due to lack of space we consider here only the following blind source separation (BSS) model

$$\mathbf{x} = \mathbf{As},$$

where $\mathbf{x}$ is the observable p-variate random vector, $\mathbf{A}$ an unknown $p \times p$ full rank mixing matrix and $\mathbf{s}$ the unknown source. The goal in BSS is to find an unmixing matrix $\mathbf{W}$ such that $\mathbf{z} = \mathbf{Wx}$ recovers $\mathbf{s}$ up to some indeterminancies, depending on the assumptions made on $\mathbf{s}$. Two cases are considered here.

3.1   Independent Component Analysis

In the independent component analysis (ICA) it is assumed that the components of $\mathbf{s}$ are independent and that at most one of them is gaussian. There are many algorithms for ICA (see e.g. [5,6]). Most of them first whiten the data using the covariance matrix and then search for an orthogonal transformation to recover the independent components. The covariance matrix is highly nonrobust and therefore using it makes all these algorithms nonrobust too. However, one can not just plug in any robust scatter functional to robustify the whitening step - only scatter functionals having the independence property can be used [4].

The problem of nonrobustness in ICA is well-known. For example Ollila [7] shows that deflation-based fastICA has an unbounded influence function. Thus even one outlier can reduce the quality of the estimated unmixing matrix hugely.

Two approaches have been considered to robustify ICA.

1. Precleaning the data: Here the goal is to remove all atypical observations before actually performing ICA. This approach was used for example in [8,9]. The problem in this approach is how to define atypical observations. In order to do that, some model assumptions are required for typical observations. That, however, is counterintuitive in ICA as often the "extreme" observations are influential in the algorithms to maximize non-gaussianity. Thus extreme observations and atypical observations should be very well separated.

2. Using robust procedures: In [10] a standard ICA algorithm is robustified by replacing each nonrobust step by a robust step. There, however, a discussion about when the same population quantities or population values of the functional with similar properties are estimated, is omitted. For example, the minimum covariance determinant scatter estimate is wrongly recommended for the whitening step, even though it does not posses the independence property, and hence the assumption of at most one skew source is made implicitly. In [11] two robust scatter functionals, both having the independence property, are used and a robust generalized version of FOBI is obtained.

3.2   Second Order Blind Source Separation

In second order blind source separation the data points are not assumed to be iid as in ICA - instead the sources are assumed to be stationary time series. The model is then written as

$$\mathbf{x}(t) = \mathbf{A}\mathbf{s}(t), \quad t = 1, 2, 3 \ldots$$

Let $\mathbf{s}_i(t)$ denote the ith time series. We assume that $\mathbf{E}(\mathbf{s}(t)) = \mathbf{0}$ for every t, and that, for $\tau = 0, 1, 2, \ldots$, $\mathbf{E}(\mathbf{s}_i(t)\mathbf{s}_j(t+\tau)) = 0$ for $i \neq j$ and $\mathbf{E}(\mathbf{s}_i(t)\mathbf{s}_i(t+\tau)) = \gamma_i(\tau)$ with $\gamma_i(\tau) \neq \gamma_j(\tau)$ for $i \neq j$. Furthermore we assume that $\mathbf{s}(t)$ is symmetric.

Not many robustness considerations have been made in this model so far. Already in univariate time series analysis, the notion of an outlier gets a bit complicated. According to [1] three different types of outliers should be addressed: (i) isolated outliers, which are single gross errors; (ii) patchy outliers, which occur in a succession and destroy completely the covariance structure; (iii) complete level shift in mean value, which is a structural break with a moderate mean increase without effect on the covariance structure.

Thinking about different approaches to robustify BSS it is obvious that the approach based on first removing atypical observations is not possible here. Since the correlation structure is important, single observations cannot be removed. An alternative approach would be to filter the observed series using robust filters in order to get rid of the outlying data points and then use a standard BSS method for the filtered data. We are not aware of such approaches used in second order blind source separation and we also believe that prefiltering might be counterproductive to the applied BSS. Therefore we prefer an approach that uses directly robust methods.

The most popular approach in this model is SOBI [12] algorithm:

1. Whitening the series: $\mathbf{x}(t) \leftarrow \text{COV}(\mathbf{x}(t))^{-1/2}(\mathbf{x}(t) - \mathbf{E}(\mathbf{x}(t)))$, where $\mathbf{A}^{1/2}$ is a symmetric square root of $\mathbf{A}$.
2. Computing K autocovariance matrices: $\mathbf{R}_\tau = \mathbf{E}(\mathbf{x}(t)\mathbf{x}(t + \tau)')$ for $\tau = 1, \ldots, K$.
3. Symmetrizing the autocovariance matrices: $\mathbf{R}_\tau = (\mathbf{R}_\tau + \mathbf{R}'_\tau)/2$.
4. Finding an orthogonal matrix $\mathbf{U} = (\mathbf{u}_1, \ldots, \mathbf{u}_p)$ that jointly diagonalizes $\mathbf{R}_1, \ldots, \mathbf{R}_K$.

There are many variants of SOBI which mainly differ on the way the matrix $\mathbf{U}$ is computed, see for example [13–15]. The original SOBI uses Jacobi rotations but for instance [13] suggest a deflation-based approach, where the directions $\mathbf{u}_1, \ldots, \mathbf{u}_p$ are found one after another by maximizing $\sum_{k=1}^K G(\mathbf{u}'\mathbf{R}_k\mathbf{u})$ under the constraint that $\mathbf{u}'_i\text{COV}(\mathbf{x}(t))\mathbf{u}_j = \delta_{ij}$ $i = 1, \ldots, j - 1$. G can in this context be any increasing twice differentiable function.

SAM-SOBI [2] was recently proposed as a robustifaction of the original SOBI by modifying it as follows:

1. Robust centering of the data: $\mathbf{x}(t) \leftarrow (\mathbf{x}(t) - \mathbf{T}(\mathbf{x}(t)))$, where $\mathbf{T}(\mathbf{x}(t))$ is the spatial median.
2. Robust uncorrelating of the data: $\mathbf{x}(t) \leftarrow \mathbf{S}(\mathbf{x}(t))^{-1/2}(\mathbf{x}(t))$, where $\mathbf{S}(\mathbf{x}(t))$ is the spatial sign covariance matrix $\mathbf{S}(\mathbf{x}) = \mathbf{E}\left(\frac{\mathbf{x}(t)}{\|x(t)\|}\frac{\mathbf{x}(t)}{\|x(t)\|}'\right)$.
3. Computing K robust autocovariance matrices based on the spatial sign autocovariance matrix: $\mathbf{R}_\tau = \mathbf{E}\left(\frac{\mathbf{x}(t)}{\|\mathbf{x}(t)\|}\frac{\mathbf{x}(t+\tau)}{\|x(t+\tau)\|}'\right)$ for $\tau = 1, \ldots, K$.

4. Symmetrizing the robust autocovariance matrices: $\mathbf{R}_\tau = (\mathbf{R}_\tau + \mathbf{R}'_\tau)/2$.
5. Using Jacobi rotations to find the orthogonal matrix $\mathbf{U}$ that jointly diagonalizes $\mathbf{R}_1, \ldots, \mathbf{R}_K$.

Thus, all the highly non-robust parts of SOBI, mean vector, covariance matrix
and autocovariance matrices have been replaced by robust alternatives. This
minimizes the effect of atypical observations and therefore SAM-SOBI can be
considered robust. However, it has some minor flaws. First, the spatial median is
not a location functional and the spatial sign covariance matrix is not a scatter
functional in the sense defined above. Both are only equivariant under orthogonal
transformations. Second, the spatial median centers the sources/signals similarly
as the mean vector only under symmetry and the spatial sign covariance matrix
does not posses the independence property. Figure 1 demonstrates that for symmetric data the location estimates are not distinguishable as in the case of skew
data. The spatial sign covariance matrix is also a bit tilted in the skew case.

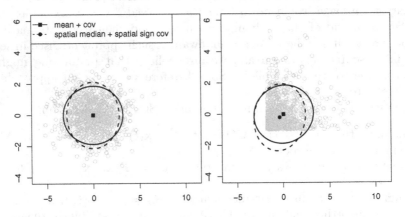

Fig. 1. Comparison of mean and spatial median together with covariance matrix and
spatial sign covariance matrix for a bivariate normal distribution (left panel) and a
distribution with two independent exponential distributions (right panel).

3.3   An Affine Equivariant SAM-SOBI

We present here an important improvement for SAM-SOBI which combines the
first two steps and uses there the affine equivariant Hettmansperger-Randles
estimate (see [3]). Therefore the new affine equivariant SAM-SOBI, eSAM-SOBI,
has the following steps:

1. Robust whitening of the data: $\mathbf{x}(t) \leftarrow \mathbf{S}^{-1/2}(\mathbf{x}(t))(\mathbf{x}(t) - \mathbf{T}(\mathbf{x}(t)))$, where
   $\mathbf{T}(\mathbf{x}(t))$ and $\mathbf{S}(\mathbf{x}(t))$ are the Hettmansperger-Randles estimates.
2. Computing K robust autocovariance matrices based on the spatial sign auto-
   covariance matrix: $\mathbf{R}_\tau = \mathbf{E}\left(\frac{\mathbf{x}(t)}{||\mathbf{x}(t)||}\frac{\mathbf{x}(t+\tau)}{||x(t+\tau)||}'\right)$ for $\tau = 1, \ldots, K$.
3. Symmetrizing the robust autocovariance matrices: $\mathbf{R}_\tau = (\mathbf{R}_\tau + \mathbf{R}'_\tau)/2$.
4. Finding an orthogonal matrix $\mathbf{U}$ that jointly diagonalizes $\mathbf{R}_1, \ldots, \mathbf{R}_K$.

This has the advantage that the starting step is affine equivariant, making the whole procedure affine equivariant. Note that we do not restrict here the joint diagonalization to using Jacobi rotations but suggest also to consider different functions G in the deflation-based approach as they might have an impact on the robustness too.

To conclude this section, note that robustness was also considered for non-stationary BSS in [16], but due to space limitation we can not discuss that here.

4   Simulation Study

To compare the three different version of SOBI discussed in this paper, we conduct a small simulation study. For that we consider SOBI, SAM-SOBI and eSAM-SOBI using Jacobi rotations and deflation-based versions of all three of them using the G functions $G(x) = |x|^a$ with $a = 1, 2, 3$ and $G(x) = \log(x)$. From these, $a = 2$ should be quite similar to the Jacobi approach, and $a = 1$ and log have a more robust flavor, see [13] for further discussion.

We sampled 2000 observations from three sources $\mathbf{S}$. The first source was an AR(3)-process with t_3 distributed innovations. The second source was a MA(7)-process and the last source an ARMA(4,3)-process. The sources were mixed with a random matrix $\mathbf{A}$ having each element simulated from a $N(0,1)$ distribution. Having a non-trivial mixing matrix $\mathbf{A}$ is important here as only the performance of the affine equivariant methods does not depend on its value. In the statistics literature the spatial sign-covariance matrix is considered efficient only for uncorrelated data with equal scales.

To evaluate the robustness properties, we have the three following outlier scenarios (i) isolated outlier: observation $\mathbf{X}(100)$ was replaced with an isolated outlier $(50, 50, 50)'$; (ii) patchy outliers: observations $\mathbf{X}(100), \ldots, \mathbf{X}(114)$ got in each component additive error coming from a $N(10,1)$ distribution; (iii) level shift: the last one hundred observations were all added $(1, 1, 1)'$. In all three cases the contaminated data points hence do not follow anymore the model $\mathbf{X} = \mathbf{AS}$.

For all 15 estimates, the unmixing matrix $\mathbf{W}$ was estimated 1000 times for the clean data, and all three outlier scenarios using $K = 12$. To evaluate the performance we used the minimum distance index D from [17], which is given by $D = \inf_{\mathbf{P,J,D}} \frac{1}{\sqrt{p-1}} ||\mathbf{PJD\hat{W}A} - \mathbf{I}_p||_2$, where $\mathbf{P}$ ia a permutation matrix, $\mathbf{J}$ a signchange matrix and $\mathbf{D}$ a scaling matrix. The values of D are in the interval $[0, 1]$ where 0 corresponds to a perfect separation.

The results in Fig. 2 show that in the uncontaminated data SOBI and eSAM-SOBI perform equally well. However, SOBI suffers considerably under all three outlier setups. eSAM-SOBI seems almost unaffected from the isolated or patchy outliers, and suffers only a little from the level shift. It is clearly the best method also under this scenario. SAM-SOBI seems unaffected by any disturbances and the performance is stable throughout. It is however always worse than eSAM-SOBI. The reason for that is the missing affine equivariance. The difference between the methods based on the way they were jointly diagonalized are only marginal and not consistent. For example, it seems that log is worse for eSAM-SOBI, but has a robustifying effect on SOBI — but only for the isolated outlier.

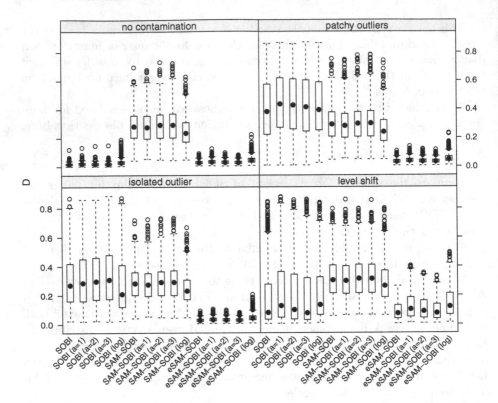

Fig. 2. Boxplots for the performance criterion D for SOBI, SAM-SOBI and eSAM-SOBI using different joint diagonalization approaches for all four scenarios.

5   Conclusion

In this paper we reviewed robust methods for BSS and suggested a new algorithm, eSAM-SOBI. Simulations demonstrated the good performance of eSAM-SOBI. Different options to relax the symmetry assumption needed here should be investigated. Also investigating the impact of different joint diagonalization approaches might be worthwhile.

Acknowledgements. The work of Klaus Nordhausen and Hannu Oja was supported by the Academy of Finland (grant 268703). The authors thank the referees for their insightful comments which helped to improve the manuscript.

References

1. Maronna, R.A., Martin, D.R., Yohai, V.J.: Robust Statistics: Theory and Methods. Wiley, Chichester (2006)
2. Theis, F.J., Müller, N.S., Plant, C., Böhm, C.: Robust second-order source separation identifies experimental responses in biomedical imaging. In: Vigneron, V., Zarzoso, V., Moreau, E., Gribonval, R., Vincent, E. (eds.) LVA/ICA 2010. LNCS, vol. 6365, pp. 466–473. Springer, Heidelberg (2010)

3. Hettmansperger, T.P., Randles, R.H.: A practical affine equivariant multivariate median. Biometrika **89**, 851–860 (2002)
4. Nordhausen, K., Tyler, D.E.: A Cautionary Note on Robust Covariance Plug-in Methods. To appear in Biometrika (2015). doi:10.1093/biomet/asv022, arXiv:1404.0860
5. Hyvärinen, A., Karhunen, J., Oja, E.: Independent Component Analysis. Wiley, New York (2001)
6. Comon, P., Jutten, C.: Handbook of Blind Source Separation: Independent Component Analysis and Applications. Academic Press, Oxford (2010)
7. Ollila E.: On the robustness of the deflation-based FastICA estimator. In: 15th IEEE/SP Workshop on Statistical Signal Processing, pp. 673–676 (2009)
8. Brys, G., Hubert, M., Rousseeuw, P.J.: A robustification of independent component analysis. J. Chemom. **19**, 364–375 (2005)
9. Anderson, M., Adali, T.: A general approach for robustification of ICA algorithms. In: Vigneron, V., Zarzoso, V., Moreau, E., Gribonval, R., Vincent, E. (eds.) LVA/ICA 2010. LNCS, vol. 6365, pp. 295–302. Springer, Heidelberg (2010)
10. Baloch, S.H., Krim, H., Genton, M.: Robust independent component analysis. In: 13th IEEE/SP Workshop on Statistical Signal Processing, pp. 61–64 (2005)
11. Nordhausen, K., Oja, H., Ollila, E.: Robust independent component analysis based on two scatter matrices. Austrian J. Stat. **37**, 91–100 (2008)
12. Belouchrani, A., Abed-Meraim, K., Cardoso, J.-F., Moulines, E.: A blind source separation technique using second order statistics. IEEE Trans. Sig. Process. **45**, 434–444 (1997)
13. Miettinen, J., Nordhausen, K., Oja, H., Taskinen, S.: Deflation-based separation of uncorrelated stationary time series. J. Multivar. Anal. **123**, 214–227 (2014)
14. Miettinen, J., Illner, K., Nordhausen, K. Oja, H., Taskinen, S., Theis, F.: Separation of Uncorrelated Stationary Time Series Using Autocovariance Matrices (2015, submitted). arXiv:1405.3388
15. Illner, K., Miettinen, J., Fuchs, C., Taskinen, S., Nordhausen, K., Oja, H., Theis, F.J.: Model selection using limiting distributions of second-order source separation algorithms. Sig. Process. **113**, 95–103 (2015)
16. Nordhausen, K.: On robustifying some second order blind source separation methods for nonstationary time series. Stat. Pap. **55**, 141–156 (2014)
17. Ilmonen, P., Nordhausen, K., Oja, H., Ollila, E.: A new performance index for ICA: properties, computation and asymptotic analysis. In: Vigneron, V., Zarzoso, V., Moreau, E., Gribonval, R., Vincent, E. (eds.) LVA/ICA 2010. LNCS, vol. 6365, pp. 229–236. Springer, Heidelberg (2010)

An Overview of the Asymptotic Performance of the Family of the FastICA Algorithms

Tianwen Wei[(✉)]

Laboratoire de Mathématiques de Besancon, Université de Franche-Comté,
16 Route de Gray, 25000 Besancon, France
tianwen.wei.2014@ieee.org

Abstract. This contribution summarizes the results on the asymptotic performance of several variants of the FastICA algorithm. A number of new closed-form expressions are presented.

Keywords: Independent component analysis · Symmetric FastICA · Deflationary FastICA · Data whitening · Data centering · Asymptotic performance

1 Introduction

In what follows, we denote scalars by lowercase letters $(a, b, c, \ldots)$, vectors by boldface lowercase letters $(\mathbf{a}, \mathbf{b}, \mathbf{c}, \ldots)$ and matrices by boldface uppercase letters $(\mathbf{A}, \mathbf{B}, \mathbf{C}, \ldots)$. Greek letters $(\alpha, \beta, \gamma, \ldots)$ are reserved for particular scalar quantities. We denote by $\mathbf{A}^{\mathsf{T}}$ the matrix transpose of $\mathbf{A}$ and by $\|\cdot\|$ the Euclidean norm.

1.1 ICA Data Model

We consider the following noiseless linear ICA model:

$$\mathbf{y}(t) = \mathbf{H}\mathbf{s}(t), \quad t = 1, \ldots, N,$$

where

1. $\mathbf{s}(t) \overset{\text{def}}{=} (s_1(t), \ldots, s_d(t))^{\mathsf{T}}$ denotes the tth realization of the unknown *source signal*. The components $s_1(t), \ldots, s_d(t)$ are mutually statistically independent, have unit variance and at most one of them is Gaussian. Furthermore, $\mathbf{s}(1), \ldots, \mathbf{s}(N)$ denote N independent realizations of $\mathbf{s}$.
2. $\mathbf{y}(t) \overset{\text{def}}{=} (y_1(t), \ldots, y_d(t))^{\mathsf{T}}$ denotes the tth realization of the *observed signal*.
3. $\mathbf{H} \in \mathbb{R}^{d \times d}$ is a full rank square matrix, called the mixing matrix.

1.2 Data Preprocessing

Most ICA methods require the observed signal $\{\mathbf{y}(t)\}$ to be standardized [1–3]. The standardization of $\{\mathbf{y}(t)\}$ consists of the data centering and data whitening,

© Springer International Publishing Switzerland 2015
E. Vincent et al. (Eds.): LVA/ICA 2015, LNCS 9237, pp. 336–343, 2015.
DOI: 10.1007/978-3-319-22482-4_39

which involve the estimation of $\mathbb{E}[\mathbf{y}]$ and $\mathrm{Cov}(\mathbf{y})$. In practice, $\mathbb{E}[\mathbf{y}]$ and $\mathrm{Cov}(\mathbf{y})$ are usually estimated by the sample mean and sample variance:

$$\bar{\mathbf{y}} \overset{\mathrm{def}}{=} \sum_{t=1}^{N} \frac{1}{N} \mathbf{y}(t), \quad \widehat{\mathbf{C}} \overset{\mathrm{def}}{=} \frac{1}{N} \sum_{t=1}^{N} (\mathbf{y}(t) - \bar{\mathbf{y}})(\mathbf{y}(t) - \bar{\mathbf{y}})^{\mathsf{T}}.$$

In this work, we shall consider several different data preprocessing scenarios. Denote

$$\widetilde{\mathbf{C}} = \frac{1}{N} \sum_{t=1}^{N} (\mathbf{y}(t) - \mathbb{E}[\mathbf{y}])(\mathbf{y}(t) - \mathbb{E}[\mathbf{y}])^{\mathsf{T}}.$$

The following data preprocessing scenarios will be studied:

1. Theoretical whitening and theoretical centering.

$$\mathbf{x}(t) \overset{\mathrm{def}}{=} \mathrm{Cov}(\mathbf{y})^{-\frac{1}{2}} (\mathbf{y}(t) - \mathbb{E}[\mathbf{y}]). \tag{1}$$

2. Theoretical whitening and empirical centering.

$$\mathbf{x}(t) \overset{\mathrm{def}}{=} \mathrm{Cov}(\mathbf{y})^{-\frac{1}{2}} (\mathbf{y}(t) - \bar{\mathbf{y}}). \tag{2}$$

3. Empirical whitening and theoretical centering.

$$\mathbf{x}(t) \overset{\mathrm{def}}{=} \widetilde{\mathbf{C}}^{-\frac{1}{2}} (\mathbf{y}(t) - \mathbb{E}[\mathbf{y}]). \tag{3}$$

4. Empirical whitening and empirical centering.

$$\mathbf{x}(t) \overset{\mathrm{def}}{=} \widehat{\mathbf{C}}^{-\frac{1}{2}} (\mathbf{y}(t) - \bar{\mathbf{y}}). \tag{4}$$

In the sequel, $\mathbf{x}(t)$ will always stand for the standardized signal under one of the scenarios defined above. The specific data preprocessing scenario will be stated explicitly when necessary.

1.3 Variants of the FastICA Algorithm

Before proceeding further, we need to introduce some notations first. We denote by $\mathcal{S}$ the unit sphere in $\mathbb{R}^d$. We denote by $g(\cdot) : \mathbb{R} \to \mathbb{R}$ the nonlinearity function, and by $G(\cdot)$ its primitive. The nonlinearity function g is usually supposed to be non-linear, non-quadratic and smooth. For any function $f : \mathbb{R}^d \to \mathbb{R}^m$, we write $\widehat{\mathbb{E}}_{\mathbf{x}}[f(\mathbf{x})] \overset{\mathrm{def}}{=} \frac{1}{N} \sum_{t=1}^{N} f(\mathbf{x}(t))$ for conciseness.

The Deflationary FastICA Algorithm. This version of the FastICA algorithm extracts the sources sequentially. It consists of the following steps [3]:

- **Input:** $\mathbf{x}(1), \ldots, \mathbf{x}(N)$.
  1. Set $p = 1$.
  2. Choose an arbitrary initial iterate $\mathbf{w} \in \mathcal{S}$;

3. Run iteration

$$\mathbf{w} \leftarrow \widehat{\mathbb{E}}_{\mathbf{x}}[g'(\mathbf{w}^{\mathsf{T}}\mathbf{x})\mathbf{w} - g(\mathbf{w}^{\mathsf{T}}\mathbf{x})\mathbf{x}] \tag{5}$$

$$\mathbf{w} \leftarrow \mathbf{w} - \sum_{i=1}^{p-1}(\mathbf{w}_i^{DFL})^{\mathsf{T}}\mathbf{w} \tag{6}$$

$$\mathbf{w} \leftarrow \frac{\mathbf{w}}{\|\mathbf{w}\|} \tag{7}$$

until convergence[1]. The limit is stored as $\mathbf{w}_p^{DFL}$.

4. Break if $p = d$. Otherwise $p \leftarrow p + 1$ then go to step 2).

- **Output:** $\mathbf{W}^{DFL} = (\mathbf{w}_1^{DFL}, \dots, \mathbf{w}_d^{DFL})$.

The Symmetric FastICA Algorithm. The symmetric version of FastICA extracts all the sources simultaneously. It can be described as follows:

- **Input:** $\mathbf{x}(1), \dots, \mathbf{x}(N)$.

  1. Choose an arbitrary orthonormal matrix $\mathbf{W} = (\mathbf{w}_1, \dots, \mathbf{w}_d) \in \mathbb{R}^{d \times d}$.
  2. Run

$$\mathbf{w}_1 \leftarrow \widehat{\mathbb{E}}_{\mathbf{x}}[g'(\mathbf{w}_1^{\mathsf{T}}\mathbf{x})\mathbf{w}_1 - g(\mathbf{w}_1^{\mathsf{T}}\mathbf{x})\mathbf{x}] \tag{8}$$

$$\vdots$$

$$\mathbf{w}_d \leftarrow \widehat{\mathbb{E}}_{\mathbf{x}}[g'(\mathbf{w}_d^{\mathsf{T}}\mathbf{x})\mathbf{w}_1 - g(\mathbf{w}_1^{\mathsf{T}}\mathbf{x})\mathbf{x}] \tag{9}$$

$$\mathbf{W} \leftarrow \left(\mathbf{W}\mathbf{W}^{\mathsf{T}}\right)^{-1/2}\mathbf{W} \tag{10}$$

until convergence. The limit is denoted by $\mathbf{W}^{SYM}$.

- **Output:** $\mathbf{W}^{SYM} = (\mathbf{w}_1^{SYM}, \dots, \mathbf{w}_d^{SYM})$.

2   Asymptotic Performance

Let us introduce the notion of gain matrix:

$$\mathbf{G}^{DFL} \stackrel{\text{def}}{=} (\mathbf{W}^{DFL})^{\mathsf{T}}\mathbf{C}^{-1/2}\mathbf{H}, \quad \mathbf{G}^{SYM} \stackrel{\text{def}}{=} (\mathbf{W}^{SYM})^{\mathsf{T}}\mathbf{C}^{-1/2}\mathbf{H},$$

where $\mathbf{C}^{-1/2}$ stands for the sphering matrix used in the data preprocessing stage, i.e. $\mathbf{C} = \mathrm{Cov}(\mathbf{y})$ in scenarios (1) and (2), $\mathbf{C} = \widetilde{\mathbf{C}}$ in scenario (3) and $\mathbf{C} = \widehat{\mathbf{C}}$ in scenario (4). Without loss of generality, we shall omit the permutation and sign ambiguities of ICA. Then, $\mathbf{G}^{DFL} \approx \mathbf{I}$ and $\mathbf{G}^{SYM} \approx \mathbf{I}$, hence $\mathbf{C}^{-1/2}\mathbf{W}^{DFL}$ and $\mathbf{C}^{-1/2}\mathbf{W}^{SYM}$ can be considered as estimators of $\mathbf{B} \stackrel{\text{def}}{=} (\mathbf{H}^{-1})^{\mathsf{T}}$. In the sequel, we will study the asymptotic errors of $N^{1/2}(\mathbf{C}^{-1/2}\mathbf{W}^{DFL} - \mathbf{B})$ and $N^{1/2}(\mathbf{C}^{-1/2}\mathbf{W}^{SYM} - \mathbf{B})$ under proposed data preprocessing scenarios.

The proofs of the results presented below are based on the method of M-estimators. However, all proofs will be omitted due to the lack of space. A complete version of this work can be provided upon request. The readers are also referred to [4] for a more detailed account of this subject.

---

[1] We impose the number of iterations to be even, so that the well known sign-flipping phenomenon disappears.

2.1   The Asymptotic Error of Deflationary FastICA

Assume that the following mathematical expectations exist for $i = 1, \ldots, d$:

$$\alpha_i \stackrel{\text{def}}{=} \mathbb{E}[g'(z_i) - g(z_i)z_i]$$

$$\beta_i \stackrel{\text{def}}{=} \mathbb{E}[g(z_i)^2]$$

$$\gamma_i \stackrel{\text{def}}{=} \mathbb{E}[g(z_i)z_i]$$

$$\eta_i \stackrel{\text{def}}{=} \mathbb{E}[g(z_i)]$$

$$\tau_i \stackrel{\text{def}}{=} (\mathbb{E}[z_i^4] - 1)/4,$$

where $z_i = s_i - \mathbb{E}[s_i]$ for $i = 1, \ldots, d$.

Theorem 1. *Let $\mathbf{b}_i$ denote the ith column of $\mathbf{B}$. Under some mild regularity conditions, we have*

$$N^{1/2}(\mathbf{C}^{-1/2}\mathbf{w}_i^{DFL} - \mathbf{b}_i) \xrightarrow[N \to \infty]{\mathscr{D}} \mathcal{N}(0, \mathbf{R}_{(k)}^{DFL}),$$

where $k \in \{1, 2, 3, 4\}$ is the label of the underlying data preprocessing scenario (see (1)–(4)) and $\mathbf{R}_{(k)}^{DFL}$ is given as follows:

$$\mathbf{R}_{(1)}^{DFL} = \sum_{j=1}^{i-1} \frac{\beta_j^2}{\alpha_j^2} \mathbf{b}_j \mathbf{b}_j^{\mathsf{T}} + \sum_{\substack{p,q=1 \\ p \neq q}}^{i-1} \frac{\eta_p \eta_q}{\alpha_p \alpha_q} \mathbf{b}_p \mathbf{b}_q^{\mathsf{T}} + \frac{\beta_i^2}{\alpha_i^2} \sum_{j=i+1}^{d} \mathbf{b}_j \mathbf{b}_j^{\mathsf{T}}, \tag{11}$$

$$\mathbf{R}_{(2)}^{DFL} = \sum_{j=1}^{i-1} \frac{\beta_j - \eta_j^2}{\alpha_j^2} \mathbf{b}_j \mathbf{b}_j^{\mathsf{T}} + \frac{\beta_i - \eta_i^2}{\alpha_i^2} \sum_{j=i+1}^{d} \mathbf{b}_j \mathbf{b}_j^{\mathsf{T}}, \tag{12}$$

$$\mathbf{R}_{(3)}^{DFL} = \sum_{j=1}^{i-1} \frac{\beta_j - \gamma_j^2 + \alpha_j^2}{\alpha_j^2} \mathbf{b}_j \mathbf{b}_j^{\mathsf{T}} + \sum_{\substack{p,q=1 \\ p \neq q}}^{i-1} \frac{\eta_p \eta_q}{\alpha_p \alpha_q} \mathbf{b}_p \mathbf{b}_q^{\mathsf{T}} + \tau_i \mathbf{b}_i \mathbf{b}_i^{\mathsf{T}}$$

$$+ \frac{\beta_i - \gamma_i^2}{\alpha_i^2} \sum_{j=i+1}^{d} \mathbf{b}_j \mathbf{b}_j^{\mathsf{T}} - \sum_{j=1}^{i-1} \frac{\mathbb{E}[s_i^3] \eta_j}{\alpha_j} (\mathbf{b}_j \mathbf{b}_i^{\mathsf{T}} + \mathbf{b}_i \mathbf{b}_j^{\mathsf{T}}), \tag{13}$$

$$\mathbf{R}_{(4)}^{DFL} = \sum_{j=1}^{i-1} \frac{\beta_j - \gamma_j^2 + \alpha_j^2 - \eta_j^2}{\alpha_j^2} \mathbf{b}_j \mathbf{b}_j^{\mathsf{T}} + \tau_i \mathbf{b}_i \mathbf{b}_i^{\mathsf{T}} + \frac{\beta_i - \gamma_i^2 - \eta_i^2}{\alpha_i^2} \sum_{j=i+1}^{d} \mathbf{b}_j \mathbf{b}_j^{\mathsf{T}}$$

$$- \sum_{j=1}^{i-1} \frac{\mathbb{E}[s_i^3] \eta_j}{\alpha_j} (\mathbf{b}_j \mathbf{b}_i^{\mathsf{T}} + \mathbf{b}_i \mathbf{b}_j^{\mathsf{T}}). \tag{14}$$

Corollary 2. *There holds $N^{1/2}(\mathbf{G}_{ij}^{DFL} - \delta_{ij}) \xrightarrow[N \to \infty]{\mathscr{D}} \mathcal{N}(0, V_{(k)}^{DFL})$, where $\mathbf{G}_{ij}^{DFL}$ denotes the (i, j)th entry of $\mathbf{G}^{DFL}$ and $V_{(k)}^{DFL}$ is given as follows:*

1. *Case $j < i$:*

$$V_{(1)}^{DFL} = \frac{\beta_j^2}{\alpha_j^2}$$

$$V_{(2)}^{DFL} = \frac{\beta_j - \eta_j^2}{\alpha_j^2}$$

$$V_{(3)}^{DFL} = \frac{\beta_j - \gamma_j^2 + \alpha_j^2}{\alpha_j^2}$$

$$V_{(4)}^{DFL} = \frac{\beta_j - \gamma_j^2 + \alpha_j^2 - \eta_j^2}{\alpha_j^2}.$$

2. *Case $j = i$:*

$$V_{(1)}^{DFL} = V_{(2)}^{DFL} = 0, \qquad V_{(3)}^{DFL} = V_{(4)}^{DFL} = \tau_i.$$

3. *Case $j > i$:*

$$V_{(1)}^{DFL} = \frac{\beta_i}{\alpha_i^2} \tag{15}$$

$$V_{(2)}^{DFL} = \frac{\beta_i - \eta_i^2}{\alpha_i^2} \tag{16}$$

$$V_{(3)}^{DFL} = \frac{\beta_i - \gamma_i^2}{\alpha_i^2} \tag{17}$$

$$V_{(4)}^{DFL} = \frac{\beta_i - \gamma_i^2 - \eta_i^2}{\alpha_i^2}. \tag{18}$$

2.2   The Asymptotic Error of Symmetric FastICA

Theorem 3. *Under some mild regularity conditions, we have* $N^{1/2}(\mathbf{C}^{-1/2} \mathbf{w}_i^{SYM} - \mathbf{b}_i) \xrightarrow[N \to \infty]{\mathscr{D}} \mathcal{N}(0, \mathbf{R}_{(k)}^{SYM})$, *where*

$$\mathbf{R}_{(1)}^{SYM} = \sum_{j \neq i}^{d} \frac{\beta_i + \beta_j - 2\gamma_i\gamma_j - 2\eta_j^2}{(|\alpha_i| + |\alpha_j|)^2} \mathbf{b}_j\mathbf{b}_j^{\mathsf{T}} + 2\sum_{j \neq i}^{d} \frac{\eta_j \mathbf{b}_j}{|\alpha_i| + |\alpha_j|} \sum_{j \neq i}^{d} \frac{\eta_j \mathbf{b}_j^{\mathsf{T}}}{|\alpha_i| + |\alpha_j|}, \tag{19}$$

$$\mathbf{R}_{(2)}^{SYM} = \sum_{j \neq i}^{d} \frac{\beta_i + \beta_j - 2\gamma_i\gamma_j - 2\eta_i^2}{(|\alpha_i| + |\alpha_j|)^2} \mathbf{b}_j\mathbf{b}_j^{\mathsf{T}}, \tag{20}$$

$$\mathbf{R}_{(3)}^{SYM} = \sum_{j \neq i}^{d} \frac{\beta_i - \gamma_i^2 + \beta_j - \gamma_j^2 + \alpha_j^2 - \eta_j^2}{(|\alpha_i| + |\alpha_j|)^2} \mathbf{b}_j\mathbf{b}_j^{\mathsf{T}} + \sum_{j \neq i}^{d} \frac{\eta_j \mathbf{b}_j}{(|\alpha_i| + |\alpha_j|)} \sum_{j \neq i}^{d} \frac{\eta_j \mathbf{b}_j^{\mathsf{T}}}{(|\alpha_i| + |\alpha_j|)}$$

$$+ \tau_i \mathbf{b}_i\mathbf{b}_i^{\mathsf{T}} - \sum_{j \neq i}^{d} \frac{\mathbb{E}[s_i^3]\eta_j}{2(|\alpha_i| + |\alpha_j|)}(\mathbf{b}_j\mathbf{b}_i^{\mathsf{T}} + \mathbf{b}_i\mathbf{b}_j^{\mathsf{T}}), \tag{21}$$

$$\mathbf{R}_{(4)}^{SYM} = \sum_{j \neq i}^{d} \frac{\beta_i - \gamma_i^2 + \beta_j - \gamma_j^2 + \alpha_j^2 - \eta_i^2 - \eta_j^2}{(|\alpha_i| + |\alpha_j|)^2} \mathbf{b}_j\mathbf{b}_j^{\mathsf{T}} + \tau_i \mathbf{b}_i\mathbf{b}_i^{\mathsf{T}}. \tag{22}$$

Corollary 4. *For* $i, j = 1, \ldots, d$, *there holds* $N^{1/2}(\mathbf{G}_{ij}^{SYM} - \delta_{ij}) \xrightarrow[N \to \infty]{\mathscr{D}}$ $\mathcal{N}(0, V_{(k)}^{SYM})$, *where*

1. *Case* $j = i$:

$$V_{(1)}^{SYM} = V_{(2)}^{SYM} = 0, \qquad V_{(3)}^{SYM} = V_{(4)}^{SYM} = \tau_i.$$

2. *Case* $j \neq i$:

$$V_{(1)}^{SYM} = \frac{\beta_i + \beta_j - 2\gamma_i\gamma_j}{(|\alpha_i| + |\alpha_j|)^2}, \tag{23}$$

$$V_{(2)}^{SYM} = \frac{\beta_i + \beta_j - 2\gamma_i\gamma_j - 2\eta_i^2}{(|\alpha_i| + |\alpha_j|)^2}, \tag{24}$$

$$V_{(3)}^{SYM} = \frac{\beta_i - \gamma_i^2 + \beta_j - \gamma_j^2 + \alpha_j^2}{(|\alpha_i| + |\alpha_j|)^2}, \tag{25}$$

$$V_{(4)}^{SYM} = \frac{\beta_i - \gamma_i^2 + \beta_j - \gamma_j^2 + \alpha_j^2 - \eta_i^2 - \eta_j^2}{(|\alpha_i| + |\alpha_j|)^2}. \tag{26}$$

Remark 5. Although the asymptotic error of the FastICA algorithm has already been studied by quite a few researchers [5–8], many of the results presented in this contribution, notably expressions (11)–(13) established in Theorem 1 and (19)–(22) in Theorem 3, are new.

Example 1. The validity of formulas (15)-(18) and (23)–(26) is verified in computer simulations, see Figs. 1 and 2. The simulations are configured as follows: $d = 3$, $N = 5000$, all three sources have identical bimodal Gaussian distribution with asymmetrical density. Both deflationary FastICA and symmetric FastICA have been tested with different data preprocessing (1)–(4) in 5000 independent trials.

2.3   Discussion

First, comparing the expressions in Corollary 2 and Corollary 4, we find that for the (i, j)th entry of the gain matrix,

$$V_{(1)}^{DFL} - V_{(2)}^{DFL} = V_{(3)}^{DFL} - V_{(4)}^{DFL} = \frac{\eta_j^2}{\alpha_j^2}, \quad j < i,$$

$$V_{(1)}^{DFL} - V_{(2)}^{DFL} = V_{(3)}^{DFL} - V_{(4)}^{DFL} = \frac{\eta_i^2}{\alpha_i^2}, \quad j > i,$$

$$V_{(1)}^{SYM} - V_{(2)}^{SYM} = \frac{2\eta_i^2}{(|\alpha_i| + |\alpha_j|)^2}, \quad i \neq j,$$

$$V_{(3)}^{SYM} - V_{(4)}^{SYM} = \frac{\eta_i^2 + \eta_j^2}{(|\alpha_i| + |\alpha_j|)^2}, \quad i \neq j.$$

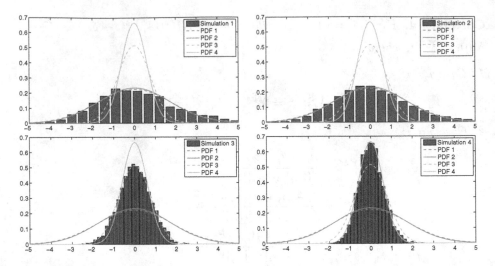

Fig. 1. Asymptotic error of the deflationary FastICA in each preprocessing scenario. We plotted the histograms of an (upper) off-diagonal entry of $N^{1/2}\mathbf{G}^{DFL}$ in 5000 independent trials versus the theoretical curves of the Gaussian PDFs with variances given by (15)–(18).

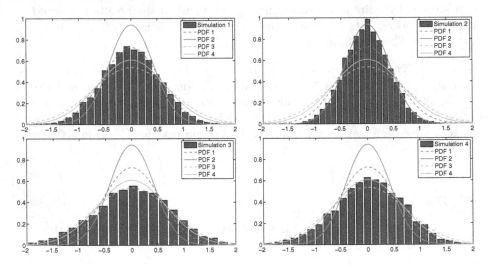

Fig. 2. Asymptotic error of the symmetric FastICA in each preprocessing scenario. We plotted the histograms of an off-diagonal entry of $N^{1/2}\mathbf{G}^{SYM}$ in 5000 independent trials versus the theoretical curves of the Gaussian PDFs with variances given by (23)–(26).

Since all the differences above are non-negative[2], we assert that the empirical data centering generally leads to a better asymptotic performance.

---

[2] They become zero if η_i and/or η_j vanish. This is the case if, e.g. g is pair and the involved sources have symmetric distributions.

3 Conclusion

The contribution of this work is twofold. First, we derived explicit formulas for the asymptotic error of the two most important variants of the FastICA algorithm, the deflationary FastICA and the symmetric FastICA, under four different data preprocessing scenarios. Many of the presented formulas are novel. Second, we assessed the impact of empirical data preprocessing procedure on the asymptotic performance of the algorithms. We showed that, compared to the theoretical data centering, the empirical data centering generally leads to a better separation performance.

References

1. Comon, P.: Independent component analysis: a new concept? Sig. Process. **36**(3), 287–314 (1994)
2. Cardoso, J.F., Souloumiac, A.: Blind beamforming for non-gaussian signals. IEEE Proc.-F **140**(6), 362–370 (1993)
3. Hyvärinen, A.: Fast and robust fixed-point algorithms for independent component analysis. IEEE Trans. Neural Netw. **10**(3), 626–634 (1999)
4. Wei, T.: A convergence and asymptotic analysis of the generalized symmetric fastica algorithm (2015, submitted)
5. Hyvärinen, A.: One-unit contrast functions for independent component analysis: a statistical analysis. In: Proceedings of IEEE NNSP Workshop 1997. Neural Networks for Signal Processing VII (1997)
6. Shimizu, S., Hyvärinen, A., Kano, Y., Hoyer, P.O., Kerminen, A.J.: Testing significance of mixing and demixing coefficients in ICA. In: Rosca, J.P., Erdogmus, D., Príncipe, J.C., Haykin, S. (eds.) ICA 2006. LNCS, vol. 3889, pp. 901–908. Springer, Heidelberg (2006)
7. Nordhausen, K., Ilmonen, P., Mandal, A., Oja, H., Ollila, E.: Deflation-based FastICA reloaded. In: 19th European Signal Processing Conference (EUSIPCO 2011), Barcelona, Spain, September 2011
8. Tichavsky, P., Koldovsky, Z., Oja, E.: Performance analysis of the FastICA algorithm and cramer-rao bounds for linear independent component analysis. IEEE Trans. Sig. Process. **54**(4), 1189–1203 (2006)

A MAP-Based Order Estimation Procedure
for Sparse Channel Estimation

Sajad Daei[1]([✉]), Massoud Babaie-Zadeh[1], and Christian Jutten[2]

[1] Department of Electrical Engineering, Sharif University of Technology, Tehran, Iran
{sajad2008se,mbzadeh}@yahoo.com
[2] GIPSA-Lab, Institut Universitaire de France, Grenoble, France
christian.jutten@gipsa-lab.grenoble-inp.fr

Abstract. Recently, there has been a growing interest in estimation of sparse channels as they are observed in underwater acoustic and ultra-wideband channels. In this paper we present a new Bayesian sparse channel estimation (SCE) algorithm that, unlike traditional SCE methods, exploits noise statistical information to improve the estimates. The proposed method uses approximate maximum a posteriori probability (MAP) to detect the non-zero channel tap locations while least square estimation is used to determine the values of the channel taps. Computer simulations shows that the proposed algorithm outperforms the existing algorithms in terms of normalized mean squared error (NMSE) and approaches Cramér-Rao lower bound of the estimation. In addition, it has low computational cost when compared to the other algorithms.

Keywords: Bayesian · Sparse channel estimation · Cramér-Rao lower bound

1 Introduction

Fast and accurate channel estimation at the receiver is often of much importance due to the need for optimal demodulation and decoding in limited time. Sparse channels, those whose time domain impulse response has much less non-zero taps than their length, have been observed in many practical scenarios such as acoustic underwater [1], ultrawideband propagation [2] and seismic exploration [3]. Since traditional channel estimation methods, such as the least squares method, fail to exploit sparsity of these channels, in the last decade, several sparse channel estimation (SCE) methods have been proposed to improve the estimates [4–10].

In [4,10], two iterative approaches called ITD-SE and MIDE are reported which utilize thresholds to detect the channel support[1] followed by a structured

---

C. Jutten—This work has been partly funded by ERC project 2012-ERC-AdG-320684 CHESS.

[1] Non-zero channel tap locations.

E. Vincent et al. (Eds.): LVA/ICA 2015, LNCS 9237, pp. 344–351, 2015.
DOI: 10.1007/978-3-319-22482-4_40

least square (LS) estimate to determine the values of the channel taps. Simple structure, low complexity and no dependency on the channel order along with acceptable accuracy are the advantages of these threshold-based support detection methods. Moreover, in [5], an iterative MAP-based approach is introduced to jointly estimate the location and the values of the taps. For this purpose, three algorithms have been examined: L2MAP with Threshold, LASSO-MAP with Threshold and Backward-Detection MAP. These methods have very low complexity and near-optimal performance at high SNRs; nevertheless they presume the channel sparsity level is known a priori while it's rarely known in practice. Furthermore, they have limitations on the sparsity rate of the channel.

The algorithms mentioned above, neglect noise statistical information and posterior information of the channel support, which can explain their limited performance. In this paper, to overcome the problems mentioned, as in [4], we present a two stage Bayesian procedure, based on support detection and then channel estimation. For the former part, following [11], we propose a MAP-based tap detection approach which not only considers sparsity of the channel but also exploits noise statistics and posterior information of the channel support to improve the estimates; and for the latter, a structured least square estimation is applied. Unlike Bayesian approaches in SCE algorithms that usually assume Gaussian distribution for the channel, regarding the procedure in [11], the channel distribution in our algorithm is arbitrary. Experimental results demonstrate that our algorithm approaches the Cramér-Rao lower bound of the estimation based on knowing the true channel support (called CRB-S in [4]) at high SNRs. Besides, it has a low computational cost. Note that as our main contribution to SCE algorithms is a support detection approach using approximate MAP, we call it Support Detection using Maximum A posteriori Probability (SDMAP) in this paper.

The paper is organized as follows: System Model and MAP Setup are given in Sect. 2. In Sect. 3, SDMAP algorithm is proposed. Experimental results are investigated in Sect. 4 in order to compare the performance of SDMAP with existing algorithms in term of normalized mean square error (NMSE) and computational complexity. Finally, we conclude the paper in Sect. 5.

Notation: Throughout the paper, we denote scalars with lowercase letters (e.g., x), vectors with lowercase boldfaced letters (e.g., $\mathbf{x}$) and matrices with uppercase boldfaced letters (e.g., $\mathbf{X}$). $\mathbf{x}_i$ stands for the ith column of the matrix $\mathbf{X}$. Sets are designated by uppercase calligraphic letters; the cardinality of the set $\mathcal{S}$ is $|\mathcal{S}|$. We use $\mathbf{x}_\mathcal{S}$ to denote the $|\mathcal{S}|$- dimensional vector of the entries in the vector $\mathbf{x}$ indexed by S. Also, for any $m \times n$ matrix $\mathbf{X}$, we use $\mathbf{X}_\mathcal{S}$ to denote the $m \times |\mathcal{S}|$ matrix corresponding to the columns of $\mathbf{X}$ indexed by $\mathcal{S}$ and $\mathbf{X}(\mathcal{S}_1, \mathcal{S}_2)$ to denote the $|\mathcal{S}_1| \times |\mathcal{S}_2|$ matrix corresponding to the rows and columns of $\mathbf{X}$ indexed by $\mathcal{S}_1$ and $\mathcal{S}_2$ respectively. $\mathbf{I}_m$ is denoted for the $m \times m$ identity matrix. $\|\mathbf{x}\|$ means the 2-norm of the vector $\mathbf{x}$. Finally, $|\mathbf{x}|$ stands for a vector whose elements are the absolute values of the corresponding elements of $\mathbf{x}$.

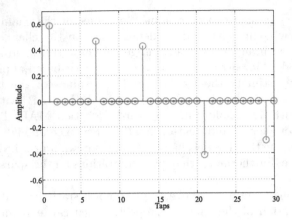

Fig. 1. Time domain discrete impulse response of a sparse channel

2 System Model and MAP Setup

2.1 System Model

Typically, channel estimation is accomplished by sending a training sequence and processing the channel output. Mathematically, let $\{u_n\}_{i=1}^L, L \in \mathbb{N}$ denote a training sequence and $\mathbf{h} \in \mathbb{R}^N, N \in \mathbb{N}$ be the finite discrete impulse response of the channel (See Fig. 1). The resulting observations $\mathbf{y} \in \mathbb{R}^M, M = L + N - 1$ are the convolution of the training signal $\mathbf{u} = [u_1, u_2, u_3, ..., u_L]^T$ and the impulse response $\mathbf{h} = [h_1, h_2, h_3, ..., h_N]^T$ corrupted by an additive noise vector $\mathbf{n}$. In matrix form, we have

$$\mathbf{y} = \mathbf{U}\mathbf{h} + \mathbf{n} = \mathbf{U}_{\mathcal{S}}\mathbf{h}_{\mathcal{S}} + \mathbf{n}, \tag{1}$$

where, $\mathbf{n} \sim \mathcal{N}(\mathbf{0}, \sigma^2 \mathbf{I}_M)$ is an $M \times 1$ Gaussian noise vector, $\mathcal{S}$ is the true support set of $\mathbf{h}$ and $\mathbf{U}$ is the $M \times N$ training Toeplitz matrix with the first column $[u_1, u_2, u_3, ..., u_L, 0, ..., 0]^T$ as in [4].

We assume that the sparse channel vector $\mathbf{h}$ is modeled as $\mathbf{h} = \mathbf{h}_B \odot \mathbf{h}_G$, in which $\odot$ is element-wise Hadamard multiplication, $\mathbf{h}_B$ is an $N \times 1$ vector, whose elements are independent and identically distributed (i.i.d) Bernoulli random variables with success probability $P_a = \frac{|\mathcal{S}|}{N}$ and the elements of $\mathbf{h}_G$ are drawn from an arbitrary distribution. Clearly, $\mathbf{h}_B$ models the support of $\mathbf{h}$, with a sparsity level equal to P_a.

2.2 MAP Setup

The goal is to estimate $\mathbf{h}$ from knowledge of the observation vector $\mathbf{y}$ and the training signal $\mathbf{u}$. To achieve this goal, first, we obtain an estimate of the channel support, $\mathcal{S}$, via MAP detection procedure, which is given by,

$$\hat{\mathcal{S}}_{\text{MAP}} = \underset{\mathcal{S}}{\arg\max} \; \mathbb{P}\{\mathbf{y}|\mathcal{S}\}\mathbb{P}\{\mathcal{S}\}, \tag{2}$$

in which, $\mathbb{P}$ denotes the probability distribution. Since each element of $\mathbf{h}$ is active according to a Bernoulli distribution with success probability P_a, $\mathbb{P}(\mathcal{S})$ is given by,

$$\mathbb{P}\{\mathcal{S}\} = P_a^{|\mathcal{S}|}(1-P_a)^{N-|\mathcal{S}|}. \tag{3}$$

Rather than obtaining the probability of $\mathbf{y}$ conditioned on the support, $\mathbb{P}\{\mathbf{y}|\mathcal{S}\}$, directly, we serve the approach in [11] to make our algorithm independent of the channel distribution. For this purpose, we project $\mathbf{y}$ onto the orthogonal complement of $\mathbf{U}_\mathcal{S}$ via multiplying (1) by $\mathbf{\Pi}_\mathcal{S}^\perp = \mathbf{I}_M - \mathbf{U}_\mathcal{S}(\mathbf{U}_\mathcal{S}^T\mathbf{U}_\mathcal{S})^{-1}\mathbf{U}_\mathcal{S}^T$ which leads to $\mathbf{\Pi}_\mathcal{S}^\perp\mathbf{y} = \mathbf{\Pi}_\mathcal{S}^\perp\mathbf{n} \sim \mathcal{N}(\mathbf{0}, \sigma^2\mathbf{\Pi}_\mathcal{S}^\perp)$. Ignoring constant multiplicative factors, we have,

$$\mathbb{P}\{\mathbf{y}|\mathcal{S}\} \varpropto \mathbb{P}\{\mathbf{\Pi}_\mathcal{S}^\perp\mathbf{y}|\mathcal{S}\}$$
$$\varpropto \exp\left(-\frac{1}{2}(\mathbf{\Pi}_\mathcal{S}^\perp\mathbf{y})^T(\sigma^2\mathbf{\Pi}_\mathcal{S}^\perp)^{-1}(\mathbf{\Pi}_\mathcal{S}^\perp\mathbf{y})\right), \tag{4}$$

in which, $\varpropto$ denotes approximate proportion[2]. Since evaluation of the support in (2) leads to prohibitive computational task, alternatively, we propose a support detection procedure in the next section that requires a fitness function which is defined by,

$$\mu(\mathcal{S}) \triangleq \ln\left(\mathbb{P}\{\mathbf{y}|\mathcal{S}\}\mathbb{P}\{\mathcal{S}\}\right)$$
$$= \ln\left(\exp(-\frac{1}{2\sigma^2}(\mathbf{y}^T\mathbf{\Pi}_\mathcal{S}^\perp\mathbf{y}))\right) + \ln\left(P_a^{|\mathcal{S}|}(1-P_a)^{N-|\mathcal{S}|}\right)$$
$$\varpropto \frac{1}{\sigma^2}\left(\mathbf{y}^T\mathbf{U}_\mathcal{S}(\mathbf{U}_\mathcal{S}^T\mathbf{U}_\mathcal{S})^{-1}\mathbf{U}_\mathcal{S}^T\mathbf{y}\right) + 2|\mathcal{S}|\ln\left(\frac{|\mathcal{S}|}{N-|\mathcal{S}|}\right). \tag{5}$$

After finding dominant channel support using the SDMAP scheme of the next section, it only remains to determine the values of the channel taps at the obtained support. To accomplish this, structured least square estimation is applied as follows,

$$\hat{\mathbf{h}} = (\mathbf{U}_\mathcal{S}^T\mathbf{U}_\mathcal{S})^{-1}\mathbf{U}_\mathcal{S}^T\mathbf{y}. \tag{6}$$

3   Proposed Algorithm for Support Detection (SDMAP)

In this section, we introduce our algorithm to detect the channel support. This algorithm is presented in two steps, first support candidates selection and then estimation of the channel order.

3.1   Support Candidates Selection

To obtain support candidates, first, we compute unstructured least square estimate, $\hat{\mathbf{h}} = (\mathbf{U}^T\mathbf{U})^{-1}\mathbf{U}^T\mathbf{y}$, and sort the absolute value of the elements in $\hat{\mathbf{h}}$, $|\hat{\mathbf{h}}|$, in descending order and keep the respective indices $\mathcal{S}$.

---

[2] A justification of (4) for Gaussian channels is given in [11].

3.2 Order Estimation Procedure

To obtain the channel order, first, we initialize the channel order by,

$$P = \#\{|\hat{\mathbf{h}}_i| > \frac{\max(|\hat{\mathbf{h}}|)}{2}, 1 \leqslant i \leqslant N, i \in \mathbb{N}\}, \tag{7}$$

in which, $\max(|\mathbf{h}|)$ stands for the largest element in the vector $|\mathbf{h}|$. Regarding the initial order P, we determine the direction toward which the current support, $\mathcal{S}_P = \{\mathcal{S}(i), i = 1, ..., P\}$, is inclined. In this regard, to determine the move direction we use some criteria which will be discussed further. After finding the direction toward which the current support tends to move, the support order changes until some stopping criteria are satisfied or the number of maximum move steps is exceeded.

Forming the direction and stopping criteria suitably, requires the knowledge of the noisy part of the fitness function (5). To choose suitable stopping rules, we can exploit the pure noisy part of the fitness function which is given by,

$$\mu_{\mathbf{n}}(\mathcal{S}) = \frac{1}{\sigma^2}\Big(\mathbf{n}^T \underbrace{\mathbf{U}_{\mathcal{S}}(\mathbf{U}_{\mathcal{S}}^T\mathbf{U}_{\mathcal{S}})^{-1}\mathbf{U}_{\mathcal{S}}^T}_{\mathbf{H}}\, \mathbf{n}\Big). \tag{8}$$

Since $\mathbf{H}$ in (8) is a symmetric, idempotent matrix with $\mathrm{rank}(\mathbf{H}) = |\mathcal{S}|$, $\mu_{\mathbf{n}}(\mathcal{S})$ is chi-squared distributed with $\nu = |\mathcal{S}|$ degrees of freedom, i.e. $\chi^2(\nu)$, in which ν is the parameter of the chi-squared distribution [12, Theorem A.87]. As the mean and variance of this chi-squared distributed random variable are ν and 2ν, respectively, we can obtain tolerance limits[3] of $\mu_{\mathbf{n}}(\mathcal{S})$ as follow,

$$\text{Lower-bound:}\quad lb(\nu) = \nu - \alpha\sqrt{2\nu};$$
$$\text{Upper-bound:}\quad ub(\nu) = \nu + \beta\sqrt{2\nu}, \tag{9}$$

in which, α and β are chosen such that about 10 % of the distribution occurs outside the bounds in (9). In our simulations, we used $\alpha = 1.1$ and $\beta = 1.3$. Considering the effect of noise on the fitness function (5) and trying to reduce it, we define the direction criterion as follows,

$$\left\{\begin{array}{l}
\text{Move backward (i.e. keep } P - 1) \text{ if,} \\
\mu(\mathcal{S}_{P-1}) - \mu(\mathcal{S}_P) > lb(P-1) - ub(P) \quad \& \\
\mu(\mathcal{S}_{P+1}) - \mu(\mathcal{S}_P) < ub(P+1) - lb(P); \qquad (10a) \\
\text{Move forward (i.e. keep } P + 1) \text{ if,} \\
\mu(\mathcal{S}_{P-1}) - \mu(\mathcal{S}_P) < lb(P-1) - ub(P) \quad \& \\
\mu(\mathcal{S}_{P+1}) - \mu(\mathcal{S}_P) > ub(P+1) - lb(P); \qquad (10b) \\
\text{Don't move if,} \\
\text{neither of the two above is satisfied.} \qquad (10c)
\end{array}\right.$$

---

[3] Tolerance limits are the bounds that the probability of random variable occurrence outside them is a certain value.

Qualitatively spoken, moving backward simultaneously requires that the tendency of the current support to change to the previous support (i.e. $\mu(\mathcal{S}_{P-1}) - \mu(\mathcal{S}_P)$) be greater than the lower limit of the noisy part and the tendency to change to the next support (i.e. $\mu(\mathcal{S}_{P+1}) - \mu(\mathcal{S}_P)$) be less than the upper limit of the noisy part. Likewise, moving forward simultaneously requires that the tendency to change to the previous support be less than the lower limit of the noisy part and the tendency to change to the next support be greater than the upper limit of the noisy part.

After finding the direction using (10), the algorithm moves in the obtained direction until approximately no change is observed in the fitness function. This is accomplished by using a stopping rule which is given by:

$$\mu(\mathcal{S}_{P_{\text{new}}}) - \mu(\mathcal{S}_{P_{\text{pre}}}) < ub(P_{\text{new}}) - lb(P_{\text{pre}}), \tag{11}$$

Algorithm 1. SDMAP Algorithm

1: **procedure** SDMAP($\mathbf{U}, \mathbf{y}, \sigma^2$)
2:     $\text{corr} = \mathbf{y}^T\mathbf{U}, \mathbf{A} = \mathbf{U}^T\mathbf{U}$
3:     $\hat{\mathbf{h}} = \mathbf{A}^{-1}\text{corr}$
4:     Sort the elements in $|\hat{\mathbf{h}}|$ in descending order and save the respective indexes in $\mathcal{S}_c$
5:     $P_{\text{init}} \leftarrow \#\{|\hat{\mathbf{h}}_i| > \frac{\max(|\hat{\mathbf{h}}|)}{2}, 1 \leqslant i \leqslant N, i \in \mathbb{N}\}$
6:     **if** (10a) is satisfied **then**
7:         $P \leftarrow P_{\text{init}}$
8:         $\mu_{\text{new}} \leftarrow \mu(\mathcal{S}_{P-1})$
9:         **repeat**
10:             $P \leftarrow P - 1$
11:             $\mu_{\text{old}} \leftarrow \mu_{\text{new}}$
12:             $\mu_{\text{new}} \leftarrow \mu(\mathcal{S}_{P-1})$
13:         **until** $\mu_{\text{new}} - \mu_{\text{old}} < ub(P-1) - lb(P) \vee P = 1$
14:     **else if** (10b) is satisfied **then**
15:         $P \leftarrow P_{\text{init}}$
16:         $\mu_{\text{new}} \leftarrow \mu(\mathcal{S}_{P+1})$
17:         **repeat**
18:             $P \leftarrow P + 1$
19:             $\mu_{\text{old}} \leftarrow \mu_{\text{new}}$
20:             $\mu_{\text{new}} \leftarrow \mu(\mathcal{S}_{P+1})$
21:         **until** $\mu_{\text{new}} - \mu_{\text{old}} < ub(P+1) - lb(P) \vee P = N$
22:     **end if**
23:     $\mathcal{S}_f = \mathcal{S}_c(1:P)$
24:     $\hat{\mathbf{h}}_f = \mathbf{A}(\mathcal{S}_f, \mathcal{S}_f)^{-1}\text{corr}(\mathcal{S}_f)$
25:     **function** $\mu(\mathcal{S})$
26:         $F = \frac{1}{\sigma^2}(\text{corr}(\mathcal{S})\mathbf{A}(\mathcal{S}, \mathcal{S})^{-1}\text{corr}(\mathcal{S})^T) + 2|\mathcal{S}|\ln(\frac{|\mathcal{S}|}{N-|\mathcal{S}|})$
27:         **return** F
28:     **end function**
29:     **Output** $\hat{\mathbf{h}}_f$
30: **end procedure**

---

in which, P_{pre} and P_{new} are the previous and new orders, respectively. (11) qualitatively expresses that the algorithm stops when the tendency of the previous support $\mathcal{S}_{P_{\text{pre}}}$ to change to the new support $\mathcal{S}_{P_{\text{new}}}$ (i.e. $\mu(\mathcal{S}_{P_{\text{new}}}) - \mu(\mathcal{S}_{P_{\text{pre}}})$) is less than the upper bound of the noisy part. The final pseudocode form of our algorithm is given in Algorithm 1.

4 Computer Simulations

In this section, we investigate the performance of our algorithm (SDMAP) in comparison with five algorithms: L2MAP [5], Backward-MAP [5], LASSO-MAP [5], MIDE [10] and ITD-SE [4]. For this purpose, we consider a sparse channel with length $N = 30$ and support size $|\mathcal{S}| = 5$ (see Fig. 1) and draw the elements of the training matrix, $\mathbf{U}_{M \times N}, M = 50$, from a zero-mean i.i.d. Gaussian distribution ($\mathcal{N}(0, \frac{1}{N})$). The estimation efficiency is evaluated using normalized mean squared error (NMSE) which is defined as,

$$\text{NMSE} = \frac{1}{N_{\text{MC}}} \sum_{n=1}^{N_{\text{MC}}} \frac{\|\mathbf{h} - \hat{\mathbf{h}}_n\|^2}{\|\mathbf{h}\|^2}, \tag{12}$$

where, N_{MC} is the number of Monte Carlo iterations, $\hat{\mathbf{h}}_n$ is the channel estimator in the nth experiment and $\mathbf{h}$ is the true channel. To compare computational complexity of the proposed algorithm with other methods, we use CPU time as a simple metric. Our simulation is implemented using MATLAB 2012 on a laptop computer with 2.4 GHz Intel i5 processor and 4 GB memory running the Windows 7 64 bit operating system.

From Fig. 2(a), we observe that SDMAP algorithm outperforms all the other compared algorithms in the sense of NMSE and approaches the theoretic lower bound CRB-S at high SNRs. Figure 2(b) demonstrates the computational efficacy of our algorithm over the other methods.

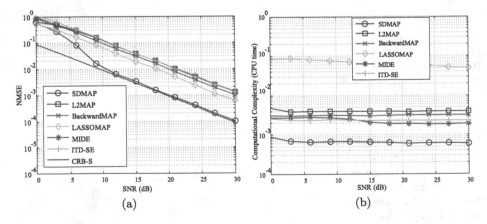

Fig. 2. Performance comparison. (a) NMSE versus SNR. (b) Computational complexity versus SNR

5   Conclusion

In this paper, we proposed a new Bayesian strategy for channel estimation called SDMAP. As it can be seen from simulation results, SDMAP has strengths in terms of NMSE and low computational cost. The reason is that our algorithm utilizes a priori information of noise in the support detection stage. The use of noise statistics provides the posterior information of the channel support, and finally leads to reducing the misdetection.

References

1. Kilfoyle, D.B., Baggeroer, A.B.: The state of the art in underwater acoustic telemetry. IEEE J. Oceanic Eng. **25**(1), 4–27 (2000)
2. Molisch, A.F.: Ultrawideband propagation channels-theory, measurement, and modeling. IEEE Trans. Veh. Technol. **54**(5), 1528–1545 (2005)
3. Claerbout, J.F., Muir, F.: Robust modeling with erratic data. Geophysics **38**(5), 826–844 (1973)
4. Carbonelli, C., Vedantam, S., Mitra, U.: Sparse channel estimation with zero tap detection. IEEE Trans. Wirel. Commun. **6**(5), 1743–1763 (2007)
5. Niazadeh, R., Babaie-Zadeh, M., Jutten, C.: An alternating minimization method for sparse channel estimation. In: Vigneron, V., Zarzoso, V., Moreau, E., Gribonval, R., Vincent, E. (eds.) LVA/ICA 2010. LNCS, vol. 6365, pp. 319–327. Springer, Heidelberg (2010)
6. Subramanian, S.: Compressed sensing for sparse underwater channel estimation: Some practical considerations (2010). arXiv preprint arXiv:1002.2677
7. Sharp, M., Scaglione, A.: Estimation of sparse multipath channels. In: Military Communications Conference, MILCOM, pp. 1–7. IEEE (2008)
8. Cotter, S.F., Rao, B.D.: Sparse channel estimation via matching pursuit with application to equalization. IEEE Trans. Commun. **50**(3), 374–377 (2002)
9. Karabulut, G.Z., Yongacoglu, A.: Sparse channel estimation using orthogonal matching pursuit algorithm. In: Vehicular Technology Conference, VTC, vol. 6, pp. 3880–3884. IEEE (2004)
10. Wan, F., Mitra, U., Molisch, A.: The modified iterative detector/estimator algorithm for sparse channel estimation. In: OCEANS, pp. 1–6. IEEE (2010)
11. Masood, M., Al-Naffouri, T.Y.: Sparse reconstruction using distribution agnostic Bayesian matching pursuit. IEEE Trans. Sig. Process. **61**(21), 5298–5309 (2013)
12. Rao, C.R., Toutenburg, H., Shalabh, H.C., Schomaker, M.: Linear Models and Generalizations: Least Squares and Alternatives, 3rd edn. Springer, Heidelberg (2008)

Bayesian Blind Source Separation
with Unknown Prior Covariance

Ondřej Tichý[1,2]([✉]) and Václav Šmídl[1]

[1] Institute of Information Theory and Automation, Pod Vodárenskou věží 4,
18208 Prague 8, Czech Republic
otichy@utia.cas.cz
[2] Faculty of Nuclear Sciences and Physical Engineering, Břehová 7,
Prague 1, Czech Republic

Abstract. The task of blind source separation (BSS) is to recover original signal sources which are observed only via their superposition with unknown weights. Since we are interested in estimation of the number of relevant sources in noisy observation, we use the Bayesian formulation which automatically removes spurious sources. A tool for this behavior is joint estimation of the unknown prior covariance matrix of the sources in tandem with the sources. In this work, we study the effect of various choices of the covariance matrix structure. Specifically, we compare models using the automatic relevance determination (ARD) principle on the first and the second diagonal, as well as full covariance matrix with Wishart prior. We obtain five versions of the variational BSS algorithm. These are tested on synthetic data and on a selected dataset from dynamic renal scintigraphy. MATLAB implementation of the methods is available for download.

Keywords: Blind source separation · Covariance model · Variational bayes approximation · Non-negative matrix factorization

1 Introduction

The blind source separation (BSS) problem arises in situations where several sources are observed only via their superposition such as in case of audio or medical signal processing [8] or hyperspectral imaging [10]. The task is to separate original sources, e.g. in the form of images and their related weights. The classical separation methods include principal or independent component analysis [6], non-negative matrix factorization [7], or projection methods [1,4].

In this work, we are focused on the Bayesian approach to the BSS problem which has advantages under poor signal to noise conditions and is capable to provides an estimate of the number of relevant sources. Another advantage for further processing of the results is the availability of uncertainty bounds around the estimate in the form of full probability distribution function. The ability to estimate the number of relevant sources is available due to a specific choice of the prior structure, typically unknown covariance matrix [9,12]. In this paper,

© Springer International Publishing Switzerland 2015
E. Vincent et al. (Eds.): LVA/ICA 2015, LNCS 9237, pp. 352–359, 2015.
DOI: 10.1007/978-3-319-22482-4_41

we study various choices of the structure of the prior covariance matrix and their effects on the behavior of the resulting separation algorithm. Specifically, we study three different assumptions leading to different covariance structures: (i) the source weights are most likely sparse which can be modeled using automatic relevance determination (ARD) approach [2,14], (ii) the source weights are smooth with occasional abrupt changes which can be modeled by sparse differences of the weights, and (iii) both weights and their differences can be sparse, which can be modeled by bi-diagonal covariance matrix. Since evaluation of exact posterior densities is not tractable, we apply the Variational Bayes method to obtain approximate posterior densities [11]. The first two structures are standard and the algorithms are well known, however, the last model is computationally intractable even under the Variational Bayes approach. Therefore, we propose to derive the posterior distribution for a full prior covariance matrix of the source weights using Wishart distribution. The introduced overparametrization is mitigated by the use technique known as matrix localization [5]. This heuristics is very successful in atmospheric modeling. Similar approach has been also applied for model of convolution kernels in blind deconvolution [13].

The resulting variants of the variational BSS algorithm are tested on a synthetic dynamic image data where advantages and disadvantages of the tested priors are demonstrated. The advantages of the proposed method were also observed on a real data set from dynamic renal scintigraphy, where the proposed method compares favorably with competing approaches such as the NMF algorithm [7]. Matlab implementations of the algorithms are freely available for download.

2   Bayesian Blind Source Separation

We introduce the Bayesian model of blind source separation. Prior models for all parameters of the model are described here except the prior for source weights which is described in the next section.

2.1   Observation Model

A sequence of recorded data vectors, $\mathbf{d}_t \in \mathbf{R}^{p \times 1}$, $t = 1, \ldots, n$, is stored columnwise in matrix $D \in \mathbf{R}^{p \times n}$. The assumed decomposition is

$$D = AX^T + E, \tag{1}$$

where matrix $A \in \mathbf{R}^{p \times r}$ represents the source images in its columns, matrix $X \in \mathbf{R}^{n \times r}$ represents source weights in its columns, matrix $E \in \mathbf{R}^{p \times n}$ represents noise term of the observation model, and symbol $()^T$ denotes transposition of a vector or a matrix in this paper.

We assume that all elements of the matrices D, A, X, and E are positive; however, modification to full support is straightforward.

2.2   Noise Model

We use the isotropic Gaussian noise model [15] with zero mean and common variance for all pixels, $e_{i,j} \sim \mathcal{N}_{e_{i,j}}(0, \omega^{-1})$. Then, the observation model can be rewritten as

$$f(D|A, X, \omega) = \prod_{t=1}^{n} \mathcal{N}_{\mathbf{d}_t} \left(A\overline{\mathbf{x}}_t^T, \omega^{-1} I_p \right), \quad f(\omega) = \mathcal{G}_\omega(\vartheta_0, \rho_0), \qquad (2)$$

where symbol $\mathcal{N}$ denotes normal distribution and symbol I_p denotes identity matrix of the given size. In the Bayesian methodology, all unknown parameters have their prior distribution. The prior distribution for the precision of the noise, ω, has a conjugate prior in the form of the Gamma distribution, denoted as $\mathcal{G}$, with selected prior constants ϑ_0, ρ_0.

2.3   Prior Model of Source Images

The prior model of the source images is common for all methods in the paper. Each source image, i.e. column of the matrix A, $\mathbf{a}_k$, has prior in the form of the normal distribution with unknown precision parameter related to each source image as

$$f(\mathbf{a}_k|\xi_k) = t\mathcal{N}_{\mathbf{a}_k} \left(\mathbf{0}_{p,1}, \xi_k^{-1} I_p, [0, \infty] \right), \quad f(\xi_k) = \mathcal{G}_{\xi_k}(\phi_0, \psi_0), \qquad (3)$$

where $t\mathcal{N}$ denotes truncated normal distribution with given support and ξ_k is an unknown precision parameter with the Gamma prior for $k = 1, \ldots, r$ where ϕ_0, ψ_0 are selected prior constants. This parameter acts as the automatic relevance determination (ARD) term [14].

3   Prior Models of Covariance Matrix of Source Weights

3.1   Isotropic Prior

The only assumption in this case is that the elements of each weights vector are isotropic [11], i.e. that their covariance matrix is identity matrix as $f(\mathbf{x}_k) = t\mathcal{N}_{\mathbf{x}_k}(\mathbf{0}_{n,1}, I_n, [0, \infty])$ for $k = 1, \ldots r$.

3.2   Sparse Prior

The key assumption of this prior is that the source weights are most likely sparse. Once again, we employ the ARD principle; however, in a different way than in Sect. 2.3. Here, each element of source weight, $x_{k,j}$, has its own ARD prior with relevance parameter, $v_{k,j}$, which can be written in vector form as

$$f(\mathbf{x}_k|\boldsymbol{v}_k) = t\mathcal{N}_{\mathbf{x}_k} \left(\mathbf{0}_{n,1}, \mathrm{diag}(\boldsymbol{v}_k^{-1}), [0, \infty] \right), \quad f(v_{k,j}) = \mathcal{G}_{v_{k,j}}(\alpha_0, \beta_0), \qquad (4)$$

$\forall j = 1, \ldots, n$, where $\mathrm{diag}()$ denotes square matrix with argument vector in its diagonal and zeros otherwise and α_0, β_0 are selected prior constants.

The purpose of this approach is to favor zeros in estimates of the elements of the weights.

3.3   Sparse Differences Prior

Sparse prior from Sect. 3.2 could possibly lead to very non-smooth solutions. If smooth solutions are preferred, a model of sparse differences instead of sparse elements could be more appropriate. The differences of $\mathbf{x}_k$ can be expressed using ∇ operator as $\nabla \mathbf{x}_k$, where $\nabla \in \mathbf{R}^{n \times n}$ is the matrix with ones on its diagonal, -1s on its superdiagonal, and zeros otherwise. We employ the ARD principle on each element of $\nabla \mathbf{x}_k$ with relevance parameter $\boldsymbol{v}_k^\nabla$. This can be formulated equally using full vector $\mathbf{x}_k$ as

$$f(\nabla \mathbf{x}_k | \boldsymbol{v}_k^\nabla) \leftrightarrow f(\mathbf{x}_k | \boldsymbol{v}_k) = t\mathcal{N}_{\mathbf{x}_k}\left(\mathbf{0}_{n,1}, \nabla^{-1}\mathrm{diag}(\boldsymbol{v}_k^{-1})(\nabla^{-1})^T, [0, \infty]\right), \quad (5)$$

$$f(\boldsymbol{v}_{k,j}) = \mathcal{G}_{\boldsymbol{v}_{k,j}}(\alpha_0, \beta_0), \quad \forall j = 1, \ldots, n, \quad (6)$$

with selected prior constants α_0, β_0.

3.4   Wishart Prior

Till this moment, we have modeled only selected diagonals of the covariance matrix. However, it is possible to model the full covariance matrix. For this task, we use vectorized form of the matrix X as $\overrightarrow{\mathbf{x}} = \mathrm{vec}(X) = [\mathbf{x}_1^T, \ldots, \mathbf{x}_r^T]^T \in \mathbf{R}^{nr \times 1}$ where the covariance between all elements is a full covariance matrix $\Upsilon \in \mathbf{R}^{nr \times nr}$. Prior distribution on an unknown full covariance matrix is usually chosen in the form of Wishart matrix distribution,

$$f(\overrightarrow{\mathbf{x}} | \Upsilon) = t\mathcal{N}_{\overrightarrow{\mathbf{x}}}\left(\mathbf{0}_{nr,1}, \Upsilon^{-1}, [0, \infty]\right), \quad f(\Upsilon) = \mathcal{W}_\Upsilon(\alpha_0 I_n, \beta_0), \quad (7)$$

where $\mathcal{W}()$ denotes the Wishart matrix distribution and α_0, β_0 are selected prior constants.

The weak point of this prior model is that $n^2 r^2$ additional parameters have to be estimated which makes this problem very ill-posed.

3.5   Wishart Prior with Localization

We assume that the most relevant prior knowledge is located only in several diagonals of the covariance matrix and its sub-matrices. This idea originates in data assimilation of atmospheric models [5]. Therefore, we replace the remaining entries in the estimate by zeros. Formally, we use the Hadamard matrix product which is defined between two matrices of the same size as $C = A \circ B$ where $c_{i,j} = a_{i,j}b_{i,j}$. Then, the localization of the posterior estimate of the full covariance matrix from Sect. 3.4 is

$$\widehat{\Upsilon}_{\mathrm{loc}} = \widehat{\Upsilon} \circ L, \quad (8)$$

where $\widehat{\Upsilon}$ denotes estimate of Υ and L is the localization matrix of the same size as the matrix Υ. There could be many possible localization matrices L [3] however their study is out of the scope of this paper. Here, we will show results with two localization matrices (i) matrix of ones, i.e. without any localization (denoted as Wishart), and (ii) localization matrix L with ones on the first and the second diagonals of all sub-matrices and zeros otherwise (denoted as Localized Wishart).

3.6   Approximate Solution Using Variational Bayes Method

The whole probabilistic prior model is formed using equations (2)–(3) and prior model from Sects. 3.1–3.5. Estimation algorithm for each prior model was derived using the Variational Bayes method [11] where equations for shaping parameters of the posterior probability densities of the model parameters are found in the form of a set of implicit equations which has to be solved iteratively. Solutions for the first three models are available from previous publications, equation for the proposed version with Wishart prior and localization are given in the Appendix A. This yields five different versions of the variational BSS algorithm (two versions with Wishart prior, with and without localization). All prior parameters (with subscript 0) are set to $10^{\pm 10}$ in order to yield non-informative priors while all algorithms are not sensitive to this selection.

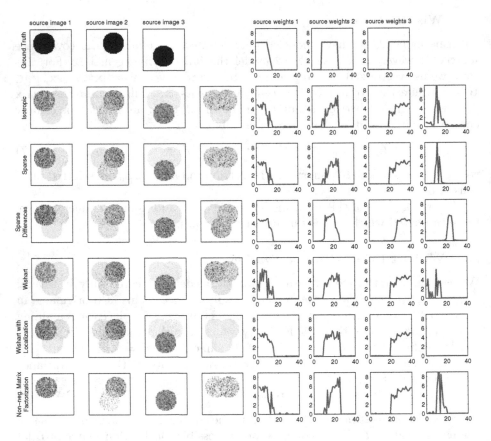

Fig. 1. The results of the five studied methods (the second to the sixth rows) together with NMF algorithm results (the seventh row) in synthetically generated data (the first row).

4 Experiments

4.1 Experiment on Synthetic Dataset

All five derived algorithms are now being tested on synthetic dataset in order to study the impact of covariance matrix models on resulting estimates. The data are generated according to model (1) using three sources with different time-dependent weights as displayed in Fig. 1, top row, degenerated by homogeneous Gaussian noise. All algorithms run with the same conditions such as starting point of iterations and expected number of sources which is set to $r = 4$ in order to study the ability of algorithms to recognize the correct number of sources since the modeled number of sources is 3.

The results from all tested algorithms are given in Fig. 1, rows 2–6, together with the state of the art non-negative matrix factorization (NMF) algorithm [7], row 7. There are estimated source images and source weights in row-wise schema where four images in each row are accompanied with related four weights vectors. It can be seen that all algorithms are capable to correctly estimate source images. The main differences between the algorithms is in estimates of the source weights. The fourth redundant source from BSS with isotropic prior of NMF has been estimated such that its activity is taken from the first and the second source. The same behavior can bee seen on the result of BSS with the Sparse prior; however, the tendency to favor zeros in source weights can be nicely observed here. The BSS with sparse differences prior provides smooth estimates of the source weights; however, the algorithm estimated the fourth source as a combination of the second and the third source. The BSS with the Wishart prior does not penalizes redundant sources and the activity in fourth source is taken from the first. Only the BSS with Wishart prior and localization achieves suppression of the redundant source. It is still estimated, however, with negligible activity which is under the displayed resolution.

4.2 Experiment on Dynamic Scintigraphy Dataset

In this experiment, we will use a selected data from dynamic renal scintigraphy[1] to demonstrate the performance of the methods on real data. The data has original resolution 128×128 pixels; however, we select a region with one kidney of the size 37×47 where medically relevant sources (kidney pelvis and parenchyma) are located. The whole sequence is composed of 100 images with sampling period of 10 s.

We compare only BSS algorithms based on the Sparse differences prior and the Wishart prior with localization with the NMF algorithm. The $r = 3$ for the tested algorithms. The results are summarized in Fig. 2. Estimated source images and source weights are in a row-wise schema. Two methods, BSS with Sparse differences prior and the NMF, estimate threes significant sources where sources 2 and 3 correspond to biological activity of the pelvis. Only the BSS with

---

[1] www.dynamicrenalstudy.org.

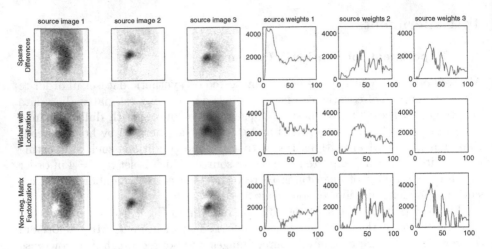

Fig. 2. Results of selected BSS algorithms on dynamic renal scintigraphy data. Source images are in the first three columns and related TACs are in the second three columns.

Wishart prior and localization estimates only two sources which correspond very well with the expected biological function of pelvis and parenchyma.

5   Discussion and Conclusion

The problem of blind source separation (BSS) is generally ill-posed, especially under the conditions such as noisy observations or unknown number of sources. Bayesian approach is generally valuable for its ability to estimate the number of relevant sources using hierarchical priors. In this work, we study various choices of prior covariance structure of the source weights. Covariance structures with ARD and ARD principle of the differences were already published. We propose another model using Wishart prior and develop Variation Bayes estimation algorithm with non-standard step of covariance localization. The proposed algorithm was found to have superior ability to suppress redundant sources in blind source separation of noisy image sequences. All versions of the variational BSS algorithm are implemented in Matlab and freely available for download from http://www.utia.cz/AS/softwaretools/image_sequences/.

Acknowledgement. This work was supported by the Czech Science Foundation, grant No. 13-29225S, and by the Grant Agency of the Czech Technical University in Prague, grant No. SGS14/205/OHK4/3T/14.

A   Shaping Parameters of Posterior Distributions

Posterior distributions are $\tilde{f}(A|D) = t\mathcal{N}_A(\mu_A, I_p \otimes \Sigma_A)$, $\tilde{f}(\xi_k|D) = \mathcal{G}_{\xi_k}(\phi_k, \psi_k)$, $\tilde{f}(\mathbf{x}|D) = t\mathcal{N}_\mathbf{x}(\mu_\mathbf{x}, \Sigma_\mathbf{x})$, $\tilde{f}(\Upsilon|D) = \mathcal{W}_{\Upsilon,nr}(\Sigma_\Upsilon, \beta)$, $\tilde{f}(\omega|D) = \mathcal{G}_\omega(\vartheta, \rho)$, with shaping parameters $\Sigma_A^{-1} = \left(\omega \widehat{X^T X} + \widehat{\Xi}\right)$, $\mu_A = \left(\omega D\widehat{X}\right)\Sigma_A$, $\phi = \phi_0 + \frac{p}{2}\mathbf{1}_{r,1}$, $\psi = \psi_0 + \frac{1}{2}\text{diag}\left(\widehat{A^T A}\right)$, $\Sigma_{\mathbf{x}}^{-1} = \left((\widehat{\omega A^T A}) \otimes I_n + \widehat{\Upsilon} \circ L\right)$ $\mu_{\mathbf{x}} = \Sigma_{\mathbf{x}}\left(\widehat{\omega}\text{vec}\left(D^T\widehat{A}\right)\right)$ $\Sigma_{\Upsilon}^{-1} = \left(\widehat{\mathbf{x}\mathbf{x}^T} + \alpha_0^{-1}I_{nr}\right)$ $\beta = \beta_0 + 1$ $\vartheta = \vartheta_0 + \frac{pn}{2}$, $\rho = \rho_0 + \frac{1}{2}\text{tr}\left((D - \widehat{AX^T})(D - \widehat{AX^T})^T\right)$.

References

1. Araújo, M.C.U., Saldanha, T.C.B., Galvão, R.K.H., Yoneyama, T., Chame, H.C., Visani, V.: The successive projections algorithm for variable selection in spectroscopic multicomponent analysis. Chemometr. Intell. Lab. Syst. **57**(2), 65–73 (2001)
2. Bishop, C.M.: Variational principal components. In: IET Conference Proceedings, pp. 509–514(5), January 1999
3. Gaspari, G., Cohn, S.E.: Construction of correlation functions in two and three dimensions. Q. J. Roy. Meteorol. Soc. **125**(554), 723–757 (1999)
4. Gillis, N.: Successive nonnegative projection algorithm for robust nonnegative blind source separation. SIAM J. Imaging Sci. **7**(2), 1420–1450 (2014)
5. Hamill, T.M., Whitaker, J.S., Snyder, C.: Distance-dependent filtering of background error covariance estimates in an ensemble kalman filter. Mon. Weather Rev. **129**(11), 2776–2790 (2001)
6. Hyvärinen, A., Karhunen, J., Oja, E.: Independent Component Analysis, vol. 46. Wiley, New York (2004)
7. Lee, D.D., Seung, H.S.: Algorithms for non-negative matrix factorization. In: Proceedings of Advances in neural information processing systems, pp. 556–562 (2001)
8. Margadán-Méndez, M., Juslin, A., Nesterov, S.V., Kalliokoski, K., Knuuti, J., Ruotsalainen, U.: ICA based automatic segmentation of dynamic cardiac PET images. IEEE Trans. Inf. Technol. Biomed. **14**(3), 795–802 (2010)
9. Miskin, J.W.: Ensemble learning for independent component analysis. Ph.D. thesis, University of Cambridge (2000)
10. Moussaoui, S., Hauksdottir, H., Schmidt, F., Jutten, C., Chanussot, J., Brie, D., Douté, S., Benediktsson, J.A.: On the decomposition of mars hyperspectral data by ica and bayesian positive source separation. Neurocomputing **71**(10), 2194–2208 (2008)
11. Šmídl, V., Quinn, A.: The Variational Bayes Method in Signal Processing. Springer, Heidelberg (2006)
12. Šmídl, V., Quinn, A.: On bayesian principal component analysis. Comput. Stat. Data Anal. **51**(9), 4101–4123 (2007)
13. Tichý, O., Šmídl, V.: Non-parametric bayesian models of response function in dynamic image sequences. Pre-print submitted to Computer Vision and Image Understanding (arXiv:1503.05684 [stat.ML]) (2015)
14. Tipping, M.E.: Sparse bayesian learning and the relevance vector machine. J. Mach. Learn. Res. **1**, 211–244 (2001)
15. Tipping, M.E., Bishop, C.M.: Probabilistic principal component analysis. J. Roy. Stat. Soc. B (Stat. Methodol.) **61**(3), 611–622 (1999)

Convex Recovery of Tensors Using Nuclear Norm Penalization

Stéphane Chrétien and Tianwen Wei[⊠]

Laboratoire de Mathématiques de Besancon, Université de Franche-Comté,
16 Route de Gray, 25000 Besancon, France
wei.lille1@gmail.com

Abstract. The subdifferential of convex functions of the singular spectrum of real matrices has been widely studied in matrix analysis, optimization and automatic control theory. Convex analysis and optimization over spaces of tensors is now gaining much interest due to its potential applications to signal processing, statistics and engineering. The goal of this paper is to present an applications to the problem of low rank tensor recovery based on linear random measurement by extending the results of Tropp [6] to the tensors setting.

1 Introduction

1.1 Background

Tensors have been recently a subject of great interest in the applied mathematics community. We refer to [3,4] for a modern reference on this subject. Many applications of tensors are based on solving tensor related optimization problems, such as minimizing certain norms under linear constraints. Such problems have been recently successfully addressed in the 2D setting, i.e. for matrices, by the statistics, signal processing, inverse problems and automatic control communities in particular. Two of the reasons for this rapid growth of interest in the application of matrix norms to penalized estimation problems is that some norms promote spectral sparsity and that much work had been done in the fields of matrix analysis and convex analysis to analyze the subdifferential of such norms; see for example [5,7]. Our goal in the present paper is to extend previous results on matrix norms to the tensor setting. In particular, we propose a general study of the subdifferential of certain convex functions of the spectrum of real tensors and apply our results to the computation of the subdifferential of useful and natural matrix norms. We also present an application of our formulas to the problem of low rank tensor recovery using sparsity promoting norm minimization under random linear constraints, a natural extension of previous works by Tropp [6].

1.2 Notations

For any convex function $f : \mathbb{R}^n \mapsto \mathbb{R} \cup \{+\infty\}$, the conjugate function f^* associated to f is defined by

$$f^*(g) \stackrel{\text{def}}{=} \sup_{x \in \mathbb{R}^n} \quad \langle g, x \rangle - f(x).$$

© Springer International Publishing Switzerland 2015
E. Vincent et al. (Eds.): LVA/ICA 2015, LNCS 9237, pp. 360–367, 2015.
DOI: 10.1007/978-3-319-22482-4_42

The subdifferential of f at $x \in \mathbb{R}^n$ is defined by

$$\partial f \overset{\text{def}}{=} \{g \in \mathbb{R}^n \mid \forall y, \in \mathbb{R}^n \quad f(y) \geq f(x) + \langle g, y - x \rangle\}.$$

Moreover, it is well known (see e.g. [2]) that $g \in \partial f(x)$ if and only if

$$f(x) + f^*(g) = \langle g, x \rangle.$$

In the present paper, a tensor represented by a multi-dimensional array in $\mathbb{R}^{d_1 \times \cdots \times d_D}$. Let D and $n_1, \ldots, n_D$ be positive integers. Let $\mathcal{X} \in \mathbb{R}^{n_1 \times \cdots \times n_D}$ denote a D-dimensional tensor. If $n_1 = \cdots = n_D$, then we say that $\mathcal{X}$ is cubic. The set of D-mode cubic tensors will be denoted by $\mathbb{R}^{n \times \cdots \times n}$, where D will stay implicit. For any index set $C \subset \{1, \ldots, n_1\} \times \cdots \times \{1, \ldots, n_D\}$, $\mathcal{X}_C$ will denote the subarray $(\mathcal{X}_{i_1, \ldots, i_D})_{(i_1, \ldots, i_D) \in C}$.

2 Basics on Tensors

2.1 Tensor Norms

The Spectrum of a Tensor. Let us define the spectrum as the mapping which to any tensor $\mathcal{X} \in \mathbb{R}^{n \times \cdots \times n}$ associates the vector $\sigma(\mathcal{X})$ given by

$$\sigma(\mathcal{X}) \overset{\text{def}}{=} \frac{1}{\sqrt{D}} \left(\sigma^{(1)}(\mathcal{X}), \ldots, \sigma^{(D)}(\mathcal{X})\right),$$

where $\sigma^{(d)}(\mathcal{X})$ denotes the vector consisting of the singular values of the mode-d matricization of $\mathcal{X}$.

Norms of Tensors. Let $\mathcal{X} = (\mathcal{X}_{ijk})$ and $\mathcal{Y} = (\mathcal{Y}_{ijk})$ be tensors in $\mathbb{R}^{n_1 \times \cdots \times n_D}$. We can define several tensor norms on $\mathbb{R}^{n_1 \times \cdots \times n_D}$. The first one is a natural extension of the Frobenius norm or Hilbert-Schmidt norm from matrices to tensors. We start by defining the following scalar product on $\mathbb{R}^{n_1 \times \cdots \times n_D}$:

$$\langle \mathcal{X}, \mathcal{Y} \rangle \overset{\text{def}}{=} \sum_{i_1=1}^{n_1} \cdots \sum_{i_D=1}^{n_D} \mathcal{X}_{i_1, \ldots, i_D} \mathcal{Y}_{i_1, \ldots, i_D}.$$

Using this scalar product, we can define the following norm, which we call the Frobenius norm

$$\|\mathcal{X}\|_F \overset{\text{def}}{=} \sqrt{\langle \mathcal{X}, \mathcal{X} \rangle}.$$

One may also define an "operator norm" in the same manner as for matrices as follows

$$\|\mathcal{X}\| \overset{\text{def}}{=} \max_{\substack{u^{(d)} \in \mathbb{R}^{n_d}, \\ \|u^{(d)}\|_2 = 1, d = 1, \ldots, D}} \langle \mathcal{X}, u^{(1)} \otimes \cdots \otimes u^{(D)} \rangle$$

We also define

$$\|\mathcal{X}\|_* \overset{\text{def}}{=} \frac{1}{D} \sum_{d=1}^{D} \|\sigma^{(d)}\|_1.$$

2.2   Orthogonally Decomposable Tensors

The Orthogonally decomposable (ODEC) tensors are defined as follows

Definition 2.1. *Let $\mathcal{X}$ be a tensor in $\mathbb{R}^{n_1 \times \cdots \times n_D}$. If*

$$\mathcal{X} = \sum_{i=1}^{r} \alpha_i \cdot u_i^{(1)} \otimes \cdots \otimes u_i^{(D)}, \qquad (2.1)$$

where $r \leqslant n_1 \wedge \cdots \wedge n_D$, $\alpha_1 \geqslant \cdots \geqslant \alpha_r > 0$ and $\{u_1^{(d)}, \ldots, u_r^{(d)}\}$ is a family of orthonormal vectors for $d = 1, \ldots, D$, then we say (2.1) is an orthogonal decomposition of $\mathcal{X}$.

Denote $\alpha = (\alpha_1, \ldots, \alpha_r, 0, \ldots, 0)$ in $\mathbb{R}^{n_1 \wedge \cdots \wedge n_D}$. For each $d \in \{1, \ldots, D\}$, we may complete $\{u_1^{(d)}, \ldots, u_r^{(d)}\}$ with $\{u_{r+1}^{(d)}, \ldots, w_{n_d}^{(d)}\}$ so that matrix $U^{(d)} = (u_1^{(d)}, \ldots, u_{n_d}^{(d)}) \in \mathbb{R}^{n_d \times n_d}$ is orthogonal. Using $U^{(1)}, \ldots, U^{(D)}$, we may write (2.1) as

$$\mathcal{X} = \mathcal{D}(\alpha) \times_1 U^{(1)} \times_2 U^{(2)} \cdots \times_D U^{(D)}. \qquad (2.2)$$

where $\mathcal{D} = \mathrm{diag}(\alpha)$ is a diagonal tensor with the ith diagonal being α_i for $i = 1, \ldots, r$ and the other diagonal entries being zero. Note that representation (2.2) is generally not unique unless $n_1 = \cdots = n_D$ and $\alpha_1, \ldots, \alpha_r$ are all distinct.

It is easy to calculate the norms of ODEC tensors.

Proposition 2.2. *Let $\mathcal{X}$ be an orthogonally decomposable tensor and let*

$$\mathcal{X} = \sum_{i=1}^{r} \alpha_i \cdot u_i^{(1)} \otimes \cdots \otimes u_i^{(D)},$$

be an orthogonal decomposition of $\mathcal{X}$. Then

$$\|\mathcal{X}\| = \alpha_1 \quad \text{and} \quad \|\mathcal{X}\|_* = \sum_{i=1}^{r} \alpha_i.$$

3   Further Results on the Spectrum

In this section, we will present some further results on the spectrum such as the question of characterizing the image of the spectrum and the subdifferential of a function of the spectrum.

3.1   A Technical Prerequisite: Von Neumann's Inequality for Tensors

Von Neumann's inequality says that for any two matrices X and Y in $\mathbb{R}^{n_1 \times n_2}$, we have

$$\langle X, Y \rangle \leq \langle \sigma(X), \sigma(Y) \rangle,$$

with equality when the singular vectors of X and Y are equal, up to permutations when the singular values have multiplicity greater than one. This result has proved useful for the study of the subdifferential of unitarily invariant convex functions of the spectrum in the matrix case in [5]. In order to study the subdifferential of the norms of certain type of tensors, we will need a generalization this result to higher orders. This was worked out in [1]. Let us recall the containt of the main result of [1].

Definition 3.1. *We say that a tensor S is blockwise decomposable if there exists an integer B and if, for all $d = 1, \ldots, D$, there exists a partition $I_1^{(d)} \cup \ldots \cup I_B^{(d)}$ into disjoint index subsets of $\{1, \ldots, n_d\}$, such that $X_{i_1, \ldots, i_D} = 0$ if for all $b = 1, \ldots, B$, $(i_1, \ldots, i_D) \notin I_b^{(1)} \times \ldots \times I_b^{(D)}$.*

An illustration of this block decomposition can be found in Fig. 1. The following result is a generalization of von Neumann's inequality from matrices to tensors. It is proved in [1].

Theorem 3.2. *Let $X, Y \in \mathbb{R}^{n_1 \times \cdots \times n_D}$ be tensors. Then for all $d = 1, \ldots, D$, we have*

$$\langle X, Y \rangle \leqslant \langle \sigma^{(d)}(X), \sigma^{(d)}(Y) \rangle. \tag{3.3}$$

Equality in (3.3) holds simultaneously for all $d = 1, \ldots, D$ if and only if there exist orthogonal matrices $W^{(d)} \in \mathbb{R}^{n_d \times n_d}$ for $d = 1, \ldots, D$ and tensors $D(X), D(Y) \in \mathbb{R}^{n_1 \times \cdots \times n_D}$ such that

$$X = D(X) \times_1 W^{(1)} \cdots \times_D W^{(D)},$$
$$Y = D(Y) \times_1 W^{(1)} \cdots \times_D W^{(D)},$$

where $D(X)$ and $D(Y)$ satisfy the following properties:

(i) $D(X)$ and $D(Y)$ are block-wise decomposable with the same number of blocks, which we will denote by B,
(ii) the blocks $\{D_b(X)\}_{b=1,\ldots,B}$ (resp. $\{D_b(Y)\}_{b=1,\ldots,B}$) on the diagonal of $D(X)$ (resp. $D(Y)$) have the same sizes,
(iii) for each $b = 1, \ldots, B$ the two blocks $D_b(X)$ and $D_b(Y)$ are proportional.

3.2 Subdifferential for ODEC Tensors

Theorem 3.3. *Let $f : \mathbb{R}^n \times \cdots \times \mathbb{R}^n \mapsto \mathbb{R}$ satisfy property*

$$f(s_1, \ldots, s_D) = f(s_{\tau(1)}, \ldots, s_{\tau(D)}) \tag{3.4}$$

for all $\tau \in \mathfrak{S}_S$. Then for all ODEC tensors X, we have

$$(f \circ \sigma)^*(X) = f^*(\sigma(X)) \tag{3.5}$$

Using this result combined with von Neumann's inequality for tensors, one easily obtains the following corollary.

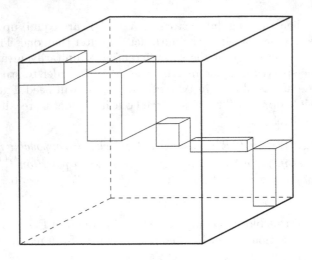

Fig. 1. A block-wise diagonal tensor.

Corollary 3.4. *Let* $f : \mathbb{R}^n \times \cdots \times \mathbb{R}^n \mapsto \mathbb{R}$ *satisfy property*

$$f(s_1, \ldots, s_D) = f(s_{\tau(1)}, \ldots, s_{\tau(D)}) \qquad (3.6)$$

for all $\tau \in \mathfrak{S}_S$. *Let* $\mathcal{X}$ *be an ODEC tensor. Then necessary and sufficient conditions for an ODEC tensor* $\mathcal{Y}$ *to belong to* $\partial(f \circ \sigma)(\mathcal{X})$ *are*

1. $\mathcal{Y}$ *has the same mode-d singular spaces as* $\mathcal{X}$ *for all* $d = 1, \ldots, D$,
2. $\sigma(\mathcal{Y}) \in \partial f(\sigma(\mathcal{X}))$.

Corollary 3.5. *Let* $\mathcal{X} = \mathcal{D}(\alpha) \times_1 U^{(1)} \times_2 \cdots \times_D U^{(D)}$ *be an ODEC tensor. Then the subdifferential* $\partial \| \cdot \|_*(\mathcal{X})$ *includes the following set*

$$\Omega = \left\{ \mathcal{D}(\mathbf{1}) \times_1 U^{(1)} \times_2 \cdots \times_D U^{(D)} + \mathcal{V} \mid \|\mathcal{V}\| \leq 1, \ \mathcal{V} \times_i U^{(i)\mathsf{T}} = 0, \ i = 1, \ldots, D \right\}.$$

4   Application to Tensor Recovery with Gaussian Measurements

Let $\mathcal{X}^\# \in \mathbb{R}^{n_1 \times n_2 \times n_3}$ be an unknown true signal, $\Phi(\cdot) : \mathbb{R}^{n_1 \times n_2 \times n_3} \mapsto \mathbb{R}^m$ be a known linear measurement mapping and

$$y = \Phi(\mathcal{X}^\#) + \xi \qquad (4.7)$$

be a noised vector of measurements in $\mathbb{R}^m$.

We focus on the following optimization problem:

$$\min_{\mathcal{X}} \|\mathcal{X}\|_* \quad \text{subject to} \quad \|\Phi(\mathcal{X}) - y\| \leq \eta. \qquad (4.8)$$

Let $\hat{\mathcal{X}}$ be any solution of optimization problem (4.8). We are interested in giving a bound for

$$\|\hat{\mathcal{X}} - \mathcal{X}^{\#}\|_F.$$

The main tool of this section is the following result by Tropp [6]:

Theorem 4.1. *Assume that $\|\xi\| \leqslant \eta$. Then with probability at least $1 - e^{-t^2/2}$, we have*

$$\|\hat{\mathcal{X}} - \mathcal{X}^{\#}\|_F \leqslant \frac{2\eta}{[\sqrt{m-1} - w(\mathscr{D}(\|\cdot\|_*, \mathcal{X}^{\#})) - t]_+},$$

where $[a]_+ = \max\{a, 0\}$ for any $a \in \mathbb{R}$.

The quantity $w(\mathscr{D}(\|\cdot\|_*, \mathcal{X}^{\#}))$ denotes the conic Gaussian width $w(\cdot)$ of the descent cone $\mathscr{D}(\|\cdot\|_*, \mathcal{X}^{\#})$. The definitions of these notions are given as follows:

Definition 4.2. *Let $K \in \mathbb{R}^d$ be a cone, the conic Gaussian width $w(K)$ is defined as*

$$w(K) = \mathbb{E}[\sup_{u \in K \cap \mathcal{S}^{d-1}} \langle g, u \rangle],$$

where $g \sim \mathcal{N}(0, I)$ is a standard Gaussian vector and $\mathcal{S}^{d-1}$ denotes the unit sphere in $\mathbb{R}^d$.

Definition 4.3. *Let $f : \mathbb{R}^d \mapsto \bar{\mathbb{R}}$ be a proper convex function. The descent cone $\mathscr{D}(f, x)$ of the function f at a point $x \in \mathbb{R}^d$ is defined as*

$$\mathscr{D}(f, x) \overset{\text{def}}{=} \{\lambda u \mid \lambda > 0, u \in \mathbb{R}^d, f(x + u) \leqslant f(x)\}.$$

According to Theorem 4.1, the error bound of $\|\hat{\mathcal{X}} - \mathcal{X}^{\#}\|_F$ depends on the conic Gaussian width $w(\cdot)$ of the descent cone $\mathscr{D}(\|\cdot\|_*, \mathcal{X}^{\#})$. The following result reveals that the latter is then closely related to the subdifferential of $\|\cdot\|_*$ at $\mathcal{X}^{\#}$.

Proposition 4.4. *Assume that $\partial\|\mathcal{X}^{\#}\|$ is nonempty and does not contain the origin. Then*

$$w^2(\mathscr{D}(\|\cdot\|_*, \mathcal{X}^{\#})) \leqslant \mathbb{E} \inf_{\tau \geqslant 0} \text{dist}_F^2(\mathcal{G}, \tau \partial\|\mathcal{X}^{\#}\|_*),$$

where $\mathcal{G} \in \mathbb{R}^{n_1 \times n_2 \times n_3}$ is a tensor with i.i.d. random Gaussian entries and

$$\text{dist}_F(\mathcal{G}, \tau\partial\|\mathcal{X}^{\#}\|_*) \overset{\text{def}}{=} \inf_{\mathcal{Y} \in \tau\partial\|\mathcal{X}^{\#}\|_*} \|\mathcal{G} - \mathcal{Y}\|_F,$$

i.e. the distance between $\mathcal{G}$ and the set $\tau\partial\|\mathcal{X}^{\#}\|_*$.

To derive a bound for $\|\hat{\mathcal{X}} - \mathcal{X}^{\#}\|_F$, we need to give an upper bound for

$$\mathbb{E} \inf_{\tau \geqslant 0} \text{dist}_F^2(\mathcal{G}, \tau\partial\|\mathcal{X}^{\#}\|_*).$$

The following result establishes such a bound in the case that $\mathcal{X}^{\#}$ is odec.

Proposition 4.5. *If $\mathcal{X}^{\#}$ is odec, then we have the following bound:*

$$\mathbb{E}\inf_{\tau\geqslant 0}\operatorname{dist}_F^2(\mathcal{G},\tau\partial\|\mathcal{X}^{\#}\|_*)\leqslant r^3 + r + 3r(n_1 + n_2 + n_3 - 3r) + r(n_1 n_2 + n_2 n_3 + n_1 n_3)$$

$$-r^2(n_1 + n_2 + n_3).$$

Proof. If $\mathcal{X}^{\#}$ is orthogonally decomposable, i.e.

$$\mathcal{X}^{\#} = \sum_{i=1}^{r}\sigma_i u_i^{(1)}\otimes u_i^{(2)}\otimes u_i^{(3)}$$

$$= \mathcal{D}(\sigma)\times_1 U^{(1)}\times_2 U^{(2)}\times_3 U^{(3)},$$

where $\mathcal{D}(\sigma)$ is a diagonal tensor with diagonal elements $\sigma = (\sigma_1,\ldots,\sigma_r)$ and $U^{(j)} = (u_1^{(j)},\ldots,u_r^{(j)})$ for $j = 1, 2, 3$, then the subdifferential $\partial\|\cdot\|_*(\mathcal{X}^{\#})$ includes the following set

$$\Omega = \left\{\sum_{i=1}^{r}u_i^{(1)}\otimes u_i^{(2)}\otimes u_i^{(3)} + \mathcal{V} \,\bigg|\, \|\mathcal{V}\|\leqslant 1,\ \mathcal{V}\times_i U^{(i)} = 0,\ i = 1, 2, 3.\right\}. \quad (4.9)$$

Hence

$$\mathbb{E}\inf_{\tau\geqslant 0}\operatorname{dist}_F^2(\mathcal{G},\tau\partial\|\mathcal{X}^{\#}\|_*)\leqslant\mathbb{E}\inf_{\tau\geqslant 0}\operatorname{dist}_F^2(\mathcal{G},\tau\Omega) = \mathbb{E}\inf_{\tau\geqslant 0}\inf_{\mathcal{Y}\in\Omega}\|\mathcal{G} - \tau\mathcal{Y}\|_F^2.$$

Note that $\mathcal{V}$ in (4.9) can also be characterized by

$$\mathcal{V} = \mathcal{T}\times_1 U_\perp^{(1)}\times_2 U_\perp^{(2)}\times_3 U_\perp^{(3)}, \quad (4.10)$$

where $\mathcal{T}\in\mathbb{R}^{(n_1-r)\times(n_2-r)\times(n_3-r)}$ is a tensor such that $\|\mathcal{T}\|\leqslant 1$ and $U_\perp^{(i)}\in\mathbb{R}^{n_i\times(n_i-r)}$ is a matrix such that $\tilde{U}^{(i)} = (U^{(i)}|U_\perp^{(i)})$ is orthogonal for $i = 1, 2, 3$. In view of (4.9) and (4.10), we assert that any $\mathcal{Y}\in\Omega$ can be written as

$$\mathcal{Y} = \mathcal{C}\times_1\tilde{U}^{(1)}\times_2\tilde{U}^{(2)}\times_3\tilde{U}^{(3)}.$$

where tensor $\mathcal{C}$ is block-wise diagonal with two diagonal blocks $\mathcal{C}_1 = \operatorname{diag}(\mathbf{1})\in\mathbb{R}^{r\times r\times r}$ and $\mathcal{C}_2 = \mathcal{T}\in\mathbb{R}^{(n_1-r)\times(n_2-r)\times(n_3-r)}$.

Because $\mathcal{G}\in\mathbb{R}^{n_1\times n_2\times n_3}$ is a tensor with i.i.d. random standard Gaussian entries, for any orthogonal matrices $W^{(1)}, W^{(2)}, W^{(3)}$ with appropriate size, tensor $\mathcal{G}\times_1 W^{(1)}\times_2 W^{(2)}\times_3 W^{(3)}$ still has i.i.d. standard Gaussian entries. Therefore, we may choose a coordinate system such that

$$\mathbb{E}\inf_{\tau\geqslant 0}\inf_{\mathcal{Y}\in\Omega}\|\mathcal{G} - \tau\mathcal{Y}\|_F^2 = \mathbb{E}\inf_{\tau\geqslant 0}\inf_{\mathcal{C}\in\tilde\Omega}\|\mathcal{G} - \tau\mathcal{C}\|_F^2,$$

where $\tilde\Omega$ denotes the set of block-wise diagonal tensors with two diagonal blocks $\mathcal{C}_{111} = \mathcal{D}(\mathbf{1})\in\mathbb{R}^{r\times r\times r}$ and $\mathcal{C}_{222}\in\mathbb{R}^{(n_1-r)\times(n_2-r)\times(n_3-r)}$ verifying $\|\mathcal{C}_2\|\leqslant 1$. Partitioning $\mathcal{G}$ in the same manner, we obtain

$$\|\mathcal{G} - \tau\mathcal{C}\|_F^2 = \|\mathcal{G}_{111} - \tau\mathcal{D}(\mathbf{1})\|_F^2 + \|\mathcal{G}_{222} - \tau\mathcal{T}\|_F^2 + \sum_{\substack{i,j,k=1 \\ i,j,k\text{ are not equal}}}^{2}\|\mathcal{G}_{i,j,k}\|_F^2.$$

Since $\mathcal{G}$ is a tensor with independent Gaussian entries, it follows that

$$\mathbb{E} \sum_{\substack{i,j,k=1 \\ i,j,k \text{ are not equal}}}^{2} \|\mathcal{G}_{ijk}\|_F^2 = r(n_1 n_2 + n_2 n_3 + n_1 n_3) - r^2(n_1 + n_2 + n_3).$$

Thus

$$\mathbb{E} \inf_{\tau \geqslant 0} \inf_{\mathcal{C} \in \tilde{\Omega}} \|\mathcal{G} - \tau \mathcal{C}\|_F^2 = \mathbb{E} \inf_{\tau \geqslant 0} \inf_{\|\mathcal{C}_2\| \leqslant 1} \left(\|\mathcal{G}_{111} - \operatorname{diag}(\tau)\|_F^2 + \|\mathcal{G}_{222} - \tau \mathcal{C}_2\|_F^2 \right)$$
$$+ r(n_1 n_2 + n_2 n_3 + n_1 n_3) - r^2(n_1 + n_2 + n_3).$$

Choosing $\tau = \|\mathcal{G}_2\|$, we get

$$\mathbb{E} \inf_{\tau \geqslant 0} \inf_{\|\mathcal{C}_2\| \leqslant 1} \left(\|\mathcal{G}_1 - \operatorname{diag}(\tau)\|_F^2 + \|\mathcal{G}_2 - \tau \mathcal{C}_2\|_F^2 \right) \leqslant \mathbb{E} \|\mathcal{G}_1 - \operatorname{diag}(\|\mathcal{G}_2\|)\|_F^2$$

Since

$$\mathbb{E}\|\mathcal{G}_1 - \operatorname{diag}(\|\mathcal{G}_2\|)\|_F^2 = r^3 + r\mathbb{E}\|\mathcal{G}_2\|^2 \leqslant r^3 + r + r\left(\sqrt{n_1 - r} + \sqrt{n_2 - r} + \sqrt{n_3 - r}\right)^2$$
$$\leqslant r^3 + r + 3r(n_1 + n_2 + n_3 - 3r),$$

It follows that

$$\mathbb{E} \inf_{\tau \geqslant 0} \operatorname{dist}_F^2(\mathcal{G}, \tau \partial \|\mathcal{X}^{\#}\|_*) \leqslant r^3 + r + 3r(n_1 + n_2 + n_3 - 3r) + r(n_1 n_2 + n_2 n_3 + n_1 n_3)$$
$$- r^2(n_1 + n_2 + n_3).$$

If the tensor is cubic, i.e. $n_i = n$ for $i = 1, 2, 3$, then we have with at least probability $1 - e^{-t^2/2}$ that

$$\|\hat{\mathcal{X}} - \mathcal{X}^{\#}\|_F \leqslant \frac{2\eta}{[\sqrt{m-1} - \sqrt{(r^3 + r + 9r(n-r) + 3rn(n-r))} - t]_+}.$$

References

1. Chrétien, S., Wei, T.: Von neumann's inequality for tensors, arXiv preprint arXiv:1502.01616 (2015)
2. Hiriart-Urruty, J.-B., Lemaréchal, C.: Convex analysis and minimization algorithms i: Part 1: Fundamentals, vol. 305, Springer Science and Business Media (1996)
3. Kolda, T.G., Bader, B.W.: Tensor decompositions and applications. SIAM Rev. **51**(3), 455–500 (2009)
4. Landsberg, J.M.: Tensors: geometry and applications, vol. 128. American Mathematical Soc, Providence (2012)
5. Lewis, A.S.: The convex analysis of unitarily invariant matrix functions. J. Convex Anal. **2**(1), 173–183 (1995)
6. Tropp, J.A.: Convex recovery of a structured signal from independent random linear measurements, arXiv preprint arXiv:1405.1102 (2014)
7. Watson, G.A.: Linear Algebra and its Applications. Charact. Subdifferential Matrix Norms **170**, 33–45 (1992)

Linear Discriminant Analysis with Adherent Regularization

Hideitsu Hino$^{(\boxtimes)}$

University of Tsukuba,
1-1-1 Tennoudai, Tsukuba, Ibaraki 305–8573, Japan
hinohide@cs.tsukuba.ac.jp

Abstract. Sparse modeling have become one of the standard approaches for latent variable analysis in the literature of statistics, machine learning and signal processing. This paper considers a supervised dimension reduction, which is a fundamental problem in data science. Particularly, the problem of linear discriminant analysis is considered. Extending the previous attempt to impose sparsity invoking regularization for Fisher's discriminant model, the proposed method bridges two different formulations of linear discriminant analysis, namely, the Fisher's discriminant model and the normal model, via a particular form of regularization. The proposed discriminant problem is efficiently solved by using the proximal point algorithm. The proposed method is shown to work well through experiments using both artificial and real-world datasets.

Keywords: Linear discriminant analysis · Sparse regularization · Classification

1 Introduction

One of the characteristics of modern data is its high dimensionality with small number of samples. It is known that traditional models in statistics usually perform poorly on such data. For example, based on the theory of random matrix [1], linear discriminant analysis (LDA [2]), which is widely used for classification in many fields, is both theoretically and experimentally shown not to work well in high dimensional settings [3].

Over the years, there have been a number of efforts to make linear discriminant analysis in high dimension reliable and to enable it to find meaningful features. Latent variable analysis assumes that only small number of intrinsic parameter play key role in explaining the observed data and describing the underlying data generating structure. Currently, sparse modeling is regarded as a promising approach for finding the latent structure in linear models. Algorithms imposing sparsity promoting regularization to the classification axes for LDA are proposed in [4,5]. In [6], the Penalized LDA(PLDA) is proposed, which minimizes between class variance with the ℓ_1-norm regularization, under the constraint that the within class variance is smaller than one.

© Springer International Publishing Switzerland 2015
E. Vincent et al. (Eds.): LVA/ICA 2015, LNCS 9237, pp. 368–375, 2015.
DOI: 10.1007/978-3-319-22482-4_43

In this paper, based on the penalized LDA framework proposed in [6], we consider an elastic-net type penalization [7] that takes account both sparsity and *adherency* to the normal discriminant model. The proposed penalization avoids too sparse solution, which sometimes cause problems in penalized classification, and it also shows *adherence* to the normal model solution. An algorithm using proximal point algorithm is derived to solve the proposed problem.

The rest of the paper is organized as follows. In Sect. 2, we define notations to describe the problem considered in this paper, and introduce two different models for linear discriminant analysis. Then, in Sect. 3, we explain the penalized linear discriminant analysis, which is the basis of the proposed method in this work. Section 4 introduces our approach for penalized linear discriminant analysis, and an efficient optimization algorithm for the problem is derived. Sections 5 is devoted to show the experimental results, and our conclusion is drawn in Sect. 6.

2 Notation and Preliminary

Let $X \in \mathbb{R}^{n \times p}$ be the data matrix of p-dimensional vectors $x_1, \ldots, x_n$, where n is the number of samples and $x_i \in \mathbb{R}^p$. Each sample is supposed to belong to one of two classes C_1 or C_2 of sizes n_1 and n_2, respectively. We consider the two class classification problem in this paper. Let μ_1, μ_2 be the mean vectors of samples in class C_1 and class C_2, respectively, and μ and ξ be the mean vectors of whole samples and difference of two mean vectors $\xi = \mu_1 - \mu_2$, respectively. Σ_w and Σ_b are the within and the between class covariance matrices, respectively, and empirically estimated as

$$\hat{\Sigma}_w = \frac{1}{n} \sum_{k=1}^{2} \sum_{i \in C_k} (x_i - \hat{\mu}_k)(x_i - \hat{\mu}_k)^\top, \quad \hat{\Sigma}_b = \frac{1}{n} X^\top X - \hat{\Sigma}_w, \quad (1)$$

where $\hat{\mu}_k = \frac{1}{n_k} \sum_{i \in C_k} x_i, k = 1, 2$ are empirical estimates of the mean vectors of samples in class $C_k, k = 1, 2$. We also define a positive-definite estimate of the within class covariance matrix Σ_w as $\tilde{\Sigma}_w$, which has the same diagonal elements as $\hat{\Sigma}_w$, but other off-diagonal elements are set to zero. The empirical estimate of μ and ξ are defined as $\hat{\mu} = \frac{1}{n} \sum_{k=1}^{2} n_k \hat{\mu}_k$ and $\hat{\xi} = \hat{\mu}_1 - \hat{\mu}_2$, respectively.

In the following, we introduce two different formulations for linear discriminate analysis, namely, *Fisher's Discriminant Model Formulation* and *Normal Model Formulation*. In their idealized situations, the classification axes obtained by these methods are identical. In our proposed method described in Sect. 4, an elastic-net type regularization is imposed to Fisher's discriminant model to keep the solution close to that of the normal model.

2.1 Fisher's Discriminant Model Formulation

Fisher's Discriminant Model Formulation of linear discriminant problem aims at finding the classification vector (axis) maximizing the between class variance while minimizing the within class variance. This is often formulated as a minimization problem of the Fisher's criterion $J_F(\boldsymbol{\beta}) = \boldsymbol{\beta}^\top \Sigma_w \boldsymbol{\beta} / \boldsymbol{\beta}^\top \Sigma_b \boldsymbol{\beta}$, and it is equivalently formulated as the following minimization problem

$$\min_{\boldsymbol{\beta}}\{-\boldsymbol{\beta}^\top \hat{\Sigma}_b \boldsymbol{\beta}\}, \quad \text{s.t.} \quad \boldsymbol{\beta}^\top \tilde{\Sigma}_w \boldsymbol{\beta} = 1. \tag{2}$$

Recalling the definition of the between class covariance matrix, the solution of the problem is shown to be $\tilde{\boldsymbol{\beta}} = \tilde{\Sigma}_w^{-1}(\boldsymbol{\mu}_1 - \boldsymbol{\mu}_2)$.

2.2   Normal Model

The *Normal Model Formulation* is another well-known form of linear discriminant analysis, which is supported by probabilistic model perspective. Suppose the distributions of data in two classes C_1 and C_2 are Gaussians with means $\boldsymbol{\mu}_1$ and $\boldsymbol{\mu}_2$ and the same covariance matrix $\Sigma = \Sigma_w$ for both classes. The Bayesian classifier defined by

$$f(\boldsymbol{x}) = \text{sign}\left(\log \frac{P(C_1|\boldsymbol{x})}{P(C_2|\boldsymbol{x})}\right)$$

is obtained by simple calculation, under the above assumption, as

$$\log \frac{P(C_1|\boldsymbol{x})}{P(C_2|\boldsymbol{x})} = \{\Sigma_w^{-1}(\boldsymbol{\mu}_1 - \boldsymbol{\mu}_2)\}^\top \boldsymbol{x} - \frac{1}{2}\boldsymbol{\mu}_1 \Sigma_w^{-1} \boldsymbol{\mu}_1 + \frac{1}{2}\boldsymbol{\mu}_2 \Sigma_w^{-1} \boldsymbol{\mu}_2 + \log \frac{P(C_1)}{P(C_2)}.$$

That is, the Bayes optimal classifier is given by a linear classifier and its projection vector is the same as that obtained by Fisher's criterion, and its empirical estimate is given by $\hat{\boldsymbol{\beta}} = \tilde{\Sigma}_w^{-1}(\hat{\boldsymbol{\mu}}_1 - \hat{\boldsymbol{\mu}}_2)$.

3   Penalized Linear Discriminant Analysis

In [6], the authors adopted Fisher's discriminant formulation

$$\min_{\boldsymbol{\beta}}\{-\boldsymbol{\beta}^\top \hat{\Sigma}_b \boldsymbol{\beta} + P_1(\boldsymbol{\beta})\}, \quad \text{s.t.} \quad \boldsymbol{\beta}^\top \tilde{\Sigma}_w \boldsymbol{\beta} \leq 1, \tag{3}$$

with weighted ℓ_1-norm regularization defined by

$$P_1(\boldsymbol{\beta}) = \lambda_1 \sum_{i=1}^{p} |\delta_i \beta_i|, \quad \lambda_1 \geq 0, \tag{4}$$

where $\delta_i, i = 1, \ldots, p$ are sample standard deviation of elements of $\boldsymbol{x}$. The problem (3) is named Penalized Linear Discriminant Analysis (PLDA).

Since the objective function $-\boldsymbol{\beta}^\top \hat{\Sigma}_b \boldsymbol{\beta} + P_1(\boldsymbol{\beta})$ to be minimized is not convex, an iterative optimization approach called the Majorization-Minimization (MM) algorithm [8] is used, which is briefly explained as follows for the sake of self-containedness. Consider the following non-convex optimization problem:

$$\min_{\boldsymbol{\beta}} \quad f(\boldsymbol{\beta}). \tag{5}$$

MM algorithm is a general algorithmic framework, which first upper bounds the objective function $f(\boldsymbol{\beta})$ by a function $g(\boldsymbol{\beta}|\boldsymbol{\beta}^{(m)})$, which is said to majorize the function $f(\boldsymbol{\beta})$ at the point $\boldsymbol{\beta}^{(m)}$ as

$$f(\boldsymbol{\beta}) \leq g(\boldsymbol{\beta}|\boldsymbol{\beta}^{(m)}), \quad f(\boldsymbol{\beta}^{(m)}) = g(\boldsymbol{\beta}^{(m)}|\boldsymbol{\beta}^{(m)}). \tag{6}$$

From an initial $\boldsymbol{\beta}^{(0)}$, the MM algorithm solves (5) by iteratively minimizing the majorized objective function as

$$\boldsymbol{\beta}^{(m+1)} = \arg \min_{\boldsymbol{\beta}} g(\boldsymbol{\beta}|\boldsymbol{\beta}^{(m)}). \tag{7}$$

The MM approach is applied to solve the PLDA problem (3). Let $f(\boldsymbol{\beta}) = -\boldsymbol{\beta}^\top \hat{\Sigma}_b \boldsymbol{\beta}$. For a fixed $\boldsymbol{\beta}^{(m)}$, $f(\boldsymbol{\beta})$ is majorized as

$$f(\boldsymbol{\beta}) \leq f(\boldsymbol{\beta}^{(m)}) + (\boldsymbol{\beta} - \boldsymbol{\beta}^{(m)})^\top \nabla f(\boldsymbol{\beta}^{(m)}) = \boldsymbol{\beta}^{(m)\top} \hat{\Sigma}_b \boldsymbol{\beta}^{(m)} - 2\boldsymbol{\beta}^\top \hat{\Sigma}_b \boldsymbol{\beta}^{(m)}.$$

Therefore, the PLDA problem is solved by iteratively minimizing the sub-problem

$$\min_{\boldsymbol{\beta}} \boldsymbol{d}^{(m)\top} \boldsymbol{\beta} + P_1(\boldsymbol{\beta}), \quad \text{s.t.} \quad \boldsymbol{\beta}^\top \tilde{\Sigma}_w \boldsymbol{\beta} \leq 1, \tag{8}$$

where $\boldsymbol{d}^{(m)} = -\hat{\Sigma}_b \boldsymbol{\beta}^{(m)}$. The solution to the sub-problem is given by the soft-thresholding operator [9].

4   Proposed Linear Discriminant Analysis Formulation

4.1   Adherent Penalization

We consider the equation constraint in penalized Fisher's problem

$$\min_{\boldsymbol{\beta}} \{-\boldsymbol{\beta}^\top \hat{\Sigma}_b \boldsymbol{\beta} + P(\boldsymbol{\beta})\}, \quad \text{s.t.} \quad \boldsymbol{\beta}^\top \tilde{\Sigma}_w \boldsymbol{\beta} = 1, \tag{9}$$

which can be shown to be equivalent to the problem with inequality constraint [6].

Then, we consider the penalization $P(\boldsymbol{\beta}) = P_1(\boldsymbol{\beta}) + P_2(\boldsymbol{\beta})$, where $P_1(\boldsymbol{\beta})$ is the weighted ℓ_1-norm term (4) which is the same as in PLDA, while

$$P_2(\boldsymbol{\beta}) = \lambda_2 \|\tilde{\Sigma}_w \boldsymbol{\beta} - \hat{\boldsymbol{\xi}}\|_2^2, \ \lambda_2 \geq 0, \tag{10}$$

which we call the *adherent penalization*. Note that $P_2(\boldsymbol{\beta})$ is a quadratic form in $\boldsymbol{\beta}$ and the penalization $P_1(\boldsymbol{\beta}) + P_2(\boldsymbol{\beta})$ is similar to the elastic-net [7]. The rational behind this penalization is in the fact that the optimal LDA classification axis in normal model is $\boldsymbol{\beta} = \Sigma_w^{-1}(\boldsymbol{\mu}_1 - \boldsymbol{\mu}_2)$, and it is reasonable to keep $\|\tilde{\Sigma}_w \boldsymbol{\beta} - \hat{\boldsymbol{\xi}}\|$ as small as possible. Here $\hat{\boldsymbol{\xi}} = \hat{\boldsymbol{\mu}}_1 - \hat{\boldsymbol{\mu}}_2$ and $\|\cdot\|$ is certain vector norm. In this work, considering that $P_1(\boldsymbol{\beta})$ is a weighted ℓ_1-norm penalization, and the elastic net-type penalization is shown to find less sparse solution that can be overlooked by ℓ_1-norm penalization, we adopt the ℓ_2-norm and defined $P_2(\boldsymbol{\beta}) =$

$\lambda_2\|\tilde{\Sigma}_w\boldsymbol{\beta} - \hat{\boldsymbol{\xi}}\|_2^2$, $\lambda_2 \geq 0$. We call the problem (9) with $P(\boldsymbol{\beta}) = P_1(\boldsymbol{\beta}) + P_2(\boldsymbol{\beta})$ the *Adherently Penalized Linear Discriminant Analysis (APLDA)* henceforth.

The sparsity penalty $P_1(\boldsymbol{\beta})$ and adherency penalty $P_2(\boldsymbol{\beta})$ complement each other. The sparsity penalty prefers a sparse structure in the general sense to control its complexity while the adherency penalty prefers a discriminant vector $\boldsymbol{\beta}$ as close as possible to the solution of the normal model. By combining these two penalties, we obtain the following objective function to be minimized:

$$J(\boldsymbol{\beta}) = -\boldsymbol{\beta}^\top \hat{\Sigma}_b \boldsymbol{\beta} + P_1(\boldsymbol{\beta}) + P_2(\boldsymbol{\beta})$$

$$= -\boldsymbol{\beta}^\top \hat{\Sigma}_b \boldsymbol{\beta} + \lambda_1 \sum_{i=1}^{p} |\delta_i \beta_i| + \lambda_2(\boldsymbol{\beta}^\top \tilde{\Sigma}_w \boldsymbol{\beta} - 2\hat{\boldsymbol{\xi}}^\top \tilde{\Sigma}_w \boldsymbol{\beta}) + const.$$

$$= -\boldsymbol{\beta}^\top \hat{\Sigma}_b \boldsymbol{\beta} + \lambda_1 \sum_{i=1}^{p} |\delta_i \beta_i| + \lambda_2 - 2\lambda_2 \hat{\boldsymbol{\xi}}^\top \tilde{\Sigma}_w \boldsymbol{\beta} + const.$$

$$\leq \boldsymbol{\beta}^{(m)\top} \hat{\Sigma}_b \boldsymbol{\beta}^{(m)} - 2\boldsymbol{\beta}^{(m)\top} \hat{\Sigma}_b \boldsymbol{\beta} - 2\lambda_2 \hat{\boldsymbol{\xi}}^\top \tilde{\Sigma}_w \boldsymbol{\beta} + \lambda_1 \sum_{i=1}^{p} |\delta_i \beta_i| + const.$$

$$= \boldsymbol{c}_{\lambda_2}^{(m)\top} \boldsymbol{\beta} + \lambda_1 \sum_{i=1}^{p} |\delta_i \beta_i| + const.$$

In the above derivation, we used the equality constraint $\boldsymbol{\beta}^\top \tilde{\Sigma}_w \boldsymbol{\beta} = 1$ and the inequality is the result of majorization. In the final line of the above equations, $\boldsymbol{c}_{\lambda_2}^{(m)} = -2(\boldsymbol{\beta}^{(m)\top} \hat{\Sigma}_b + \lambda_2 \boldsymbol{\xi}^\top \tilde{\Sigma}_w)$. At the m-th iteration step, the majorized objective function $J^{(m)}(\boldsymbol{\beta}) = \boldsymbol{c}_{\lambda_2}^{(m)\top} \boldsymbol{\beta} + \lambda_1 \sum_{i=1}^{p} |\delta_i \beta_i|$ is minimized by solving

$$\min_{\boldsymbol{\beta}} \quad J^{(m)}(\boldsymbol{\beta}) \quad \text{s.t.} \quad \boldsymbol{\beta}^\top \tilde{\Sigma}_w \boldsymbol{\beta} = 1. \tag{11}$$

This sub-problem is iteratively solved until convergence. The convergence of this procedure is guaranteed by the property of the MM algorithm. Since the problem (9) is not convex, we need to select a reasonable initial value $\boldsymbol{\beta}^{(0)}$ for this algorithm. Following the method in [6], we use the leading eigenvector of $\tilde{\Sigma}_w^{-1}\hat{\Sigma}_b$ as an initial value of $\boldsymbol{\beta}$. To solve the sub-problem, we propose an algorithm based on the proximal point algorithm as shown in next section.

4.2   Proximal Gradient Method

The proximal operator of a convex function h is defined by

$$\text{prox}_{h/L}(\boldsymbol{x}) = \arg\min_{\boldsymbol{u}} \left(h(\boldsymbol{u}) + \frac{L}{2}\|\boldsymbol{u} - \boldsymbol{x}\|_2^2 \right), \tag{12}$$

where $L > 0$ is a parameter. Proximal method is a class of algorithms for solving a convex optimization problem that uses the proximal operators of the objective function [10]. We solve the sub-problem (11) by proximal point algorithm [11], which iteratively updates the estimate of $\boldsymbol{\beta}$ by $\tilde{\boldsymbol{\beta}}^{t+1} = \text{prox}_{J^{(m)}/L}(\boldsymbol{\beta}^t)$ followed by the projection $\beta^{t+1} = \Pi(\tilde{\beta})^{t+1}$. Here, the projection operator is defined by $\Pi(\beta) = \frac{1}{\sqrt{\beta^\top \tilde{\Sigma}_w \beta}}\beta$, which is to satisfy the equality constraint $\beta^\top \tilde{\Sigma}_w \beta = 1$.

With a simple manipulation, the minimizer of the proximal operator of the objective function $J^{(m)}$ is shown to be given by

$$\tilde{\beta}_i^{t+1} = \begin{cases} \frac{L\beta_i^t - c_i^{(m)} - \lambda_1 \delta_i}{\delta_i L}, & (\frac{L\beta_i^t - c_i^{(m)}}{\delta_i L} \geq \lambda_1), \\ 0, & (-\lambda_1 < \frac{L\beta_i^t - c_i^{(m)}}{\delta_i L} < \lambda_1), \\ \frac{L\beta_i^t - c_i^{(m)} + \lambda_1 \delta_i}{\delta_i L}, & (\frac{L\beta_i^t - c_i^{(m)}}{\delta_i L} \leq \lambda_1). \end{cases} \tag{13}$$

5   Experimental Results

5.1   Artificial Dataset

In this subsection, with a simple artificial dataset, we show that the proposed method could improve the classification accuracy compared to conventional LDA and PLDA. In conventional LDA, the classification axis is obtained by solving the generalized eigen-value problem $\hat{\Sigma}_b \beta = \lambda \hat{\Sigma}_w \beta$. In PLDA, the classification axis is obtained by solving the problem (3), and in APLDA, it is obtained by solving the problem (9).

The dataset is composed of samples from two Gaussian distributions with means $\mu_1 = (0.5, 0.5, 0.4, 0.4, \ldots, 0.4) \in \mathbb{R}^p$ and $\mu_2 = (-0.5, -0.5, 0.4, 0.4, \ldots, 0.4)$, and a common covariance matrix Σ. The covariance matrix is a symmetric band matrix with diagonal element 1, and its first and second neighbors are 0.09 and 0.03, respectively. An example of the covariance matrix when $p = 5$ is $\Sigma = \begin{pmatrix} 1.00 & 0.09 & 0.03 & 0.00 & 0.00 \\ 0.09 & 1.00 & 0.09 & 0.03 & 0.00 \\ 0.03 & 0.09 & 1.00 & 0.09 & 0.03 \\ 0.00 & 0.03 & 0.09 & 1.00 & 0.09 \\ 0.00 & 0.00 & 0.03 & 0.09 & 1.00 \end{pmatrix}$. Varying the dimensionality $p = 5, 10, 100, 300$, and 500, we generated 50 samples (i.e., 25 samples for each class) and estimated the classification vectors with LDA, PLDA, and the proposed APLDA. The classification accuracy is evaluated by using the 1000 test samples.

From Fig. 1 and Table 1, we can see that the accuracy of ordinal LDA starts to degenerate between $p = 100$ and 300, and the accuracy of LDA in high dimension setting is just slightly better than random guess. On the other hand, Penalized LDA (PLDA) is not harmed by the increase of dimensionality. APLDA is also robust to the increase of the dimensionality, and basically it shows superior accuracy compared to PLDA.

5.2   Real-World Dataset

In the following, we show empirical results using three UCI-classification datasets [12] with high dimensional feature vectors (isolet with $p = 617$, secom with $p = 591$, and USPS with $p = 256$). For these datasets, since the classification accuracy of LDA was around 50%, LDA is removed from the list of comparison. Instead, naïve Bayes discriminant analyses proposed in [3] (labeled nB),

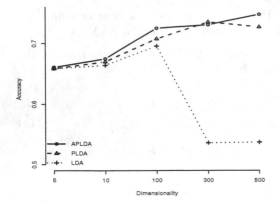

Table 1. A table of classification accuracy for different methods with different sample dimensionalities for estimating the classification axes.

p	LDA	PLDA	APLDA
5	0.65877	0.65873	0.66095
10	0.66376	0.66962	0.67457
100	0.69469	0.70690	0.72463
300	0.53518	0.73457	0.73002
500	0.53610	0.72646	0.74752

Fig. 1. Classification accuracy vs sample size

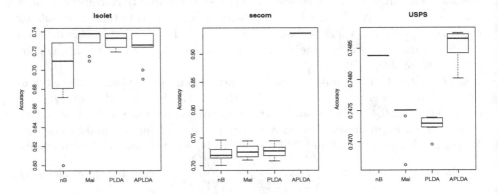

Fig. 2. Classification accuracies for real-world datasets.

sparse discriminant analyses proposed by Mai et al. [5] (labeled `Mai`), and by Cai &Liu [4](labeled `Cai`) are compared to `PLDA` and the proposed `APLDA`. The classification accuracies are summarized in Fig. 2. From Fig. 2, although there is no single method consistently outperforms others, APLDA is comparable to or better than other conventional methods.

6   Conclusion

For the Fisher's discriminant model, we introduced the adherent penalization, which is a combination of ℓ_1-norm penalization and ℓ_2-norm penalization similar to the elastic-net [7]. The rational behind the proposed penalization is that in addition to the sparsity promoting ℓ_1-norm penalization, it is reasonable to keep the solution of the penalized Fisher's problem close to the optimal solution for the normal model of the linear discriminant analysis. The adherent penalization is imposed in the form of ℓ_2-norm, and by using the equality constraint of the Fisher's problem, the penalized objective function is reduced to a simple sub-problem by using the MM algorithm approach. Then the sub-problem is solved by using the proximal point algorithm followed by projection to satisfy the equality constraint.

The aim of this paper is in introducing an elastic-net type regularization to linear discriminant problem, which we called the adherent regularization. In this work, we experimentally see that the proposed method work well in high dimensionality setting and comparable to or sometimes improves PLDA. It is our important future work to develop asymptotic theory or to show consistency of APLDA.

Acknowledgement. Part of this work is supported by KAKENHI No.26120504, 25870811, and 15H01678.

References

1. Marcenko, V., Pastur, L.: Distribution of eigenvalues for some sets of random matrices. Math. USSR Sb. **1**(4), 457 (1967)
2. Fisher, R.A.: The use of multiple measurements in taxonomic problems. Ann. Eugen. **7**, 179–188 (1936)
3. Bickel, P.J., Levina, E.: Some theory for Fisher's linear discriminant function, 'naive Bayes', and some alternatives when there are many more variables than observations. Bernoulli **10**(6), 989–1010 (2004)
4. Cai, T., Liu, W.: A direct estimation approach to sparse linear discriminant analysis. JASA. J. Am. Stat. Assoc. **106**(496), 1566–1577 (2011)
5. Mai, Q., Zou, H., Yuan, M.: A direct approach to sparse discriminant analysis in ultra-high dimensions. Biometrika **99**(1), 29–42 (2012)
6. Witten, D.M., Tibshirani, R.: Penalized classification using fisher's linear discriminant. J. Roy. Stat. Soc. B (Stat. Methodol.) **73**(5), 753–772 (2011)
7. Zou, H., Hastie, T.: Regularization and variable selection via the elastic net. J. Roy. Stat. Soc. B **67**, 301–320 (2005)
8. Hunter, D.R., Lange, K.: A tutorial on MM algorithms. Am. Stat. **58**, 30–37 (2004)
9. Donoho, D., Johnstone, I.: Ideal spatial adaptation via wavelet shrinkage. Biometrika **81**, 425–455 (1994)
10. Parikh, N., Boyd, S.: Proximal algorithms. Found. Trends Optim. **1**(3), 123–231 (2013)
11. Rockafellar, R.: Monotone operators and the proximal point algorithm. SIAM J. Control Optim. **14**(5), 877–898 (1976)
12. Murphy, P.M., Aha, D.W.: UCI Repository of machine learning databases. Technical report, University of California, Department of Information and Computer Science, Irvine, CA, US. (1994)

Split Gradient Method for Informed Non-negative Matrix Factorization

Robert Chreiky[1,2], Gilles Delmaire[1(✉)], Matthieu Puigt[1], Gilles Roussel[1], Dominique Courcot[3], and Antoine Abche[2]

[1] LISIC, ULCO, Université Lille Nord de France, Calais, France
{robert.chreiky,gilles.delmaire,matthieu.puigt,
gilles.roussel}@lisic.univ-littoral.fr
[2] University of Balamand, Koura, Lebanon
antoine.abche@balamand.edu.lb
[3] UCEIV, ULCO, Université Lille Nord de France, Dunkerque, France
courcot@univ-littoral.fr

Abstract. Recently, some informed Non-negative Matrix Factorization (NMF) methods were introduced in which some a priori knowledge (i.e., expert's knowledge) were taken into account in order to improve the separation process. This knowledge was expressed as known components of one factor, namely the profile matrix. Also, the sum-to-one property of the profile matrix was taken into account by an appropriate sequential normalization. However, our previous approach was unable to check both constraints at the same time. In this work, a new parametrization is proposed which takes into consideration both constraints simultaneously by incorporating a new unconstrained matrix. From this parameterization, new updates rules are introduced which are based on the framework of the Split Gradient Method by Lantéri *et al.* The cost function is defined in terms of a weighted Frobenius norm and the developed rules involve a new shift in order to ensure the non-negativity property. Simulations on a noisy source apportionment problem show the relevance of the proposed method.

Keywords: Informed source separation · Non-negative matrix factorization · Split gradient · Source apportionment

1 Introduction

Source apportionment consists of estimating the particulate matter sources (and their relative concentrations) which are present in the ambiant air. A source is fully characterized by a *profile* which gathers the m chemical species proportions (expressed in ng/ng) that constitute it. In practice, n data samples—collected from a chemical sampler—can be written as a mixture of p profiles, with different concentrations (expressed in ng/m^3). Mathematically, if we respectively denote by X, G, and F the non-negative $n \times m$ data matrix, $n \times p$ contribution matrix, and $p \times m$ profile matrix, the collected data reads

$$X \approx G \cdot F. \tag{1}$$

E. Vincent et al. (Eds.): LVA/ICA 2015, LNCS 9237, pp. 376–383, 2015.
DOI: 10.1007/978-3-319-22482-4_44

G and F are usually unknown and their estimation from X can be referred as a Blind Source Separation (BSS) problem [1] which can be solved, e.g., by Non-negative Matrix Factorization (NMF) [1, Ch. 13]. However, Viana et al. [12] show that the state-of-the-art blind NMF approaches do not provide a consistent performance.

In our recent work [7,8], several *informed* NMF methods were proposed which incorporate a priori knowledge about the source separation problem: (i) some entries of F are known [7] (or bounded [8]) by experts, and (ii) the rows of F are summed to one [7,8]. In order to tackle the first information, we introduced a specific parameterization which allows to take into account the known values of F. Then, a normalization step was applied after the updates of F, thus resulting in a somewhat inelegant yet efficient strategy, where all the above constraints were never satisfied within iterations, except after convergence. The informed methods in [7,8] should be seen as extensions of the well-known multiplicative-update NMF introduced by Lee and Seung [6].

Recently, many NMF approaches have been introduced to take into account the sum-to-one constraints, e.g., for remote sensing applications [9]. However, such constraints are applied to the columns of F and the source separation problem can be addressed by solving separately independent sub-problems, each one is related to a column of F, using the Sum-to-one Constraint Least Squares (SCLS) algorithm [3].

But, this strategy is not appropriate for the application under consideration in this work since the sum-to-one constraints are applied along the rows of F, thus leading to dependent sub-problems. To the best of the authors' knowledge, the only exception is the work proposed by Lantéri et al. [5], where the rows of both G and F are normalized. In this paper, an extension of the work in both [5,7] is developed. Indeed, the proposed method considers some known values of the profile matrix while satisfying the sum-to-one constraints within iterations. Therefore, it is far more elegant than [7,8] and provides a better source separation performance in the tests provided in this paper.

The remainder of the paper is structured as follows. In Sect. 2, the method in [5] is presented. In Sect. 3, the proposed informed method is introduced and its performance is investigated in Sect. 4.

2 Split Gradient NMF Methods

In [5], the authors introduced an approach with sum constraints on both G and F. More specifically, they assumed[1] the rows of F to be summed to 1 and the rows of G to be summed to the corresponding sums of the rows of X, i.e.,

$$F_{p \times m} \cdot 1_{m \times m} = 1_{p \times m}, \tag{2}$$

$$G_{n \times p} \cdot 1_{p \times p} = X_{n \times m} \cdot 1_{m \times p}, \tag{3}$$

---

[1] Actually, Lantéri et al. summed the rows of G to 1 and the rows of F to the sums of the rows of X. When expressed in the source apportionment application, it can be equivalently written as Eqs. (2) and (3).

where $1_{p\times m}$ is the $(p \times m)$ matrix of ones. Lantéri *et al.* [5] took into account the sum-to-one constraint by introducing unconstrained matrices Z and T such that

$$F = \frac{Z}{Z \cdot 1_{m\times m}} \quad \text{and} \quad G = \frac{T \circ (X \cdot 1_{m\times p})}{T \cdot 1_{p\times p}}, \tag{4}$$

where $\circ$ and $/$ denote the Hadamard product and division, respectively. Denoting $\mathcal{D}(.,.)$ a discrepancy measure, NMF with sum-to-one constraints aims to estimate

$$\arg \min_{G\succeq 0, F\succeq 0} \mathcal{D}(X, GF) \quad \text{s.t.} \quad \begin{cases} F \cdot 1_{m\times m} = 1_{p\times m}, \\ G_{n\times p} \cdot 1_{p\times p} = X_{n\times m} \cdot 1_{m\times p}. \end{cases} \tag{5}$$

Using Eq. (4), the problem (5) then reads

$$\{\hat{T}, \hat{Z}\} = \arg \min_{T\succeq 0, Z\succeq 0} \mathcal{D}\left(X, \frac{T \circ (X \cdot 1_{m\times p})}{T \cdot 1_{p\times p}} \cdot \frac{Z}{Z \cdot 1_{m\times m}}\right). \tag{6}$$

The approach proposed in [5] is relying on the differentiation of the cost function with respect to T and Z and provides update rules from a heuristic way [2] by using non-negative gradient descent formulations.

Due to the presence of noise and of potential outliers in the data X, only Eq. (2) is satisfied in the application considered in this paper and Eq. (3) is thus not taken into account below. In this paper, we only consider the squared Frobenius norm denoted $\|.\|_f^2$ but an extension to other parametric divergences is possible. Therefore, the source apportionment problem may be outlined as

$$\{\hat{G}, \hat{Z}\} = \arg \min_{G\succeq 0, Z\succeq 0} \|X - G \cdot \left(\frac{Z}{Z \cdot 1_{m\times m}}\right)\|_f^2. \tag{7}$$

Additionally, we would like to include some set components to the profile matrix according to a specific parameterization which is developed in the next section.

3   Informed Split Gradient NMF

3.1   NMF Parameterization

In many applications, the values of some components of the profile matrix may be provided by experts. Here we detail the parameterization—proposed in [7] and outlined hereafter for the sake of clarity—which takes into account this knowledge. Let Ω be a $p \times m$ binary matrix which informs the presence or the absence of constraints on each element f_{ij} of the matrix F, i.e., $\omega_{ij} = 1$ if f_{ij} is known and 0 otherwise. We then define the $p \times m$ binary matrix $\overline{\Omega}$ as $\overline{\Omega} \triangleq 1_{p\times m} - \Omega$. We denote by Φ the $p \times m$ sparse matrix of set values, i.e.,

$$\Phi \triangleq F \circ \Omega. \tag{8}$$

By construction, φ_{ij}—the (i,j)-th element of Φ—is equal to zero when $\omega_{ij} = 0$. We can easily prove that

$$\Phi \circ \Omega = \Phi, \qquad \Phi \circ \overline{\Omega} = 0. \tag{9}$$

From [7], we define ΔF as the free part of the matrix profile under the form

$$\Delta F \triangleq F - \Phi \circ \Omega. \tag{10}$$

Following the general procedure in [7]—which combines Eqs. (10), (8), and (9)—we obtain the matrix form:

$$F = \Omega \circ \Phi + \overline{\Omega} \circ \Delta F. \tag{11}$$

It should be noticed that the rows of F are summed to 1, which implies that

$$(\overline{\Omega} \circ \Delta F) \cdot 1_{m \times m} = 1_{p \times m} - \Phi \cdot 1_{m \times m}. \tag{12}$$

As a consequence—and as in [5]—we define an unconstrained matrix Z, of the same size as the profile matrix F, such that

$$F \circ \overline{\Omega} = \overline{\Omega} \circ \Delta F = \frac{\overline{\Omega} \circ Z}{(\overline{\Omega} \circ Z) \cdot 1_{m \times m}} \cdot (1_{p \times m} - \Phi \cdot 1_{m \times m}). \tag{13}$$

We thus derive from Eq. (11) our new parameterization, i.e.,

$$F = \Omega \circ \Phi + \frac{\overline{\Omega} \circ Z}{(\overline{\Omega} \circ Z) \cdot 1_{m \times m}} \cdot (1_{p \times m} - \Phi \cdot 1_{m \times m}). \tag{14}$$

As an example, let us assume F is a 1×4 profile matrix and $\Phi = [0, 0.1, 0, 0.3]$. It leads to $1_{1 \times 4} - \Phi \cdot 1_{4 \times 4} = [0.6, 0.6, 0.6, 0.6]$. As a consequence, the parameterization reads $F = \left[\dfrac{Z_{11} \cdot 0.6}{Z_{11} + Z_{13}}, 0.1, \dfrac{Z_{13} \cdot 0.6}{Z_{11} + Z_{13}}, 0.3 \right]$. The parameterization (14) is the basis for the development of new update rules.

3.2   Proposed Method

In this subsection, we focus on the update rules for F. Indeed, as G is unconstrained in this paper, we do not need to develop extended update rules. The informed problem may be formulated as a weighted Frobenius norm between X and its estimation $G \cdot F$ with additional constraints, i.e.,

$$\min \mathcal{J}(G, F) = \min_{G \succeq 0, F \succeq 0} \| X - GF \|_{f,W}^2 \text{ s.t. } F \text{ satisfies Eq. (14)}, \tag{15}$$

where W is built with expert's individual uncertainties σ_{ij}, i.e., $W_{ij} = \frac{1}{\sigma_{ij}^2}$. Expressing $\mathcal{J}$ with respect to a matrix trace, its differentiation reads

$$\frac{\partial \mathcal{J}}{\partial F} = 2G^T ((GF - X) \circ W). \tag{16}$$

At this stage, the sum-to-one constraints on F were not taken into account yet. For that purpose—and as in [5]—we consider an unconstrained matrix Z which satisfies Eq. (14) and we derive $\mathcal{J}$ with respect to its general term, which is applied to the i-th row of F:

$$\frac{\partial \mathcal{J}}{\partial Z_{ik}} = \sum_{j=1}^{m} \frac{\partial \mathcal{J}}{\partial F_{ij}} \cdot \frac{\partial F_{ij}}{\partial Z_{ik}}. \tag{17}$$

Firstly, we need to compute the derivative of the general term of the matrix profile with respect to Z_{ik}, i.e.,

$$\frac{\partial F_{ij}}{\partial Z_{ik}} = 0 + \left(1 - \sum_{l=1}^{m} \Phi_{il}\right) \frac{\partial}{\partial Z_{ik}} \left(\frac{\overline{\Omega}_{ij} Z_{ij}}{\sum_{l=1}^{m} \overline{\Omega}_{il} Z_{il}}\right). \tag{18}$$

Noticing that $\frac{\partial}{\partial Z_{ik}}(\sum_{l=1}^{m} \overline{\Omega}_{il} Z_{il}) = \overline{\Omega}_{ik}$, the fraction to derive becomes:

$$\frac{\partial}{\partial Z_{ik}} \left(\frac{\overline{\Omega}_{ij} Z_{ij}}{\sum_{l=1}^{m} \overline{\Omega}_{il} Z_{il}}\right) = \frac{\overline{\Omega}_{ik} \delta_{jk}}{\sum_{l=1}^{m} \overline{\Omega}_{il} Z_{il}} + \overline{\Omega}_{ij} Z_{ij} \frac{\partial}{\partial Z_{ik}} \left(\frac{1}{\sum_{l=1}^{m} \overline{\Omega}_{il} Z_{il}}\right), \tag{19}$$

$$= \frac{\overline{\Omega}_{ik} \delta_{jk}}{\sum_{l=1}^{m} \overline{\Omega}_{il} Z_{il}} - \frac{\overline{\Omega}_{ij} Z_{ij}}{\left(\sum_{l=1}^{m} \overline{\Omega}_{il} Z_{il}\right)^2} \overline{\Omega}_{ik}, \tag{20}$$

where δ_{jk} is the Kronecker function equal to 1 when $j = k$. Equation (18) then reads:

$$\frac{\partial F_{ij}}{\partial Z_{ik}} = \left(1 - \sum_{l=1}^{m} \Phi_{il}\right) \frac{\overline{\Omega}_{ij}}{\sum_{l=1}^{m} \overline{\Omega}_{il} Z_{il}} \left[\delta_{jk} - \frac{\overline{\Omega}_{ik} Z_{ij}}{\sum_{l=1}^{m} \overline{\Omega}_{il} Z_{il}}\right]. \tag{21}$$

Secondly, we need to differentiate $\mathcal{J}$ with respect to Z_{ik} using Eq. (17). Replacing Eq. (21) into Eq. (17) leads to

$$\frac{\partial \mathcal{J}}{\partial Z_{ik}} = \frac{1 - \sum_{l} \Phi_{il}}{\left(\sum_{l=1}^{m} Z_{il} \overline{\Omega}_{il}\right)} \left(\frac{\partial \mathcal{J}}{\partial F_{ik}} . \overline{\Omega}_{ik} - \frac{\overline{\Omega}_{ik}}{\left(\sum_{l=1}^{m} Z_{il} \overline{\Omega}_{il}\right)} \sum_{j=1}^{m} \left(\frac{\partial \mathcal{J}}{\partial F_{ij}} . Z_{ij} . \overline{\Omega}_{ij}\right)\right), \tag{22}$$

which can be transformed into a matrix form by noticing that the sums in the above equation correspond to the right multiplication by $1_{m \times m}$. Noticing that

$$(B \circ (A \cdot 1_{m \times m}))1_{m \times m} = (B \cdot 1_{m \times m}) \circ (A \cdot 1_{m \times m}) \tag{23}$$

for any matrices A and B, and that

$$\frac{\overline{\Omega} \circ Z}{(\overline{\Omega} \circ Z) \cdot 1_{m \times m}} = \frac{\overline{\Omega} \circ F}{1_{p \times m} - \Phi \cdot 1_{m \times m}}, \tag{24}$$

the matrix form of Eq. (22)—not explicited for space consideration—reads

$$\frac{\partial \mathcal{J}}{\partial Z} = \frac{1_{p \times m} - \Phi \cdot 1_{m \times m}}{(Z \circ \overline{\Omega}) \cdot 1_{m \times m}} \circ \left[\frac{\partial \mathcal{J}}{\partial F} \circ \overline{\Omega} - \frac{\overline{\Omega}}{1_{p \times m} - \Phi \cdot 1_{m \times m}} \circ \left(\left(\frac{\partial \mathcal{J}}{\partial F} \circ F \circ \overline{\Omega}\right) \cdot 1_{m \times m}\right)\right].$$
$$\tag{25}$$

Let $U \triangleq -\frac{\partial J}{\partial F}$. The opposite of Eq. (25) can be written with respect to U as

$$-\frac{\partial J}{\partial Z} = \overline{\Omega} \circ \left(U - \frac{(U \circ \overline{\Omega} \circ F) \cdot 1_{p \times m}}{1_{p \times m} - \Phi \cdot 1_{m \times m}} \right) \circ \left(\frac{1_{p \times m} - \Phi \cdot 1_{m \times m}}{(Z \circ \overline{\Omega}) \cdot 1_{m \times m}} \right). \qquad (26)$$

The third KKT condition with respect to Z (which is a necessary condition to get a stationary point) is expressed as $Z \circ \frac{\partial J}{\partial Z} = 0$. This expression may be written with respect to F by using the property (13), i.e.,

$$F \circ \overline{\Omega} \circ \frac{\partial J}{\partial Z} = 0. \qquad (27)$$

In the case of a stationary point $(F = F^k = F^{k+1})$, combining Eqs. (26) and (27) yields

$$\overline{\Omega} \circ F^k \circ U - \frac{(U \circ \overline{\Omega} \circ F^k) \cdot 1_{m \times m}}{1_{p \times m} - \Phi \cdot 1_{m \times m}} \circ F^{k+1} \circ \overline{\Omega} = 0. \qquad (28)$$

Since $(F^{k+1} \circ \overline{\Omega})$ is the free part of F^{k+1}, it turns out that the new updated rule for the profile matrix is given by

$$F^{k+1} = \Omega \circ \Phi + \overline{\Omega} \circ \frac{F^k \circ U}{[U \circ \overline{\Omega} \circ F^k] \cdot 1_{m \times m}} \circ (1_{p \times m} - \Phi \cdot 1_{m \times m}). \qquad (29)$$

Equation (29) stands for the general form of the update rules of our informed NMF method. However, we need to make sure that non-negativity is preserved. In the general case, it means that U should be positive. The authors in [5] proposed to shift U by adding $\eta \cdot 1_{p \times m}$, where η is a parameter estimated at each iteration. In this paper, we propose to add a shift equal to the sum of the negative part of $-\frac{\partial J}{\partial F}$ in Eq. (16). The shifted matrix—denoted U_s hereafter—then reads

$$U_s = G^T ((X - GF) \circ W) + \left[G^T (GF \circ W) \right] 1_{m \times m}. \qquad (30)$$

The actual update rule for our proposed Split-Gradient Method for Constrained Weighted NMF (SG-CWNMF) consists of applying U_s, defined in Eq. (30), to Eq. (29). These update rules satisfy (i) the non-negativity of F, (ii) sum-to-one constraints, and (iii) the third KKT condition. However, it cannot be shown in this paper for space consideration.

4   Experimental Validation

In this section, we investigate the enhancement of our proposed SG-CWNMF method in a simulation of source apportionment. The data matrix consists of 50 samples and 7 species, with a known uncertainty measure σ_{ij}—provided by a chemical expert—associated to each data point x_{ij}. A uniform noise ranging in $[-\min(\lambda \sigma_{ij}; x_{ij}); \lambda \sigma_{ij}]$ is added while keeping the data positive. Note also

Table 1. Positions and values of the constraints used in the informed NMF methods. X means no constraint.

	Fe	Ca	SO$_4$	Zn	Mg	Al	Cr
Source 1	700	X	X	X	X	X	0
Source 2	X	400	5	0	X	75	X
Source 3	400	X	X	0	X	X	0

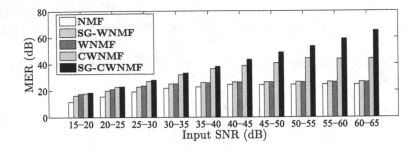

Fig. 1. Performance of the tested NMF methods with respect to the input SNR. Perf. criterion: MER (in dB).

that λ is related to an input Signal-to-Noise Ratio (SNR). The profile matrix—whose initialization is provided by chemical experts—consists of three (partially correlated) industrial profiles. The contribution matrix G is then initialized as the solution of a weighted and constrained least-square cost function [11].

In addition to our proposed SG-CWNMF method, we test blind NMF [6], WNMF [4], SG-WNMF (a weighted version of [5] with our proposed shift), and our previous informed CWNMF [7]. Nine constraints on the profile matrix are used in the tests, as shown in Table 1. The performance criterion is the Mixing Error Ratio (MER) [13], computed as the average[2] of the MERs estimated on each column of the contribution matrix G. The tested methods were stopped after $5 \cdot 10^5$ iterations. Figure 1 shows the MERs of the tested methods with respect to the input SNR. Our proposed SG-CWNMF outperforms all the other methods (around 2 dB in the noisiest scenarios). Additionally, in a noiseless case—not shown in Fig. 1—the SG-CWNMF method provides MERs at least 200 dB higher than those from the other tested methods.

5   Conclusion

In this paper, we proposed a new NMF parameterization to combine the presence of known components in the profile matrix together with the sum-to-one constraint of the rows. We derived new updates rules—consistent with the KKT conditions—which can be viewed as multiplicative update rules applied to the free part of the profile matrix. In that sense, the proposed approach is far

---

[2] Please note that in our previous work [7,8], we were computing the sums of the MERs estimated on each column of G.

more elegant than our previous method [7]—which sequentially tackles these constraints—and can be easily extended to a gradient-like technique, as in [5]. Our proposed method was shown to outperform four state-of-the-art approaches tested on simulated mixtures of industrial particulate matter sources, with various input SNR conditions. In future work, we will search for alternatives to multiplicative update rules and other kinds of constraints to inform the NMF, e.g., soft constraints [10].

Acknowledgements. This work was funded by the "ECUME" project granted by the DREAL Nord Pas de Calais Agency.

References

1. Comon, P., Jutten, C.: Handbook of Blind Source Separation, Independent Component Analysis and Applications. Academic Press, Oxford (2010)
2. Févotte, C., Idier, J.: Algorithms for nonnegative matrix factorization with the beta-divergence. Neural Comput. **23**(9), 2421–2456 (2011)
3. Heinz, D.C., Chang, C.: Fully constrained least squares linear mixture analysis for material quantification in hyperspectral imagery. IEEE Trans. Geosci. Remote Sens. **39**, 529–545 (2001)
4. Ho, N.D.: Non negative matrix factorization algorithms and applications. Ph.D. thesis, Université Catholique de Louvain (2008)
5. Lantéri, H., Theys, C., Richard, C., Févotte, C.: Split gradient method for non-negative matrix factorization. In: Proceedings of EUSIPCO (2010)
6. Lee, D., Seung, H.: Learning the parts of objects by non negative matrix factorization. Nature **401**(6755), 788–791 (1999)
7. Limem, A., Delmaire, G., Puigt, M., Roussel, G., Courcot, D.: Non-negative matrix factorization under equality constraints–a study of industrial source identification. Appl. Numer. Math. **85**, 1–15 (2014)
8. Limem, A., Puigt, M., Delmaire, G., Roussel, G., Courcot, D.: Bound constrained weighted NMF for industrial source apportionment. In: Proceedings of MLSP (2014)
9. Ma, W.K., Bioucas-Dias, J.M., Chan, T.H., Gillis, N., Gader, P., Plaza, A.J., Ambikapathi, A., Chi, C.Y.: A signal processing perspective on hyperspectral unmixing. IEEE Sig. Proc. Mag. **31**, 67–81 (2014)
10. Peng, C., Wong, K., Rockwood, A., Zhang, X., Jiang, J., Keyes, D.: Multiplicative algorithms for constrained non-negative matrix factorization. In: 12th IEEE International Conference on Data Mining, ICDM 2012, 10–13 December 2012, Brussels, Belgium, pp. 1068–1073 (2012)
11. Plouvin, M., Limem, A., Puigt, M., Delmaire, G., Roussel, G., Courcot, D.: Enhanced NMF initialization using a physical model for pollution source apportionment. In: Proceedings of ESANN, pp. 261–266 (2014)
12. Viana, M., Pandolfi, A., Minguillo, M.C., Querol, X., Alastuey, A., Monfort, E., Celades, I.: Inter-comparison of receptor models for PM source apportionment: case study in an industrial area. Atmos. Environ. **42**, 3820–3832 (2008)
13. Vincent, E., Araki, S., Bofill, P.: The 2008 signal separation evaluation campaign: A community-based approach to large-scale evaluation. In: Adali, T., Jutten, C., Romano, J.M.T., Barros, A.K. (eds.) ICA 2009. LNCS, vol. 5441, pp. 734–741. Springer, Heidelberg (2009)

Audio Applications

The 2015 Signal Separation Evaluation Campaign

Nobutaka Ono[1]([✉]), Zafar Rafii[2], Daichi Kitamura[3], Nobutaka Ito[4],
and Antoine Liutkus[5]

[1] National Institute of Informatics, Tokyo, Japan
onono@nii.ac.jp
[2] Media Technology Lab, Gracenote, Emeryville, USA
[3] SOKENDAI (The Graduate University for Advanced Studies), Hayama, Japan
[4] NTT Communication Science Laboratories, NTT Corporation, Kyoto, Japan
[5] INRIA, Villers-lès-Nancy, France

Abstract. In this paper, we report the 2015 community-based Signal Separation Evaluation Campaign (SiSEC 2015). This SiSEC consists of four speech and music datasets including two new datasets: "Professionally produced music recordings" and "Asynchronous recordings of speech mixtures". Focusing on them, we overview the campaign specifications such as the tasks, datasets and evaluation criteria. We also summarize the performance of the submitted systems.

1 Introduction

Sharing datasets and evaluating methods with common tasks and criteria has recently become a general and popular methodology to accelerate the development of new technologies. Aiming to evaluate signal separation methods, the Signal Separation Evaluation Campaign (SiSEC) has been held about every one-and-half year in conjunction with the LVA/ICA conference since 2008. The tasks, datasets, and evaluation criteria in the past SiSECs are still available online with the results of the participants. They have been referred to and utilized for comparison and further evaluation by researchers in the source separation community, not limited to the past participants, as shown in Fig. 1.

In this fifth SiSEC, two new datasets were added: A new music dataset for a large-scale evaluation was provided in "Professionally produced music recordings" and another new dataset including real recording was provided in "Asynchronous recordings of speech mixtures". For further details, the readers are referred to the web page of SiSEC 2015 at https://sisec.inria.fr/. In Sect. 2, we specify the tasks, datasets and evaluation criteria, with a particular focus on these new datasets. Section 3 summarizes the evaluation results.

2 Specifications

SiSEC 2015 focused on the following source separation tasks and datasets.

© Springer International Publishing Switzerland 2015
E. Vincent et al. (Eds.): LVA/ICA 2015, LNCS 9237, pp. 387–395, 2015.
DOI: 10.1007/978-3-319-22482-4_45

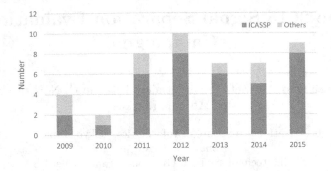

Fig. 1. The number of papers referring SiSEC datasets found by full-text-search on all ICASSP proceedings (ICASSP) and by abstract-search on IEEE Xplore (Others).

T1 Single-channel source estimation
T2 Multichannel source image estimation

D1 Underdetermined speech and music mixtures
D2 Two-channel mixtures of speech and real-world background noise
D3 Professionally produced music recordings
D4 Asynchronous recordings of speech mixtures

T1 aims to estimate single-channel source signals observed by a specific reference microphone, whereas T2 aims to estimate multichannel source images observed by the microphone array.

In D1 and D2, we utilized the same datasets as in SiSEC 2013, which permits easy comparison. Their specifications are given in details in [1].

The new D3 dataset, the Mixing Secret Dataset 100 (MSD100) is designed to evaluate the separation of multiple sources from professionally-produced music recordings. MSD100 consists of 100 full-track songs of different styles, and includes both the stereophonic mixtures and the original stereo sources images. The data is divided into a development set and a test set, each consisting of 50 songs, so that algorithms which need supervised learning can be trained on the development set and tested on the test set. The duration of the songs ranges from 2 min and 22 s to 7 min and 20 s, with an average duration of 4 min and 10 s.

For each song, MSD100 includes 4 stereo sources corresponding to the bass, the drums, the vocals and "other" (i.e., the other instruments). The sources were created using stems from selected raw multitrack projects downloaded from the 'Mixing Secrets' Free Multitrack Download Library[1]. Stems corresponding to a given source were summed together and the result was normalized, then scaled so that the mixture would also be normalized. The mixtures were then generated by summing the sources together. For a given song, the mixture and the sources have the same duration; however, while the mixture is always stereo, some sources can be mono (typically, the vocals). In that case, it appears identical in the left and right channels of the mixture. All items are WAV files sampled at 44.1 kHz.

The D4 dataset aims to evaluate the separation of mixtures recorded with asynchronous devices. A new dataset added to D4 contains real recordings

---

[1] www.cambridge-mt.com/ms-mtk.htm.

of three or four speakers using four different stereo IC recorders (8 channels in total). A standard way to make datasets for BSS evaluation is to record each source image first, which is used as the ground truth, and then to make a mixture by summing them up. Unlike conventional synchronized recording, it is not easy in an asynchronous setting because the time offset (time of recording start) of each device is unknown and because there is a sampling frequency mismatch between channels. To obtain consistent source images and real mixtures, a chirp signal was played back from a loudspeaker for time-marking, and the time offsets at the different devices were aligned precisely at a sub-sample level. It is assumed that the sampling frequency of each device is invariant over the whole recording. This dataset consists of three types of mixing: *realmix*, *sumrefs* and *mix*. The realmix is a recording of the real mixture, the sumrefs is the summation of the source images, and the mix is the simulated mixture generated by convolving impulse responses with the dry source and applying resampling for the artificial sampling frequency mismatch.

The BSS Eval toolbox [2] was used to evaluate the following four power-based criteria: the signal to distortion ratio (SDR), the source image to spatial distortion ratio (ISR), the signal to interference ratio (SIR), and signal to artifacts ratio (SAR). The version 2.0 of the PEASS toolbox [3] was used to evaluate the following four perceptually-motivated criteria: the overall perceptual score (OPS), the target-related perceptual score (TPS), the interference-related perceptual score (IPS), and the artifact-related perceptual score (APS). More specifically, T1 was evaluated by `bss_eval_source_denoising.m` for D2 or `bss_eval_source.m` for others. T2 on D3 and D4 was evaluated with `bss_eval_image.m`. For D1 and D2, the PEASS toolbox was used for the comparison with previous SiSEC.

3   Results

We evaluated 27 algorithms in total: 3, 2, 19, and 3 algorithms for D1, D2, D3 and D4, respectively. The average performance of the systems is summarized in Tables 1, 2, 3, and Figs. 2 and 3. Because of the space limitation, only part of the results is shown.

Three algorithms were submitted to D1 as shown in Table 1. Sgouros's method [4] for instantaneous mixtures is based on direction of arrival (DOA) estimation by fitting a mixture of directional Laplacian distributions. The other two algorithms are for convolutive mixtures. Bouafif's method [5] exploits a detection of glottal closure instants in order to estimate the number of speakers and their time delays of arrival (TDOA). It also aims at separation with less artifacts and distortion. Indeed, it shows higher SARs and APSs. However, the SIRs and IPSs are lower. This fact illustrates the well known trade-off between SIR and SAR in BSS. Nguyen's method is similar to [6] and the permutation problem is solved by multi-band alignment [25]. Overall, the performance is almost equivalent to the past SiSEC, which indicates that underdetermined BSS for convolutive mixtures is still a tough problem.

Two algorithms were submitted to D2 as shown in Table 2. López's method [7] designs the demixing matrix and the post-filters based on a single-channel

Table 1. Results for the D1 dataset: (a) The performance of T1 for the instantaneous mixtures averaged over datasets "test" and "test2" in 2 mics and the over dataset "test3" in 3 mics. (b) The performance of T2 for the convolutive mixtures averaged over "test" dataset in 2 mics and over "test3" dataset in 3 mics. SP and MU represents speech and music data, respectively.

(a)

System	2mic/3src (SP) SDR SIR SAR	2mic/3src (MU) SDR SIR SAR	2mic/4src (SP) SDR SIR SAR	3mic/4src (SP) SDR SIR SAR
Sgouros [4]	7.6 18.8 8.6	8.3 18.4 9.4	5.6 15.6 6.5	6.6 19.1 7.0

(b)

System	2mic/3src (SP) SDR ISR SIR SAR OPS TPS IPS APS	2mic/4src (SP) SDR ISR SIR SAR OPS TPS IPS APS	3mic/4src (SP) SDR ISR SIR SAR OPS TPS IPS APS
Bouafif [5]	-4.3 1.4 -1.9 8.6 8.4 67.0 1.4 85.1	-5.7 1.6 -3.6 8.2 8.4 55.1 1.0 83.3	– – – – – – – –
Nguyen	7.0 11.6 11.6 9.2 40.9 65.3 55.9 58.0	4.5 8.3 8.0 6.4 36.9 62.2 51.0 48.7	4.3 7.2 6.6 8.0 35.6 62.2 53.3 47.0

Table 2. Results for the D2 dataset (only for task T1)

Systems	Criteria	dev			test					
		Ca1	Sq1	Su1	Ca1	Ca2	Sq1	Sq2	Su1	Su2
López [7]	SDR	-	-	-	4.0	4.5	5.1	11.0	−3.8	3.9
	SIR	-	-	-	14.9	16.1	9.6	16.3	−1.6	8.8
	SAR	-	-	-	4.7	5.0	8.6	13.0	4.3	6.3
Ito [8]	SDR	7.2	8.9	4.9	8.1	7.8	10.8	13.8	6.7	7.6
	SIR	25.9	23.7	15.3	25.7	27.7	26.8	28.6	21.0	27.9
	SAR	7.2	9.2	5.6	8.2	7.8	11.0	14.0	6.9	7.7

source separation method. In this submission, they used spectral subtraction as the single-channel source separation method. Note that the performance may vary depending on the choice of the single-channel method. Ito's method is based on full-band clustering of the time-frequency components [8]. Thanks to a frequency-independent time-varying source presence model, the method robustly solves the permutation problem and shows good denoising performance even though it does not explicitly include spectral modeling of speech and noise.

Similarly to the previous SiSEC, D3 attracted most participants. The evaluated methods includes 5 methods available online (not submitted by participants) and are as follows.

– CHA: system using a two-stage Robust Principal Component Analysis (RPCA)[2], with an automatic vocal activity detector and a melody detector [9].

---

[2] http://mac.citi.sinica.edu.tw/ikala/.

Table 3. Results of T2 for the D4 dataset

Systems	Criteria	3src			4src		
		realmix	sumrefs	mix	realmix	sumrefs	mix
Wang [25]	SDR	4.4	4.4	4.6	3.0	3.0	2.5
	ISR	4.8	4.9	5.2	3.5	3.6	3.3
	SIR	20.8	20.7	18.6	18.0	17.9	16.8
	SAR	12.8	12.9	13.9	11.0	11.2	10.9
Miyabe [26]	SDR	6.9	6.8	10.6	4.0	3.8	3.3
	ISR	11.2	11.1	15.1	8.8	8.5	7.3
	SIR	11.0	10.9	14.9	6.7	6.4	6.0
	SAR	11.7	11.6	15.5	7.8	7.6	7.4
Murase	SDR	2.7	2.6	2.4	0.9	0.8	1.0
	ISR	7.0	6.8	7.0	5.2	5.1	5.3
	SIR	5.2	4.6	4.2	1.7	1.6	2.3
	SAR	5.0	5.3	5.5	4.2	4.2	3.6

– DUR1, DUR2: systems using a source-filter model for the voice and a Non-negative Matrix Factorization (NMF) model for the accompaniment[3], without (DUR1) and with (DUR2) unvoiced vocals model [10].
– HUA1, HUA2: systems using RPCA[4], with binary (HUA1) and soft (HUA2) masking [11].
– KAM1, KAM2, KAM3: systems using Kernel Additive Modelling (KAM), with light kernel additive modelling (KAM1)[5], a variant with only one iteration (KAM2), and a variant where the energy of the vocals is adjusted at each iteration (KAM3) [12,13].
– NUG1, NUG2, NUG3: systems using spatial covariance models and Deep Neural Networks (DNN) for the spectrograms, with one set of four DNNs for the four sources for all the iterations (NUG1), one set for the first iteration and another set for the subsequent iterations (NUG2), and one DNN for all the sources (NUG3) [14].
– OZE: system using the Flexible Audio Source Separation Toolbox (FASST) (version 1)[6] [15,16].
– RAF1, RAF2, RAF3: systems using the REpeating Pattern Extraction Technique (REPET)[7], with the original REPET with segmentation (RAF1) [17–20], the adaptive REPET (RAF2) [18,20], and REPET-SIM (RAF3) [19,20].

---

[3] http://www.durrieu.ch/research/jstsp2010.html.
[4] https://sites.google.com/site/singingvoiceseparationrpca/.
[5] http://www.loria.fr/~aliutkus/kaml/.
[6] http://bass-db.gforge.inria.fr/fasst/.
[7] http://zafarrafii.com/repet.html.

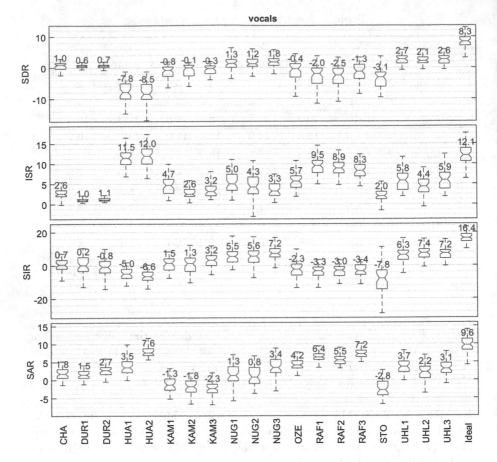

Fig. 2. Results of T2 for the D3 dataset (vocals).

- STO: system using a predominant pitch extraction and an efficient comb filtering[8] [21,22].
- UHL1, UHL2, UHL3: systems using DNN, with an independent training material, with four DNNs for the four sources (UHL1) [23], then augmented with an extended training material (UHL2) [23], then using a phase-sensitive cost function (UHL3) [23,24].
- Ideal: system using the ideal soft masks computed from the mixtures and the sources.

Figures 2 and 3 show the box plots for the SDR, ISR, SIR, and SAR (in dB), for the vocals and the accompaniment, respectively, for the test subset. Outliers are not shown, median values are displayed, and higher values are better. As we can see, the separation performance is overall better for the accompaniment, as many songs feature weak vocals. Also, supervised systems typically

---

[8] http://www.audiolabs-erlangen.de/resources/2014-DAFx-Unison/.

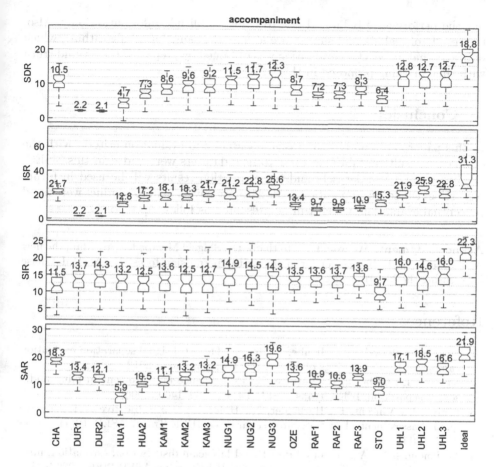

Fig. 3. Results of T2 for the D3 dataset (accompaniment).

achieved better results compared to unsupervised systems. Finally, depending on the systems, more or less large statistical dispersions are observed, meaning that different methods lead to different performances, depending on the songs, hence the need for a large-scale evaluation for music source separation.

Three methods were submitted to D4. Wang's method consists of an exhaustive search for estimating the sampling frequency mismatch and a state-of-the-art source separation technique [25]. Their results show the highest SIR but ISR is not so high. Miyabe's method consists of the maximum likelihood estimation of the sampling frequency mismatch [26] followed by auxiliary function based independent vector analysis [27]. Their results show the highest ISR. So, this combination would be interesting. Murase's system does not include the compensation of sampling frequency mismatch. It directly designs the time-frequency mask based on non-negative matrix factorization in the time-channel domain with sparse penalty added to [28]. It is robust to the sampling frequency mismatch, but the performance is limited due to using amplitude information only. Also, the results of realmix and simrefs are almost the same for all algorithms, which indicates that an effective evaluation was obtained by preparing the ground truth with time marking proposed in this task.

4 Conclusion

In this paper, we reported the tasks, datasets and evaluation criteria with the evaluation results in SiSEC 2015. Two new datasets were added in this SiSEC. We hope that these datasets and the evaluation results will be used in future research of the source separation field. Also, we have a plan to conduct web-based perceptual evaluation, which will be presented as follow-up report.

Acknowledgment. We would like to thank Dr. Shigeki Miyabe for providing the new ASY dataset, and Mike Senior for giving us the permission to use the MSD database for creating the MSD100 corpus.

References

1. Ono, N., Koldovsky, Z., Miyabe, S., Ito, N.: The 2013 signal separation evaluation campaign. In: Proceedings of MLSP, pp. 1–6, September 2013
2. Vincent, E., Griboval, R., Févotte, C.: Performance measurement in blind audio source separation. IEEE Trans. ASLP **14**(4), 1462–1469 (2006)
3. Emiya, V., Vincent, E., Harlander, N., Hohmann, V.: Subjective and objective quality assessment of audio source separation. IEEE Trans. ASLP **19**(7), 2046–2057 (2011)
4. Mitianoudis, N.: A generalised directional laplacian distribution: estimation, mixture models and audio source separation. IEEE Trans. ASLP **20**(9), 2397–2408 (2012)
5. Bouafif, M., Lachiri, Z.: Multi-sources separation for sound source localization. In: Proceedings of Interspeech, pp. 14–18, September 2014
6. Sawada, H., Araki, S., Makino, S.: Underdetermined convolutive blind source separation via frequency bin-wise clustering and permutation alignment. IEEE Trans. ASLP **19**(3), 516–527 (2011)
7. López, A.R., Ono, N., Remes, U., Palomäki, K., Kurimo, M.: Designing multichannel source separation based on single-channel source separation. In: Proceedings of ICASSP, pp. 469–473, April 2015
8. Ito, N., Araki, S., Nakatani, T.: Permutation-free convolutive blind source separation via full-band clustering based on frequency-independent source presence priors. In: Proceedings of ICASSP, pp. 3238–3242, May 2013
9. Chan, T.-S., Yeh, T.-C., Fan, Z.-C., Chen, H.-W., Su, L., Yang, Y.-H., Jang, R.: Vocal activity informed singing voice separation with the iKala dataset. In: Proceedings of ICASSP, pp. 718–722, April 2015
10. Durrieu, J.-L., David, B., Richard, G.: A musically motivated mid-level representation for pitch estimation and musical audio source separation. IEEE J. Sel. Top. Sign. Process. **5**(6), 1180–1191 (2011)

11. Huang, P.-S., Chen, S.D., Smaragdis, P., Hasegawa-Johnson, M.: Singing-voice separation from monaural recordings using robust principal component analysis. In: Proceedings of ICASSP, pp. 57–60, March 2012
12. Liutkus, A., FitzGerald, D., Rafii, Z., Pardo, B., Daudet, L.: Kernel additive models for source separation. IEEE Trans. SP **62**(16), 4298–4310 (2014)
13. Liutkus, A., FitzGerald, D., Rafii, Z., Daudet, L.: Scalable audio separation with light kernel additive modelling. In: Proceedings of ICASSP, pp. 76–80, April 2015
14. Nugraha, A.A., Liutkus, A., Vincent, E.: Multichannel audio source separation with deep neural networks. Research report RR-8740, Inria (2015)
15. Ozerov, A., Vincent, E., Bimbot, F.: A general flexible framework for the handling of prior information in audio source separation. IEEE Trans. ASLP **20**(4), 1118–1133 (2012)
16. Salaün, Y., Vincent, E., Bertin, N., Souviraà-Labastie, N., Jaureguiberry, X., Tran, D.T., Bimbot, F.: The flexible audio source separation toolbox version 2.0. In: Proceedings of ICASSP, 4–9 May 2014
17. Rafii, Z., Pardo, B.: REpeating pattern extraction technique (REPET): a simple method for music/voice separation. IEEE Trans. ASLP **21**(1), 71–82 (2013)
18. Liutkus, A., Rafii, Z., Badeau, R., Pardo, B., Richard, G.: Adaptive filtering for music/voice separation exploiting the repeating musical structure. In: Proceedings of ICASSP, pp. 53–56, March 2012
19. Rafii, Z., Pardo, B.: Music/voice separation using the similarity matrix. In: Proceedings of ISMIR, pp. 583–588, October 2012
20. Rafii, Z., Liutkus, A., Pardo, B.: REPET for background/foreground separation in audio. In: Naik, G.R., Wang, W. (eds.) Blind Source Separation, Chap. 14. Signals and Communication Technology, pp. 395–411. Springer, Heidelberg (2014)
21. Salamon, J., Gómez, E.: Melody extraction from polyphonic music signals using pitch contour characteristics. IEEE Trans. ASLP **20**(6), 1759–1770 (2012)
22. Stöter, F.-R., Bayer, S., Edler, B.: Unison Source Separation. In: Proceedings of DAFx, September 2014
23. Uhlich, S., Giron, F., Mitsufuji, Y.: Deep neural network based instrument extraction from music. In: Proceedings of ICASSP, pp. 2135–2139, April 2015
24. Erdogan, H., Hershey, J.R., Watanabe, S., Le Roux, J.: Phase-sensitive and recognition-boosted speech separation using deep recurrent neural networks. In: Proceedings of ICASSP, pp. 708–712, April 2015
25. Wang, L.: Multi-band multi-centroid clustering based permutation alignment for frequency-domain blind speech separation. Digit. Signal Process. **31**, 79–92 (2014)
26. Miyabe, S., Ono, N., Makino, S.: Blind compensation of interchannel sampling frequency mismatch for ad hoc microphone array based on maximum likelihood estimation. Elsevier Signal Process. **107**, 185–196 (2015)
27. Ono, N.: Stable and fast update rules for independent vector analysis based on auxiliary function technique. In: Proceedings of WASPAA, pp. 189–192, October 2011
28. Chiba, H., Ono, N., Miyabe, S., Takahashi, Y., Yamada, T., Makino, S.: Amplitude-based speech enhancement with nonnegative matrix factorization for asynchronous distributed recording. In: Proceedings of IWAENC, pp. 204–208, September 2014

A Geometrically Constrained Independent Vector Analysis Algorithm for Online Source Extraction

Affan H. Khan[✉], Maja Taseska, and Emanuël A.P. Habets

International Audio Laboratories Erlangen, Erlangen, Germany
affanhasan@gmail.com

Abstract. In this paper, an online constrained independent vector analysis (IVA) algorithm that extracts the desired speech signal given the direction of arrival (DOA) of the desired source and the array geometry is proposed. The far-field array steering vector calculated using the DOA of the desired source is used to add a penalty term to the standard cost function of IVA. The penalty term ensures that the speech signal originating from the given DOA is extracted with small distortion. In contrast to unconstrained IVA, the proposed algorithm can be used to extract the desired speech signal online when the number of interferers is unknown or time varying. The applicability of the algorithm in various scenarios is demonstrated using simulations.

1 Introduction

Many modern communication systems require a high-quality handsfree capture of speech using one or more microphones. The signal received at the microphones is usually a mixture of desired and undesired source signals. One common approach to extract the desired source signal from the received mixture is through the use of beamforming algorithms. Beamformers can either be fixed, which require the knowledge of the DOA of the desired source, or data-dependent, which require an accurate estimate of the second-order statistics (SOS) of the desired and the undesired signals. The estimation of the SOS is a challenging task that can be accomplished, for example, by detecting the activity of the desired sound sources [1,14].

Independent component analysis (ICA) provides an alternative approach to source extraction in which sound sources are separated based on the assumption that their signals are mutually statistically independent. Some common methods to obtain independent components include maximization of non-Gaussianity of the separated signals [10,11], minimization of mutual information, [6] and maximum likelihood-based signal estimation [4]. ICA algorithms however suffer from scaling and permutation ambiguities and non-uniqueness of solution in underdetermined scenarios [6].

A joint institution of the University of Erlangen-Nuremberg and Fraunhofer IIS.

© Springer International Publishing Switzerland 2015
E. Vincent et al. (Eds.): LVA/ICA 2015, LNCS 9237, pp. 396–403, 2015.
DOI: 10.1007/978-3-319-22482-4_46

To improve the quality of the extracted signal and to mitigate the inherent problems in ICA, various researchers have proposed to incorporate prior information by adding constraints to the optimization problem of ICA. DOA-based constraints were first introduced in ICA in [13]. The authors performed source separation through joint diagonalization of SOS [12] with a soft DOA-based constraint on each separation filter. Similarly in [9], a hard constraint on one of the separation filters was applied to ensure an undistorted response in the direction of the desired source. Both of these are batch algorithms and require prior knowledge of the number of sources. In contrast, the authors in [15] used a soft constraint to extract the desired source without prior knowledge of the number of sources. The use of DOA-based prior information mitigates the permutation ambiguity in ICA as each separation filter is constrained to extract an independent component originating from a given direction. However, if the sources are close to each other, permutation ambiguity might occur nonetheless.

To solve the permutation ambiguity, IVA was proposed in [8] as a generalization of ICA. In IVA, statistical dependence between the output signals is jointly minimized across all frequency bins such that permutation ambiguity does not occur. An online variant of IVA was proposed in [7]. In this paper, we propose an online geometrically constrained IVA (CIVA) algorithm that works in the frequency domain to extract the desired source whose DOA is known. We augment the standard cost function of IVA with a penalty term that restricts the Euclidean angle between one of the separation filters and the far-field steering vector calculated using the desired source DOA. This ensures that the desired speech signal is always delivered at the output of the corresponding separation filter with small distortion and without the knowledge of number of interferers. In contrast, the unconstrained IVA algorithm introduces higher distortion of the desired speech signal in non-determined and reverberant scenarios.

2   Problem Formulation

We consider a scenario where a sound field composed of L speech signals and background noise is captured by M microphones. The L speech signals are assumed to be mutually statistically independent. The signal received at the m-th microphone can be described in the short-time Fourier transform (STFT) domain with sufficiently long time-frames as follows

$$Y_m(n, k) = \sum_{l=1}^{L} A_{m,l}(k)\, S_l(n, k) + V_m(n, k), \tag{1}$$

where n and k represent the time and frequency indices, $S_l(n, k)$ denotes the signal of the l-th source, $V_m(n, k)$ denotes the background noise component, and $A_{m,l}(k)$ denotes the acoustic transfer function (ATF) between l-th source and m-th microphone. The M microphone signals can be expressed in vector notation as follows

$$\mathbf{y}(n, k) = \mathbf{A}(k)\mathbf{s}(n, k) + \mathbf{v}(n, k), \tag{2}$$

where $\mathbf{s}(n,k) = [S_1(n,k) \cdots S_L(n,k)]^{\mathrm{T}}$, $\mathbf{y}(n,k) = [Y_1(n,k) \cdots Y_M(n,k)]^{\mathrm{T}}$, and $\mathbf{A}(k) = [\mathbf{a}_1(k) \cdots \mathbf{a}_L(k)]$ with $\mathbf{a}_l(k) = [A_{1,l}(k) \cdots A_{M,l}(k)]^T$. In this paper, we consider the problem where only one of the L sources is desired. Without loss of generality, we assume source 1 to be desired and rewrite (2) as follows

$$\mathbf{y}(n,k) = \mathbf{a}_1(k)\, S_1(n,k) + \sum_{u=2}^{L} \mathbf{a}_u(k)\, S_u(n,k) + \mathbf{v}(n,k). \tag{3}$$

The extraction of source signals from the received mixture in a standard blind source separation (BSS) algorithm is achieved by a demixing matrix $\mathbf{W}(k)$ as follows

$$\hat{\mathbf{s}}(n,k) = \mathbf{W}(k)\, \mathbf{y}(n,k), \tag{4}$$

where $\hat{\mathbf{s}}(n,k)$ is a vector of estimated sources at the output of the BSS algorithm, and each row of the demixing matrix represents a filter. For $L = M$, the demixing matrix $\mathbf{W}(k)$ can then be written out as follows

$$\mathbf{W}(k) = [\mathbf{w}_1(k)\ \mathbf{w}_2(k)\ \mathbf{w}_3(k)\ \cdots\ \mathbf{w}_M(k)]^{\mathrm{H}}. \tag{5}$$

The aim in this paper is as follows: given the DOA of the desired source, compute a demixing matrix $\mathbf{W}(k)$, by minimizing the statistical dependence among the output signals $\hat{\mathbf{s}}(n,k)$, while ensuring that $\mathbf{w}_1(k)$ extracts the desired source signal.

3   Geometrically Constrained Independent Vector Analysis

The optimization criterion employed in this paper to estimate the demixing matrix is based on minimization of mutual information. In Sect. 3.1, we review the concept of IVA proposed in [8]. In Sect. 3.2, we present the proposed geometrically constrained IVA algorithm for online extraction of the desired source. For simplicity of derivation, we assume $M = L$ and use the index m for sources and microphones. In the performance evaluation, we demonstrate the applicability of the proposed algorithm to scenarios where $M \neq L$ as well.

3.1   Unconstrained IVA

The derivation of unconstrained IVA in this section follows from [8]. In standard IVA, the source signals are modelled as multivariate random variables $\mathbf{S}_m(n) = [S_m(n,1) \cdots S_m(n,K)]^{\mathrm{T}}$, where K denotes the total number of frequency bins. The cost function of IVA based on mutual information between the multivariate random variables $\hat{\mathbf{S}}_m(n)$ is then given by

$$J_{\mathrm{iva}}(\mathbf{W}) = -\sum_{m=1}^{M} \mathrm{E}\left\{\log p\left[\hat{\mathbf{S}}_m(n)\right]\right\} - \sum_{k=1}^{K} \log|\det[\mathbf{W}(k)]|. \tag{6}$$

Gradient-based iterative algorithms are used to find a demixing matrix $\mathbf{W}(k)$ that minimizes (6). The iterative update for $\mathbf{W}(k)$ is given by

$$\mathbf{W}_b(k) = \mathbf{W}_{b-1}(k) - \eta \frac{\partial J_{\text{iva}}}{\partial \mathbf{W}_{b-1}(k)} = \mathbf{W}_{b-1}(k) - \eta \nabla \mathbf{W}_b(k), \tag{7}$$

where η ($\eta \geq 0$) is the learning rate of the algorithm and b is the iteration index. The probability density function (PDF) of the output signals $\hat{\mathbf{S}}_m(n)$, required in (6) to compute the gradient, can be estimated using the data or modeled based on prior knowledge [6]. Since speech signals are known to have a supergaussian PDF, they are modelled in IVA using a multivariate Laplacian distribution as follows

$$p\left[\mathbf{S}_m(n)\right] = p\left[S_m(n,1), \cdots, S_m(n,K)\right] = \alpha \exp\left(-\sqrt{\sum_{k=1}^{K} |S_m(n,k)|^2}\right). \tag{8}$$

For the m-th output signal, score functions are then calculated as the following partial derivatives

$$\varphi^{(k)}\left[\hat{\mathbf{S}}_m(n)\right] = \frac{\partial \log p\left[\hat{S}_m(n,1) \cdots \hat{S}_m(n,K)\right]}{\partial \hat{S}_m(n,k)} = \frac{\hat{S}_m(n,k)}{\sqrt{\sum_{k=1}^{K} |\hat{S}_m(n,k)|^2}}. \tag{9}$$

Using (9), the gradient $\nabla \mathbf{W}_{\text{iva}}(k)$ of the cost function in (6) is given by

$$\nabla \mathbf{W}_{\text{iva}}(k) = \frac{\partial J_{\text{iva}}}{\partial \mathbf{W}(k)} = \mathrm{E}\left\{\varphi^{(k)}(n)\mathbf{y}^H(n,k)\right\} - \mathbf{W}^{-H}(k), \tag{10}$$

where

$$\varphi^{(k)}(n) = \left[\varphi^{(k)}\left[\hat{\mathbf{S}}_1(n)\right] \quad \varphi^{(k)}\left[\hat{\mathbf{S}}_2(n)\right] \cdots \varphi^{(k)}\left[\hat{\mathbf{S}}_M(n)\right]\right]^{\mathrm{T}}. \tag{11}$$

3.2   Geometrically Constrained IVA

Due to the inherent scaling ambiguity, the estimated signals in a BSS system must be normalized to a reference microphone. We can therefore replace the desired source ATF vector $\mathbf{a}_1(k)$ in (3) by the relative transfer function (RTF) vector of the desired source with respect to a reference microphone. Assuming far field propagation, the RTF with respect to the first microphone can be written as

$$\mathbf{g}_1(k) = \left[1 \ e^{j(2\pi f/c)[\mathbf{r}_2 - \mathbf{r}_1]^{\mathrm{T}}\mathbf{q}_1} \ \cdots \ e^{j(2\pi f/c)[\mathbf{r}_M - \mathbf{r}_1]^{\mathrm{T}}\mathbf{q}_1}\right]^{\mathrm{T}}, \tag{12}$$

where $\mathbf{r}_m$ is the location of the m-th microphone, $\mathbf{q}_1$ represents a unit-norm vector pointing in the direction of the desired source, c is the speed of sound and $f = k \, F_s(2\,K)^{-1}$ is the frequency in Hertz with F_s being the sampling frequency.

We define a penalty function to restrict the Euclidean angle between $\mathbf{w}_1(k)$ and $\mathbf{g}_1(k)$. The Euclidean angle between $\mathbf{w}_1(k)$ and $\mathbf{g}_1(k)$ is defined as

$$\cos \Theta(k) = \frac{\mathrm{Re}\left\{\mathbf{w}_1^{\mathrm{H}}(k)\,\mathbf{g}_1(k)\right\}}{||\mathbf{w}_1(k)||\,||\mathbf{g}_1(k)||}. \tag{13}$$

The proposed penalty function that steers the filter of interest $\mathbf{w}_1$ in the direction of the desired source is then given by

$$J_{\mathrm{p}}(\mathbf{w}_1) = \sum_{k=1}^{K} \left[\cos \Theta(k) - 1\right]^2. \tag{14}$$

The cost function for the geometrically constrained IVA algorithm is then obtained by augmenting the IVA cost function in (6) by J_{p}, i.e.,

$$J_{\mathrm{civa}}(\mathbf{W}) = J_{\mathrm{iva}}(\mathbf{W}) + \lambda\, J_{\mathrm{p}}(\mathbf{w}_1), \tag{15}$$

where λ ($\lambda \geq 0$) is the penalty parameter. The gradient of J_{civa} with respect to the elements of the demixing matrix can be expressed as

$$\nabla \mathbf{W}_{\mathrm{civa}}(k) = \frac{\partial J_{\mathrm{civa}}}{\partial \mathbf{W}(k)} = \nabla \mathbf{W}_{\mathrm{iva}}(k) + \lambda \nabla \mathbf{W}_{\mathrm{p}}(k), \tag{16}$$

where $\nabla \mathbf{W}_{\mathrm{iva}}(k)$ is the gradient of J_{iva} given in (10) and $\nabla \mathbf{W}_{\mathrm{p}}(k)$ is the gradient of J_{p}. Since the penalty function J_{p} is only a function of $\mathbf{w}_1(k)$, the gradient of J_{p} with respect to filters $\mathbf{w}_u(k)$ ($u = 2, 3 \ldots M$) is zero, i.e.,

$$\nabla \mathbf{W}_{\mathrm{p}}(k) = \frac{\partial J_{\mathrm{p}}}{\partial \mathbf{W}(k)} = \begin{bmatrix} \nabla \mathbf{w}_1^{\mathrm{H}}(k) \\ \mathbf{0}_{M-1 \times M} \end{bmatrix}, \tag{17}$$

where the gradient of the proposed penalty function with respect to $\mathbf{w}_1^*(k)$ is derived based on the theorems in [3]. It is given by

$$\nabla \mathbf{w}_1(k) = C \cdot \left[(\cos \Theta(k) - 1) \left(\mathbf{g}_1(k) - \frac{\mathbf{w}_1(k)}{||\mathbf{w}_1(k)||^2} \, \mathrm{Re}\left\{\mathbf{w}_1^{\mathrm{H}}(k)\mathbf{g}_1(k)\right\} \right) \right], \tag{18}$$

where $C = 1/\left(||\mathbf{w}_1(k)|| \cdot ||\mathbf{g}_1(k)||^2\right)$. Using (10) and (17), the gradient matrix for geometrically constrained IVA algorithm is given by

$$\nabla \mathbf{W}_{\mathrm{civa}}(k) = \underbrace{\mathrm{E}\{\varphi^{(k)}(n)\mathbf{y}^{\mathrm{H}}(n,k)\} - \mathbf{W}^{-\mathrm{H}}(k)}_{\nabla \mathbf{W}_{\mathrm{iva}}(k)} + \lambda \underbrace{\begin{bmatrix} \nabla \mathbf{w}_1^{\mathrm{H}}(k) \\ \mathbf{0}_{M-1 \times M} \end{bmatrix}}_{\nabla \mathbf{W}_{\mathrm{p}}(k)}. \tag{19}$$

Similar to the online IVA algorithm derived in [7], we obtain an online variant of the CIVA algorithm by omitting the expectation operator in (19). To avoid divergence of the algorithm due to source signal fluctuations, we normalize the gradient matrix at each frame by its Frobenius norm $||\cdot||_{\mathrm{F}}$ and update as follows

$$\mathbf{W}_n(k) = \mathbf{W}_{n-1}(k) - \eta\, \frac{\nabla \mathbf{W}_{n,\mathrm{civa}}(k)}{||\nabla \mathbf{W}_{n,\mathrm{civa}}(k)||_{\mathrm{F}}}. \tag{20}$$

Finally, scaling ambiguity is mitigated by multiplying $\mathbf{W}_n^{-1}(k) \odot \mathbf{I}$ at each frame, where $\odot$ denotes the element-wise product.

4   Performance Evaluation

The quality of the desired speech signal at the output of the proposed algorithm was evaluated using simulated audio data. To obtain the microphone signals, clean speech signals sampled at 16 kHz were convolved with simulated room impulse responses. Room impulse responses were generated using [5]. A circular microphone array with a diameter of 2.5 cm was employed. The STFT frame size was 1024 samples with 50 % overlap. In all experiments, a diffuse noise with 30 dB signal-to-noise (SNR) ratio and a sensor noise with 40 dB SNR was added to the microphone signals. The segmental signal-to-interference (segSIR) ratio and segmental speech distortion index (segSD) [2] were used to measure the performance and the desired source signal at the reference microphone was used as ground truth.

The performance of the proposed online CIVA algorithm was compared to the online IVA algorithm proposed in [7]. The learning rate η was set to 150 and λ was set to 10. The filter $\mathbf{w}_1(k)$ that extracts the desired signal in CIVA was initialized with $\mathbf{g}_1(k)$ while all the other filters were initialized with columns of an identity matrix. In the first scenario, we evaluated the quality of the extracted desired signal when the number of interferers was fixed. The simulations were repeated for $T_{60} = 150$ ms and $T_{60} = 300$ ms on a 20 s speech segment with source positions as depicted in Fig. 1. Source 1 to 3 were selected as desired one by one and the results were averaged over the three simulations. The performance measures for $M > L$, $M = L$ and $M < L$ are given in Table 1. With $M = L$, unconstrained IVA provided better interferer suppression as the solution in this case is unique. Nevertheless, constrained IVA still resulted in lower speech distortion. When M was increased from 3 to 4, the CIVA algorithm provided a gain in segSIR of 5.3 dB at $T_{60} = 150$ ms and 3.5 dB at $T_{60} = 300$ ms along with a decrease in desired speech distortion, while the performance of the IVA algorithm deteriorated due to non-uniqueness of solution with $M = 4, L = 3$. The performance of the IVA algorithm deteriorated further when L was increased to 5, while the CIVA algorithm provided a segSIR of 8.7 dB and 3.6 dB for $T_{60} = 150$ ms and $T_{60} = 300$ ms respectively. Moreover, it can be noted that the

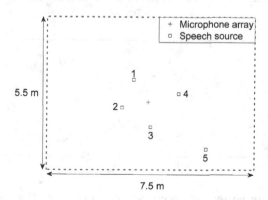

Fig. 1. Room geometry

CIVA algorithm maintained a very low desired speech distortion in all cases as the proposed penalty restricts the Euclidean angle between $\mathbf{w}_1(k)$ and $\mathbf{g}_1(k)$.

Table 1. Performance measures for fixed number of sources

Algorithms	$T_{60} = 150\,\text{ms}$		$T_{60} = 300\,\text{ms}$	
	segSD	segSIR (dB)	segSD	segSIR (dB)
Unprocessed mixture		1.9		1.5
IVA [7] (M = 3, L = 3)	0.21	**14.9**	0.31	4.7
CIVA (M = 3, L = 3)	**0.12**	11.7	**0.24**	**5.2**
Unprocessed mixture		1.9		1.5
IVA [7] (M = 4, L = 3)	0.27	8.2	0.39	3.8
CIVA (M = 4, L = 3)	**0.04**	**17.0**	**0.07**	**8.7**
Unprocessed mixture		0.7		−0.3
IVA [7] (M = 4, L = 5)	0.54	0.2	0.57	0.1
CIVA (M = 4, L = 5)	**0.05**	**8.7**	**0.08**	**3.6**

In the second scenario, the proposed algorithm was evaluated with a time varying number of interferers with $M = 6$ and $T_{60}=150\,\text{ms}$. The number of active sources over time is plotted in Fig. 2 (Top). Source 1 was selected as desired. The segSIR improvement calculated for segments of 1 s is plotted in Fig. 2 (Bottom). The experiment showed the effectiveness of the proposed algorithm with an unknown, time varying number of interferers. It must be noted that the online IVA algorithm did not converge as M was greater than L. The proposed online CIVA algorithm, however, converges in all cases.

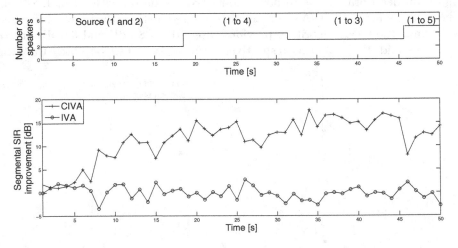

Fig. 2. Top: Activity pattern of speech sources over time; Bottom: Segmental SIR improvement over time. $M = 6$, $T_{60} = 150\,\text{ms}$.

5   Conclusions

An online constrained IVA algorithm was developed to extract the desired speech signal given the desired source DOA. The DOA was used to obtain the array steering vector. A penalty function was then added to IVA to penalize the Euclidean angle between one separation filter and the array steering vector. Simulations demonstrated the applicability of the algorithm to scenarios with fixed and time-varying number of interferers. Future work includes comparison of the algorithm against beamforming algorithms and evaluation with measured data.

References

1. Araki, S., Sawada, H., Makino, S.: Blind speech separation in a meeting situation with maximum SNR beamformers. In: IEEE International Conference on Acoustics, Speech and Signal Processing (ICASSP) (2007)
2. Benesty, J., Chen, J., Huang, Y.: Microphone Array Signal Processing. Springer, Berlin (2008)
3. Brandwood, D.: A complex gradient operator and its application in adaptive array theory. IEE Proc. F Commun. Radar Signal Process. **130**, 11–16 (1983)
4. Cardoso, J.F.: Infomax and maximum likelihood for blind source separation. IEEE Signal Process. Lett. **4**, 112–114 (1997)
5. Habets, E.A.P.: Room Impulse Response Generator. Technical report, Technische Universisteit Eindhoven (2006)
6. Hyvärinen, A., Karhunen, J., Oja, E.: Independent Component Analysis. Wiley, New York (2001)
7. Kim, T.: Real-time independent vector analysis for convolutive blind source separation. IEEE Trans. Circuits Syst. **1**, 1431–1438 (2010)
8. Kim, T., Lee, S.Y.: Blind source separation exploiting higher-order frequency dependencies. IEEE Trans. Audio Speech Lang. Process. **15**, 70–79 (2006)
9. Knaak, M., Araki, S., Makino, S.: Geometrically constrained independent component analysis. IEEE Trans. Audio Speech Lang. Process. **15**, 715–726 (2007)
10. Li, H., Adali, T.: A class of complex ICA algorithms based on the kurtosis cost function. IEEE Trans. Audio Speech Lang. Process. **19**, 408–420 (2008)
11. Novey, M., Adali, T.: Complex ICA by negentropy maximization. IEEE Trans. Neural Networks **19**, 596–609 (2008)
12. Parra, L., Spence, C.: Convolutive blind separation of non-stationary sources. IEEE Trans. Speech Audio Process **8**, 320–327 (2000)
13. Parra, L.C., Alvino, C.V.: Geometric source separation: merging convolutive source separation with geometric beamforming. IEEE Trans. Speech Audio Process. **10**, 352–362 (2002)
14. Taseska, M., Habets, E.A.P.: Spotforming using distributed microphone arrays. In: IEEE workshop on Applications of Signal Processing to Audio and Acoustics (WASPAA) (2013)
15. Zhang, W., Rao, B.D.: Combining independent component analysis with geometric information and its application to speech processing. In: IEEE International Conference on Acoustics, Speech and Signal Processing (ICASSP) (2009)

On-line Multichannel Estimation of Source Spectral Dominance

Francesco Nesta[1]([✉]), Trausti Thormundsson[1], and Zbyněk Koldovský[2]

[1] Conexant System, 1901 Main Street, Irvine, CA, USA
francesco.nesta@gmail.com, trausti.thormundsson@conexant.com
[2] Technical University of Liberec, Studentská 2, 461 17 Liberec, Czech Republic
zbynek.koldovsky@tul.cz

Abstract. Despite its popularity, multichannel source demixing is intrinsically limited in real-world applications due to the model mismatch between the convolutive mixing model and the actual recordings. Varying number of sources, reverberation, diffuseness and spatial changes are common uncertainties that need to be handled. Post-processing is commonly adopted to compensate for these mismatches, generally in the form of non-linear spectral filtering. In this work we analyze the property of the normalized differences between the output magnitudes of a linear spatial filter. We show that thanks to the time-frequency sparsity of acoustic signals, such distributions can be approximatively modeled by a bimodal Gaussian mixture model. An on-line bimodal constrained GMM fitting is proposed, in order to estimate the posterior probability of source spectral dominance. It is shown that the estimated posteriors can be used to produce a filtered output with very low distortion, outperforming traditional non-linear methods.

Keywords: Source separation · GMM · Binary masking · Speech enhancement

1 Introduction

Multichannel spatial filtering has shown to be effective with the enhancement of a given sound source of interest from the remaining noise. Supervised methods exploit prior geometrical information in the form of source position, leading to classical beamformer [2] or to more sophisticated cancellation filter bank (CFB)-based methods [3]. On the other hand, unsupervised methods are traditionally based on separation frameworks using Independent Component Analisys (ICA) [9] or spatial/spectral clustering [1]. Differently from traditional single channel enhancement methods, multichannel filtering exploits spatial cues to discriminate between multiple sources and do not necessarily need any strong assumption on the nature of the sound signal. As a main advantage such methods are able

F. Nesta and T. Thormundsson—The work of Zbyněk Koldovský was partially supported by California Community Foundation through Project No. DA-15-114599.

to deal with the separation of highly non-stationary signals such as concurrent speech sounds.

Despite their popularity, spatial filtering methods have intrinsic limitations due to the approximated mixing system modeling. Mixtures are generally approximated as a linear combination of signals generated by a finite number of spatially localized sources, often referred to as coherent sources. However, this condition is only partially fulfilled in real-world since the noise spatial covariance can be highly time-varying. Furthermore, in presence of high reverberation, linear deximing with short filters is suboptimal and leads to a large cross-source output signal leakage.

For the above limitations, spatial demixing is rarely used alone for source separation and is usually complemented by post-filtering methods exploiting other spectral cues. For instance, in classical beamforming a GSC structure is employed to remove the residual noise in the target channel [2]. In source separation systems with a limited number of microphones, spectral masking is generally adopted in the form of binary masks [6] or Wiener-like gains [3]. However, these methods do not explicitly model the uncertainty of the spatial filter and as a result, they require heuristic tuning hyperparameter optimized to avoid distortion in the target signal.

In this work we discuss on the meaning of the normalized cross-output-channel magnitude differences, i.e. the normalized differences of magnitudes measured at the outputs of the spatial filter, and show how its pdf can be used to predict the posterior probability of source dominance, compared to the Ideal Binary Mask (IdBM) [4,11]. Then, we propose an on-line fitting of a constrained GMM model whose posteriors are used to generate spectral gains for the filtering.

2   Models for Multichannel Observations

In this work we limit the analysis to the case of recordings made by 2 microphones but the discussion can be easily extended to a generic multichannel case. We indicate with $s(t)$ the time-domain signal generated by a target speech and with $x_1(t)$ and $x_2(t)$ the signal sampled at the first and second microphones which can be modeled as $x_i(t) = s_i(t) + n_i(t)$, where $s_i(t)$ and $n_i(t)$ $\forall i = 1, 2$ indicates the reverberant image of the target source and the noise contributions to each microphone (which can be viewed as generated by a multiplicity of coherent noise sources). We assume that a generic spatial filtering system is trained to produce an estimate of $s_i(t)$ and of $n_i(t)$. If $s(t)$ is a coherent source, regardless of the spatial characteristic of $n(t)$, the output of the system can be approximatively modeled as

$$\hat{s}_i(t) = s_i(t) + \alpha[n_i(t)], \quad \hat{n}_i(t) = \gamma[n_i(t)] + \beta[s_i(t)] \tag{1}$$

where $\alpha[\cdot]$ and $\beta[\cdot]$ are time-varying convolutive transformations modeling the residual of noise and speech in the corresponding channels, and $\gamma[\cdot]$ models the distortion of the estimated noise due to the approximated linear demixing. This model comes from the application of the Minimal Distortion Principle (MDP)

[5] to a generic inverse multichannel filter where the noise sources might exceed the number of microphones (see [7] Sect. 5.2 for details).

By means of a time-frequency analysis, e.g. a weighted short-time Fourier transform (STFT), each signal can be transformed from time-domain to a discrete time-frequency representation. Therefore, let $S_i(k,l)$ and $N_i(k,l)$ (with $i = 1, 2$) be the downsampled subband representation of the time-domain signals where k and l indicates the frequency bin and subband frame, respectively. Assuming that the convolutive transformations are approximatively stationary within the STFT analysis window and that the target speech and noise are uncorrelated, we can model the output magnitude of the spatial filter as

$$|\widehat{S}_i(k,l)| \simeq |S_i(k,l)| + \alpha(k,l)|N_i(k,l)| \qquad (2)$$
$$|\widehat{N}_i(k,l)| \simeq \gamma(k,l)|N_i(k,l)| + \beta(k,l)|S_i(k,l)|$$

where $\alpha(k,l)$, $\beta(k,l)$ and $\gamma(k,l)$ are positive constants. In oracle conditions, if the magnitudes of the target source and noise were available the Ideal Binary Mask extracting the target source could be estimated as

$$IdBM(k,l) = 1, \text{if } |S_i(k,l)| > LC \cdot |N_i(k,l)|, \qquad IdBM(k,l) = 0, \text{otherwise} \quad (3)$$

where LC is the local signal-to-noise ratio (SNR) in linear scale, typically set to 1 (i.e. 0 dB) [4,11]. In our case we only observe $\widehat{S}_i(k,l)$ and $\widehat{N}_i(k,l)$ from which we cannot directly infer the IdBM. However, by modeling the statistical distribution of features derived from $|\widehat{S}_i(k,l)|$ and $|\widehat{N}_i(k,l)|$ it is possible to estimate the probability that the IdBM is 1 in a particular time-frequency point.

3   Normalized Cross-output-channel Magnitude Differences as Discriminative Features for T-F Source Dominance

First, we define the normalized cross-output-channel magnitude differences as

$$f_k(l) = \frac{|\widehat{S}(k,l)| - |\widehat{N}(k,l)|}{|\widehat{S}(k,l)| + |\widehat{N}(k,l)|} \qquad (4)$$

where we remove the dependence on the channel i to simplify the notation. By substituting (2) in (4) we get

$$f_k(l) \simeq \frac{[1 - \beta(k,l)]\sqrt{SNR(k,l)} + [\alpha(k,l) - \gamma(k,l)]}{[1 + \beta(k,l)]\sqrt{SNR(k,l)} + [\gamma(k,l) + \alpha(k,l)]} \qquad (5)$$

where $SNR(k,l)$ is the true instantaneous signal-to-noise ratio (in linear scale) in the (k,l) T-F point. Due to the sparseness of acoustic signals in the T-F domain, the SNR will assume values close to 0 or close to infinity according to which source is dominant/active for a particular time-frequency point. As a

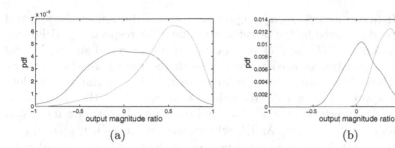

Fig. 1. Kernel density of $f(k, l)$ for a specific subband-based k (a) and for the entire frame (b).

consequence, the gains are expected to cluster around the centers $\frac{\alpha(k,l)-\gamma(k,l)}{\gamma(k,l)+\alpha(k,l)}$ and $\frac{1-\beta(k,l)}{1+\beta(k,l)}$ for T-F points dominated by the noise and target source, respectively. Therefore, we can expect the density of $f_k(l)$, $\forall k$, being approximatively bimodal with each component displaced and spread according to the statistic of the variables $\alpha(k,l), \gamma(k,l), \beta(k,l)$.

To help the understanding of this discussion, we consider the case where $\widehat{S}(k,l)$ and $\widehat{N}(k,l)$ are estimated through a two-channel spatial filter based on a geometrically constrained Independent Component Analysis [8]. By using the true oracle images $S(k,l)$ and $N(k,l)$, we define the IdBM and cluster T-F points in two classes, target source and noise source dominated points. Then, the kernel density estimate of $f_k(l)$ is computed for each class separately. In Fig. 1(a) the resulting densities for the two classes are shown with different colors. It can be observed that the full distribution resembles a mixture of exponential components which can be approximatively modeled with two Gaussian components. Note, this is only a convenient approximation since the observations in $f_k(l)$ are bounded in the range $[-1; 1]$ and in general each component cannot be symmetric. However, in practice, a simple Gaussian model is enough accurate to describe the uncertainty of $f_k(l)$ in representing the dominance classes.

Figure 1(b) shows the density obtained with the $f_k(l)$ of all the subbands. On average a bimodal GMM fits well the empirical distribution, which implies that the model can be also used for a frame-based multichannel target source activity detection. Note, function in Eq. (4) produces a convenient 1-dimensional discriminative representation correlated to the IdBM class probabilities. However, an alternative function could be used as long as a suitable model is available for describing its pdf.

4   Constrained On-line GMM Parameter Fitting

According to the above analysis, in each T-F point, we model the density of $f_k(l)$ with a bimodal GMM as

$$p[f_k(l)] = w_{1,k}(l) \cdot N[\mu_{1,k}(l), \sigma^2_{1,k}(l)] + w_{2,k}(l) \cdot N[\mu_{2,k}(l), \sigma^2_{2,k}(l)] \qquad (6)$$

where the $N(\mu_1, \sigma_1^2)$ and $N(\mu_2, \sigma_2^2)$ represent the distribution of the spectral gains for points dominated by the target source and noise respectively. Following the interpretation of $f_k(l)$, $\mu_{1,k}(l)$ is expected to be larger than $\mu_{2,k}(l)$ and therefore proper constraints need to be added to the model. In this work we fit the model variables using a sequential approximation of the expectation-maximization approach as in [12]. We define with $c \in \{1, 2\}$ the class labels, where 1= "target speech dominant", 2= "noise dominant". We are interested in the probability $p[c = 1|f_k(l), \lambda_k(l)]$, where $\lambda_k(l) = [\mu_{1,k}(l), \sigma_{1,k}^2(l), w_{1,k}(l), \mu_{2,k}(l), \sigma_{2,k}^2(l), w_{2,k}(l)]$ is the parameter vector for the target speech and noise component models, estimated at the frame l. The probability of T-F target speech dominance $p_k(l)$ can be computed using the Bayes formula as

$$p_k(l) = p[c = 1|f_k(l), \lambda_k(l)] = \frac{w_{1,k}(l)p[f_k(l)|c = 1, \lambda_k(l)]}{\sum_{c=1}^{2} w_{c,k}(l)p[f_k(l)|c, \lambda_k(l))} \tag{7}$$

In the sequential on-line learning, within the frame l, the mixture parameters are updated $\forall c = 1, 2$ and $\forall k$ as

$$w_{c,k}(l) = (1 - \eta_c) \cdot w_{c,k}(l - 1) + \eta_c \cdot p[c|f_k(l), \lambda_k(l - 1)] \tag{8}$$

$$\mu_{c,k}(l) = \frac{(1 - \eta_c) \cdot \mu_{c,k}(l - 1)}{w_{c,k}(l)} + \frac{\eta_c \cdot p[c|f_k(l), \lambda_k(l - 1)]f_k(l)}{w_{c,k}(l)} \tag{9}$$

$$\sigma_{c,k}(l) = \frac{(1 - \eta_c) \cdot \sigma_{c,k}(l - 1)}{w_{c,k}(l)} + \frac{\eta_c \cdot p[c|f_k(l), \lambda_k(l - 1)](f_k(l) - \mu_{c,k}(l))^2}{w_{c,k}(l)} \tag{10}$$

where η_c is a learning rate step-size. Iterating equations (7)-(10) the GMM parameters are updated on-line with the incoming data. To avoid divergence in trivial solutions some constraints are necessary. First, the weights $w_{1,k}(l)$ or $w_{2,k}(l)$ can approach zero if either the target or the noise signal is absent for a long time. To avoid this divergence the values of the weights are constrained within the iteration as

$$w_{1,k}(l) = \min[\max(w_{1,k}(l), \epsilon), 1 - \epsilon], \quad w_{2,k}(l) = 1 - w_{1,k}(l) \tag{11}$$

where ϵ is set to a small value (e.g. 0.05). To guarantee that the estimated components are in the correct order, for each frequency bin k we impose:

$$\mu_1(k, l) > \mu_2(k, l), \tag{12}$$

and to avoid $\sigma_{1,k}^2(l)$ and $\sigma_{2,k}^2(l)$ approach 0, the following constraint is applied:

$$\sigma_{c,k}^2(l) = \min(\sigma_{c,k}^2(l), \epsilon_{\sigma^2}), \forall c \tag{13}$$

where ϵ_{σ^2} is a small value (e.g. 0.0001).

5   Proposed System Architecture

Figure 2(b) shows the block architecture of the proposed filtering scheme. The multichannel recordings are sent to the input of a convolutive spatial filter which decomposes the input in target speech and noise signal components. In this work we evaluate the performance with both a blind spatial filter learned with a geometrically constrained ICA [8] and with a semi-blind filter based on a pre-trained cancellation filterbank [3]. The time-domain outputs are transformed in T-F representation by means of a STFT analysis with Hanning windows of 2048 points with increments of 128 samples.

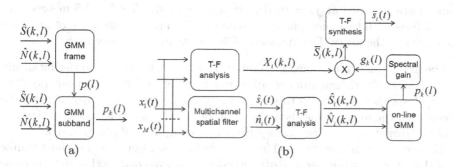

Fig. 2. Block diagram of the proposed filtering structure.

To improve the robustness of the on-line fitting, a two level hierarchy structure is proposed as shown in Fig. 2(a). In the top level, a GMM model is used to fit all the subband features $f_k(l)$, $\forall k$ in a single model and estimate the probability of speech dominance $p(l)$ in the entire frame. This block can be considered as an adaptive multichannel Voice Activity Detector (VAD). The frame-based posterior dominance probability $p(l)$ is then used to weight the learning rate of the subband-based GMM fitting, as follow:

$$\eta_1 = \eta \cdot p(l), \quad \eta_2 = \eta \cdot (1 - p(l)) \tag{14}$$

where η is the maximum step-size used for the subband parameter tracking. This approach prevents the GMM to update the target model parameters during speech pauses, which improves the convergence speed of the on-line learning. The subband posterior probabilities $p_k(l)$ are then used to compute spectral gains used to filter the multichannel input. Two alternative filtering approaches are proposed:

- M1: use the posteriors to directly compute the gains as $g_k(l) = p_k(l)$, if $p_k(l) > 0.5$ (0 otherwise). This approach is justified by the interpretation of the posteriors learned by the GMM model, which are related to the probability of IdBM(k, l) being equal to 1.
- M2: use the posteriors indirectly to estimate the expectation of the noise power $P(k, l)$ as

$$P(k, l) = (1 - a_k(l)) \cdot P(k, l) + a_k(l) \cdot |\hat{N}(k, l)|^2] \tag{15}$$
$$a_k(l) = 1 - p_k(l), \text{ if } p_k(l) < 0.5, \text{ (0 otherwise)}$$

and then use the noise power to estimate the spectral gains with conventional single-channel methods (in this work we used a standard spectral subtraction [10]).

6   Experimental Evaluation

Sources are recorded at $f_s = 16$ kHz, in a room of size $5 \times 5 \times 2.5$ meters with $T_{60}=300$ ms with two microphones spaced of $0.2m$ and at a distance of 2 m from the center of the array. Two datasets of 100 mixtures are generated:

- The first dataset is obtained by combining a target and noise interfering speakers (randomly chosen from a collection of male and female speakers). The target speaker is assumed to be in an angular region of $+/-10^o$ around the center of the array, while the noise is randomly displaced in any direction out of the target region. The average SNR at the input is of about -2 dB.
- The second dataset is obtained from the first dataset but reducing the dynamic of the interfering speaker by 6 dB and adding a stereo real-world cafeteria noise to the mixture. The average SNR at the input is of about -0.5 dB.

Tables 1 and 2 show the SNR and SDR improvement obtained for the two datasets with the blind and the semi-bind spatial filter. Performance are compared to conventional spectral masking methods such as binary masking

Table 1. Mean and (standard deviation) SNRi and SDRi performance when using a blind ICA-based spatial filter.

Dataset1	BM	GW	M1	M2
SNRi	5.46 (3.21)	6.46 (3.11)	8.42 (2.90))	**11.27** (4.51)
SDRi	5.04 (1.55)	5.42 (1.45)	**5.62** (1.60)	5.47 (1.63)
Dataset2	BM	GW	M1	M2
SNRi	2.20 (1.01)	3.09 (1.29)	5.40 (1.99)	**9.49** (2.62)
SDRi	3.59 (0.96)	4.03 (0.88)	**4.09** (1.26)	3.94 (1.38)

Table 2. Mean and (standard deviation) SNRi and SDRi performance when using a semi-blind CFB-based spatial filter.

Dataset1	BM	GW	M1	M2
SNRi	6.97 (3.90)	6.26 (2.28)	10.23 (3.01)	**12.08** (4.04)
SDRi	5.14 (1.68)	5.46 (1.35)	**6.21** (2.17)	5.64 (2.18)
Dataset2	BM	GW	M1	M2
SNRi	1.35 (0.50)	2.68 (1.18)	8.83 (1.32)	**11.78** (1.72)
SDRi	3.60 (0.82)	3.95 (0.70)	**5.64** (1.44)	5.37 (1.46)

$BM(k,l) = |\hat{S}(k,l)| > c \cdot |\hat{N}(k,l)|$ and parametric wiener-like filter $GW(k,l) = \frac{|\hat{S}(k,l)|}{|\hat{S}(k,l)| + c \cdot |\hat{N}(k,l)|}$ where the hyper parameter c was tuned to optimize the SDR scores. It can be noted that method $M1$ achieves the best overall scores which highlight that the GMM model fits well the IdBM class probabilities as long as the noise can be considered enough sparse in the time-frequency domain. On the other hand, in real-world diffuse noise, the indirect use of the posteriors in $M2$ delivers the best overall results since the noise has a lower degree of sparsity.

7   Conclusions

In this paper we discuss on the property of the normalized cross-output-channel magnitude difference of a spatial filter. It is shown that under certain conditions, its statistic can be approximatively modeled with a bimodal GMM whose posteriors can predict the probability of target source dominance. An on-line constrained GMM learning structure together with a filtering scheme is then proposed. Through an experimental evaluation with both coherent and real-world reverberant challenging recordings, it is shown that the proposed method can generate masks able to enhance the target source with a limited amount of distortion.

Future works may concern the use of better density models to generate more accurate posteriors, in combination with more advanced spectral filtering structures.

References

1. Araki, S., Nakatani, T., Sawada, H., Makino, S.: Stereo source separation and source counting with MAP estimation with dirichlet prior considering spatial aliasing problem. In: Adali, T., Jutten, C., Romano, J.M.T., Barros, A.K. (eds.) ICA 2009. LNCS, vol. 5441, pp. 742–750. Springer, Heidelberg (2009)
2. Brandstein, M., Ward, D.: Microphone Arrays. Springer, Berlin (2001)
3. Koldovský, Z., Málek, J., Tichavský, P., Nesta, F.: Semi-blind noise extraction using partially known position of the target source. IEEE Trans. Audio Speech Lang. Process. **21**(10), 2029–2041 (2013)
4. Loizou, P., Kim, G.: Reasons why current speech-enhancement algorithms do not improve speech intelligibility and suggested solutions. IEEE Trans. Audio Speech Lang. Process. **19**(1), 47–56 (2011)
5. Matsuoka, K., Nakashima, S.: Minimal distortion principle for blind source separation. In: Proceedings of International Symposium on ICA and Blind Signal Separation, San Diego, CA, USA, December 2001
6. Mori, Y., Saruwatari, H., Takatani, T., Ukai, S., Shikano, K., Hiekata, T., Morita, T.: Real-time implementation of two-stage blind source separation combining simo-ica and binary masking. In: IWAENC, pp. 229–232 (2005)
7. Nesta, F., Matassoni, M.: Blind source extraction for robust speech recognition in multisource noisy environments. Comput. Speech Lang. **27**(3), 703–725 (2013)
8. Nesta, F., Matassoni, M., Astudillo, R.F.: A flexible spatial blind source extraction framework for robust speech recognition in noisy environments. In: Proceedings of CHiME, pp. 33–40 (2013)

9. Pedersen, M.S., Larsen, J., Kjems, U., Parra, L.C.: A survey of convolutive blind source separation methods. In: Benesty, J., Huang, Y., Sondhi, M. (eds.) Springer Handbook of Speech. Springer, Berlin (2007)

10. Tashev, I., Lovitt, A., Acero, A.: Unified framework for single channel speech enhancement. In: IEEE Pacific Rim Conference on Communications, Computers and Signal Processing, August 2009

11. Wang, D.: Time-frequency masking for speech separation and its potential for hearing aid design. Trends Amplification 12(4), 332–353 (2008)

12. Ying, D., Yan, Y., Dang, J., Soong, F.: Voice activity detection based on an unsupervised learning framework. IEEE Trans. Audio Speech Lang. Process. 19(8), 2624–2633 (2011)

Component-Adaptive Priors for NMF

Julian M. Becker[✉] and Christian Rohlfing

Institut für Nachrichtentechnik, RWTH Aachen University, 52056 Aachen, Germany
{becker,rohlfing}@ient.rwth-aachen.de
http://www.ient.rwth-aachen.de

Abstract. Additional priors for nonnegative matrix factorization (NMF) are a powerful way of adapting NMF to specific tasks, such as for example audio source separation. For this application, priors supporting sparseness or temporal continuity have been proposed. However, these priors are not helpful for all kinds of signals and should therefore only be used when needed. For some mixtures, only some components of the mixtures should be supported by these priors. We present an easy, but efficient method of adapting priors to different components. We show, that the separation results are improved, while the computational complexity is even slightly reduced. We also show, that our method is a helpful modification for the combination of different priors.

Keywords: NMF · Audio source separation · Temporal continuity · Sparseness

1 Introduction

In the past, several extensions to nonnegative matrix factorization (NMF) have been proposed to adapt it to the task of source separation. Some extensions use convolutive bases instead of multiplicative ones [9,10], others introduce additional constraints such as sparsity [6,13], temporal continuity [13] or spectral continuity [2]. An overview over different versions of NMF can be found in [4].

We focus on NMF with additional constraints. Most existing methods [2,6,13] add these constraints to all of the components of the NMF, weighting them equally. However, these priors often only make sense for some of the components. Recent work on only applying the constraints to some components either require an additional training step [8] or are restricted to specific tasks, such as separation of harmonic and percussive sources [3]. We propose a way of individually adapting priors to the components, so that they are used stronger on the components where they are more helpful. Our method can be used on any kind of mixture and does not require training or prior information about the sources.

The paper is structured as follows: In Sect. 2, we provide basic information about NMF for source separation. In Sect. 3 we describe NMF with additional temporal continuity criterion as an example for additional priors. We then introduce our method of adapting this prior to the NMF components and generalize this method for other priors in Sect. 4. In Sect. 5 we evaluate our method by experiment, before closing the paper with our conclusions in Sect. 6.

© Springer International Publishing Switzerland 2015
E. Vincent et al. (Eds.): LVA/ICA 2015, LNCS 9237, pp. 413–420, 2015.
DOI: 10.1007/978-3-319-22482-4_48

2   NMF for Monaural Source Separation

We assume an audio mixture $\mathbf{x}$ in time domain, consisting of M sources $\mathbf{s}_m$. $\underline{\mathbf{X}}$ is the complex valued result of the short time Fourier transform (STFT) of $\mathbf{x}$. For source separation, NMF can be applied to the magnitude spectrogram $\mathbf{X} = |\underline{\mathbf{X}}|$.

NMF approximates a matrix $\mathbf{X} \in \mathbb{R}_+^{K \times N}$ by a product of two matrices $\mathbf{B}$ and $\mathbf{G}$ as $\mathbf{X} \approx \tilde{\mathbf{X}} = \mathbf{B}\mathbf{G}$, $\mathbf{B} \in \mathbb{R}_+^{K \times I}$, $\mathbf{G} \in \mathbb{R}_+^{I \times N}$. I defines the number of NMF components. $\mathbf{B}$ and $\mathbf{G}$ are iteratively calculated, minimizing a cost term $c(\mathbf{B}, \mathbf{G})$ between $\mathbf{X}$ and $\tilde{\mathbf{X}}$. Usually $c(\mathbf{B}, \mathbf{G})$ only consists of a reconstruction term $c_r(\mathbf{B}, \mathbf{G})$, where commonly used terms are the Euclidean distance, the Kullback-Leibler (KL) divergence and the Itakura-Saito (IS) distance. Lee and Seung [7] introduced multiplicative update rules for the squared Euclidean distance as well as for the KL divergence, resulting in convergence to a local minimum of the cost term. These update rules can be calculated using the gradient of $c(\mathbf{B}, \mathbf{G})$ with respect to $\mathbf{B}$, $\nabla_\mathbf{B} c(\mathbf{B}, \mathbf{G}) = \nabla_\mathbf{B}^+ c(\mathbf{B}, \mathbf{G}) - \nabla_\mathbf{B}^- c(\mathbf{B}, \mathbf{G})$, where $\nabla_\mathbf{B}^+ c(\mathbf{B}, \mathbf{G})$ and $\nabla_\mathbf{B}^- c(\mathbf{B}, \mathbf{G})$ are elementwise nonnegative terms, and the equivalently defined gradient with respect to $\mathbf{G}$, $\nabla_\mathbf{G} c(\mathbf{B}, \mathbf{G})$. The update rules are

$$\mathbf{B} \leftarrow \mathbf{B} \otimes \frac{\nabla_\mathbf{B}^- c(\mathbf{B}, \mathbf{G})}{\nabla_\mathbf{B}^+ c(\mathbf{B}, \mathbf{G})} \quad \text{and} \quad \mathbf{G} \leftarrow \mathbf{G} \otimes \frac{\nabla_\mathbf{G}^- c(\mathbf{B}, \mathbf{G})}{\nabla_\mathbf{G}^+ c(\mathbf{B}, \mathbf{G})}, \tag{1}$$

where $\otimes$ denotes elementwise multiplication and the divisions are also elementwise. For the methods presented in this paper, this generalized formulation of the update rules is sufficient. Exact update rules for KL-divergence and squared Euclidean distance can be found in [7] and for the IS-distance in [4]. Figure 1 shows the factorization of a spectrogram of a mixture of a harmonic and a percussive source with NMF with $I = 2$ with the resulting matrices $\mathbf{B}$ (on the left) and $\mathbf{G}$ (on top). The columns of $\mathbf{B}$ capture the spectral shape of the acoustical events and can be interpreted as spectral bases. The rows of $\mathbf{G}$ can be interpreted as temporal activations. After performing NMF, phase information and finer structures of the spectrograms are restored using a filtering step, which is usually done by Wiener filtering (see e.g. [5,11]). If I is higher than the number of sources M, the components have to be assigned to the sources in a clustering step, resulting in M spectrograms corresponding to the estimated sources. Finally, these spectrograms are transformed back to time domain by inverse STFT.

3   NMF with Temporal Continuity

An example for an additonal prior for NMF is the temporal continuity, proposed by Virtanen [13] to prevent incorrect factorizations. In the factorization in Fig. 1 it can be observed, that temporal gaps appear in the activation vector of the harmonic note where it is overlapped by the percussive tone. To prevent this, Virtanen proposed to add a temporal continuity term $c_t(\mathbf{G})$ to the reconstruction error term $c_r(\mathbf{B}, \mathbf{G})$, resulting in a cost function $c(\mathbf{B}, \mathbf{G}) = c_r(\mathbf{B}, \mathbf{G}) + \alpha_t c_t(\mathbf{G})$,

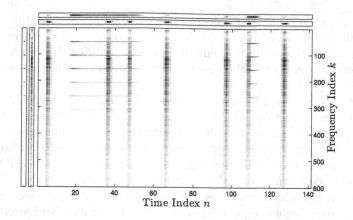

Fig. 1. Factorization of an audio mixture using NMF.

where α_t is a weight to adjust the influence of $c_t(\mathbf{G})$. For $\alpha_t = 0$ this model equals the standard NMF. The update rule for $\mathbf{G}$ transforms to

$$\mathbf{G} \leftarrow \mathbf{G} \otimes \frac{\nabla_{\mathbf{G}}^- c_r(\mathbf{B}, \mathbf{G}) + \alpha_t \nabla_{\mathbf{G}}^- c_t(\mathbf{G})}{\nabla_{\mathbf{G}}^+ c_r(\mathbf{B}, \mathbf{G}) + \alpha_t \nabla_{\mathbf{G}}^+ c_t(\mathbf{G})}, \tag{2}$$

while the update term for $\mathbf{B}$ stays the same as in Eq. (1). Virtanen proposes a temporal squared difference (TSD) cost term,

$$c_t(\mathbf{G}) = \sum_{i=1}^{I} \frac{1}{\sigma_i^2} \sum_{n=2}^{N} (g_{i,n} - g_{i,n-1})^2, \tag{3}$$

with $\sigma_i = \sqrt{(1/N) \sum_{n=1}^{N} g_{i,n}^2}$ being the standard deviation of each row of $\mathbf{G}$ and $g_{i,n}$ denoting one element of the matrix $\mathbf{G}$ at indizes i and n. The negative and positive gradient terms of this cost function are

$$[\nabla_{\mathbf{G}}^- c_t(\mathbf{G})]_{i,n} = \frac{2N(g_{i,n-1} + g_{i,n+1})}{\sum_{l=1}^{N} g_{i,l}^2} + \frac{2N g_{i,n} \sum_{l=2}^{N}(g_{i,l} - g_{i,l-1})^2}{\left(\sum_{l=1}^{N} g_{i,l}^2\right)^2} \tag{4}$$

and

$$[\nabla_{\mathbf{G}}^+ c_t(\mathbf{G})]_{i,n} = \frac{4N g_{i,n}}{\sum_{l=1}^{N} g_{i,l}^2}. \tag{5}$$

4   Component-Adaptive Priors

The assumption of temporal continuity does not hold for all signals. In the example in Fig. 1 it only holds for the first (harmonic) component, whereas the second (percussive) component is not continuous in time. In fact, percussive sources usually have an impulse-like behaviour in time. Therefore, using the additional cost term is not advisable for the second component, as it deteriorates the resulting temporal activation vector. The percussive component is being smeared in time when using the temporal continuity prior. However, Virtanen only proposed a method to either use the temporal continuity term on all or no components.

In [3] it was proposed to only use priors on some components. Priors supporting harmonic structures are used on one half of the components, priors for percussive structures on the other half, assuming that harmonic and percussive structures will then automatically develop in the corresponding components. This approach has several disadvantages: First of all, it is not possible to use a structured initialization (e.g. SVD) with this approach, since these initializations already define which structures will develop in which component. Secondly, different mixtures might need a different number of percussive or harmonic components, which is not considered in this approach. The method also has the downside of only being applicable for mixtures of harmonic and percussive sources.

In the following, we will introduce a method for adapting the temporal continuity prior in a way that it is used stronger for components that need an additional temporal continuity term, while it is used weaker for components that do not. With this method, structured initializations are possible and the priors are automatically adapted to the different components, solving the problem of the fixed harmonic and percussive components and making it applicable to any kind of mixture. We also describe, how this method can be used for other priors.

4.1 Finding Harmonic Components

Temporal continuity is mostly only desirable for harmonic components. To adapt the prior to the components, harmonic components have to be identified. Performing harmonic/percussive classification has two downsides: First of all, harmonic and percussive signals are not perfectly distinguishable. A harmonic signal might have a percussive onset, or a percussive signal a harmonic decay. This makes classification difficult and might lead to additional errors. Secondly, this step produces additional computational complexity, which might be unwanted.

An easier way is to use the cost term $c_t(\mathbf{G})$. For undistorted signals we expect it to be low for signals that are continuous in time (e.g. harmonic) and high for others (e.g. percussive). Assuming that separation distortions are small, we can use this term as information about the behaviour of the different components.

4.2 Adapting the Prior

With this motivation, it seems reasonable to adapt the prior by multiplying the gradient terms $\nabla_{\mathbf{G}}^{-} c_t(\mathbf{G})$ and $\nabla_{\mathbf{G}}^{+} c_t(\mathbf{G})$ in Eq. (2) with a factor $1/c_t(\mathbf{G})$. Thus, the effect of the additional cost term is amplified for components with a low value of c_t (e.g. harmonic components) compared to components with high c_t. The update rule for $\mathbf{G}$ transforms to

$$\mathbf{G} \leftarrow \mathbf{G} \otimes \frac{\nabla_{\mathbf{G}}^{-} c_r(\mathbf{B}, \mathbf{G}) + \frac{\alpha_t \nabla_{\mathbf{G}}^{-} c_t(\mathbf{G})}{c_t(\mathbf{G})}}{\nabla_{\mathbf{G}}^{+} c_r(\mathbf{B}, \mathbf{G}) + \frac{\alpha_t \nabla_{\mathbf{G}}^{+} c_t(\mathbf{G})}{c_t(\mathbf{G})}}. \tag{6}$$

The gradient terms for the TSD prior, weighted with this factor are

$$\left[\frac{\nabla_{\mathbf{G}}^{-} c_t(\mathbf{G})}{c_t(\mathbf{G})}\right]_{i,n} = \frac{2(g_{i,n-1} + g_{i,n+1})}{\sum_{l=2}^{N}(g_{i,l} - g_{i,l-1})^2} + \frac{2g_{i,n}}{\sum_{l=1}^{N} g_{i,l}^2} \tag{7}$$

and

$$\left[\frac{\nabla_{\mathbf{G}}^{+} c_t(\mathbf{G})}{c_t(\mathbf{G})}\right]_{i,n} = \frac{4g_{i,n}}{\sum_{l=2}^{N}(g_{i,l} - g_{i,l-1})^2}. \tag{8}$$

Note, that Eqs. (7) and (8) are less computationally complex than Eqs. (4) and (5) because of a reduced number of multiplications.

Figure 2 shows a comparison of the temporal activations for factorization of the same example as in Fig. 1 with standard NMF, Virtanens TSD ($\alpha_t = 250$) and the presented adaptive prior ($\alpha_t = 431$, the equivalent value to $\alpha_t = 250$ for Virtanens TSD with respect to the harmonic component). Figure 2a shows the resulting temporal activations for the harmonic component. The positive effect of the temporal continuity is preserved with the proposed method, the gaps in the temporal activation of the harmonic component are avoided. However, the negative effect, the temporal smearing of the percussive component (Fig. 2b), is reduced compared to the factorization with NMF with Virtanens TSD.

NMF
$c_t(\mathbf{G})$
$\tilde{c}_t(\mathbf{G})$

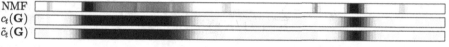

(a) Comparison of the temporal activation vectors for the harmonic component. The proposed method ($\tilde{c}_t(\mathbf{G})$) preserves the advantages of Virtanens TSD ($c_t(\mathbf{G})$).

NMF
$c_t(\mathbf{G})$
$\tilde{c}_t(\mathbf{G})$

(b) Comparison of the temporal activation vectors for the percussive component. The proposed method ($\tilde{c}_t(\mathbf{G})$) reduces the temporal smearing of Virtanens TSD ($c_t(\mathbf{G})$).

Fig. 2. Comparison of temporal activation vectors $\mathbf{G}$ of standard NMF, NMF with Virtanens temporal continuity ($c_t(\mathbf{G})$, $\alpha_t=250$) and the proposed method ($\tilde{c}_t(\mathbf{G}),\alpha_t=431$).

4.3 Mathematical Description and Generalization

Comparing Eqs. (2) and (6) it can be observed, that the proposed method can be interpreted as using a new cost function $\tilde{c}_t$ with $\nabla_{\mathbf{G}}\tilde{c}_t(\mathbf{G}) = \frac{\nabla_{\mathbf{G}} c_t(\mathbf{G})}{c_t(\mathbf{G})}$. With the properties of the natural logarithm, it is obvious, that $\tilde{c}_t(\mathbf{G}) = \ln(c_t(\mathbf{G}))$.

Thus, we can generalize our model for any cost function c with the requirement, that the prior only makes sense for components with a low value of c. Then, using $\ln(c)$ as prior leads to an adaption of the cost function c to the components.

5   Experimental Results

We performed source separation as described in Sect. 2. To evaluate the separation quality of the NMF without being affected by errors of a clustering algorithm, we used a non-blind clustering with knowledge of the original signals, as described in [13]. As measure for separation quality, we used the signal-to-distortion ratio (SDR), signal-to-inference ratio (SIR) and signal-to-artifacts ratio (SAR), as proposed in [12]. All given values are averaged over the complete testset.

We compared our method to standard NMF and to NMF with Virtanens TSD. To show, that our method is applicable to other priors and that it is beneficial for the combination of different priors, we also tested it on a combination of the TSD prior and a sparseness prior on **B**. Sparse spectral basis vectors can be assumed for some components (e.g. harmonic notes), but not for all (e.g. impulse-like or noisy components). Therefore, the prior should be used stronger on the components, that are relatively sparse and the requirement for our method is fulfilled. We used the same sparseness prior as was used in [13], with the difference, that we used the prior on **B** and not on **G**. The cost function of this prior was

$$c_s(\mathbf{B}) = \sum_{i=1}^{I} \frac{1}{\sqrt{(1/K)\sum_{l=1}^{K} b_{l,i}^2}} \sum_{k=2}^{K} b_{k,i} \tag{9}$$

We compared a combination of c_t and c_s to one of $\tilde{c}_t$ and $\tilde{c}_s = \ln(c_s)$, using the proposed adaptive weighting. We chose $\alpha_{s,c_s} = 1.4$ and $\alpha_{s,\tilde{c}_s} = 130$, the weights with the best separation quality, for the spectral priors.

5.1   Testset and Setup

The testset consists of 60 audio signals, including harmonic and percussive signals, speech, vocals and noise, each being sampled with 44.1 kHz. These signals were mixed in every possible two-source combination, resulting in 1770 mixtures. The testset is identical to the one used in [11].

For the STFT, we used a window size of $s_w = 2^{12}$ and a hop size of $s_h = 2^{11}$ samples. I was set to 20 for every mixture, since this had shown to be a suitable number of components for this testset. As reconstruction error term, we used KL-divergence, as this produced the best separation results. We initialized the NMF by performing an SVD on the complex spectrogram $\underline{\mathbf{X}}$ as proposed in [1].

5.2   Results

A comparison of the SDR using NMF with Virtanens TSD and our adaptive version of it for different values of α_t, is shown in Fig. 3a. Note, that $\alpha = 0$ equals the standard NMF without priors. The SDR results for the combination of priors are shown in Fig. 3b. Our method reached higher SDR values over a broad range of α_t in both cases. An overview over the highest reached values

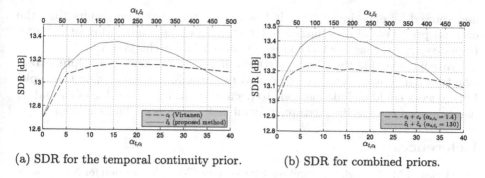

(a) SDR for the temporal continuity prior. (b) SDR for combined priors.

Fig. 3. Experimental results: Adaptive priors lead to a higher maximum separation quality.

Table 1. Maximum SDR, SIR and SAR for different priors.

	NMF	c_t	$\tilde{c}_t$	$c_t + c_s$	$\tilde{c}_t + \tilde{c}_s$
SDR [dB]	12.69	13.17	**13.36**	13.24	**13.47**
SIR [dB]	18.76	19.10	**19.34**	19.32	**19.56**
SAR [dB]	15.64	16.65	**16.66**	16.87	**17.00**

of SDR, SIR and SAR is given in Table 1. The separation quality is improved for all measures, when using the adaptive cost terms $(\tilde{c}_t, \tilde{c}_s)$ compared to the original ones (c_t, c_s).

Comparing our method with the method in [3] is difficult because of the different preconditions. Thus, we used a random initialization and only evaluated the mixtures of harmonic and percussive signals in our testset, since these are the limitations of [3]. We evaluated the methods for Virtanens TSD (Eq. (3)) as harmonic prior and an equivalent spectral squared difference as percussive prior. Those are two of the priors that are used in [3], where the harmonic prior is put on one half of the components and the percussive prior on the other half. For our method we used the natural logarithm of the two priors on all components to adapt the priors to the components. We performed the methods with different combinations of the weights, choosing the best combination for evaluation. We also evaluated NMF without priors. The average SDR over all 440 harmonic/percussive mixtures was 14.84 dB for the standard NMF. Method [3] reached a maximum of 15.28 dB, our method reached a maximum of 15.76 dB.

6 Conclusion

In this paper, we introduced a way of adapting a temporal continuity prior to the NMF components so that the prior is used stronger on the components, where it is more helpful. Our method does not need any additional computational steps, but only changes the cost function, leading to less computational complex update rules. We showed, that our method can be generalized for other priors and showed by experiment, that it leads to better separation results than the original prior. We also evaluated our method for the combination of different priors, verifying, that our method is also beneficial for this scenario.

We conclude, that the adaption to the different components is a helpful extension to existing priors. Our results should motivate future research on this topic. Future work could include methods to decide for the optimal value of α_t and α_s depending on the specific mixture, or combinations of more different priors.

References

1. Becker, J.M., Menzel, M., Rohlfing, C.: Complex SVD initialization for NMF source separation on audio spectrograms. In: DAGA 2015. Nürnberg, Germany (2015)
2. Becker, J.M., Sohn, C., Rohlfing, C.: NMF with spectral and temporal continuity criteria for monaural sound source separation. In: 2013 Proceedings of the 22nd European Signal Processing Conference (EUSIPCO), pp. 316–320. IEEE (2014)
3. Canadas-Quesada, F.J., Vera-Candeas, P., Ruiz-Reyes, N., Carabias-Orti, J., Cabanas-Molero, P.: Percussive/harmonic sound separation by non-negative matrix factorization with smoothness/sparseness constraints. EURASIP J. Audio Speech Music Process. **2014**(1), 1–17 (2014)
4. Cichocki, A., Zdunek, R., Phan, A.H., Amari, S.I.: Nonnegative Matrix and Tensor Factorizations: Applications to Exploratory Multi-way Data Analysis and Blind Source Separation. Wiley, New York (2009)
5. Jaiswal, R., FitzGerald, D., Barry, D., Coyle, E., Rickard, S.: Clustering NMF basis functions using shifted NMF for monaural sound source separation. In: 2011 IEEE International Conference on Acoustics, Speech and Signal Processing (ICASSP), pp. 245–248. IEEE (2011)
6. Joder, C., Weninger, F., Virette, D., Schuller, B.: A comparative study on sparsity penalties for nmf-based speech separation: beyond lp-norms. In: 2013 IEEE International Conference on Acoustics, Speech and Signal Processing (ICASSP), pp. 858–862. IEEE (2013)
7. Lee, D.D., Seung, H.S.: Algorithms for non-negative matrix factorization. In: Leen, T.K., Dietterich, T.G., Tresp, V. (eds.) Advances in Neural Information Processing Systems, pp. 556–562. MIT Press, Cambridge (2001)
8. Marxer, R., Janer, J.: Study of regularizations and constraints in NMF-based drums monaural separation. In: International Conference on Digital Audio Effects Conference (DAFx-13) (2013)
9. Schmidt, M.N., Mørup, M.: Nonnegative matrix factor 2-D deconvolution for blind single channel source separation. In: Rosca, J.P., Erdogmus, D., Príncipe, J.C., Haykin, S. (eds.) ICA 2006. LNCS, vol. 3889, pp. 700–707. Springer, Heidelberg (2006)
10. Smaragdis, P.: Non-negative matrix factor deconvolution; extraction of multiple sound sources from monophonic inputs. In: Puntonet, C.G., Prieto, A.G. (eds.) ICA 2004. LNCS, vol. 3195, pp. 494–499. Springer, Heidelberg (2004)
11. Spiertz, M., Gnann, V.: Beta divergence for clustering in monaural blind source separation. In: Audio Engineering Society Convention 128. Audio Engineering Society (2010)
12. Vincent, E., Gribonval, R., Févotte, C.: Performance measurement in blind audio source separation. IEEE **14**, 1462–1469 (2006)
13. Virtanen, T.: Monaural sound source separation by nonnegative matrix factorization with temporal continuity and sparseness criteria. IEEE **15**, 1066–1074 (2007)

Estimating Correlation Coefficient Between Two Complex Signals Without Phase Observation

Shigeki Miyabe[1](✉), Notubaka Ono[2], and Shoji Makino[1]

[1] University of Tsukuba, 1-1-1 Tennodai, Tsukuba, Ibaraki 305-8573, Japan
{miyabe,maki}@tara.tsukuba.ac.jp
[2] National Institute of Informatics (NII)/The Graduate University for Advanced Studies (SOKENDAI), 2-1-2 Hitotsubashi, Chiyoda-ku, Tokyo 101-8430, Japan
onono@nii.ac.jp

Abstract. In this paper, we propose a method to estimate a correlation coefficient of two correlated complex signals on the condition that only the amplitudes are observed and the phases are missing. Our proposed method is based on a maximum likelihood estimation. We assume that the original complex random variables are generated from a zero-mean bivariate complex normal distribution. The likelihood of the correlation coefficient is formulated as a bivariate Rayleigh distribution by marginalization over the phases. Although the maximum likelihood estimator has no analytical form, an expectation-maximization (EM) algorithm can be formulated by treating the phases as hidden variables. We evaluate the accuracy of the estimation using artificial signal, and demonstrate the estimation of narrow-band correlation of a two-channel audio signal.

Keywords: Correlation · Complex signal · Maximum likelihood · EM algorithm

1 Introduction

Correlation of complex sequences plays an important role in array signal processing and almost all of its application [1]. The estimation of correlation is easily obtained by a simple product sum of the signal sequences. However, the correlation estimation by the product sum cannot be used when the phase observation is unreliable or unavailable. For example, in asynchronous distributed acoustic sensing systems which gather signals observed by multiple independent recording devices, the biases of the sampling frequencies of the individual devices cause drift of the phases [2,3]. The effect of the drift on the amplitude is not serious, but the phase is strongly affected [4]. Also, when we analyze the output of the nonlinear signal processing in the amplitude domain, such as nonnegative matrix factorization (NMF) [5], the phases are often missing. Although many phase estimation methods are studied [6], the accurate estimation of the correlation cannot be guaranteed.

The goal of this paper is to estimate a correlation coefficient of two complex signal channels from the observation of amplitude without information of the

© Springer International Publishing Switzerland 2015
E. Vincent et al. (Eds.): LVA/ICA 2015, LNCS 9237, pp. 421–428, 2015.
DOI: 10.1007/978-3-319-22482-4_49

phase. Since the correlation coefficient is expressed by a nonnegative number, it cannot be used for estimation of phase difference, and the usage is somewhat limited. Still, the correlation coefficient gives important information, and the correlation estimation from amplitude is useful for specific purposes. For example, the estimation can be used to reduce the computational cost of the maximum likelihood compensation of drift [3], which requires large computational power and memory. As discussed in [3], only the limited frequency bins with high correlation contributes much to the observation, and estimation of the correlation only from the amplitude is informative for the efficient analysis discarding the useless frequency bins. Also, the correlation estimation from amplitude is expected to be useful for evaluation of channel capacity [7], for the estimation of SNR to optimize of the coefficients of the signal enhancement such as SS or Wiener filter, and so on.

To estimate a correlation coefficient between two complex random variables without phase observation, we propose a maximum likelihood estimation assuming that the original complex variables are generated from a zero-mean bivariate complex normal distribution. We show that the likelihood is given by the marginalization of the arguments, appearing as a *bivariate Rayleigh distribution* [7], whose parameter estimation algorithm has not been derived to the best of our knowledge. We derive an expectation-maximization (EM) algorithm to maximize the likelihood of the bivariate Rayleigh distribution, and obtain the maximum likelihood estimator of the correlation coefficient.

2   Statement of Problem

Suppose there are two correlated complex random variables $X_i \in \mathbb{C}$, $i = 1, 2$, which have means of zero, variances σ_i^2, correlation ρ and the uniform arguments Θ_i as

$$E[X_i] = 0, \tag{1}$$

$$E\left[|X_i|^2\right] = \sigma_i^2, \tag{2}$$

$$E[X_1 X_2^*] = \sigma_1 \sigma_2 \rho, \tag{3}$$

$$f_{\Theta_i}(\theta_i) = \frac{1}{2\pi}, \quad -\pi \le \theta_i < \pi, \tag{4}$$

where $E[\cdot]$ is the expectation operator, $|\cdot|$ is the absolute value, $\{\cdot\}^*$ is the complex conjugate, σ_i^2 is the variance of X_i, $\Theta_i = \angle X_i$, $\angle\{\cdot\}$ is the argument, and $f_A(a)$ denotes the probability density of a random variable A whose sample is denoted as a. Hereafter, we denote random variables and the samples with upper and lower case letters, respectively.

Suppose the complex random variables X_1, X_2 are unavailable, but we can observe their absolute values Y_1, Y_2:

$$Y_i = |X_i|. \tag{5}$$

Our goal is to estimate the correlation coefficient $|\rho|$, the absolute value of the cross correlation ρ between X_1 and X_2, under the condition that only the absolute values are observed. It is obvious that the maximum likelihood estimator of the variance σ_i^2 can be obtained as the average of the square of the observation:

$$\sigma_i^2 = E\left[|X_i|^2\right] \leftarrow \frac{1}{N}\sum_{n=1}^{N} y_i(n)^2, \tag{6}$$

where (n), $n = 1,\ldots,N$ denotes the index of the N observations. Note that we omit the sample index (n) when the explicit declaration is unnecessary. In contrast to the estimation of variance, the correlation coefficient $|\rho|$ is different from the correlation of the absolute observations:

$$|\rho| = \frac{|E[X_1 X_2^*]|}{\sigma_1\sigma_2} \neq \frac{E[Y_1 Y_2]}{\sigma_1\sigma_2}. \tag{7}$$

Thus the following mean of the product of the observed absolute samples does not give a good estimation:

$$|\rho| \leftarrow \frac{1}{N\sigma_1\sigma_2}\sum_{n=1}^{N} y_1(n)\, y_2(n). \tag{8}$$

Therefore, the estimation of the correlation coefficient with the absolute observation is not trivial.

3   Correlation Estimation Assuming Bivariate Complex Normal Distribution

3.1   Probabilistic Model

In this section, we discuss the estimation of the correlation coefficient $|\rho|$ from the absolute samples y_1, y_2 assuming that the unobserved complex samples x with the statistics given by (1)–(3) are generated from a zero-mean bivariate complex normal distribution as

$$f_{X_1,X_2}(x_1,x_2;\rho) = \frac{\exp\left(-\frac{\sigma_2^2|x_1|^2+\sigma_1^2|x_2|^2-2\sigma_1\sigma_2\mathrm{Re}[\rho^* x_1 x_2^*]}{\sigma_1^2\sigma_2^2(1-|\rho|^2)}\right)}{\pi^2\sigma_1^2\sigma_2^2\left(1-|\rho|^2\right)}. \tag{9}$$

Then, the joint density of the observation Y_i and the unobserved argument Θ_1, Θ_2 is expressed as

$$f_{Y_1,Y_2,\Theta_1,\Theta_2}(y_1,y_2,\theta_1,\theta_2;\rho) = \frac{y_1 y_2 \exp\left(-\frac{\sigma_2^2 y_1^2+\sigma_1^2 y_2^2-2\sigma_1\sigma_2|\rho|y_1 y_2\cos(\theta_1-\theta_2-\angle\rho)}{\sigma_1^2\sigma_2^2(1-|\rho|^2)}\right)}{\pi^2\sigma_1^2\sigma_2^2\left(1-|\rho|^2\right)}, \tag{10}$$

by applying the polar coordinate conversion to (9). By the marginalization of the uniform distribution of the arguments Θ_1, Θ_2, the likelihood of the absolute observation Y_1, Y_2 is given as a bivariate Rayleigh distribution, which can be found in many papers, e.g., [7] as a special case of multivariate Nakagami-m distributions:

$$
f_{Y_1,Y_2}(y_1, y_2; |\rho|) = \int_{-\pi}^{\pi} \int_{-\pi}^{\pi} f_{Y_1,Y_2,\Theta_1,\Theta_2}(y_1, y_2, \theta_1, \theta_2; \rho)\, d\theta_1 d\theta_2
$$

$$
= \frac{4 y_1 y_2}{\sigma_1^2 \sigma_2^2 \left(1 - |\rho|^2\right)} I_0 \left(\frac{2|\rho|\, y_1 y_2}{\sigma_1 \sigma_2 \left(1 - |\rho|^2\right)} \right) \exp\left(- \frac{\sigma_2^2 y_1^2 + \sigma_1^2 y_2^2}{\sigma_1^2 \sigma_2^2 \left(1 - |\rho|^2\right)} \right),
\tag{11}
$$

where $I_\nu(\cdot)$ denotes the modified Bessel function of the first kind with the order ν. It can be seen that the density of Y_1 and Y_2 depends on the correlation coefficient $|\rho|$ but not on the argument $\angle\rho$ of the correlation. Therefore, the maximization of the likelihood gives the estimation of the correlation coefficient $|\rho|$. However, the maximum likelihood estimator does not have the analytical form.

3.2   Maximum Likelihood Estimation by EM Algorithm

Here we describe the maximum likelihood estimation of the correlation coefficient $|\rho|$ by the iterative procedure. By treating the observed samples $y_1(n)$ and $y_2(n)$ together with the unobserved arguments $\theta_1(n)$ and $\theta_2(n)$, we can formulate the EM algorithm. We treat the arguments Θ_1, Θ_2 as hidden variables, and the posterior density of the hidden variables is given by

$$
f_{\Theta_1,\Theta_2|Y_1,Y_2}(\theta_1, \theta_2 | y_1, y_2; \rho) = \frac{f_{Y_1,Y_2,\Theta_1,\Theta_2}(y_1, y_2, \theta_1, \theta_2; \rho)}{f_{Y_1,Y_2}(y_1, y_2; \rho)}
$$

$$
= \frac{\exp\left(\frac{2|\rho| y_1 y_2 \cos(\theta_1 - \theta_2 - \angle\rho)}{\sigma_1 \sigma_2 (1 - |\rho|^2)} \right)}{2\pi I_0 \left(\frac{2|\rho| y_1 y_2}{\sigma_1 \sigma_2 (1 - |\rho|^2)} \right)}.
\tag{12}
$$

Then, the auxiliary function $Q(|\rho|, |\bar\rho|)$ to maximize in each iteration of the EM algorithm is obtained as

$$
Q(|\rho|; |\bar\rho|) = \sum_{n=1}^{N} \left\langle \log f_{Y_1,Y_2,\Theta_1,\Theta_2}(y_1(n), y_2(n), \theta_1(n), \theta_2(n); \rho) \right\rangle_{\theta_1(n),\theta_2(n)|y_1(n),y_2(n);|\bar\rho|}
$$

$$
\propto \frac{2|\rho|}{\sigma_1 \sigma_2 \left(1 - |\rho|^2\right)} \sum_{n=1}^{N} y_1(n) y_2(n) \lambda(n) - N \log\left(1 - |\rho|^2\right),
\tag{13}
$$

$$
\langle g(a) \rangle_{a|b;c} = \int_{\mathcal{D}(a)} f_{A|B}(a|b; c)\, g(a)\, da,
\tag{14}
$$

where $\mathcal{D}(a)$ is the domain of the variable a, and $\lambda(n)$ is the posterior expectation of $\cos(\theta_1(n) - \theta_2(n) - \angle\rho)$ with the current parameter estimation $|\bar{\rho}|$, given by

$$\lambda(n) = \langle \cos(\theta_1(n) - \theta_2 - \angle\rho) \rangle_{\theta_1(n),\theta_2(n)|y_1(n),y_2(n);|\bar{\rho}|}$$

$$= L\left(\frac{2|\bar{\rho}|\,y_1(n)y_2(n)}{\sigma_1\sigma_2\left(1 - |\bar{\rho}|^2\right)}\right), \tag{15}$$

$$L(x) = I_1(x)/I_0(x) \tag{16}$$

Since $L(x)$ is a monotonically increasing function giving $L(0) = 0$ and $\lim_{x\to\infty} L(x) = 1$, $\lambda(n)$ acts as weighting in the update of the estimation of $|\rho|$ in the M-step:

$$|\rho| = \frac{1}{N\sigma_1\sigma_2}\sum_{n=1}^{N} y_1(n)\,y_2(n)\,\lambda(n). \tag{17}$$

By iterating the updates of E- and M-steps given by (15) and (17), respectively, the estimation of $|\rho|$ converges to a local optimal. Note that σ_1^2 and σ_2^2 are estimated by (6), which can also be derived as the maximization of the auxiliary function, although the related terms are omitted in (13). Also note that $L(x)$ can be calculated by one-dimensional table lookup.

4 Experimental Results

4.1 Evaluation with Artificial Signal

To evaluate the performance of the proposed method, we conducted a numerical simulation to estimate the correlation coefficients of artificial two-channel complex signals generated from pseudorandom numbers. To show the baseline, we also evaluated the performance of absolute correlation given by (8). In addition, to show the upper limit with the ideal condition, we evaluated the standard maximum likelihood estimation under the condition where the original complex sequences is available, given by

$$|\rho| \leftarrow \frac{1}{N\sigma_1\sigma_2}\left|\sum_{n=1}^{N} x_1(n)^* x_2(n)\right|. \tag{18}$$

As the evaluation criterion, we calculated the root mean squared errors (RMSEs).

The correlated data are generated by the linear mixture of the two independent pseudorandom numbers with the same variance. We controlled the correlation by manipulating the linear mixture. To evaluate the robustness of the proposed method against the mismatch of the Gaussian assumption, we also evaluated the artificial signals generated by super-Gaussian pseudorandom numbers. The super-Gaussian data are generated from the circular generalized normal distribution [8], whose density of the random variables X_1, X_2 is given by

$$f_{X_i}(x_i) = \frac{c_i \exp\left(-\left(\frac{|x_i|}{\xi_i}\right)^{c_i}\right)}{2\pi\xi_i^2\Gamma\left(\frac{2}{c_i}\right)}, \tag{19}$$

where $\xi_i > 0$ is the scale parameter, $c_i > 0$ is the shape parameter, and $\Gamma(\cdot)$ is a gamma function. The shape parameters are set as $c_1 = 0.5$ and $c_2 = 0.8$, and scale parameters are adjusted to give the unit variance. With such shape parameter setting, the sequences have super Gaussian property with long tail, and their kurtosis, a well-used Gaussianity measure, are about 7.4 and 2.3. Note that the kurtosis is changed after the mixing to give correlation.

We show an example of the estimation in Fig. 1. For both Gaussian and super-Gaussian data, the absolute correlation does not give accurate estimation. Although the variance of the proposed method is slightly larger than the estimation with arguments, the estimated correlation coefficients distribute near the true correlation coefficients and the accuracy of the proposed method is match better than the baseline. The estimation of the correlation coefficients of super-Gaussian data tends to underestimate the correlation, but the RMSEs are similar to the Gaussian case without the model mismatch. Thus it is confirmed that the proposed method can effectively estimate the correlation coefficients of the complex sequences without the observation of arguments.

To examine the effect of the bias and variances of the estimation, we evaluate RMSEs of various numbers of observed samples. The result is shown in Fig. 2. Although the proposed method has larger variance than the ideal estimation with phase, we can see that the estimation accuracy of the proposed method improves according to the increase of the number of the samples when the data is Gaussian. Thus the proposed method can estimate the correlation of Gaussian data effectively with small bias. However, we can see the saturation of the improvement of the accuracy under the condition mismatch with the super-Gaussian data.

4.2   Demonstration with Audio Data

As a demonstration of the correlation coefficient estimation in practical signal processing, we evaluated the estimation of the correlation coefficients of the narrow band amplitudes of a two-channel audio signal.

We analyzed observation of speech mixture by an array of two microphones. The data is chosen from the Underdetermined Test dataset of the Signal Separation Evaluation Campaign (SiSEC) [9], a benchmark of speech separation. Utterances of four female speakers were recorded in a room whose reverberation time T_{60} is 250 ms. The spacing of the microphones was 1 m. The recorded data was 10 s long, sampled with the frequency of 16 kHz. The signal was analyzed by short-time Fourier transform with the von Hann window of the length 1024 samples and the shift 128 samples. The number of the frames is 1258.

The estimated results are shown in Fig. 3. In contrast to the artificial signals, the true correlation coefficients are unknown. Thus it should be noted that the horizontal axis is not the true correlation coefficients but the estimated correlation coefficients by the ideal estimation in (18) with the original complex signal given. We can see that the proposed method is much better than the baseline. Superiority of the proposed method holds for other frame lengths, which strongly

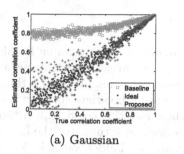 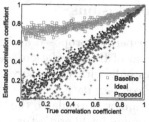

(a) Gaussian   (b) Super-Gaussian

Fig. 1. An example of the estimation results of correlation coefficients between two artificial signals generated from (a) Gaussian and (b) super-Gaussian pseudorandom numbers. The number of samples was 100 for each trial. The EM iteration number was 10. The RMSEs are about 0.42, 0.05 and 0.12 for the baseline, the ideal estimation and the proposed method, respectively for the results in (a), and about 0.05, 0.37 and 0.12 for the results in (b).

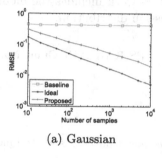 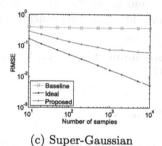

(a) Gaussian   (c) Super-Gaussian

Fig. 2. Root mean squared errors for the number of the samples $N = 10, 20, 50,$ $100, 200, 500, 1000, 2000, 5000, 10000$. The RMSEs of each condition was calculated from 100 trials. The EM iteration number was set to 200.

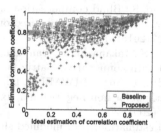

Fig. 3. Estimation results of narrow band correlation coefficients of a two-channel audio signal. The EM iteration number was 10. The RMSEs from the ideal estimation are about 0.28 and 0.20 for the baseline and the proposed method, respectively.

affects the correlation. However, the accuracy is not as good as that of the artificial signal. Thus, further analysis is required to improve the estimation accuracy with the realistic signals.

5   Conclusions

In this paper, we proposed an EM algorithm to obtain the maximum likelihood estimation of the correlation coefficient of the two correlated complex random variables only with the observation of the absolute values. Assuming that the original complex data are generated from a zero-mean bivariate complex normal distribution, we formulated the likelihood of the correlation coefficient. Although the maximum likelihood estimator is not analytical, we formulated the EM algorithm by treating the difference of the arguments as a hidden variable. We evaluated accuracy and the robustness against the model mismatch of the proposed method by the simulation using artificial signals. Also, we demonstrated the estimation of the narrow-band correlation coefficients of the two-channel audio signal. It is confirmed that the proposed method is much more accurate than the estimation with the correlation of the absolute values. However, estimation of the real signal was worse than that of the artificial signal, and the further analysis and improvement are required. Our future work includes the modified maximum likelihood estimation with a non-Gaussian distribution.

Acknowledgment. This work was supported by JSPS KAKENHI Grant Number 23240023.

References

1. Johnson, D., Dudgeon, D.: Array Signal Processing: Concepts and Techniques. Simon & Schuster, New York (1992)
2. Markovich-Golan, S., Gannot, S., Cohen, I.: Blind sampling rate offset estimation and compensation in wireless acoustic sensor networks with application to beamforming. In: Proceedings of IWAENC, pp. 1–4 (2012)
3. Miyabe, S., Ono, N., Makino, S.: Blind compensation of interchannel sampling frequency mismatch for ad hoc microphone array based on maximum likelihood estimation. Signal Process. **107**(2015), 185–196 (2015)
4. Chiba, H., Ono, N., Miyabe, S., Takahashi, Y., Yamada, T., Makino, S.: Amplitude-based speech enhancement with nonnegative matrix factorization for asynchronous distributed recording. In: Proceedings of IWAENC, pp. 204–208 (2014)
5. Lee, D., Seung, H.: Algorithms for non-negative matrix factorization. In: Proceedings of NIPS, pp. 556–562 (2001)
6. Griffin, D., Lim, J.: Signal estimation from modified short-time Fourier transform. IEEE Trans. ASSP **32**(2), 236–243 (1984)
7. Fraidenraich, G., Lévêque, O., Cioffi, J.: On the MIMO channel capacity for the nakagami-m channel. IEEE Trans. Inf. Theory **54**(8), 3752–3757 (2008)
8. Novey, M., Adali, T., Roy, A.: A complex generalized Gaussian distribution-characterization, generation and estimation. IEEE Trans. Signal Process. **58**(3), 1427–1433 (2010)
9. Ono, N., Koldovsky, Z., Miyabe, S., Ito, N.: The 2013 Signal Separation Evaluation Campaign. In: Proceedings of MLSP, pp. 1–6 (2013)

Deep Karaoke: Extracting Vocals from Musical Mixtures Using a Convolutional Deep Neural Network

Andrew J.R. Simpson$^{(\boxtimes)}$, Gerard Roma, and Mark D. Plumbley

Centre for Vision, Speech and Signal Processing,
University of Surrey, Guildford, UK
{andrew.simpson, g.roma, m.plumbley}@surrey.ac.uk

Abstract. Identification and extraction of singing voice from within musical mixtures is a key challenge in source separation and machine audition. Recently, deep neural networks (DNN) have been used to estimate 'ideal' binary masks for carefully controlled cocktail party speech separation problems. However, it is not yet known whether these methods are capable of generalizing to the discrimination of voice and non-voice in the context of musical mixtures. Here, we trained a convolutional DNN (of around a billion parameters) to provide probabilistic estimates of the ideal binary mask for separation of vocal sounds from real-world musical mixtures. We contrast our DNN results with more traditional linear methods. Our approach may be useful for automatic removal of vocal sounds from musical mixtures for 'karaoke' type applications.

Keywords: Deep learning · Supervised learning · Convolution · Source separation

1 Introduction

Much work in audio source separation has been inspired by the ability of human listeners to maintain separate auditory neural and perceptual representations of competing speech in 'cocktail party' listening scenarios [1–3]. A common engineering approach is to decompose a mixed audio signal, comprising two or more competing speech signals, into a spectrogram in order to assign each time-frequency element to the respective sources [4–7]. Hence, this form of source separation may be interpreted as a classification problem.

A benchmark for this approach is known as the 'ideal binary mask' and represents a performance ceiling on the approach by providing a fully-informed separation based on the spectrograms for each of the component source signals. Using the source spectrograms, each time-frequency element of the mixture spectrogram may be attributed to the source with the largest magnitude in the respective source spectrogram. This ideal binary mask may then be used to establish reference separation performance. In a recent approach to binary-mask based separation, the ideal binary mask was used to train a deep neural network (DNN) to directly estimate binary masks for new mixtures [7]. However, this approach was limited to a single context of two known speakers and

© Springer International Publishing Switzerland 2015
E. Vincent et al. (Eds.): LVA/ICA 2015, LNCS 9237, pp. 429–436, 2015.
DOI: 10.1007/978-3-319-22482-4_50

a sample rate of only 4 kHz. Therefore, it is not yet known whether the approach is capable of generalizing to less well controlled scenarios featuring unknown voices and unknown background sounds. In particular, it is not known whether such a DNN architecture is capable of generalizing to the more demanding task of extracting unknown vocal sounds from within unknown music [8–10].

In this paper, we employed a diverse collection of real-world musical multi-track data produced and labelled (on a song-by-song basis) by music producers. We used 63 typical 'pop' songs in total, each featuring vocals of various kinds. For each multi-track song/mix, comprising a set of component 'stems' (vocals, bass, guitars, drums, etc.), we pooled audio labeled as 'vocal' separately to all other audio (i.e., the accompanying instruments). We then obtained arbitrary mixtures for each song, simulating the process of mixing to produce 'mixes' for each song. Using the first 50 songs as training data, we trained a convolutional DNN to predict the ideal binary masks for separating the respective vocal and non-vocal signals for each song. For reference, we also trained an equivalent linear method (convolutional non-negative matrix factorization - NMF) of similar scale. We then tested the respective models on mixes of new songs featuring different musical arrangements, different singing and different production. From both models we obtained probabilistic estimates of the ideal binary mask and analyzed the resulting separation quality using objective source separation quality metrics. These results demonstrate that a convolutional DNN approach is capable of generalizing voice separation, learned in a musical context, to new musical contexts. We also illustrate the capability of the probabilistic convolutional approach [7] to be optimized for different priorities of separation quality according to the statistical interpretation employed. In particular, we highlight the differences in performance for the two respective architectures in the context of the trade-off between artefacts and separation.

2   Method

We consider a typical simulated ensemble musical performance scenario featuring a variety of musical contexts and a variety of vocal performances. In each context, which we refer to as 'a song', there are a multitude of musical accompaniment signals and at least one (often more) vocal signals. The various signals are mixed together (arbitrarily) and the resulting mixture is refered to as 'a mix'. The engineering problem is to auto-matically separate all vocal signals from the concurrent accompaniment signals. We used 63 fully produced songs, taken from the MedleyDB database [11]. The average duration of the songs was 3.7 min (standard deviation (STD): ±2.7 min). The average number of accompanying sources (stems) was 7.2 (STD: ±6.6 sources) and the average number of vocal sources was 1.8 (STD: ±0.8 sources).

For each song, the source signals were classified as either vocal or non-vocal (according to the labels assigned by the music producers). Vocal sounds included both male and female singing voice and spoken voice ('rap'). Non-vocal sounds included accompanying instruments (drums, bass, guitars, piano, etc.). Source sounds were studio recorded and featured relatively little interference from other sources. All source sounds were then peak normalized before being linearly summed into either a vocal mixture or a non-vocal mixture respectively. The two separate (vocal/non-vocal) mixtures were then peak normalized and linearly summed to provide a complete mixture (i.e., a 'final production mix'). This provided for a mixture that resembled a mix that might be produced by a human mixing engineer [12]. All sources and mixtures were monaural (i.e., we did not employ any stereo processing).

All signals were sampled at a rate of 44.1 kHz. The respective source (vocal/non-vocal) and mixture signals were transformed into spectrograms using the short-time Fourier transform (STFT) with window size of 2048 samples, overlap interval of 512 samples and a Hanning window. This provided spectrograms with 1025 frequency bins. The phase component of each spectrogram was removed and retained for later use in inversion. From the source spectrograms a binary mask was computed where each element of the mask was determined by comparing the magnitudes of the corresponding elements of the source (vocal/non-vocal) spectrograms and assigning the mask a '1' when the vocal spectrogram had greater magnitude and '0' otherwise.

The first 50 songs (taken in arbitrary order) were used as training data and the final 13 songs were used as test data. The magnitude-only mixture spectrograms computed from the first 50 songs and the respective ideal binary masks were used as training data. Note, phase was not used in training the model.

For the training data, the mixture spectrogram and the corresponding source spectrograms were cut up into corresponding windows of 20 samples (in time). The windows shifted at intervals of 60 samples (i.e., there was no overlap). Thus, for every 20-sample window, for training the models there was a mixture spectrogram matrix of size 1025×20 (frequency bins x time) samples and an ideal binary mask matrix of the same size. From the 50 songs designated as training data, this gave approximately 15,000 training examples. For the testing stage, the spectrograms for the remaining 13 songs were cut up with overlap intervals of 1 sample (which would ultimately be applied in an overlapping convolutional output stage). Prior to windowing, all spectrogram data was normalized to unit scale.

Deep Neural Network. We used a feed-forward DNN of size $20500 \times 20500 \times 20500$ units ($1025 \times 20 = 20500$). Each spectrogram window of size 1025×20 was unpacked into a vector of length 20500. The DNN was configured such that the input layer was the mixture spectrogram (20500 samples). The DNN was trained to synthesize the ideal binary mask at its output layer. The DNN employed the biased-sigmoid activation function [13] throughout with zero bias for the output layer. The DNN was trained using 100 full iterations of stochastic gradient descent (SGD). Each iteration of SGD featured a full sweep of the training data. Dropout was not used in training. After training, the model was used as a feed-forward probabilistic device.

Probabilistic Binary Mask. In the testing stage, there was an overlap interval of 1 sample. This means that the test data described the mixture spectrogram in terms of a sliding window and the output of the model described predictions of the ideal binary mask in the same sliding window format. The output layer of the DNN was sigmoidal and hence we may interpret these predictions in terms of the logistic function. Therefore, because of the sliding window, this procedure resulted in a distribution (size 20) of predictions for each time-frequency element of the mixture spectrogram [7]. We chose to summarize this distribution by taking the mean and we evaluate the result in terms of an empirical confidence estimate, separately for each source, as follows:

For each time-frequency element, of each source, we computed the mean prediction and applied a confidence threshold (α);

$$M_{t,f}^{V} = \begin{cases} 1 & for \quad \frac{1}{T}\sum_{i=0}^{T} S_{t+i,f} > \alpha \\ 0 & for \quad \frac{1}{T}\sum_{i=0}^{T} S_{t+i,f} \leq \alpha \end{cases} \tag{1}$$

where M^{V} refers to the binary mask for the vocal source, T refers to the window size (20), t is the time index, i is the window index and f is the frequency (bin) index into the estimated mask (S). The corresponding (but independent) binary mask for the non-vocal source (M^{NV}) is computed as follows;

$$M_{t,f}^{NV} = \begin{cases} 1 & for \quad \frac{1}{T}\sum_{i=0}^{T} S_{t+i,f} < (1-\alpha) \\ 0 & for \quad \frac{1}{T}\sum_{i=0}^{T} S_{t+i,f} \geq (1-\alpha) \end{cases} \tag{2}$$

Thus, by adjustment of α, masks at different levels of confidence could be constructed for both sources.

Non-negative Matrix Factorization. For comparison to the DNN approach, an equivalent non-negative matrix factorization (NMF) based approach was implemented using the same training and test data (as described above). We used the same unpacking strategy, which has been tested before for NMF-based separation of speech and music [14]. The spectrograms of the training data were sampled and unpacked analogously to the DNN approach, resulting in a large (220500 × 15000) matrix that was then decomposed using the traditional multiplicative updates algorithm with KL divergence [15]. This means that for this training matrix V, $V = WH$, where we set the number of basis vectors (columns of W) and the respective activations (rows of H) to 1500. We performed this training stage for both vocal and non-vocal mixtures, and kept the two basis vectors matrices W_{v} and W_{nv}. For the testing stage, we concatenated both matrices and initialized a corresponding H_{u} matrix randomly, so that for each unpacked spectrogram, V_{u}, of the set of test songs, $V_{u} = [W_{v}\ W_{nv}]\ H_{u}$. We then ran the same multiplicative updates algorithm but keeping the composite W_{u} matrix fixed [14], and updating H_{u}. The test spectrogram was then re-composed for either vocal $(V_{v} = W_{v}H_{v})$ or non-vocal $(V_{nv} = W_{nv}H_{nv})$ vectors, and used to define a soft mask via the element-wise division $S_{v} = V_{v}/(V_{v} + V_{nv})$. The matrix was then packed back to the original spectrogram size by averaging the consecutive frames of the soft mask. This allowed us to define an equivalent α parameter (as used in the DNN approach) so that the binary mask $B_{v} = 1$ when $S_{v} > \alpha$, 0 when $S_{v} <= \alpha$ and analogously for B_{nv} and S_{nv}.

Finally, the respective masks were resolved by multiplication with the original (complex) mixture spectrogram and the resulting masked spectrograms were inverted with a standard overlap-and-add procedure. Separation quality (for the test data) was measured using the BSS-EVAL toolbox [16] and is quantified in terms of signal-to-distortion ratio (SDR), signal-to-artefact ratio (SAR) and signal-to-interference ratio (SIR). Separation quality was assessed at different confidence levels by setting different values of α.

3 Results

Figure 1 plots spectrograms illustrating the stages of mixture and separation for a brief excerpt (~ 1.5 s) from a randomly chosen test song separated using the DNN (at $\alpha = 0.5$). The spectrograms for the source vocal and non-vocal signals are shown at the top. The middle panel plots the mixture spectrogram, illustrating the difficulty of the problem (even for an ideal binary mask). At the bottom of Fig. 1 are plotted spectrograms representing the separated audio for the vocal and non-vocal signals respectively (DNN, $\alpha = 0.5$).

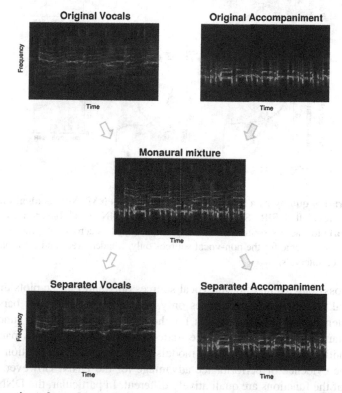

Fig. 1. Separation of vocal sounds from musical mixtures using a probabilistic convolutional deep neural network. Spectrograms for 1.5 s excerpt from test song; Upper: original sources, middle: monaural mixture, lower: separated audio (DNN, $\alpha = 0.5$). Note: frequency axis represents the range 0–22 kHz on a logarithmic axis.

The various objective source separation quality metrics (SDR/SIR/SAR) were computed for the separated sources, as estimated with each model, as a function of α. The same measures were also computed for the ideal binary mask. Figure 2 plots a summary of the respective measures. For each measure, and for each separation context (DNN/ideal binary mask/NMF), Fig. 2a plots the mean across-song performance computed by first averaging the measures across vocal/non-vocal sources. Figure 2b

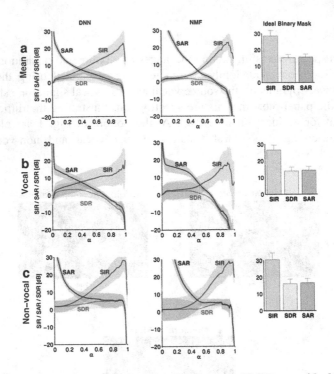

Fig. 2. Separation quality as a function of α: DNN versus NMF versus ideal binary mask. Left to right are plotted SIR, SDR and SAR for the DNN, ideal binary mask and NMF respectively. (**a**) plots across-source, across-song mean, (**b**) plots across-song mean for the vocal sources, (**c**) plots the same for the non-vocal sources only. Shaded areas and error bars represent 95 % confidence intervals.

plots the across-song average for the vocal sources only and Fig. 2c plots the same for the non-vocal (accompaniment) sources only. Shaded areas and error bars represent 95 % confidence intervals. The results for the DNN and NMF (as a function of α) feature similar functions illustrating the trade-off between the various parameters as statistical confidence is adjusted. Both models provide similar intersection points and there is some evidence of performance advantage for the DNN. However, the slopes and shapes of the functions are qualitatively different. In particular, the DNN functions for SAR and SIR more closely resemble 'ideal' sigmoid functions. In this context, SAR and SIR may be interpreted as energetic equivalents of positive hit rate (SIR) and false positive rate (SAR). Hence, if these slopes are interpreted as being analogous to cumulative density functions (indexed using α), then the DNN results might be interpreted as demonstrating a wider probability function that is closer to normally distributed. However, although these plots provide insight into the mapping of probability to performance, they do not provide a very interpretable comparison of the models. In particular, the plots do not allow us to interpret performance in like terms with respect to the critical trade-off between artefacts and separation.

In order to provide a like-for-like comparison, Fig. 3a plots mean SAR as a function of mean SIR for both models (taken from Fig. 2a) for the useable range of

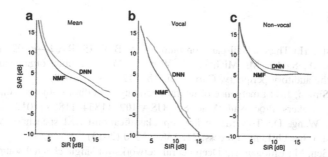

Fig. 3. Trade-off: interference versus artefacts. DNN versus NMF. mean SAR as a function of mean SIR [0.1 < α < 0.9] (**a**) overall mean (Fig. 2a), (**b**) vocals (Fig. 2b), (**c**) non-vocals (Fig. 2c).

0.1 < α < 0.9. Figure 3b plots the same for the functions of Figs. 2b and 3c plots the same for the functions of Fig. 2c. Overall (Fig. 3a), the DNN provides $\sim$3 dB better SAR performance for a given SIR index. This advantage is mostly explained by the $\sim$5 dB advantage for the vocal sources (Fig. 3b) and only a small advantage is evident for the non-vocal signals (Fig. 3c).

4   Discussion and Conclusion

We have demonstrated that a convolutional deep neural network is capable of separating vocal sounds from within typical musical mixtures. Our convolutional DNN is of nearly a billion parameters and was trained with relatively little data (and relatively few iterations of SGD). We have contrasted this performance with a like-for-like (suitably scaled) NMF approach, in the context of a trade-off between artefact and separation quality, indexed via confidence in the statistical predictions made.

The main advantage of the DNN appears to be in its general learning of what 'vocal' sounds are (Fig. 3b). Since the NMF approach is limited to linear factorization, we may at least partly attribute the advantage of the (nonlinear) DNN to abstract learning via demodulation [13]. The DNN appears to have biased it's learning towards making good predictions via correct positive identification of the vocal sounds.

Both methods feature the largest known parameterizations for this particular problem and, to some extent, both methods may be considered 'deep' [7, 13, 17]; both featured demodulated (magnitude) spectrograms produced using STFT and re-synthesis via inverse STFT. We also note that the relatively small amount of data employed in training the DNN may have been offset by the fact that the spectrograms were sampled using a Hanning window, hence minimizing aliasing/distortion in the training data that may otherwise have resulted in over-fitting [18] (and see [19]).

Acknowledgment. AJRS, GR and MDP were supported by grant EP/L027119/1 from the UK Engineering and Physical Sciences Research Council (EPSRC). Data and materials are available at doi:10.15126/surreydata.00807909.

References

1. McDermott, J.H.: The cocktail party problem. Curr. Biol. **19**, R1024–R1027 (2009)
2. Pressnitzer, D., Sayles, M., Micheyl, C., Winter, I.M.: Perceptual organization of sound begins in the auditory periphery. Curr. Biol. **18**, 1124–1128 (2008)
3. Ding, N., Simon, J.Z.: Emergence of neural encoding of auditory objects while listening to competing speakers. Proc. Natl. Acad. Sci. USA **109**, 11854–11859 (2012)
4. Wang, Y., Wang, D.: Towards scaling up classification-based speech separation. IEEE Trans. Audio Speech Lang. Process. **21**, 1381–1390 (2013)
5. Grais, E., Sen, M., Erdogan, H.: Deep neural networks for single channel source separation. In: 2014 IEEE International Conference on Acoustics, Speech and Signal Processing (ICASSP), pp. 3734–3738 (2014)
6. Huang, P.S., Kim, M., Hasegawa-Johnson, M., Smaragdis, P.: Deep learning for monaural speech separation. In: 2014 IEEE International Conference on Acoustics, Speech and Signal Processing (ICASSP), pp. 1562–1566 (2014)
7. Simpson, A.J.R.: Probabilistic Binary-Mask Cocktail-Party Source Separation in a Convolutional Deep Neural Network, arxiv.org abs/1503.06962 (2015)
8. Abrard, F., Deville, Y.: A time–frequency blind signal separation method applicable to underdetermined mixtures of dependent sources. Sig. Process. **85**, 1389–1403 (2005)
9. Ryynanen, M., Virtanen, T., Paulus, J., Klapuri, A.: Accompaniment separation and karaoke application based on automatic melody transcription. In: 2008 IEEE International Conference on Multimedia and Expo, pp. 1417–1420 (2008)
10. Raphael, C.: Music plus one and machine learning. In: Proceedings 27th International Conference on Machine Learning (ICML-2010), pp. 21–28 (2010)
11. Bittner, R., Salamon, J., Tierney, M., Mauch, M., Cannam, C., Bello, J.P.: MedleyDB: a multitrack dataset for annotation-intensive MIR research. In: 15th International Society Music Information Retrieval Conference (2014)
12. Terrell, M.J., Simpson, A.J.R., Sandler, M.: The mathematics of mixing. J. Audio Eng. Soc. **62**(1/2), 4–13 (2014)
13. Simpson, A.J.R.: Abstract Learning via Demodulation in a Deep Neural Network, arxiv.org abs/1502.04042 (2015)
14. Grais, E.M., Erdogan, H.: Single channel speech music separation using nonnegative matrix factorization and spectral masks. In: 2011 17th International Conference on Digital Signal Processing (DSP), pp. 1–6. IEEE (2011)
15. Lee, D.D., Seung, H.S.: Algorithms for non-negative matrix factorization. In: Advances in Neural Information Processing Systems, pp. 556–562 (2001)
16. Vincent, E., Gribonval, R., Févotte, C.: Performance measurement in blind audio source separation. IEEE Trans. Audio Speech Lang. Process. **14**, 1462–1469 (2006)
17. Simpson, A.J.R.: Deep Transform: Error Correction via Probabilistic Re-Synthesis, arxiv.org abs/1502.04617 (2015)
18. Simpson, A.J.R.: Over-Sampling in a Deep Neural Network, arxiv.org abs/1502.03648 (2015)
19. Hinton, G.E., Srivastava, N., Krizhevsky, A., Sutskever, I., Salakhutdinov, R.: Improving neural networks by preventing co-adaptation of feature detectors. The Computing Research Repository (CoRR), abs/1207.0580 (2012)

Evaluation of the Convolutional NMF for Supervised Polyphonic Music Transcription and Note Isolation

Stanislaw Gorlow[✉] and Jordi Janer

Music Technology Group, Universitat Pompeu Fabra,
Roc Boronat 138, 08018 Barcelona, Spain
{stanislaw.gorlow,jordi.janer}@upf.edu

Abstract. We evaluate the convolutive nonnegative matrix factorization in the context of automatic music transcription of polyphonic piano recordings and the associated problem of note isolation. Our intention is to find out whether the temporal continuity of piano notes is truthfully captured by the convolutional kernels and how the performance scales with complexity. Systematic studies of this kind are lacking in existing literature. We make use of established measures of accuracy and similarity. NMF dictionaries covering the piano's pitch range are learned from a given sample bank of isolated notes. The kernel alias patch size is varied. By using a measure of performance advantage, we show up that the improvements due to convolved bases do not justify the extra computational effort as compared to the standard NMF. In particular, this is true for the more realistic case, in which the dictionary does not fully correspond to the mixture signal. Further pertinent conclusions are drawn as well.

Keywords: Nonnegative matrix factorization · Convolution · Supervised learning · Polyphony · Automatic music transcription · Note separation

1 Introduction

Nonnegative matrix factorization (NMF) [1] meanwhile is an established tool in music processing, and music transcription has emerged as its main area of application, see [2,3]. Since a complete transcription would also include a note's velocity, this information, together with the learned bases, can be used to isolate notes from the mixture by Wiener filtering. This can be done either in a supervised or in an unsupervised task.

In this study, we seek to compare the performance of the convolutional NMF [4,5] with the standard NMF in regard to supervised learning, i.e. where the bases

S. Gorlow is now with Sony Computer Science Laboratory (CSL) in Paris, France.
This work was funded in part by the Yamaha Corporation.

E. Vincent et al. (Eds.): LVA/ICA 2015, LNCS 9237, pp. 437–445, 2015.
DOI: 10.1007/978-3-319-22482-4_51

are held fixed and their activations are updated until the modeled spectrogram is in the shortest distance from the observed spectrogram. In our evaluation, we resort to more frequently used measures, such as the root-mean-square deviation (RMSD). We also provide perception-related ratings. Above, we are interested in seeing how the superior modeling accuracy that convolutional bases are expected to bring about relates to the extra computational effort. As the convolutional NMF was designed for capturing the temporal evolution of sound patterns, we expect it to track the temporal decay of notes more faithfully than the standard NMF. For the transcription of polyphonic recordings and the related task of note isolation, temporal continuity of notes is a crucial factor. Thus, the convolutional NMF looks promising and seems to be a reasonable alternative to other variants that favor temporal continuity through additional penalty terms in the cost function. More generally speaking, our interest is in evaluating the aptitude of the convolutional NMF for musical applications.

2   Convolutional NMF

The basic idea behind the convolutional or convolutive NMF is to treat sequences of single-column bases, or multi-column bases, in the exact same manner that single-column bases are treated by the standard NMF. This is meant to better capture the temporal evolution of repeating patterns of the dominant or principal components in the mixture as compared with the standard, i.e. non-convolutional, NMF. In our case, we mean sequences of magnitude and/or power spectra when speaking of patterns and the principal components are piano notes. We will further refer to the length of such a sequence of spectra as the "patch size". The rank of the factorization is given by the number of distinct piano notes.

Now consider a Bregman distance $D_F^{\mathbf{X}}$ formally given in the form of the Kullback–Leibler (KL) divergence with $\mathbf{X}$ of size $K \times N$, $x_{kn} \in \mathbb{R}_0^+$, being approximated as

$$\mathbf{X} \approx \mathbf{Y} = \sum_{m=0}^{M-1} \mathbf{S}(m) \cdot \mathbf{A} \text{ rshift } m, \tag{1}$$

where $\mathbf{A}$ is the activations matrix, rshift is the zero-fill right-shift operator applied to the rows of $\mathbf{A}$, $\mathbf{S}$ is the bases matrix or the spectral imprint, m is the patch index and M the patch size, respectively. To show that (1) is indeed a convolution in n, we need to consider the following term which is applied to every element of $\mathbf{Y}$,

$$y_{kn} = \sum_{r=1}^{R} \sum_{m=0}^{M-1} s_{kr}(m) \cdot a_{r,n-m} = \sum_{r=1}^{R} s_{kr}(n) * a_{rn}, \tag{2}$$

where $*$ denotes convolution and R is the rank of $\mathbf{Y}$, $R \ll \min(K, N)$. The generalized KL divergence w.r.t. $\mathbf{X}$,

$$D_{\text{KL}}^{\mathbf{X}}(\mathbf{Y}, \mathbf{X}) = \sum_{k,n} y_{kn} \log \frac{y_{kn}}{x_{kn}} - \sum_{k,n} y_{kn} + \sum_{k,n} x_{kn}, \tag{3}$$

is generated from the convex function

$$F(\mathbf{X}) = \sum_{k,n} x_{kn} \log x_{kn} - \sum_{k,n} x_{kn}. \tag{4}$$

Alternatively, the KL divergence from (3) can be replaced by [2]

$$D_{\mathrm{KL'}}^{\mathbf{X}}(\mathbf{Y}, \mathbf{X}) = \|\mathbf{Y} \odot \log(\mathbf{Y} \oslash \mathbf{X}) - \mathbf{Y} + \mathbf{X}\|_{\mathrm{F}}, \tag{5}$$

where $\|\cdot\|_{\mathrm{F}}$ is the Frobenius norm, $\oslash$ stands for element-wise division and $\odot$ for element-wise multiplication, respectively. Note that for $M = 1$, (1) turns into the standard NMF. So, in supervised learning, the problem at hand can be stated as follows. Given $\mathbf{X}$ and $\mathbf{S}$, $s_{kr} \in \mathbb{R}_0^+$, $R \ll \min(K, N)$, $M \in \mathbb{N}$, find

$$\mathbf{A}_{\mathrm{opt}} = \arg\min_{\mathbf{A}} D_{\mathrm{KL'}}^{\mathbf{X}}(\mathbf{Y}, \mathbf{X}) \qquad \text{s.t. } a_{rn} \in \mathbb{R}_0^+. \tag{6}$$

2.1 Multiplicative Update Rule

$\mathbf{A}_{\mathrm{opt}}$ in (6) can be found using the convolutional update rule given in [5], which is

$$\mathbf{A} \leftarrow \mathbf{A} \odot \left[\mathbf{S}^{\mathsf{T}}(m) \cdot (\mathbf{X} \oslash \mathbf{Y}) \text{ lshift } m\right] \oslash \left[\mathbf{S}^{\mathsf{T}}(m) \cdot \mathbf{1}\right], \tag{7}$$

where $\mathbf{1}$ is a $K \times N$ all-ones matrix and lshift stands for the row-wise zero-fill left-shift operator. In [5], it is further suggested that for each $\mathbf{S}(m)$ a different $\mathbf{A}_m$ should be learned and that the final $\mathbf{A}$ should be computed as $\mathbf{A} = \langle \mathbf{A}_m \rangle$, where $\langle \cdot \rangle$ denotes the time average operator,

$$\langle \mathbf{A}_m \rangle = \frac{1}{M} \sum_{m=0}^{M-1} \mathbf{A}_m. \tag{8}$$

2.2 Dictionary Learning and Normalization

To construct an instrument's dictionary, one requires a dataset of separate note recordings. A typical piano range, e.g., would consist of $I = 88$ notes, starting with A_0 and ending with C_8, at a distance of a semitone. For every ith note, one computes the spectrogram $\mathbf{X}_i$ and learns the corresponding patch of M bases using [5]

$$\mathbf{S}_i(m) \leftarrow \mathbf{S}_i(m) \odot \left[(\mathbf{X}_i \oslash \mathbf{Y}_i) \cdot (\mathbf{A}_i \text{ rshift } m)^{\mathsf{T}}\right] \oslash \left[\mathbf{1} \cdot (\mathbf{A}_i \text{ rshift } m)^{\mathsf{T}}\right], \tag{9}$$

while alternating with (7). In a final step, the M $\mathbf{A}_i$ matrices are discarded and the convolutional bases $\mathbf{S}_i(m)$ are kept. The overcomplete dictionary, which is held stiff in (6), is obtained by stringing the note patches together to

$$\mathbf{S}(m) = [\mathbf{S}_1(m)\ \mathbf{S}_2(m) \cdots \mathbf{S}_I(m)] \qquad \in \mathbb{R}^{K \times MI}. \tag{10}$$

After each update (9), it is very common to normalize the columns of $\mathbf{S}_i(m)$ by their lengths in the Euclidean space. A reason for doing this is numerical stability. Another way of normalizing is by patch, i.e. either by relating each matrix element to the largest singular value of $\mathbf{S}_i(m)$, by taking the Euclidean matrix norm, or as an alternative by dividing each matrix element by the Frobenius norm. In this wise, the temporal decay of the notes' spectral envelopes can be tracked.

2.3 Gaussian-Additive Mixture Model

Consider the short-time Fourier transform (STFT) domain. In reference to the central limit theorem, the Fourier coefficients are approximately complex-normally distributed. We further assume that they are circularly-symmetric, i.e. that they have zero mean and zero covariance matrix. Now, if we stipulate that the note components are mutually independent, the mixture's PSD can be decomposed into a sum of notes' PSDs. In other words, the NMF can be performed on the mixture's PSD. Yet note that this model does not hold for the magnitude spectra, as the square root of a sum of squares is not equal to the sum of magnitudes.

2.4 Note Separation

With the signal model from Sect. 2.3, Wiener filtering can be used to separate the note components from the mixture. In a first step, one computes the learned spectrograms

$$\widehat{\mathbf{Y}}_i = \sum_{m=0}^{M-1} \mathbf{S}_i(m) \cdot \hat{\mathbf{A}}_i \text{ rshift } m, \qquad (11)$$

$i = 1, 2, \ldots, I$, and applies Wiener filtering to every element separately:

$$\hat{z}_{ikn} = \frac{\hat{y}_{ikn}}{\sum_{j=1}^{I} \hat{y}_{jkn}} \cdot x_{kn} e^{\jmath \phi_{kn}} \qquad \forall\, i, k, n, \qquad (12)$$

where ϕ is the phase of x in time-frequency (TF) point (n, k) and $\jmath$ is the imaginary unit. The corresponding time-domain signal is obtained by the inverse STFT on $\widehat{\mathbf{Z}}_i$.

3 Evaluation

For the purpose of evaluation, we design various dictionaries using the RWC Music Database, while each dictionary is trained for Yamaha's Pianoforte, normal playing style, and "mezzo" level of dynamics.[1] For the STFT, we apply a 4-term Blackman–Harris window of the size of the transform and overlap succeeding blocks by 87.5 %.

---

[1] https://staff.aist.go.jp/m.goto/RWC-MDB/.

As for the mixture signal, we generate it from a MIDI file taken from the Saarland Music Data (SMD) using Kontakt 5 by Native Instruments.[2] The 32-s excerpt is part of Chopin's Opus 10.[3] We generate two mixtures: one synthetic using the RWC samples and one realistic for the Berlin Concert Grand. We perform the NMF on the mixture using the NMFlib for a fixed number of 30 iterations.[4][5] The critical testing parameter is the patch size M which is increased from 1 onwards. Also, we evaluate the NMF performance for the transform lengths of 2048 and 4096 points for two different nonnegative TF representations: the magnitude spectrum and the power spectrum. Overall, we train 24 dictionaries, one for each set of configuration parameters. We normalize the basis spectra by patch using the Euclidean matrix norm.

3.1   Performance Measures

F-measure. In binary classification, the F-measure indicates the accuracy of a system under test and it is defined as the harmonic mean of precision and recall:

$$F \triangleq 2 \cdot \frac{\text{precision} \cdot \text{recall}}{\text{precision} + \text{recall}} = \frac{2 \cdot TP}{2 \cdot TP + FP + FN}, \tag{13}$$

where TP is the number of true positives, FP is the number of false positives and FN is the number of false negatives. In the case of music transcription, true positives denote those TF points that have significant contributions according to (11) in the same spots as in the perfect transcription. False positives are activations in the wrong spots and false negatives denote missing activations, respectively. The F-score attains its best value at 1 and its worst value at 0.

Root-mean-square Deviation. The root-mean-quare deviation (RMSD) is a frequently used measure of accuracy for comparing errors of different models for a particular variable. With regard to notes:

$$\text{RMSD}_i \triangleq \sqrt{\frac{1}{N} \sum_{n=1}^{N} [\hat{z}_i(n) - s_i(n)]^2}, \tag{14}$$

where N is the length (in samples) of the time-domain signal $s_i(n)$ and $\hat{z}_i(n)$ is its estimate. Lower values are preferred.

---

[2] http://www.mpi-inf.mpg.de/resources/SMD/SMD_MIDI-Audio-Piano-Music.html.
[3] The results shown are representative of what we experienced for different piano recordings.
[4] https://code.google.com/p/nmflib/.
[5] The number was chosen empirically. Above it, no significant improvement was observed.

Perceptual Similarity Measure. "PEMO-Q" [6] is a method for the objective assessment of the perceptual quality of audio. It uses the model of auditory perception by Dau et al. to predict the audio quality of a test signal relative to a reference signal. PEMO-Q aligns the levels of both signals and transforms them into so-called "internal representations" of the auditory model. The cross-correlation coefficient between the two representations serves as a measure of the perceived similarity, PSM. And so, it can be used as a measure of the test signal's degradation.

Average Performance and Performance Advantage. The major goal of this evaluation is to relate the performance of the convolutional NMF to its computational complexity in a more formal manner. We state the average performance as

$$P_{avg} \triangleq \frac{P}{T},\tag{15}$$

where P can be expressed as any of the above measures and T shall denote the execution time of the NMF. Moreover, we define the performance advantage of the convolutional NMF as the logarithm of the ratio between the performances of the convolutional and the standard NMF over time,

$$PA \triangleq \log \frac{P_M/T_M}{P_1/T_1} \approx \log \frac{P_M}{M \cdot P_1}\tag{16}$$

with $T_M \approx M \cdot T_1$, i.e. on the assumption that it takes M times longer to compute the convolutional M-basis NMF as compared to the standard single-basis NMF [5]. A PA that is above zero indicates an advantage, a disadvantage if below zero, i.e. if it is negative, and a value of zero means equality.

3.2   Music Transcription and Note Isolation

In the first part of our evaluation, we compute the accuracy of the convolutional NMF as a function of the patch size M for different configurations using the F-measure. The perfect or reference transcription is computed from the score. For each note, we obtain a waveform signal from the respective MIDI track using the Kontakt 5 sampler. For all notes, we compute the time-pitch power spectra. We compare the powers with a threshold of $-60\,$dB, and so we obtain a binary mask for the entire excerpt. The same thresholding procedure is applied to each note signal estimated according to (12). The two binary masks are then compared against each other in terms of (13). Errors are manifested in missing or superfluous positives that represent a mismatch between the signal and the model. Figure 1 summarizes the results.

In the second part, we evaluate the quality of separated notes. For this, we use the RMSD and the PSM. All note signals are normalized to 0 dBFS RMS before computing the RMSD. In Fig. 2, the average over all isolated notes is shown.

3.3   Interpretation of Results and Observations

Looking at Fig. 1, one can observe a slight improvement that is due to a greater patch size in the case of the synthetic mix. For the realistic mix, the improvement is minor. The greater patch size seems rather counterproductive when the power is used as the nonnegative representation together with a lower frequency resolution. It looks like the magnitude spectra yield a better accuracy for both the mixtures. It is also evident that a higher frequency resolution improves the transcription. The fact that some curves are not monotonically increasing might be due to random initializations in the NMFlib.

Figure 2 confirms once more that a significantly better result can be expected if the dictionary fits the mixture. In regard to the RMSD, a gain of 3 dB can be stated. Here again, a higher frequency resolution has a stronger impact on the result than a larger patch. When listening to the note samples, we would further observe that for low-pitched notes a 2048-point STFT is insufficient to discriminate neighboring partials. For high-pitched notes, this issue is less critical. For the synthetic mix, the perceptual similarity between notes is higher in the case of magnitude spectra. Yet for the realistic mixture, the power spectral representation gives comparable if not better results.

Even though a performance improvement with respect to the F-measure and also the PSM is undeniable between 1 and 4 bases in particular, the PA-curves indicate that it comes at the expense of an almost M times higher effort. For a patch size greater than 4, the improvement looks negligible in most cases. Plus,

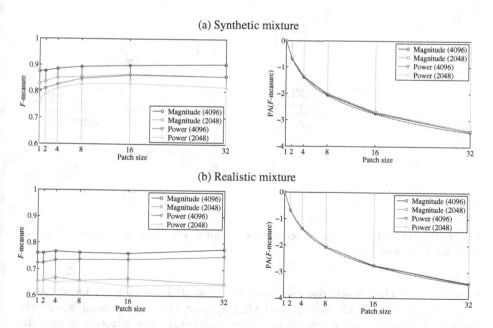

Fig. 1. F-measure values versus performance advantage of the convolutional NMF for music transcription

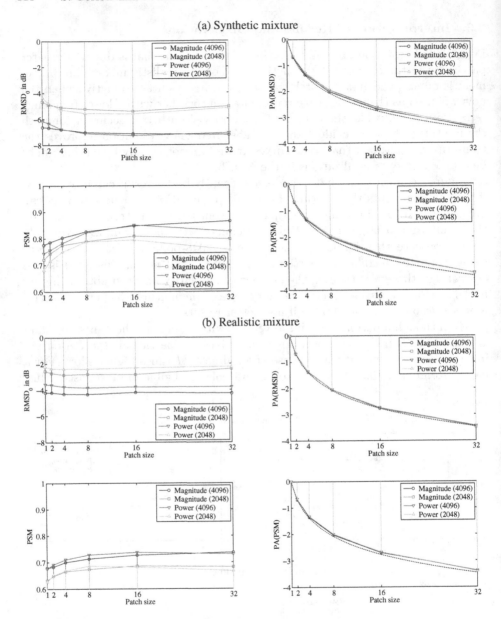

Fig. 2. RMSD and PSM values versus performance advantage of the convolutional NMF for note isolation

irrespective of the chosen test case and measure, $PA \approx -\log M$, i.e. negative (disadvantageous) for all $M > 1$. And what is more, the improvement is scarcely audible. Another negative side effect of the convolution worth noting is that the attacks of low-pitched notes are smoothed out.

4   Conclusion

We conclude that because of the large pitch range of the piano, the STFT size should be no smaller than 4096 at a sampling rate of 44.1 kHz to separate low-pitched notes. As for the spectral representation, in most test cases the magnitude spectrum is more performant than the power spectrum. At this point, we do not have an explanation for this enigma that questions the validity of the Gaussian-additive mixture model. Finally, the study shows that is it senseful to favor a single-basis NMF over a computationally intensive convolutional NMF in musical applications, especially if the runtime plays an important role. No significant sound quality loss was established in our experiments.

References

1. Lee, D.D., Seung, H.S.: Learning the parts of objects by non-negative matrix factorization. Nature **401**(6755), 788–791 (1999)
2. Smaragdis, P., Brown, J.C.: Non-negative matrix factorization for polyphonic music transcription. In: Proceedings of the WASPAA 2003, pp. 177–180, October 2003
3. Abdallah, S.A., Plumbley, M.D.: "Polyphonic music transcription by non-negative sparse coding of power spectra. In: Proceedings of the ISMIR 2004, pp. 318–325, October 2004
4. Smaragdis, P.: Non-negative matrix factor deconvolution; extraction of multiple sound sources from monophonic inputs. In: Puntonet, C.G., Prieto, A.G. (eds.) ICA 2004. LNCS, vol. 3195, pp. 494–499. Springer, Heidelberg (2004)
5. Smaragdis, P.: Convolutive speech bases and their application to supervised speech separation. IEEE Audio, Speech, Lang. Process. **15**(1), 1–12 (2007)
6. Huber, R., Kollmeier, B.: PEMO-Q – a new method for objective audio quality assessment using a model of auditory perception. IEEE Audio, Speech, Lang. Process. **14**(6), 1902–1911 (2006)

Masked Positive Semi-definite Tensor Interpolation

Dave Betts[✉]

CEDAR Audio Ltd., 20 Home End, Fulbourn, Cambridge CB21 5BS, UK
dave.betts@cedaraudio.com

Abstract. Time-frequency constrained interpolation of audio has proven to be an effective technique in removing a wide variety of acoustic disturbances. Traditionally these techniques assume that the signal is stationary for the duration of the interpolation, which limits the types of disturbances that can be addressed. In this paper we propose masked positive semi-definite tensor factorisation followed by a novel form of multi-channel spectral subtraction to solve the problem, and we demonstrate excellent results on some real-world examples. The proposed methods can remove disturbances that were previously considered highly challenging to interpolate, for example a burst of wind noise in a voice recording.

Keywords: PSTF · Masked PSTF · Multi-channel spectral subtraction · Minorisation-Maximisation

1 Introduction

Auto-regressive (AR) interpolation using a noise model combined with a signal model was introduced in [1] to remove low frequency noise pulses from an audio signal. In [2], we demonstrated how to use this AR model to interpolate time-frequency constrained disturbances. Such methods have been used to remove many different sorts of disturbances; coughs, chairs banging and a dropped wrench being just some of the successful examples. While this is an industry standard technique, it does not work well for very long duration disturbances. A good example of this problem is wind noise, where the disturbance is a burst of non-stationary noise, and the duration is long enough for the stationarity assumptions in the AR model to be invalid.

The short time Fourier transform (STFT) is a well-known technique for the processing of audio in the time frequency domain. Within the time-frequency domain, non-negative matrix factorisation (NMF) [3,4] and its multi-channel variant, positive semi-definite tensor factorisation (PSTF) [5,6] have proven to be good models for acoustic behaviour on larger time-scales. However, one well-known problem with NMF and PSTF is non-uniqueness [7]. In general, there are many equivalent solutions to the simple NMF/PSTF model including permutations and scaling, which means that the correct categorisation of the detected features as disturbance or desired signal can be problematic. Many techniques

© Springer International Publishing Switzerland 2015
E. Vincent et al. (Eds.): LVA/ICA 2015, LNCS 9237, pp. 446–453, 2015.
DOI: 10.1007/978-3-319-22482-4_52

have been used to resolve these ambiguities, such as sparsity and using training data [8,9]. These are relevant for many source separation tasks, but we would prefer to have a blind interpolation technique, as that has wider applicability.

User-defined masks have been applied to NMF for missing data interpolation and bandwidth extension [10,11]. These problems are different to those addressed in this paper as they are replacing missing data instead of reducing or removing the effect of an unwanted disturbance while perceptually preserving the underlying desired audio signal.

In this paper we use a masked PSTF model based on that given in [6] that can be solved iteratively using an improved minorisation-maximisation algorithm to give a Maximum Likelihood estimate of the model parameters. As has been shown, e.g. in [4], this maximum likelihood approach is equivalent to minimising the Itakura-Saito divergence. The use of the mask resolves the uniqueness ambiguities normally associated with NMF/PSTF approaches. Consequently there is no training or dictionary learning required in our approach.

Armed with the masked PSTF model, we can express our interpolation problem as the sum of a time-frequency varying signal process and a time-frequency varying noise process, where the user defines a mask for each process. Each process is further characterised as being the sum of one or more PSTF components. Having found an optimal solution for the model parameters, the interpolation is performed by applying gains in the STFT domain. The gains can be calculated either using a Wiener filtering approach similar to that in [3], or our proposed multi-channel spectral subtraction method which incorporates a novel 'least harm' criterion. While Wiener filtering is attractive as it provides the minimum mean square error estimator given the masked PSTF model, it tends to sound overprocessed in audio applications. Spectral subtraction, on the other hand, attempts to replace the disturbance with synthesised audio that matches the expected statistics of the desired signal, and in our application is often perceptually superior to the minimum mean square error interpolation.

2   The Masked PSTF Model

We first define some notation:

- C is the number of audio channels.
- F is the number of STFT frequencies.
- T is the number of STFT frames.
- K is the number of components in the PSTF model.
- f, t, k are indices corresponding to F, T, K.
- we express differentiation as a function $\partial f(\partial x)$ rather than a derivative $\frac{\partial f}{\partial x}$.

At each time-frequency point in the STFT we can collate the observations for each of the C channels to form an observation vector $\underline{x}_{ft}$

$$\underline{x}_{ft} = [x_{1,f,t}, \cdots x_{C,f,t}]^H.$$

We assume that $\underline{x}_{ft}$ is the sum of K unknown independent components $\underline{z}_{ftk}$. We also assume that each $\underline{z}_{ftk}$ is independently drawn from a multivariate complex circular symmetric normal distribution $\mathcal{N}$ with an unknown covariance matrix ψ_{ftk} that varies over both time and frequency. Lastly, we assume that the covariance ψ_{ftk} is the product of a mask m_{ftk}, a spatial covariance matrix U_{fk} and an activation v_{tk} with the constraints that $m_{ftk} \geq 0$, U_{fk} is a positive semi-definite matrix, and $v_{tk} \geq 0$. To summarise:

$$\underline{x}_{ft} = \sum_k \underline{z}_{ftk}, \quad \underline{z}_{ftk} \sim \mathcal{N}\left(0, \psi_{ftk}\right), \quad \psi_{ftk} = m_{ftk} U_{fk} v_{tk}. \tag{1}$$

U and v form the latent variables of the model. The mask m is defined a-priori, and is normally binary $m_{ftk} \in \{0, 1\}$, but non-binary masking could be used instead. It is principally the use of the mask that distinguishes our model from [6], and this resolves the non-uniqueness problem to allow factorisation into disturbance and signal components. In our case the mask for the disturbance components should indicate the time-frequency region of the disturbance, whereas the mask for the signal components should contain both the disturbance and some surrounding undisturbed signal.

Assuming that the observations are independent, we can define an intermediate covariance matrix σ_{ft} and write the likelihood of the observations given the latent variables can be written as:

$$\underline{x}_{ft} \sim \mathcal{N}\left(0, \sigma_{ft}\right), \quad \sigma_{ft} = \sum_k \psi_{ftk} \tag{2}$$

$$L\left(X; U, v\right) = \sum_f \sum_t -\ln \det \pi \sigma_{ft} - \underline{x}_{ft}^H \sigma_{ft}^{-1} \underline{x}_{ft}. \tag{3}$$

3   Maximum Likelihood Solution

The maximum likelihood estimates for U and v are found by maximising (3). Many techniques are available for solving this problem. We will use a minorisation-maximisation approach [6,12] because of its convergence guarantees. Minorisation-maximisation requires that we construct an auxiliary function $L^+\left(\hat{U}, \hat{v}, U, v\right)$ which has to satisfy the minorisation criterion over the search subspace [12]. We can fix $\hat{v} = v$ and then maximise the auxiliary function with respect to $\hat{U}$ to give a guaranteed improvement in the likelihood. Similarly we can fix $\hat{U} = U$ and maximise with respect to $\hat{v}$.

There are of course any number of auxiliary functions that satisfy these constraints. The art is in choosing a function that is both tractable and gives good convergence. A suitable minorisation in our case is given by:

$$\hat{\psi}_{ftk} = m_{ftk} \hat{U}_{fk} \hat{v}_{tk}, \quad \hat{\sigma}_{ft} = \sum_k \hat{\psi}_{ftk}$$

$$L^+\left(\hat{U}, \hat{v}, U, v\right) = \sum_t \sum_f -\ln \det \pi \sigma_{ft} - \operatorname{tr}\left(\hat{\sigma}_{ft} \sigma_{ft}^{-1}\right) + C - \underline{x}_{ft}^H \sigma_{ft}^{-1}\left(\sum_k \psi_{ftk} \hat{\psi}_{ftk}^{-1} \psi_{ftk}\right) \sigma_{ft}^{-1} \underline{x}_{ft}. \tag{4}$$

3.1 Optimising with Respect to U

Setting the partial differentiation of (4) with respect to $\hat{U}_{fk}$ to zero when $\hat{v}_{tk} = v_{tk}$ gives an analytically tractable solution for updating U. We define two intermediate positive semi-definite matrices A_{fk}, B_{fk} as

$$A_{fk} = \sum_t \sigma_{ft}^{-1} m_{ftk} v_{tk} \tag{5}$$

$$B_{fk} = U_{fk} \left(\sum_t \sigma_{ft}^{-1} \underline{x}_{ft} \underline{x}_{ft}^H \sigma_{ft}^{-1} m_{ftk} v_{tk} \right) U_{fk}. \tag{6}$$

The solution to $\partial L^+ \left(\partial \hat{U}_{fk} \right) = 0$ for all ∂U_{fk} is given by $\hat{U}_{fk} A_{fk} \hat{U}_{fk} = B_{fk}$. We note that it is also given by the modified equation

$$\hat{U}_{fk} A_{fk} \hat{U}_{fk}^H = B_{fk}. \tag{7}$$

The general solution set to this modified equation can be expressed in terms of square root factorisation and an arbitrary orthonormal matrix Θ_{fk} as

$$A_{fk} = A_{fk}^{1/2H} A_{fk}^{1/2}, \quad B_{fk} = B_{fk}^{1/2H} B_{fk}^{1/2}, \quad \hat{U}_{fk} = B_{fk}^{1/2H} \Theta_{fk} A_{fk}^{-1/2H}. \tag{8}$$

We choose Θ_{fk} to preserve the positive semi-definite nature of $\hat{U}_{fk}$, which can be done using singular value decomposition of $A_{fk}^{1/2} B_{fk}^{1/2H}$ by

$$A_{fk}^{1/2} B_{fk}^{1/2H} = \alpha \Sigma \beta^H, \quad \Theta_{fk} = \beta \alpha^H. \tag{9}$$

Note that the positive semi-definite solution to (7) is invariant to the choice of square root factorisation. So to update U given the current estimates of U and v we use the following algorithm:

1. Use (1) and (2) to calculate σ_{ft} for each frame and frequency.
2. For each frequency f and component k:
   (a) Use (5) and (6) to calculate A_{fk} and B_{fk}.
   (b) Use Cholesky factorisation to calculate $A_{fk}^{1/2}$ and $B_{fk}^{1/2}$.
   (c) Use (9) to calculate Θ_{fk}.
   (d) Use (8) to calculate $\hat{U}_{fk}$.
3. Copy $\hat{U} \rightarrow U$.

3.2 Optimising with Respect to v

Similarly, setting the partial derivative of (4) with respect to $\hat{v}_{tk}$ to zero when $\hat{U}_{fk} = U_{fk}$ gives an analytically tractable update for v. We define two intermediate variables $A'_{tk}, B'_{tk} \geq 0$:

$$A'_{tk} = \sum_f \text{tr} \left(\sigma_{ft}^{-1} U_{fk} \right) m_{ftk} \tag{10}$$

$$B'_{tk} = v_{tk}^2 \sum_f \underline{x}_{ft}^H \sigma_{ft}^{-1} U_{fk} \sigma_{ft}^{-1} \underline{x}_{ft} m_{ftk}. \tag{11}$$

The solution to $\partial L^+/\partial \hat{v}_{tk} = 0$ is then given by

$$\hat{v}_{tk} = \sqrt{\frac{B'_{tk}}{A'_{tk}}}. \tag{12}$$

So to update v given the current estimates of U and v we use the following algorithm:

1. Use (1) and (2) to calculate σ_{ft} for each frame and frequency.
2. For each frame t and component k:
   (a) Use (10) and (11) to calculate A'_{tk} and B'_{tk}.
   (b) Use (12) to calculate the updated $\hat{v}_{tk}$.
3. Copy $\hat{v} \to v$.

We update with respect to $\hat{U}$ and $\hat{v}$ alternately until convergence. We found that using a fixed number of iterations was found to be adequate rather than using a convergence test.

4   The Masked PSTF Interpolation

With suitably defined masks we can estimate U and v as shown in Sect. 3. We then calculate an estimate for the desired data $\tilde{x}_{ft}$ using the proposed spectral subtraction technique with a 'least harm' criterion as follows.

To define the masks, the operator graphically selects a region on a spectrogram that indicates where the disturbance occurs. Similarly an enclosing region that is representative of the desired signal can be indicated. We can use these two regions to define a disturbance mask and a desired mask. We can then split the components in our masked PSTF model into two categories, disturbance and desired, and assign the appropriate mask to each component.

We define a selection vector s_k that is zero for disturbance components and one for desired components. The covariance of the desired data $\tilde{\sigma}_{ft}$ is then

$$\tilde{\sigma}_{ft} = \sum_k \psi_{ftk} s_k. \tag{13}$$

Our interpolation task is then to create an estimate for the desired signal $\tilde{\underline{x}}_{ft}$ as a linearly transformed version of the observation via

$$\tilde{\underline{x}}_{ft} = G_{ft} \underline{x}_{ft}. \tag{14}$$

The simplest solution is to use the minimum mean square estimator which is given by the classic Wiener equation $G_{ft} = \tilde{\sigma}_{ft}\sigma_{ft}^{-1}$. Unfortunately, the Wiener filter tends to sound over processed if the disturbance is significantly larger than the desired signal. We propose a multi-channel variant of spectral subtraction to create an interpolated result that matches the desired covariance. This matching involves square root factorisations and an arbitrary orthonormal matrix Θ_{ft}:

$$\tilde{\sigma}_{ft} = G_{ft}\sigma_{ft}G_{ft}^H, \quad G_{ft} = \tilde{\sigma}_{ft}^{1/2H}\Theta_{ft}\sigma_{ft}^{-1/2H}. \tag{15}$$

Given that there is a continuum of possible solutions to (15), we introduce a least harm criterion to resolve the ambiguity; we find the solution that is closest to the original in a Euclidean sense $\left(E\left\{ \left\| \tilde{\underline{x}}_{ft} - \underline{x}_{ft} \right\|^2 \right\} \right)$. Substituting (14) and (15) shows that the closest solution is found by maximising $\text{tr}\left(\tilde{\sigma}_{ft}^{1/2H} \Theta_{ft} \sigma_{ft}^{1/2} \right)$, and Θ_{ft} can be found via singular value decomposition of $\sigma_{ft}^{1/2} \tilde{\sigma}_{ft}^{1/2H}$:

$$\sigma_{ft}^{1/2} \tilde{\sigma}_{ft}^{1/2H} = \alpha \Sigma \beta^H, \quad \Theta_{ft} = \beta \alpha^H. \tag{16}$$

Note that this makes G_{ft} positive semi-definite and invariant to the choice of square root factorisation. The algorithm is therefore:

1. For each frame t and frequency f:
   (a) For each k, use (1) to calculate ψ_{ftk} from U_{fk}, v_{tk}.
   (b) Use (2) and (13) to calculate σ_{ft} and $\tilde{\sigma}_{ft}$.
   (c) Use Cholesky factorisation to calculate $\sigma_{ft}^{1/2}$ and $\tilde{\sigma}_{ft}^{1/2}$.
   (d) Use (16) to calculate Θ_{ft}.
   (e) Use and (15) and (14) to calculate $\hat{\underline{x}}_{ft}$.

5   Results

In this section some real-data results are presented using spectrogram plots. These are accompanied by the website www.cedaraudio.com/PSTFexamples which contains further examples and the accompanying audio, including comparisons against a state-of-the-art AR interpolator. We believe that these subjective results are much the best way to assess the performance of an applied algorithm such as this.

For illustration we show the real-world example of a recording made on a windy day in Cape Cod. This extract is taken from a promotional video where the narrator is moving out of a wind shadow and back while recording on a handheld microphone. Figure 1a shows the original spectrogram, where the 2 s burst of wind noise can be seen. Figure 1b shows the same image with the disturbance mask as a grey overlay. The good mask is indicated by the outer dotted rectangle.

Figures 1c and d show the results using both Wiener filtering and spectral subtraction. The masked PSTF model used just 2 disturbance components, 2 good components and 30 iterations. The Wiener filter has lost a lot of the ambient background as well as putting a low frequency drop out in the voice. We estimate that the spectral subtraction method achieves about 20dB of reduction of the wind noise for virtually no perceptible loss in the voice and preserves the background ambience. The noise reduction is less than that for the Wiener filter, but the lack of artefacts makes the result much more palatable.

We discovered that it is generally better to underestimate the number of components required for the disturbance and the desired categories. With more components there is more opportunity for audio features to be misclassified as disturbance or desired, leading to suboptimal results.

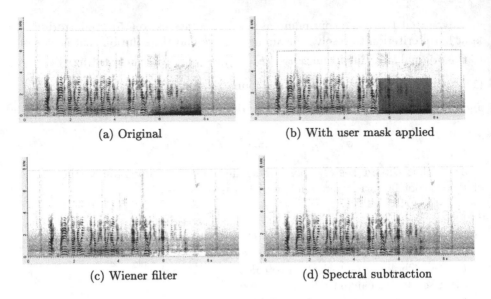

(a) Original

(b) With user mask applied

(c) Wiener filter

(d) Spectral subtraction

Fig. 1. Spectrograms showing the burst of wind noise from 5.5 to 7.9 s.

6   Conclusion

We have introduced a new method to approach the time-frequency constrained interpolation problem. The Masked PSTF technique with either Wiener Filtering or the proposed multi-channel spectral subtraction significantly extend the range of problems that can be successfully addressed in practice. It is our opinion that the spectral subtraction algorithm gives more acoustically pleasing results than the Wiener filtering approach. Future work would include investigating the effects of priors to create a Bayesian solution, and alternative optimisation schemes to minorisation-maximisation such as expectation-maximisation and Markov Chain Monte Carlo (MCMC) methods.

References

1. Godsill, S., Rayner, P.: Digital Audio Restoration - A Statistical Model Based Approach, pp. 153–163. Springer, London (1998)
2. Betts, D.A.: Method and apparatus for audio signal processing. US Patent 7 978 862 (2011)
3. Fevotte, C., Bertin, N., Durrieu, J.-L.: Nonnegative matrix factorization with the Itakura-Saito Divergence: with application to music analysis. Neural Comput. **21**(3), 793–830 (2008)
4. Févotte, C., Ozerov, A.: Notes on nonnegative tensor factorization of the spectrogram for audio source separation: statistical insights and towards self-clustering of the spatial cues. In: Aramaki, M., Jensen, K., Kronland-Martinet, R., Ystad, S. (eds.) CMMR 2010. LNCS, vol. 6684, pp. 102–115. Springer, Heidelberg (2011)

5. Ozerov, A., Fevotte, C.: Multichannel nonnegative matrix factorization in convolutive mixtures. With application to blind audio source separation. In: IEEE International Conference on Acoustics, Speech and Signal Processing. ICASSP 2009, pp. 3137–3140, April 2009

6. Sawada, H., Kameoka, H., Araki, S., Ueda, N.: Efficient algorithms for multichannel extensions of Itakura-Saito nonnegative matrix factorization. In: 2012 IEEE International Conference on Acoustics, Speech and Signal Processing (ICASSP), pp. 261–264, March 2012

7. Laurberg, H., Christensen, M.G., Plumbley, M.D., Hansen, L.K., Jensen, S.H.: Theorems on positive data: on the uniqueness of nmf. Comput. Intell. Neurosci. **2008**, 1–9 (2008)

8. King, B., Févotte, C., Smaragdis, P.: Optimal cost function and magnitude power for NMF-based speech separation and music interpolation. In: 2012 IEEE International Workshop on Machine Learning for Signal Processing (MLSP), pp. 1–6. IEEE (2012)

9. Mohammadiha, N., Dodo, S.: Transient noise reduction using nonnegative matrix factorization. In: 2014 4th Joint Workshop on Hands-free Speech Communication and Microphone Arrays (HSCMA), pp. 27–31. IEEE (2014)

10. Bansal, D., Raj, B., Smaragdis, P.: Bandwidth expansion of narrowband speech using non-negative matrix factorization. In: INTERSPEECH, pp. 1505–1508 (2005)

11. Smaragdis, P., Raj, B., Shashanka, M.: Missing data imputation for spectral audio signals. In: IEEE International Workshop on Machine Learning for Signal Processing. MLSP 2009, pp. 1–6. IEEE (2009)

12. de Leeuw, J.: Block-relaxation algorithms in statistics. In: Bock, H.-H., Lenski, W., Richter, M.M. (eds.) Information Systems and Data Analysis. Studies in Classification, Data Analysis, and Knowledge Organization, pp. 308–324. Springer, Heidelberg (1994)

On the Suppression of Noise from a Fast Moving Acoustic Source Using Multimodality

Wendyam Serge Boris Ouedraogo$^{(\boxtimes)}$, Bertrand Rivet, and Christian Jutten

Université Grenoble-Alpes, GIPSA-lab, 38000 Grenoble, France
{wendyam.ouedraogo,bertrand.rivet,
christian.jutten}@gipsa-lab.grenoble-inp.fr

Abstract. The problem of removing non stationary noise coming from a moving acoustic source in outdoor environment is investigated in this paper. By making use of the known instantaneous location of the moving source (provided by a second modality), we propose a time-domain method for removing the noise coming from this source in a mixture of acoustic sources. The proposed method consists in an irregular resampling of the mixed data recorded by the sensors, and by linearly combining the resampled data to remove the undesired source. It's similar to the beamforming approach, but instead of seeking to enhance a target source, we seek to remove an undesired one. Simulation on synthetic data show the effectiveness of the proposed method with a few number of sensors.

Keywords: Non-stationary noise suppression · Moving acoustic source · Multimodality · Delay compensation · Beamforming

1 Introduction

The task of removing noise coming from an undesired and dominant source in a mixture of several acoustic sources has application in many areas like hearing aids, speech enhancement in wireless communication, or auditory scene analysis to name a few [1,2]. This issue remains a challenge when the undesired source is moving. In this paper, we consider the problem where a target fixed acoustic source is corrupted by a dominant noise coming from a fast moving acoustic source. We consider the case where the noise source moves along a straight line at constant speed in a plane. The high speed of the moving source may cause a doppler frequency shift between the recording microphones [3]. The task of canceling the noise coming from the moving source is discussed in an outdoor environment but we consider only the direct path and we do not take into account the reflexions. This kind of problem can occur in airport platforms, on roadsides or along railway lines. The problem described above belongs to the general framework of non-stationary noise suppression. Conventional methods

W.S.B. Ouedraogo—This work was supported by the project CHESS, 2012-ERC-AvG-320684.

for the non-stationary noise suppression include spectral subtraction [4], source separation [5] and beamforming [6].

The spectral subtraction methods performs subtraction of the estimated noise spectrum to the spectrum of the recorded signal (containing both the target source and the noise) to get an estimated noiseless spectrum of the target source from which one can compute its temporal profile [7]. However, these methods require a proper estimate of the noise spectrum, which is difficult to obtain in a context of highly non-stationary noise, like the case of a fast moving source.

Another approach to remove undesired source from the mixture is to perform source separation. Most of source separation methods have been proposed for fixed sources and exploit the source statistics, especially their mutual independence. For moving sources, some methods performed a block-wise separation using statistical methods, where for each bloc the sources are assumed to be fixed [5]. However, this assumption no longer holds if the sources are moving faster so that the bloc length for which they can be considered as fixed is not enough to perform statistical processing.

Regarding the beamforming methods, they proceed by combining the signals recorded by an array of sensors in order to achieve a spatial filtering that will enhance the signal coming from the direction of the target source and remove noises coming from the other directions [6,8]. Recently, new methods coupling video and audio modalities have been proposed to unmix moving acoustic sources [9,10]. These techniques firstly estimate the instantaneous positions and the velocities of the sources by the video, and make use of the source locations to design some beamforming methods.

The method proposed in this paper is within the scope of the approaches using multimodality. Assuming that the instantaneous location of the moving source are provided by a second modality like a video system or the Global Positioning System (GPS), we develop a temporal method for removing this source in the mixture. The first step of the proposed method consists in an irregular resampling of the recorded signals, in order to compensate the difference of propagation delays of the moving source between two sensors. In a second step, we linearly combine the resampled data to remove the noise coming from the moving source. The paper is organized as follows: Sect. 2 presents the problem modelization and a simple way to instantaneously estimate the location of the moving source. Section 3 presents the proposed method for cancelling the noise coming the undesired fast moving source. In Sect. 4, we show simulation results and in Sect. 5 we derive conclusions.

2   Problem Statement

2.1   Mixing Model

Let's recall that we deal with the problem where a target fixed source s_1 is corrupted by a dominant and moving source s_2. By considering only the direct path, the mixing model is given by Eq. (1):

$$m_k(t) = a_{k1}s_1\left(t - \tau_{k1}\right) + a_{k2}\left(t - \tau_{k2}(t)\right)s_2\left(t - \tau_{k2}(t)\right), \ 1 \le k \le K. \quad (1)$$

where $m_k(t)$ is the mixed data recorded at sensor k, K is the number of sensors. $a_{kl}(t)$ and $\tau_{kl}(t)$ are the mixing coefficient and the propagation delay from source l to sensor k at time index t. The delay $\tau_{kl}(t)$ between source l and sensor k at time index t is proportional to the distance $D_{kl}(t)$ between source l and sensor k and in a far field context, we can also assume that the mixing coefficient $a_{kl}(t)$ is proportional to the inverse of the distance $D_{kl}(t)$. Denoting c the sound velocity in air, it leads to:

$$a_{kl}(t) = \frac{1}{D_{kl}(t)} \text{ and } \tau_{kl}(t) = \frac{D_{kl}(t)}{c}. \quad (2)$$

Since $s_1(t)$ does not move, then $a_{k1}(t) = a_{k1}$ and $\tau_{k1}(t) = \tau_{k1}$.

2.2 Instantaneous Localization of the Moving Source

Set $(X_{s_2}(t), Y_{s_2}(t), Z_{s_2}(t))$ the instantaneous localization of the moving source s_2. Since we consider that the source s_2 moves along a straight line at a constant speed v_2 in a plane, for example in the XY plane (see Fig. 1), the coordinates of s_2 at any time t is given by:

$$\begin{cases} X_{s_2}(t) = X_{s_2}(t_0) + v_2(t - t_0)\cos(\theta_2) \\ Y_{s_2}(t) = Y_{s_2}(t_0) + v_2(t - t_0)\sin(\theta_2) \\ Z_{s_2}(t) = Z_{s_2}(t_0) \end{cases} \quad (3)$$

where θ_2 is the angle between the x-axis and the trajectory axis of s_2 (Fig. 1).

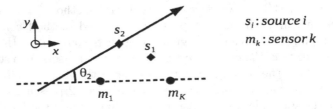

Fig. 1. Mixing configuration

Equation (3) shows that to estimate the current location of s_2, one only require the estimated position of s_2 at a reference time t_0, the estimated speed v_2 of s_2 and the estimated angle θ_2. These parameters can be obtained with a second modality like GPS in a cooperative environment, or can be estimated by processing video and/or audio signal [11,12]. In this paper, we will not address the problem of estimating the previous parameters, but we assume that they are provided by one of the methods mentioned above. Furthermore, since θ_2 is constant and without loss of generality, we will consider in the remaining of the paper that $\theta_2 = 0$, to simplify the equations.

3   Removing of the Noise Coming from the Moving Source

Below we describe a method for removing the moving source noise at a given sensor k. In this purpose we choose a reference sensor r $(r \neq k)$, the mixed data recorded at sensors k and r are respectively:

$$m_k(t) = a_{k1} s_1 (t - \tau_{k1}) + a_{k2} (t - \tau_{k2}(t)) s_2 (t - \tau_{k2}(t)) \qquad (4)$$

$$m_r(t) = a_{r1} s_1 (t - \tau_{r1}) + a_{r2} (t - \tau_{r2}(t)) s_2 (t - \tau_{r2}(t)) \qquad (5)$$

The propagation delay from source s_2 to the sensors k and r are respectively $\tau_{k2}(t)$ and $\tau_{r2}(t)$. We are going to resampled irregularly the signal recorded on sensor r such that the propagation delay from s_2 to the two sensors (k and r) becomes the same. According to equation (2), $\tau_{k2}(t) = \frac{D_{k2}(t)}{c}$. Set $(X_{m_k}, Y_{m_k}, Z_{m_k})$ the known location of the sensor m_k. Then the distance between source 2 and sensor k at time t is given by:

$$D_{k2}(t) = \sqrt{(X_{m_k} - X_{s_2}(t))^2 + (Y_{m_k} - Y_{s_2}(t))^2 + (Z_{m_k} - Z_{s_2}(t))^2}. \qquad (6)$$

According to (3) and for $\theta_2 = 0$, one can easily show that:

$$D_{k2}(t) = \sqrt{D_{k2}^2(t_0) - 2v_2(t - t_0)\left(X_{m_k} - X_{s_2(t_0)}\right) + v_2^2(t - t_0)^2}. \qquad (7)$$

Therefore:

$$\tau_{r2}(t) = \sqrt{\tau_{r2}^2(t_0) - \frac{2v_2(t - t_0)\left(X_{m_r} - X_{s_2(t_0)}\right)}{c^2} + \frac{v_2^2(t - t_0)^2}{c^2}}. \qquad (8)$$

To resample the signal recorded on sensor r, we compute $m_r(t + \varepsilon_{kr}(t))$ by Eq. (9), where $\varepsilon_{kr}(t)$ is a time shift that will allow to equalize the propagation delays of from the source 2 to the sensors k and r.

$$m_r(t + \varepsilon_{kr}(t)) = a_{r1} s_1 [t + \varepsilon_{kr}(t) - \tau_{r1}]$$
$$+ a_{r2} [t + \varepsilon_{kr}(t) - \tau_{r2}[t + \varepsilon_{kr}(t)]] s_2 [t + \varepsilon_{kr}(t) - \tau_{r2}[t + \varepsilon_{kr}(t)]]$$
$$(9)$$

We then seek of $\varepsilon_{kr}(t)$ such that:

$$\varepsilon_{kr}(t) - \tau_{r2}[t + \varepsilon_{kr}(t)] = -\tau_{k2}(t). \qquad (10)$$

One can shows that:

$$\tau_{r2}[t + \varepsilon_{kr}(t)] = \sqrt{\tau_{r2}^2(t) + \frac{v_2^2}{c^2}[\varepsilon_{kr}(t)]^2 + \frac{2v_2\varepsilon_{kr}(t)}{c^2}\left[v_2(t - t_0) - (X_{m_r} - X_{s_2(t_0)})\right]}. \qquad (11)$$

Otherwise:

$$\varepsilon_{kr}(t) - \tau_{r2}[t + \varepsilon_{kr}(t)] = -\tau_{k2}(t) \Rightarrow [\tau_{k2}(t) + \varepsilon_{kr}(t)]^2 = \tau_{r2}[t + \varepsilon_{kr}(t)]^2. \qquad (12)$$

By combining Eqs. (11) and (12), one gets the quadratic Eq. (13), from which one of the two solutions gives the desired $\varepsilon_{kr}(t)$.

$$[\varepsilon_{kr}(t)]^2 \left[1 - \frac{v_2^2}{c^2}\right] + 2\varepsilon_{kr}(t) \left[\tau_{k2}(t) - \frac{v_2^2(t - t_0) - v_2(X_{m_r} - X_{s_2(t_0)})}{c^2}\right] + \left[\tau_{k2}^2(t) - \tau_{r2}^2(t)\right] = 0.$$

(13)

It follows that the resampled signal on sensor r is given by:

$$m_r(t + \varepsilon_{kr}(t)) = a_{r1}s_1\left[t - \tau_{r1} + \varepsilon_{kr}(t)\right] + a_{r2}\left[t - \tau_{k2}(t)\right]s_2\left[t - \tau_{k2}(t)\right].$$

(14)

Thus, for removing the moving source s_2 from sensor k, we just have to linearly combine $m_k(t)$ and $m_r(t + \varepsilon_{kr}(t))$ as illustrated on Eq. (15):

$$\begin{aligned}
\tilde{m}_k(t) &= a_{r2}\left[t - \tau_{k2}(t)\right]m_k(t) - a_{k2}\left[t - \tau_{k2}(t)\right]m_r(t + \varepsilon_{kr}(t)) \\
&= a_{k1}a_{r2}\left[t - \tau_{k2}(t)\right]s_1(t - \tau_{k1}) - a_{r1}a_{k2}\left[t - \tau_{k2}(t)\right]s_1\left[t - \tau_{r1} + \varepsilon_{kr}(t)\right].
\end{aligned}$$

(15)

By construction, $\tilde{m}_k$ contains two paths of the target source s_1, from which a scaled and delayed estimation $\hat{s}_1$ of s_1 can be estimated by:

$$\hat{s}_1(t) = a_{k1}s_1(t - \tau_{k1}) - a_{r1}\frac{a_{k2}\left[t - \tau_{k2}(t)\right]}{a_{r2}\left[t - \tau_{k2}(t)\right]}s_1\left[t - \tau_{r1} + \varepsilon_{kr}(t)\right].$$

(16)

The method described above allows us to remove the noise coming from the moving source, but it also creates a second path on the target source (right term in (16)), that we deal with the multipath suppression algorithm proposed in [13]. This algorithm is not detailed here due to lack of place.

4   Simulation Results

This section shows simulation results on synthetic data. The proposed method is compared with the Linearly Constrained Minimum Variance (LCMV) beamforming method [8]. Let's recall that, in the same way than the method described in this paper, LCMV-beamforming also make use of the sources locations. It's aims at enhancing the signal coming from the direction of the target source, and reduce noises coming from the other directions.

The efficiency of the estimation of s_1 is quantified by the signal-to-interference ratio (SIR) and by the signal-to-distortion ratio (SDR), as defined in [14]. To compute the SIR and SDR, $\hat{s}_1$ is decomposed as:

$$\hat{s}_1 = s_{target} + e_{interf} + e_{artifact}.$$

(17)

where s_{target} is a scaled and delayed version of the original source s_1, and where e_{interf}, and $e_{artifact}$ are the interference, and artifact error terms, respectively [14]. The SIR and the SDR are then computed through Eqs. (18) and (19), respectively. The larger the SIR and the SDR are, the better the estimation is.

Due to the dynamic nature of the model, the SIR and the SDR are computed blockwise of length $1s$.

$$SIR = 10\log_{10}\frac{\|s_{target}\|^2}{\|e_{interf}\|^2}.\tag{18}$$

$$SDR = 10\log_{10}\frac{\|s_{target}\|^2}{\|e_{interf} + e_{noise} + e_{artifact}\|^2}.\tag{19}$$

The fixed source, s_1, is a speech while the moving source, s_2, is a tone at frequency $400\,Hz$ that moves at $50\,km/h$ parallel to the x-axis. Figure 3 shows the original source s_1 and the mixed signal recorded at microphone 1. We set the coordinates of s_1 to $(0\,m,\ 0.1\,m,\ 1.5\,m)$ and the coordinates of s_2 at the reference time t_0 to $(0\,m,\ 5\,m,\ 0\,m)$. We consider a uniform linear antenna whose sensors are distributed along the x-axis and centered at the origin of this axis. The distance between two consecutive sensors is $d = 5\,cm$ and for each sensor k, we set $Y_{m_k} = 0$ and $Z_{m_k} = 1.5\,m$. Figure 2 shows the simulation scenario.

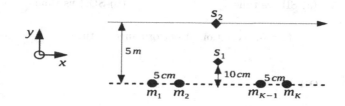

Fig. 2. Simulation scenario

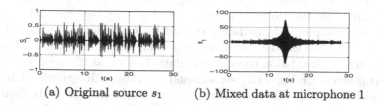

(a) Original source s_1 (b) Mixed data at microphone 1

Fig. 3. Original source s_1 and mixed data at microphone 1

Figure 4 shows the estimations of the target source, $\hat{s}_1$, by proposed method and by beamforming with different numbers of sensors. One can see from Fig. 4 that the proposed method is able to remove the noise from the moving source with only 2 microphones, therefore there is no need to increase the number of sensors. To achieve this performance the LCMV-beamforming method requires at least 3 microphones (in this example).

Figure 5 shows the performance indices evolution versus time, both for the proposed method and for LCMV-Beamforming. From the performance reported on Fig. 5, we can conclude that our method performs better than the LCMV-beamforming with 2 microphones, and achieves similar results than LCMV-beamforming with 3 microphones.

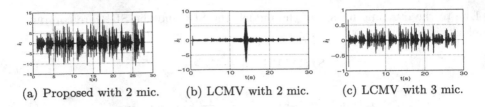

| (a) Proposed with 2 mic. | (b) LCMV with 2 mic. | (c) LCMV with 3 mic. |

Fig. 4. Results of estimation of the target source s_1

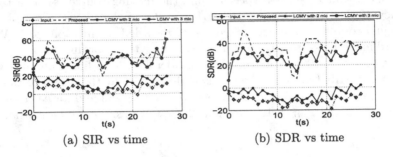

| (a) SIR vs time | (b) SDR vs time |

Fig. 5. Evolution of performance vs time

5   Conclusion

In this paper, we propose a method for removing noise coming from a fast moving acoustic source. The proposed method requires only two microphones and exploits the known position of the moving source, assumed to be provided by other modality like a video system or a GPS. The proposed method consists in an irregular resampling of the data recorded by a reference sensor, and to linearly combine it with the data recorded by second sensor in order to remove the undesired source. Simulation on synthetic data shows that the proposed method outperforms beamforming with 2 sensors and achieves similar results than beamforming with a great number of sensors. Future works include evaluation of the proposed method on actual data and to take into account localization errors. The case of multiple undesired moving sources will also be investigated.

References

1. Widrow, B., Luo, F.-L.: Microphone arrays for hearing aids: an overview. Speech Commun. **39**, 139–146 (2003)
2. Okutani, K., Yoshida, T., Nakamura, K., Nakadai, K.: Outdoor auditory scene analysis using a moving microphone array embedded in a quadrocopter. In: IEEE/RSJ International Conference on Intelligent Robots and Systems, pp. 3288–3293 (2012)
3. Australian Centre for Field Robotics, Chap. 14 - Doppler Measurement. http://www.acfr.usyd.edu.au/pdfs/training/sensorSystems/14%20Doppler%20 Measurement.pdf

4. Manohar, K., Rao, P.: Speech enhancement in nonstationary noise environments using noise properties. Speech Commun. **48**, 96–109 (2006)
5. Mukai, R., Sawada, H., Haraki, S., Makino, S.: Blind source separation for moving speech signals using blockwise ICA and residual crosstalk subtraction. IEICE Trans. Fundam. **E87**(8), 1941–1948 (2004)
6. Cohen, I., Gannot, S., Berdugo, B.: An integrated real-time beamforming and postfiltering system for nonstationary noise environments. EURASIP J. Appl. Sig. Process. **2003**, 1064–1073 (2003)
7. Verteletskaya, E., Simak, B.: Noise reduction based on modified spectral subtraction method. Int. J. Comput. Sci. **38**(1), 1–7 (2011)
8. Souden, M., Benesty, J., Affes, S.: A study of the LCMV and MVDR noise reduction filters. IEEE Trans. Signal Process. **58**(9), 4925–4935 (2010)
9. Maganti, H.K., Gatica-Perez, D., McCowan, I.: Speech enhancement and recognition in meetings with an audio-visual sensor array. IEEE Trans. Audio Speech Lang. Process. **15**(8), 2257–2269 (2007)
10. Naqvi, S.M., Wang, W., Khan, M.S., Barnard, M., Chambers, J.A.: Multimodal (audiovisual) source separation exploiting multi-speaker tracking, robust beamforming and time-frequency masking. IET Signal Process. **6**(5), 466–477 (2012)
11. Strobel, N., Spors, S., Rabenstein, R.: Joint audio-video object localization and tracking. IEEE Signal Process. Mag. **18**(1), 22–31 (2001)
12. Nishie, S., Akagi, M.: Acoustic sound source tracking for a moving object using precise Doppler-Shift measurement. In: Proceedings of the 21st European Signal Processing Conference, pp. 1–5 (2013)
13. Piper, J.E.: Multipath cancellation in the time domain. In: OCEANS Conference, pp. 1–4 (2010)
14. Vincent, E., Gribonval, R., Fevotte, C.: Performance measurement in blind audio source separation. IEEE Trans. Audio Speech Lang. Process. **14**(4), 1462–1469 (2006)

Speaker Verification Using Adaptive Dictionaries in Non-negative Spectrogram Deconvolution

Szymon Drgas[1](✉) and Tuomas Virtanen[2]

[1] Chair of Control and Systems Engineering, Poznan University of Technology,
Poznań, Poland
szymon.drgas@put.poznan.pl
[2] Department of Signal Processing, Tampere University of Technology,
Tampere, Finland

Abstract. This article presents a new method for speaker verification, which is based on the non-negative matrix deconvolution (NMD) of the magnitude spectrogram of an observed utterance. In contrast to typical methods known from the literature, which are based on the assumption that the desired signal dominates (for example GMM-UBM, joint factor analysis, i-vectors), compositional models such as NMD describe a recording as a non-negative combination of latent components. The proposed model represents a spectrogram of a signal as a sum of spectro-temporal patterns that span durations of order about 150 ms, while many state of the art automatic speaker recognition systems model a probability distribution of features extracted from much shorter excerpts of speech signal (about 50 ms). Longer patterns carry information about dynamical aspects of modeled signal, for example information about accent and articulation. We use a parametric dictionary in the NMD and the parameters of the dictionary carry information about the speakers' identity. The experiments performed on the CHiME corpus show that with the proposed approach achieves equal error rate comparable to an i-vector based system.

1 Introduction

In this article an application of the non-negative matrix deconvolution (NMD) with a parametric dictionary to the speaker recognition is considered. The specific task is the text-independent speaker verification, which is usually done by modeling the probability distribution of mel-frequency cepstral coefficients (MFCCs) [1]. MFCCs represent the spectral envelope of a signal within short-time (20 ms) frames and their probability distribution is often modeled using Gaussian mixture models (GMMs).

In GMM-based speaker recognition systems different short-time frames are assumed to be statistically independent. Thus, temporal information is lost (e.g. how a word was uttered, what was the accent and rate of speech). In order to obtain a contextual information, delta and delta-delta features are typically concatenated to the feature vectors [1]. This results with a set of approximately

© Springer International Publishing Switzerland 2015
E. Vincent et al. (Eds.): LVA/ICA 2015, LNCS 9237, pp. 462–469, 2015.
DOI: 10.1007/978-3-319-22482-4_54

decorrelated features that can carry information extracted from signal excerpts with duration about 50 ms. There are also attempts to model the distribution of spectrogram excerpts decorrelated by means of temporal DCT or 2D DCT in the standard GMM or i-vector framework [2,3].

The compositional models [4] of a spectrogram provide alternative possibility for speaker modeling. In these models, the magnitude spectrum is modeled as a sum of magnitude spectra generated by different sources or atomic parts of each source. In [5] application of compositional model in the speaker identification system was presented. This system was prepared for a closed-set speaker identification task. The dictionary was composed of groups of templates (so-called exemplars) for a specific set of speakers. Each group contained exemplars extracted from utterances of a single speaker. During the speaker recognition, the NMD algorithm determined which exemplars were activated. The recognized speaker was the one, whose exemplars were activated most often. In [6] non-negative matrix factorization was used in order to extract robust features.

In this article a method for the speaker verification using nonnegative matrix deconvolution with parametric dictionary is described. This solution gives a possibility to apply a compositional model to the speaker verification task as there is no need to know all of the tested speakers in advance. Moreover, adaptation of small number of parameters instead of whole dictionary requires less data. In the method proposed in this work transformation of the spectrogram segments to decorrelated coefficients is obtained separately for each state (phase of phone realization). Thus, for each state various spectral patterns (about 150 ms long) can be obtained by varying these uncorrelated coefficients.

2 Method

2.1 Model

The proposed system models the magnitude spectrogram $\mathbf{X} \in \mathbb{R}^{B \times N}$ of an observed utterance with non-negative matrix deconvolution [7,8], where B is the number of frequency bands and N is the number of time frames in the utterance. NMD models $\mathbf{X}$ as a weighted linear combination of temporally shifted spectrogram patches and it can be mathematically formulated as

$$\mathbf{X} \approx \sum_{t=0}^{T-1} \mathbf{D}(t) \overset{t \rightarrow}{\mathbf{A}}, \tag{1}$$

where $\mathbf{D}(t) \in \mathbb{R}^{B \times P}$ is the dictionary and $\mathbf{A} \in \mathbb{R}^{P \times N}$ is the activation matrix. P is the number of templates in the dictionary. pth columns of each $\mathbf{D}(t)$, $t = 1, \ldots, T$ form a template, which is a $B \times T$ spectrogram segment that is used as the basic building block in the spectrogram deconvolution. Operator over matrix $\mathbf{A}$ denotes the shift of columns of matrix $\mathbf{A}$ t columns to the right. The t first columns are padded with zeros. For example:

$$\mathbf{A} = \begin{bmatrix} 1 & 2 & 3 & 4 \\ 5 & 6 & 7 & 8 \end{bmatrix} \qquad \overset{2 \rightarrow}{\mathbf{A}} = \begin{bmatrix} 0 & 0 & 1 & 2 \\ 0 & 0 & 5 & 6 \end{bmatrix}. \tag{2}$$

In the present work we use parametric dictionary $\mathbf{D}(t)$ which is expressed as

$$\mathbf{D}(t) = \exp\left(\hat{\mathbf{D}}(t) + [\mathbf{V}_1(t)\mathbf{T}_1\mathbf{s} \ \cdots \ \mathbf{V}_P(t)\mathbf{T}_P\mathbf{s}]\right), \qquad (3)$$

where $\mathbf{s}$ is a length J column vector representing speaker variability of the dictionary, $\hat{\mathbf{D}}(t) \in \mathbb{R}^{B \times P}$ denotes speaker-independent (average) dictionary, and matrices $\mathbf{V}_1(t), \ldots, \mathbf{V}_P(t) \in \mathbb{R}^{B \times K}$ contain components representing speaker variability for each of the P templates. Matrices $\mathbf{T}_p \in \mathbb{R}^{K \times J}$ transform state-independent weights of components representing speaker variability $\mathbf{s}$ to state-dependent weights (see Eq. 8). Thus, each template (which refers to the columns of matrix $\mathbf{D}(t)$) is composed of speaker-independent component and a linear combination of J components modeling speaker variability. Each template corresponds to a state, while state refers to a segment of spectrogram centered on a certain stationary part of a phoneme.

The system works in development, training and test phases. From the development dataset, matrices $\hat{\mathbf{D}}(t)$, $\mathbf{V}_1(t), \ldots, \mathbf{V}_P(t)$, and $\mathbf{T}_1, \ldots, \mathbf{T}_P$ are determined. In the training phase of the speaker verification system, given an enrollment utterance, vector $\mathbf{s}_{tr}$ is estimated and stored in the database. During the test phase, vector $\mathbf{s}_{te}$ is adapted to a test utterance. The similarity of the claimed speaker and test speaker in the test utterance can be computed using a vector similarity measure between $\mathbf{s}_{tr}$ and $\mathbf{s}_{te}$. In this work the cosine similarity is used. The spectro-temporal features used in this technique are mel-frequency spectral coefficients (square root of energies at the outputs of mel-scale filterbank).

2.2 Development Phase

In order to determine speaker-independent template matrix $\hat{\mathbf{D}}(t)$ and matrices with components representing speaker variability $\mathbf{V}_1(t), \ldots, \mathbf{V}_P(t)$, development data with a phonetic annotation is needed. Excerpts of spectrogram that are specific to each state are extracted from the training corpus in the form of so-called exemplars and stored in set $E = \{\mathbf{e}_{pqs}\}_{p=1,\ldots,P;s=1,\ldots,S;q=1,\ldots,Q_{ps}}$, where p is the index of a state, q is the index of exemplar, Q_{ps} is the number of exemplars for pth state, and s is the index of the speaker. Matrices $\hat{\mathbf{D}}(t)$ and $\mathbf{V}_1(t), \ldots, \mathbf{V}_P(t)$ are computed in such a way, that by changing parameter vector $\mathbf{s}$ (which has small number of elements) all exemplars from the development can be approximated. Each exemplar is an excerpt (T consecutive frames) of a spectrogram centered at the center of a segment labeled with a desired state. Each exemplar can be considered as $B \times T$ matrix. After vectorization, vector $\mathbf{e}_{pqs}$ with BT elements is obtained, where elements of the exemplar are ordered with respect to the columns.

For each state all the exemplars are gathered, and the center point is estimated. The center point $\mathbf{m}_p$ is a point for which a sum of Kullback-Leibler divergences to the exemplars is lowest:

$$\mathbf{m}_p = \arg\min_{\mathbf{d}} \sum_{s=1}^{S}\sum_{q=1}^{Q_{ps}} D_{KL}(\mathbf{e}_{pqs}\|\mathbf{d}) = \log\left(\frac{1}{\sum_{s=1}^{S}Q_{ps}}\sum_{s=1}^{S}\sum_{q=1}^{Q_{ps}}\mathbf{e}_{qps}\right), \qquad (4)$$

where

$$D_{KL}(\mathbf{X}\|\tilde{\mathbf{X}}) = \sum_{ij}\left[\mathbf{X}_{ij}\log\left(\frac{\mathbf{X}_{ij}}{\tilde{\mathbf{X}}_{ij}}\right) - \mathbf{X}_{ij} + \tilde{\mathbf{X}}_{ij}\right]. \tag{5}$$

Center point vectors $\mathbf{m}_p$ (defined in Eq. 4) are used as the columns of speaker-independent template matrix $\hat{\mathbf{D}}(t)$; Each B-dimensional block of vector $\mathbf{m}_p$ is assigned to the pth column of matrix $\hat{\mathbf{D}}(t)$ for $t = 1, \ldots, T$.

The estimation of matrixces $\mathbf{V}_p(t)$ is done for each state p separately. Each matrix $\mathbf{V}_p(t)$ is obtained from all exemplars of state p available in the development dataset. For each state p covariance matrix $\mathbf{P}_p$ is determined as

$$\mathbf{P}_p = \sum_{s=1}^{S}\sum_{q=1}^{Q_{ps}}(\log\mathbf{e}_{pqs} - \log\mathbf{m}_p)(\log\mathbf{e}_{pqs} - \log\mathbf{m}_p)^{\mathrm{T}} \tag{6}$$

and its eigendecomposition is performed. Let us denote eigenvectors of $\mathbf{P}_p$ ordered according to respective eigenvalues as $\mathbf{v}_1^{(p)}, \ldots, \mathbf{v}_{BT}^{(p)}$. The first K vectors are used to construct $\mathbf{V}_p$ matrix as

$$\mathbf{V}_p = \begin{bmatrix}\mathbf{v}_1^{(p)} & \cdots & \mathbf{v}_K^{(p)}\end{bmatrix}. \tag{7}$$

The role of matrices $\mathbf{T}_p$ is to transform blocks of matrix $\mathbf{V}(t)$ in such a way that good approximation of exemplars can be achieved with state-independent weights $\mathbf{s}_s$. This can be written as

$$\mathbf{c}_{pqs} \approx \mathbf{T}_p\mathbf{s}_s \qquad \forall q, s, \tag{8}$$

where $\mathbf{c}_{pqs}$ are weights of components representing speaker variability $\mathbf{V}_p(t)$ for pth state, for qth example of speaker s, while $\mathbf{s}_s$ is the average vector $\mathbf{s}$ extracted for speaker s. In order to obtain matrices $\mathbf{T}_p$, vector $\mathbf{c}_{pqs}$ representing speaker variability is optimized for each exemplar $\mathbf{e}_{pqs}$ in the development dataset individually as

$$\mathbf{c}_{pqs} = \arg\min_{\mathbf{c}'}\left\{\sum_{t=0}^{T-1}\left[D_{KL}\left(\mathbf{e}_{pqs}|\exp(\hat{\mathbf{d}}_p(t) + \mathbf{V}_p(t)\mathbf{c}')\right)\right]\right\}. \tag{9}$$

Next, for each state p and for each speaker s in the development dataset, mean $\boldsymbol{\mu}_{ps}$ is calculated as

$$\boldsymbol{\mu}_{p,s} = \frac{1}{Q_{ps}}\sum_{q=1}^{Q_{ps}}\mathbf{c}_{pqs}. \tag{10}$$

These means are used to construct matrix $\mathbf{M}$ as

$$\mathbf{M} = \begin{bmatrix}\boldsymbol{\mu}_{1,1} & \boldsymbol{\mu}_{1,2} & \cdots & \boldsymbol{\mu}_{1,S} \\ \boldsymbol{\mu}_{2,1} & \boldsymbol{\mu}_{2,2} & \cdots & \boldsymbol{\mu}_{2,S} \\ \vdots & \vdots & \ddots & \vdots \\ \boldsymbol{\mu}_{P,1} & \boldsymbol{\mu}_{P,2} & \cdots & \boldsymbol{\mu}_{P,S}\end{bmatrix}. \tag{11}$$

Matrix $\mathbf{M}$ is decomposed using semi-NMF algorithm [9]. In the preliminary experiments, other techniques like PCA for this decomposition, but semi-NMF gave the best results. The semi-NMF gives approximation of matrix $\mathbf{M}$ as the product of mixed-sign matrix $\mathbf{N}$ and non-negative matrix $\mathbf{W}$ as $\mathbf{M} \approx \mathbf{NW}$. In contrast to NMF only matrix $\mathbf{W}$ is constrained to have nonnegative elements. The columns of matrix $\mathbf{N}$ are basis vectors $\mathbf{n}_1, \ldots, \mathbf{n}_K$. Matrices $\mathbf{T}_1, \ldots, \mathbf{T}_P$ are the blocks of matrix with columns $\mathbf{n}_k$

$$\begin{bmatrix} \mathbf{T}_1 \\ \vdots \\ \mathbf{T}_P \end{bmatrix} = [\mathbf{n}_1 \; \ldots \; \mathbf{n}_K]. \tag{12}$$

2.3 Training and Test Phase

During the training and test phases in the speaker verification system, the elements of matrix $\mathbf{A}$ and vector $\mathbf{s}$ have to be determined.

The parameters $\mathbf{s}$ and $\mathbf{A}$ of the model are optimized to minimize the Kullback-Leibler divergence between a given utterance $\mathbf{X}$ and its model $\tilde{\mathbf{X}} = \sum_{t=0}^{T-1} \mathbf{D}(t) \overset{t\rightarrow}{\mathbf{A}}$ as

$$\min_{\mathbf{s},\mathbf{A}} \quad D_{KL}\left(\mathbf{X} \| \tilde{\mathbf{X}}(\mathbf{s}, \mathbf{A})\right) + \lambda_s \|\mathbf{A}\|_1,$$
$$\text{s.t.} \quad \begin{aligned} \mathbf{A} &\geq \mathbf{0}, \\ \mathbf{s} &\geq \mathbf{0}. \end{aligned} \tag{13}$$

There is also norm-1 penalty $\lambda_s \|\mathbf{A}\|_1$ of the activation matrix included in the cost function, which can be changed by tuning weight λ_s. The starting matrix $\mathbf{A}$ was obtained from annotations of the development dataset. Each frame has a label that refers to one of six word-classes. Each state corresponds to on of these word-classes. The parameters of the Markov chain for these labels are estimated during the development phase. The starting matrix $\mathbf{A}$ contain likelihoods of being in a given state for a given frame. Adaptation of the model's parameters can be summarized as a sequence of the following steps:

1. Initialization.
2. Update $\mathbf{A}$ matrix using the formula

$$\mathbf{A} \leftarrow \sum_{t=0}^{T} \mathbf{A} \odot \frac{\mathbf{D}^{\mathrm{T}}(t) \left(\overset{\leftarrow t}{\frac{\mathbf{X}}{\tilde{\mathbf{X}}}}\right)}{\mathbf{D}^{\mathrm{T}}(t) \left(\mathbf{1}_{B \times N}\right) + \lambda_s \mathbf{1}_{P \times N}}, \tag{14}$$

where multiplication $\odot$ and division are element-wise, $\mathbf{1}$ denotes matrix filled with ones, and operator $\leftarrow$ works analogously to $\rightarrow$ in (1) but columns of a matrix are shifted to the left.
3. Compute partial derivatives $\frac{\partial D_{KL}}{\partial \mathbf{s}}$.
4. Perform line search with clipping of values of $\mathbf{s}$ below zero.

5. Update the dictionary parameters

$$\mathbf{s} \leftarrow \mathbf{s} - \alpha \frac{\partial D_{KL}}{\partial \mathbf{s}} \qquad (15)$$

where α refers to the step size obtained during the line-search.

6. Update the dictionary using new $\mathbf{s}$. Normalize the dictionary - divide each column of matrix $\mathbf{D}(t)$ by its norm.

7. If the iteration count is lower than the maximum number of iterations go to Step 1.

During the training phase vector $\mathbf{s}_{tr}$ is adapted to the enrollment utterance. During the test phase $\mathbf{s}_{te}$ is adapted to the test utterance. Next, these vectors are compared using cosine similarity and the score is obtained as

$$\text{score}(\mathbf{s}_{tr}, \mathbf{s}_{te}) = \frac{\mathbf{s}_{tr}^{T} \mathbf{s}_{te}}{\|\mathbf{s}_{tr}\| \|\mathbf{s}_{te}\|} . \qquad (16)$$

This score is used to decide whether speaker in the training and test utterances is the same or not. If the score is higher than a certain threshold then speaker the answer of the system is that the speaker is the same in both (training and test) utterances. The mentioned threshold can be set according to false alarm/miss error tradeoff according to an application. In this paper, the threshold is set to obtain equal error rates of two types.

3   Experiments

3.1   Data and Spectro-Temporal Representation

Experiments have been performed using recordings from the CHiME corpus [10], which contains utterances from 34 speakers. The experiments were conducted using recordings without additive noise. However, relatively small amount of reverberation is present in the CHiME recordings (RT60 300 ms). For each speaker 500 utterances with the same grammar are available. The vocabulary contains 52 words. The first 300 utterances of each speaker have been used as a development dataset, while the other 200 utterances were have been used as a test utterances. The original data are sampled with rate 16000 samples/second. The data were downsampled to 8000 samples/second in order to obtain a telephone bandwidth. The telephone data were used in many previous speaker recognition studies. The average duration of each utterance is about 1.5 s.

The proposed algorithm has been compared to state of the art systems which use MFCCs to represent the signals. First, the preemphasis filter with coefficient 0.97 was applied. Next, the signal was divided into 25 ms frames. The frame step was 10 ms. After FFT computation mel-frequency filterbank was used. The 40 filters spanned the frequency range 64–4000 Hz. After type II discrete cosine transform 20 cepstral coefficients were used.

Table 1. EER (%) for the baseline i-vector based system

UBM components	i-vector dimension		
	10	15	20
128	8.89	7.62	7.46
256	**4.41**	4.57	6.16
512	5.02	5.55	6.19

The frame-wise feature extraction in the proposed system is equal to the MFCC calculation, except the last two steps, namely the application of a logarithm and the discrete cosine transform are not used. This is because the compositional model is based on sum of magnitudes approximation and an extraction of uncorrelated components is done during estimation of matrix $\mathbf{V}(t)$.

3.2 Results

In order to compare the system proposed in this work, an i-vector based system which can be considered as the state of the art system, was used. The MSR Toolbox code was used to extract i-vectors [11]. First, the development dataset was used to train the GMM universal background model (UBM). In the experiments several number of components of the UBM were tried: 128, 256, and 512. Next, for the test dataset the sufficient statistics and i-vectors were calculated. The number of i-vector elements was 10, 15, or 20. Finally, a similarity between training and test i-vectors was computed with the cosine similarity. The scores were used to make a decision whether speaker in the training and test utterances were the same. Based on these scores, equal error rate was calculated for the test dataset. It is done by finding the threshold for which false acceptance and rejection error rates are equal. The results are presented in Table 1. The best results were obtained for the case where the number of UBM components is 256 and the dimension of i-vector is equal to 10. The results obtained for the proposed system are presented in Table 2. The system was tested for the following values of parameters: the number of dictionary templates $P = 250$, the number of components representing speaker variability $K = 10$. Parameter K was chosen according to the results of the preliminary experiments. Dimensions J of parameter vector $\mathbf{s}$ was varied. The best result was obtained for $J = 25$, where EER is 4.29.

Table 2. Results for the proposed system

Dimension of vector s	EER
25	**4.29**
20	5.59
15	7.30

4 Conclusions

In this work a new method for the speaker recognition is proposed. It is based on compositional model with a parametric dictionary. The advantage of such a model in contrast to the other known from the literature speaker recognition methods is that compositional models describe a recording as a mixture of signals (desired signal and interference signals). Additionally, the applied models represent spectrogram of the recording as a sum of spectro-temporal patterns that span duration of order about 150 ms, while typical speaker recognition systems are based on substantially shorter speech excerpts (about 40 ms). Longer patterns carry information about dynamical aspects of modeled signal.

In the best case the achieved EER is 4.29 % while the system based on i-vector performs at 4.40 %. The obtained error is comparable to the baseline, but the proposed method can be easily extended to model the interfering sources.

References

1. Kinnunen, T., Li, H.: An overview of text-independent speaker recognition: From features to supervectors. Speech Commun. **52**, 12–40 (2010)
2. McLaren, M., Scheffer, N., Ferrer, L., Lei, Y.: Effective use of DCTs for contextualizing features for speaker. In: IEEE International Conference on Acoustic, Speech and Signal Processing (ICASSP) (2014)
3. Kinnunen, T., Wei, C., Koh, E., Wang, L., Li, H., Chng, E.S.: Temporal discrete cosine transform: Towards longer term temporal features for speaker verification. In: Proceedings of the Fifth International Symposium on Chinese Spoken Language Processing (ISCSLP 2006), Singapore, pp. 547–558 (2006)
4. Virtanen, T., Gemmeke, J.F., Raj, B., Smaragdis, P.: Compositional models for audio processing. IEEE Signal Process. Mag. **32**(2), 125–144 (2015)
5. Saeidi, R., Hurmaleinen, A., Virtanen, T., van Leeuwen, D.A.: Exemplar-based sparse representation and sparse discrimination for noise robust speaker identification. In: Odyssey 2012: The Speaker and Langueage Recognition Workshop (2012)
6. Joder, C., Schuller, B.: Exploring nonnegative matrix factorization for audio classification: application to speaker recognition. In: Proceedings of Speech Communication; 10, ITG Symposium, VDE, 1–4 (2012)
7. Smaragdis, P.: Convolutive speech bases and their application to supervised speech separation. IEEE Trans. Speech and Audio Process. **15**(1), 1–12 (2005)
8. Hurmalainen, A., Virtanen, T.: Learning state labels for sparse classification of speech with matrix deconvolution. IEEE Workshop Autom. Speech Recogn. Underst. (ASRU) **2013**, 168–173 (2013)
9. Ding, C., Li, T., Jordan, M.: Convex and semi-nonnegative matrix factorizations. IEEE Trans. Pattern Anal. Mach. Intell. **32**, 45–55 (2010)
10. Christensen, H., Barker, J., Ma, N., Green, P.D.: The CHiME corpus: a resource and a challenge for computational hearing in multisource environments. In: INTERSPEECH, pp. 1918–1921 (2010)
11. Sadjadi, S.O., Slaney, M., Heck, L.: MSR identity toolbox v1.0: A MATLAB toolbox for speaker-recognition research. Speech and Language Processing Technical Committee Newsletter (2013)

Towards Individualized Spatial Audio via Latent Variable Modeling

Eric S. Schwenker[1,2](✉) and Griffin D. Romigh[2]

[1] Carnegie Mellon University, Pittsburgh, PA 15213, USA
eschwenk@andrew.cmu.edu
[2] Air Force Research Laboratory, Wright Patterson AFB, Dayton, OH 45433, USA
griffin.romigh@us.af.mil

Abstract. Currently, many virtual reality systems and augmented reality displays lack the immersiveness and fidelity of real-life auditory space. This limitation is derived from an inability to easily measure individualized head related transfer functions (HRTFs), the key technology behind high-fidelity spatial audio. This study presents an initial framework based on a joint maximization EM algorithm for affordable HRTF estimation that eliminates the need for both head-tracking and/or prior source location knowledge from the HRTF measurement process.

1 Introduction

In natural listening environments, humans use a unique set of acoustic cues to localize incoming sounds. These localization cues, collectively represented by a head related transfer function (HRTF), describe the acoustic transformations caused by direction specific interactions of a sound wave with a listener's head, shoulders, and outer ears. For high-fidelity virtual audio applications, an HRTF measured on the specific listener is essential. Headphone-based spatial rendering with non-individualized HRTFs can lead to inside-the-head localization, front-back reversals, and unwanted spectral distortions [4]. While individualized HRTF filtering can provide users of 3D audio systems with a requisite sense of immersive realism, traditional processes for measuring individualized HRTFs are both cumbersome and prohibitively expensive for the common consumer. As such, significant advances must be made in the ease with which individualized HRTFs are measured for the end user of a spatial audio system.

Traditional HRTFs are measured by playing a known broadband reference audio signal from a loudspeaker at a known location with respect to the listener's head, and recording the signals that arrive at two microphones placed inside the listener's ear canals. The reference and recorded audio signals are then used to calculate a transfer function for the linear time-invariant (LTI) systems created by the anatomical features of the listener's head, shoulders, and outer ears. These *sample* HRTFs (extracted from binaural recordings) are then measured at many spatial locations around the listener by changing the location of the loudspeaker or listener in a known way. Later, the set of *sample* HRTFs and their

© US Government (outside the US) 2015
E. Vincent et al. (Eds.): LVA/ICA 2015, LNCS 9237, pp. 470–477, 2015.
DOI: 10.1007/978-3-319-22482-4_55

corresponding head-relative locations are used to estimate a spatially-continuous HRTF (in a process sometimes referred to as HRTF interpolation or estimation).

Because both features (*sample* HRTFs) and labels (head-relative locations) are known, traditional continuous HRTF estimation can be thought of as a relatively straightforward supervised learning problem, and has been successfully accomplished using a number of methods (for a review, see [5]). However, conventional *sample* HRTF measurements utilize prohibitively expensive loudspeaker arrays or headtracking devices to obtain head-relative loudspeaker location data. This makes widespread commercialization of high-fidelity HRTFs challenging. A potential solution to this problem is to try to estimate a continuous HRTF without head-tracking or loudspeaker arrays. Consider a *sample* HRTF collected without knowledge of the location of the sound source. Viewing this location information as a latent variable (functionally) converts HRTF estimation into an unsupervised, or "blind" learning problem. With this interpretation, the algorithmic framework developed in this study is a novel first step towards the ultimate goal of commecially feasible individualized spatial audio.

The organization of the paper is as follows: Sect. 2 presents the structure of the blind HRTF estimation process, which is initialized with an average HRTF taken from an existing database. Section 3 further details the two fundamental maximization steps, Binaural Source Localization and Continuous HRTF Estimation, providing details on how these specific steps are leveraged in context of the overall process. Section 4 presents the results of the blind HRTF estimation algorithm on simulated data. Finally, Sect. 5 outlines some conclusions and suggests the logical direction for future work and development.

2   The Iterative Localization and HRTF Estimation Technique

The blind estimation process can be formulated as a coordinate ascent algorithm. In this case, the algorithm is initialized with the parameters of a continuous HRTF estimation model, and the estimation procedure is carried out in an iterative joint maximization EM procedure [2] (two alternating maximization steps): Binaural Source Localization (M-Step 1) and Continuous HRTF Estimation (M-Step 2). Both steps are given the distinction of being a maximization step because in each process the defined objective is being maximized under the assumption that the missing data from the other step is known.

Figure 1 provides a simplified schematic of the relationship between the Binaural Source Localization and Continuous HRTF Estimation M-Steps. As indicated, starting with the parameters of an average continuous HRTF model (an average HRTF taken from an existing database), and binaural recordings (extracted from real *sample* HRTFs, see Sect. 3), a binaural localizer can be used to estimate the locations (labels), of the sample data (Binaural Source Localization step). Once estimated labels are available, a method for estimating the parameters of the continuous HRTF model can then use those location labels to reapproximate the continuous HRTF for the next iteration (Continuous

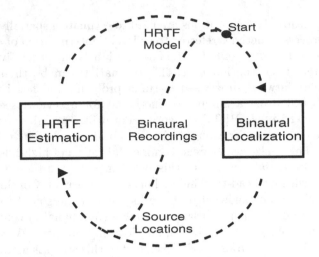

Fig. 1. The iterative localization and HRTF estimation technique (ILHET). Starting with an initial HRTF model estimate and a set of binaural recordings, traverse the circle until an optimal HRTF estimation is obtained.

HRTF Estimation step). The newly approximated continuous HRTF is expected to yield better localization results on the next iteration, and those better localization results should in turn create an incrementally better continuous HRTF estimate. This cycle continues until a convergence threshold is achieved.

In the current study, a broader view of a traditional EM algorithm is adopted where focus is placed on the overall success of the individual M-Steps rather than successful attainment of an optimal joint distribution over both the location labels and continuous HRTF parameters. This modular formulation makes the procedure more amenable to plugging-in existing algorithms for Continuous HRTF Estimation and Binaural Source Localization to accomplish each M-Step. One set of algorithms is described in detail below. For convenience, the overall two-stage framework will be hereafter referred to as the Iterative Localization and HRTF Estimation Technique or ILHET (pronounced /'aɪlət/).

3   Experimental Methods

To evaluate the effectiveness of the ILHET, two existing algorithms were implemented in context of the Binaural Source Localization and Continuous HRTF Estimation steps defined above. Note that these M-Step algorithms use pre-recorded datasets of binaural recordings in place of real-time measurements for the purpose of developing and refining the ILHET.

3.1   Binaural Source Localization

For the binaural source localization stage of the ILHET, the perceptually-based localization model from Middlebrooks [1] was used. This model works by comparing the signals that arrive at the two ears (the binaural recordings) to a database of the listener's directional transfer function (DTF) templates, a set of normalized HRTFs. Within the ILHET implementation a new set of DTF templates are calculated at each iteration by sampling the current estimate of the continuous HRTF at 600 equally distributed locations, and normalizing those sample HRTFs by the average sample HRTF spectrum across those locations. For each binaural recording and DTF, a frequency-independent interaural level difference (ILD) and right and left log-power spectral features are extracted. A single location estimate is then generated for each binaural recording based on which set of DTF features best match the features of the binaural recording. A more detailed explanation of the feature extraction and template matching procedure can be found in Middlebrooks [1].

3.2 Continuous HRTF Estimation

The continuous HRTF model used was that of Romigh *et al.* [3], based on the low order spherical harmonic (SH) representation of HRTF spectra and the spatial interaural time difference (ITD) function. With this model, *sample* HRTFs are fit with a set of spatially continuous SH basis functions, and the resulting SH coefficients (weights) form the new continuous HRTF parameters. The fidelity of the SH model is controlled by the SH representation order, a constant that determines the rate of spatial change of the basis functions over the sphere. As a result, SH order becomes an important consideration in ILHET evaluation procedure that is specific to the continuous model described in this study. Using this representation with a SH order of 4 or greater preserves all of the features relevant for sound source localization with a relatively small number of parameters, and also allows the use of a Bayesian HRTF estimation technique described by Romigh [3]. The Bayesian HRTF estimation technique has the added advantage of incorporating *a priori* knowledge of general HRTFs to significantly lower the number of sample HRTFs required to estimate a full continuous HRTF. Details of the Bayesian HRTF estimation technique and low-order SH representation can be found in Romigh *et al.* [3].

3.3 Dataset

Datasets of binaural recordings containing 277 non-redundant measurements (evenly distributed over 3D space) were created by convolving a known periodic chirp stimuli, a commonly used reference signal in HRTF measurement, with the measured HRTF set (collection of *sample* HRTFs) for 12 different listeners. All HRTF sets were recorded at the Auditory Localization Facility (ALF) of the Air Force Research Laboratory at Wright-Patterson AFB, OH, see Fig. 2.

3.4 Procedure

For each listener in the data set, a continuous HRTF was estimated using the ILHET and a limited set of the available binaural recordings N = [6, 12, 25,

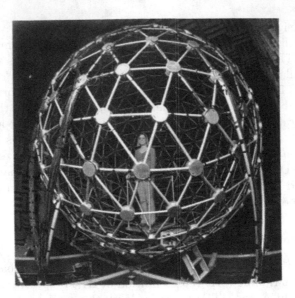

Fig. 2. Auditory Localization Facility (ALF), at Wright Patterson Air Force Base, OH.

50, 100, 150, 200, 250]. Locations were chosen from those available by randomly selecting a quasi-equally distributed set of spatial locations of the appropriate number. Every run started with a non-individualized average HRTF model and for each N, the ILHET was run for two iterations at each increasing SH order from 0 to 14, with the resultant HRTF model forming the initial estimate at the next SH-order run. This step-up procedure ensures that as the localizer becomes more accurate, the HRTF model moves to higher orders, where finer spatial variation is captured. The algorithm terminated only after the final ILHET iteration of the final SH order specified.

4   Results and Discussion

Evaluation results are summarized in Fig. 3 and broken down into two metrics that reflect the performance of the two-stages of the ILHET, localization error and the spectral distortion of the estimated HRTF. Here, localization accuracy is defined as the total angular error between the true location of a *sample* HRTF and the estimated location provided by the binaural localization step. Spectral distortion is the root mean square (RMS) error between the true and estimated spectra (in dB) from 200 Hz to 15 kHz. All results are averaged over listener and location, and plotted as a function of the number of locations used in the ILHET.

The results of Fig. 3a indicate that the average localization error (over unseen sample locations) decreases rapidly as the number of locations is increased from 0 and approaches performance with the ground truth HRTF with 50 to 100 measurements ("ground truth HRTF" refers to the HRTF set recorded for each subject in the ALF facility). This means that with 50 to 100 measurements,

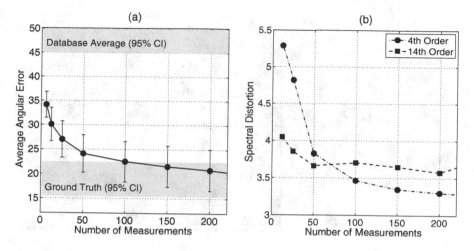

Fig. 3. (a) Average angular error (simulated localization with final c estimate) versus number of training locations (b) Spectral distortion versus number of training locations as a function of SH interpolation order.

the localization stage of the algorithm trained with unlabeled data performs as well as the localization stage with all of the labeled HRTF data available. Figure 3b above highlights how the choice of order in the continuous HRTF estimation (M-Step 2) affects spectral distortion. The results show that for a small measurement set size (<50 locations), a high (14th order) representation produces better estimate than a low-order representation (4th order). The lower 4th order representation appears to benefit most from the design of the algorithm and presents an interesting point of discussion, as Romigh *et al.* [3] find that a 4th order SH representation achieved localization accuracy at a level of performance comparable to a fully individualized HRTF, despite the fact that a low order representation induces a significant amount of both spectral and spatial smoothing. The spectral smoothing present in a 4th order estimate can be visualized in Fig. 4.

Figure 4 shows the dramatic progression of the HRTF estimation procedure for two different subjects as more locations are used in formation of the estimation. Each separate panel represents the estimated 4th order HRTF spectrum plotted as a function of angle around the median plane. The number at the top of each panel identifies the number of spatial measurements used in the estimation. In addition, note that color is used to indicate the level in decibels, where red and blue specify regions of high and low magnitude, respectively. Each time the ILHET algorithm is run, it starts with the same initial guess for the database average, given on the far left and as is shown, does begin to capture the important features of an individuals ground truth 14th order HRTF that was custom measured in an anechoic facility. As a final point, its important to recognize that the results presented here were exploratory in scope and that these insights into ideal measurement set sizes and interpolation order, will serve to help guide the algorithm towards a more refined design.

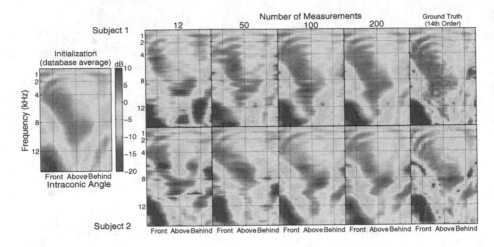

Fig. 4. 4th order HRTF magnitudes (in dB) plotted as a function of angle along the median plane (comparison between 2 different subjects for 4 different measurement set sizes).

5   Conclusion and Future Work

This study used the iterative construct of a joint-maximization EM algorithm to derive a novel method for HRTF estimation (ILHET) that eliminates both knowledge of head orientation and/or loudspeaker location knowledge from the process. Keeping to the practical problem, assumptions were made which generalized each M-Step into a more modular form; however, the overall method was successful in its estimation of a continuous HRTF across a database of 12 subjects, as both the average localization error on testing data and the spectral distortion improved over multiple iterations. To ensure robustness for measurements taken in everyday listening environments (the eventual objective), the necessary next steps involve consideration of non-anechoic and noisy recordings to see how the proposed method handles more realistic input data. Furthermore, note that currently the testing locations are evenly distributed over 3D space. Randomizing or perhaps developing realistic head-movement paths (representing the motions of a person using the technique) would present an interesting follow-up study. Finally, it is essential to begin formulating a plan for perceptual testing of the estimated continuous HRTF structures, to ensure that the metrics defined here function as suitable maximum likelihood estimators.

Acknowledgements. This research was supported in part by an appointment to the Student Research Participation Program at the U.S. Air Force Research Laboratory, 711th Human Performance Wing, Human Effectiveness Directorate, Warfighter Interface Division, Battlespace Acoustics Branch administered by the Oak Ridge Institute for Science and Education through an interagency agreement between the U.S. Department of Energy and USAFRL.

References

1. Middlebrooks, J.C.: Narrow-band sound localization related to external ear acoustics. J. Acoust. Soc. Am. **92**(5), 2607–2624 (1992)
2. Neal, R.M., Hinton, G.E.: A view of the EM algorithm that justifies incremental, sparse, and other variants. In: Proceedings of the NATO Advanced Study Institute on Learning in Graphical Models, pp. 355–368. Kluwer Academic Publishers, Norwell (1998). http://dl.acm.org/citation.cfm?id=299068.299082
3. Romigh, G.D.: Individualized Head-Related Transfer Functions: Efficient Modeling and Estimation from Small Sets of Spatial Samples. Ph.D. thesis, Carnegie Mellon University (2012)
4. Wenzel, E.M., Arruda, M., Kistler, D.J., Wightman, F.L.: Localization using non-individualized head-related transfer functions. J. Acoust. Soc. Am. **94**(1), 111–123 (1993)
5. Xie, B.: Head-Related Transfer Function and Virtual Auditory Display. J. Ross Publishing, Incorporated (2013). https://books.google.com/books?id=LEaNm QEACAAJ

Biomedical and Other Applications

Blind Separation of Surface Electromyographic Mixtures from Two Finger Extensor Muscles

Anton Dogadov[1,2]($\boxtimes$), Christine Serviere[1,2], and Franck Quaine[1,2]

[1] Univ. Grenoble Alpes, GIPSA-Lab, 38000 Grenoble, France
[2] CNRS, GIPSA-Lab, 38000 Grenoble, France
{anton.dogadov, christine.serviere, franck.quaine}
@gipsa-lab.grenoble-inp.fr

Abstract. Blind source separation (BSS) was performed to reduce the crosstalk in the surface electromyografic signals (SEMG) for the muscle force estimation applications. A convolutive mixture model was employed to separate the SEMG signals from two finger extensor muscles using a frequency-domain approach. It was assumed that the tension of each muscle varies independently and the independence of the SEMG was replaced by minimization of the covariance of muscle forces represented by integrated SEMG. This covariance was also used to resolve the permutation ambiguity inherent to the frequency-domain BSS. The forces estimated by the reconstructed sources were compared with the measured forces to calculate the crosstalk reduction efficiency. The proposed algorithm was shown to be more effective in frequency domain than an ICA algorithm for extensor muscles crosstalk reduction.

Keywords: Blind source separation · Convolutive mixture · Surface electromyography · Muscle crosstalk reducing

1 Introduction

Surface electromyografic signals (SEMG) are widely used in medicine, prosthesis control and biomechanical studies [1]. Integrated SEMG (IEMG) is commonly used in biomechanics as an estimator of muscle force [2]. However, crosstalk or interference from neighbor muscles is a widespread problem in SEMG measurements [3]. It appears when two or more muscles situated close to each other are active during a SEMG recording. This effect may cause precision decrease of IEMG-based force estimations. Blind source separation (BSS) methods may be performed to reduce this crosstalk [2, 4] and thus improve the performance of muscle force estimation.

BSS is a method of source signal recovering from several mixtures when no a priori information is available on source properties (source spatial position etc.). To reduce the crosstalk from neighbor muscles BSS is applied to mixtures, in which each muscle is thought to be a source, and mixture signals are thought to be transformed vectors of source signals. Most of mixing transformations are supposed to be linear, instantaneous or convolutive. A linear instantaneous model can be used in the case of small muscles

E. Vincent et al. (Eds.): LVA/ICA 2015, LNCS 9237, pp. 481–488, 2015.
DOI: 10.1007/978-3-319-22482-4_56

located close to each other [4]. However, validity of the instantaneity hypothesis is very sensitive to electrode location [2]. Merletti [5] explained the limitations of the instantaneous model by a convolutive effect of a volume conductor and by action potential propagation. Jiang and Farina [6] proposed an extension of a BSS technique based on second-order moments (SOBI) for the case of sources being delayed in the mixtures. However, little information is available on convolutive mixture separation of SEMG signals.

The purpose of this study was to perform convolutive BSS of SEMG mixtures to reduce the crosstalk in the SEMG signals and to improve the precision of the muscle force estimation. We first discus the hypothesis of independent sources for SEMG signals. As independence of the SEMG sources is not always verified, we propose to replace it by a criterion calculated on the integrated SEMG. The criterion characterizes the fact that the variations of the forces produced by the two muscles are biomechanically independent. This criterion was also used to resolve the permutation ambiguity inherent to the frequency-domain BSS.

2   Blind Source Separation of SEMG Signals

2.1   Mixture Model

Let us consider the standard convolutive mixing model of two EMG signals:

$$x_k(t) = \sum_{j=1}^{2} \sum_{l=0}^{L-1} h_{kj}(l) s_j(t-l), \tag{1}$$

where $x_1(t)$, $x_2(t)$ denote the observation mixtures, $s_1(t)$, $s_2(t)$ the source signals. $h_{kj}(l)$ are the elements of the impulse response matrix from source j to sensor k, L is the length of the impulse response. A noise-free model was considered as the noise can be previously filtered before SEMG separation. The FIR convolutive mixing model can be reformulated into an instantaneous one in the frequency domain. N-point short-time Fourier transform was performed (STFT), which transformed the observation sequences $\{x_k(t)\}$ to the time-frequency domain $\{X_k(t, f)\}$. For each frequency bin f Eq. (1) becomes:

$$X_k(t,f) = \sum_{j=1}^{2} H_{kj}(f) S_j(t,f), \tag{2}$$

or in the matrix form:

$$\mathbf{X}(t,f) = \mathbf{H}(f)\mathbf{S}(t,f). \tag{3}$$

2.2   Blind Source Separation of SEMG

BSS is generally based on the assumption of independent sources and the idea is to adjust the separating filters $\mathbf{G}(f)$ such that the outputs $\mathbf{y}$ are as mutually independent as possible in each frequency bin f:

$$\mathbf{Y}(t,f) = \mathbf{G}(f)\mathbf{X}(t,f), \tag{4}$$

Usually in ICA algorithms, $\mathbf{G}(f)$ is decomposed as:

$$\mathbf{Y}(t,f) = \mathbf{R}(f)\mathbf{V}(f)\mathbf{X}(t,f), \tag{5}$$

where $\mathbf{R}$ is a rotation matrix, $\mathbf{V}$ is a whitening matrix. As sources are supposed to be uncorrelated, the whitening matrix is calculated by eigenvalue decomposition of the covariance matrix in each frequency. To calculate the rotation matrix $\mathbf{R}$ an independence criterion must be chosen, for example, with the help of fourth-order cumulants [7].

For SEMG signals, the hypothesis of mutually independent sources has to be discussed. A muscle is composed of Muscle Fibres (MFs) organized into Motor Units (MUs). This functional unit is composed of an alpha-motoneuron, innervating several MFs (from tens to hundreds MFs depending on the MU and on the muscle characteristics). Activation of the MU produces an electric field generated by each MF into the MU. The summation of these electric fields provides a specific waveform called Motor Unit Action Potential (MUAP). This full electrical activity is called electromyogram (EMG) and can be recorded using electrodes located on the skin surface (SEMG). The contraction of Muscle fibres requires a train of action potentials which induces a MUAP Train (MUAPT).

Finally a MUAPT (i.e. one source) writes as a train of a specific waveform. The waveforms issued from two distinct muscles are very close as the physiological process is the same and so are not independent. Some differences may appear only if the muscle widths (and so the number of Fibers) are very different. Recall that an SEMG signal is a sparse signal composed of a train of this waveform. If the two trains are not synchronized or if a great part of the waveforms from two sources are not temporally overlapped, then a BSS method based on independence may work as proved in [2, 4, 6]. The MUAP train produces a frequency discharge which varies with the force produced by the muscle. The period of the train decreases when the force increases. It means that in the case of low force level, the sources will be less temporally overlapped than in case of high force level.

Consequently, it may appear that SEMG sources are not independent and even not uncorrelated. We cannot use the whitening step and we must find a new criterion to separate the sources. We focused on the signal nature to choose this criterion. Biomechanics studies [8] showed that in isometric contraction, a linear relation exists between the IEMG, which represents muscle activity, and the tension (or force) produced by the muscles. The IEMG is calculated from the estimated outputs by:

$$Y_j^{\text{int}}(t,f) = \frac{1}{\tau} \int\limits_{t-\tau}^{t} |Y_j(z,f)|^2 dz, \tag{6}$$

where Y_j^{int} is an IEMG of a j-th estimated sources, τ is an integration window varying between 25 ms and 200 ms. Here we assume that the tension of each muscle varies independently and we propose a new separation criterion where the minimization of the covariance of the integrated SEMG replaces the independence of the SEMG.

$$\text{cov}\left(Y_1^{\text{int}}(t,f), Y_2^{\text{int}}(t,f)\right) \to \min. \tag{7}$$

The separating matrices $\mathbf{G}(f)$ were researched by adding the constraint as follows. As we looked for normalized sources the main diagonal elements of the mixture matrix $\mathbf{H}(f)$ are ones and the secondary diagonal elements are complex numbers h_{ij} with a modulus in [0; 1]. By putting ones to the main diagonal we assume that the vector of the sources is a vector of the contributions from the i-th source to the i-th electrode.

To reduce the estimation error we directly calculated the matrix $\mathbf{G}(f)$, which is inverse to $\mathbf{H}(f)$. Firstly we looked for a 2-by-2 matrix whose main diagonal elements are ones and the secondary diagonal elements are complex numbers equal to $-h_{ij}$. After finding the matrix satisfying the criterion (7) it was divided by its determinant to make this matrix inverse to $\mathbf{H}(f)$.

The minimization was performed using Nelder-Mead algorithm.

2.3   Permutation Problem

The main problem in a frequency approach is to resolve the permutation ambiguity. As we had only two sources, only two permutations of estimated sources are possible in each frequency bin f. We assume that the force profile of one source is similar in all frequency bins, hence the IEMG of each estimated source in the f frequency bin is highly correlated with the IEMG of the same estimated source in the $f-1$ frequency bin. So, we choose the permutation which maximizes the sum of the covariances:

$$\sum_{j=1}^{2} \text{cov}\left(Y_j^{\text{int}}(t,f), Y_j^{\text{int}}(t,f-1)\right) \to \max. \tag{8}$$

2.4   Source Reconstruction

The sources $\mathbf{Y}(t,f)$ were calculated in each frequency bin by applying the corresponding separating matrix $\mathbf{G}(f)$ to the mixture $\mathbf{X}(t,f)$ in the same frequency bin.

Obtained sources were used to reconstruct the source signal using inverse STFT:

$$\hat{s}_j(t) = \text{ISTFT}\left(Y_j(t,f)\right), \tag{9}$$

3 Results

3.1 Experimental Setting

Bipolar SEMG recordings (Fig. 1) were performed as described in [2] on one male volunteer with no prior known condition or trauma on his right forearm. The SEMG signals from extensor indicis (EI) and extensor digiti minimi (EDM) were acquired during the experimental task of the extension of the index and the little finger by two pairs of electrodes, which were placed over the muscles. These two fingers have been chosen because they are known to be biomechanically independent. Five extensions were produced. The first four extensions were alternating, i.e. only one muscle was extended at the same time, and the fifth trial was a simultaneous extension of the index and little fingers. Finger extension forces were measured at the same time by the KISTLER 9017B force sensors. The SEMG and force recordings were performed by mean of the BIOPAC MP150 acquisition system. Both force and SEMG recordings were sampled synchronously at 2 kHz with a 500 Hz anti-aliasing filter. After acquiring the SEMG signals were filtered forwards and backwards with an order eight Butterworth band-pass digital filer in [20; 500] Hz. The force signals were filtered forwards and backwards with an order four Butterworth low-pass digital filer with $f_s = 5$ Hz because only low-frequency variations of the force have biomechanical application [9]. Integration time τ was fixed to 100 ms to calculate IEMG.

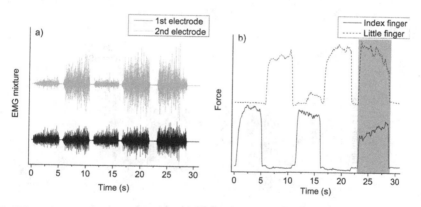

Fig. 1. The reordered SEMG signals (a) and the forces (b) of the index and the little fingers. The movement when the both fingers were active simultaneously is highlighted

3.2 Blind Source Separation

Two BSS algorithms, one ICA based (JADE) [7] and the algorithm based on proposed criterion (IEMG-based), were performed in frequency domain using the convolutive mixture model. BSS were performed by two different ways. At first, BSS algorithms were applied to the whole signal mixtures. Afterwards, the BSS algorithms were performed as follows. The separation matrices were calculated only for the part of the mixtures corresponding to the simultaneous extension of the index and little fingers, i.e.

to the fifth movement (Fig. 1) and then applied to the whole signal mixtures in each frequency bin.

The permutation problem, which is typical for the frequency approach, was solved using the fact that the IEMG profiles of the reconstructed sources remained the same from one frequency bin to another (Fig. 2).

The sources were reconstructed (Fig. 3) and used to estimate the finger forces. The crosstalk reduction efficiency was estimated as described below.

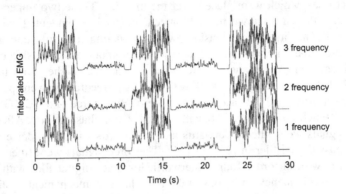

Fig. 2. The IEMG of the 3 first frequency bins of the estimated signal from extensor indicis muscle

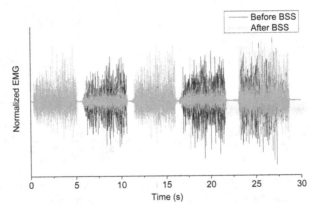

Fig. 3. The SEMG signal from extensor indicis muscle before and after BSS

3.3 Force Estimation and Crosstalk Reduction Efficiency Measure

The forces of EI and EDM muscles were estimated by calculating IEMG using the reconstructed source signals. The IEMG was calculated by applying (6) to the reconstructed source signals. As the force measurements, the calculated IEMG signals were filtered forwards and backwards with an order four Butterworth low-pass digital filer with $f_s = 5$ Hz.

Signal-to-Crosstalk Ratio. The first four trials, which are corresponding to the alternating finger movements, were used to calculate the Signal-to-crosstalk ratio (SCR). As only one finger was active at the same time during these finger movements, the signal of the source corresponding to the active muscle is a useful signal, and the signal coming from the other source could be thought as a crosstalk. The SCR was calculated as:

$$SCR(dB) = 10\log_{10}\left(\frac{P_1}{P_2}\right), \tag{10}$$

where P_1 is a power of the signal of interest and P_2 is the crosstalk.

The resulting SCR was calculated as a mean of SCRs in both initial or both reconstructed signals.

Root-Mean-Square Deviation. The RMS deviation between the measured force and the force estimated by IEMG was calculated after normalization of the measured and estimated force to unit power.

3.4   Discussion

The crosstalk reduction efficiencies of JADE and IEMG-based algorithm were compared for two different ways of separation matrices calculation (Table 1).

Table 1. Comparison of IEMG-based algorithm with JADE algorithm. The separation matrices were calculated from the whole mixtures and from the fifth movement

Separation method	SCR (dB)	RMS
No method (signal mixture)	5,34	0,74
The matrices $\mathbf{G}(f)$ *were calculated from the whole mixtures*		
IEMG-based method	12,98	0,47
Frequency JADE	12,84	0,47
The matrices $\mathbf{G}(f)$ *were calculated from the fifth movement*		
IEMG-based method	11,12	0,54
Frequency JADE	8,30	0,56

The findings lead us to believe that the crosstalk reduction in SEMG of finger extensor muscles seems to be efficient when a convolutive mixture model is used. We can also assume that the permutation problem was properly solved for the both methods.

The performances of the both methods are close when $\mathbf{G}(f)$ were calculated from the whole mixtures, because first four movements were alternating that makes the sources unsynchronized and independent during the main part of the mixtures.

For the fifth movement the sources are strongly temporally overlapped that could make the sources no more independent, because the MUAPs' waveforms are similar. That may explain the decrease of JADE performance when $G(f)$ were calculated from the fifth movement.

However, it would be beneficial to replicate the outlined approach for different electrode positions and for SEMG signals from other muscles.

4   Conclusion

We focused on the separation of FIR convolutive mixtures of Surface electromyografic signals (SEMG) to reduce the crosstalk in the SEMG signals and to improve the precision of the muscle force estimation. As independence of the SEMG sources is not always verified, we proposed to replace it by a criterion calculated on the integrated SEMG. The criterion characterizes the fact that the variations of the forces produced by the two muscles are biomechanically independent.

Acknowledgement. The authors thank Dana Lahat for her valuable comments.

References

1. Holobar, A., Farina, D.: Blind source identification from the multichannel surface electromyogram. Physiol. Meas. **35**, R143–R165 (2014)
2. Léouffre, M., Quaine, F., Servière, C.: Testing of instantaneity hypothesis for blind source separation of extensor indicis and extensor digiti minimi surface electromyograms. J. Electromyogr. Kinesiol. **23**, 908–915 (2013)
3. Mesin, L., Smith, S., Hugo, S., Viljoen, S., Hanekom, T.: Effect of spatial filtering on crosstalk reduction in surface EMG recordings. Med. Eng. Phys. **31**, 374–383 (2009)
4. Farina, D., Févotte, C., Doncarli, C., Merletti, R.: Blind separation of linear instantaneous mixtures of nonstationary surface myoelectric signals. IEEE Trans. Biomed. Eng. **51**, 1555–1567 (2004)
5. Merletti, R., Holobar, A., Farina, D.: Analysis of motor units with high-density surface electromyography. J. Electromyogr. Kinesiol. **18**, 879–890 (2008)
6. Jiang, N., Farina, D.: Covariance and Time-scale methods for blind separation of delayed sources. IEEE Trans. Biomed. Eng. **58**(3), 550–556 (2011)
7. Cardoso, J.F., Souloumiac, A.: Blind beamforming for non-gaussian signals. In: Radar and Signal Processing. IEE Proceedings F, vol. 140, pp. 362–370. IET (1993)
8. Lippold, O.C.J.: The relation between integrated action potentials in a human muscle and its isometric tension. J. Physiol. **117**, 492–499 (1952)
9. Hoozemans, M.J.M., Van Dieën, J.H.: Prediction of handgrip forces using surface EMG of forearm muscles. J. Electromyogr. Kinesiol. **15**, 358–366 (2005)

Multivariate Fusion of EEG and Functional MRI Data Using ICA: Algorithm Choice and Performance Analysis

Yuri Levin-Schwartz[1]([✉]), Vince D. Calhoun[2,3], and Tülay Adalı[1]

[1] Department of Computer Science and Electrical Engineering,
University of Maryland Baltimore County, Baltimore, MD 21250, USA
ylevins1@umbc.edu
[2] The Mind Research Network, Albuquerque, NM 87106, USA
[3] Department of Electrical and Computer Engineering, University of New Mexico,
Albuquerque, NM 87131, USA

Abstract. It has become common for neurological studies to gather data from multiple modalities, since the modalities examine complementary aspects of neural activity. Functional magnetic resonance imaging (fMRI) and electroencephalogram (EEG) data, in particular, enable the study of functional changes within the brain at different temporal and spatial scales; hence their fusion has received much attention. Joint independent component analysis (jICA) enables symmetric and fully multivariate fusion of these modalities and is thus one of the most widely used methods. In its application to jICA, Infomax has been the widely used, however the relative performance of Infomax is rarely shown on real neurological data, since the ground truth is not known. We propose the use of number of voxels in physically meaningful masks and statistical significance to assess algorithm performance of ICA for jICA on real data and show that entropy bound minimization (EBM) provides a more attractive solution for jICA of EEG and fMRI.

Keywords: Data fusion · Independent component analysis · fMRI · EEG · Medical imaging

1 Introduction

In neurological studies, the collection of data using more than one modality is becoming increasingly common since each modality provides a complementary view of neural activity [12]. Thus, effective utilization of all such joint information forms the main motivation for performing a combined analysis on multimodality data. Two functional modalities, fMRI and EEG, are more frequently fused [2,19] due to the fact that the high spatial resolution of fMRI and the high temporal resolution of EEG [10] augment each other, providing a more detailed view

Y. Levin-Schwartz—This work was supported in part by NSF grant NSF-IIS 1017718 and NIH grant R01 EB 005846.

© Springer International Publishing Switzerland 2015
E. Vincent et al. (Eds.): LVA/ICA 2015, LNCS 9237, pp. 489–496, 2015.
DOI: 10.1007/978-3-319-22482-4_57

of functional changes within the brain. JICA is a straightforward, yet effective method for coupling these two modalities in a symmetric and fully multivariate manner and has become one of the most popular data multivariate fusion methods. Though any independent component analysis (ICA) algorithm can be used to perform jICA, Infomax [3] is currently the most widely used technique. This is due to several factors: first, Infomax has been shown to perform well on fMRI data [17], a frequently fused modality in fusion studies, second, Infomax was the first algorithm used to perform jICA [4], and third, Infomax is the default method in the fusion ICA toolbox (FIT, http://mialab.mrn.org/software/fit/).

Despite this popularity, the comparative advantages of Infomax over other ICA algorithms for fusion using jICA have not been studied for real neurological data due to the lack of a ground truth. In this paper, we show that since little is known about the nature of fMRI or EEG latent sources and their interactions, it is preferable to use an algorithm with a flexible nonlinearity that can fit a wide range of source distributions. In particular, we present a unified framework for the comparison of ICA algorithms for jICA. We show that an algorithm employing a dynamic nonlinearity that can better match latent sources drawn from both super- and sub-Gaussian distributions, such as EBM [13], provides more desirable performance than Infomax, whose fixed nonlinearity favors super-Gaussian distributions [3].

Since the goal of fusing neurological data is to estimate interpretable results and, where applicable, to determine components that differentiate between two groups, we use the number of voxels in physically meaningful masks and statistical significance values to assess the performance of jICA algorithms on real data. The masks are created with the widely used WFU PickAtlas toolbox [14,15] to correspond to activation regions for the well known auditory oddball (AOD) task, see e.g. [16]. The measure of statistical significance is defined in terms of p-values corresponding to a two-sample t-test run on the estimated component loadings.

The remainder of the paper is organized as follows. In Sect. 2, we introduce the jICA model, AOD task, and mask creation. Then we describe the experimental results in Sect. 3. Lastly, in Sect. 4 we present our conclusions.

2   Methods

The inherent dissimilarities between different modalities make their direct fusion difficult. For this reason, rather than jointly analyzing the multimodality datasets directly, it is often beneficial to reduce each modality to a feature, i.e., a lower-dimensional representation of the dataset, for each subject within the study. This allows for the exploration of associations across these feature sets through an analysis of variations across individuals [1,5,6,18]. This investigation of the variations between subjects through fused features provides a natural way to find associations across modalities as this allows for the construction of a common dimension for datasets of otherwise different dimensionality as well as ease of identification and interpretation of biomarkers of disease [8,19].

2.1   JICA

We begin with an extension of the generative model for ICA to K datasets

$$\mathbf{X}^{[k]} = \mathbf{A}^{[k]}\mathbf{S}^{[k]}, \ k = 1, \ldots, K, \tag{1}$$

where each dataset, $\mathbf{X}^{[k]}$, is a linear mixture of M sources, $s_m^{[k]} \ 1 \leq m \leq M$, via mixing matrices, $\mathbf{A}^{[k]}$. For solving (1), ICA can be performed on each dataset separately, which while straightforward, is not a desirable solution because it does not exploit interactions between datasets from different modalities.

By assuming that each dataset is mixed with the same mixing matrix, $\mathbf{A}$, we can extend ICA to the fusion of multiple datasets thus allowing us to convert the problem posed in (1) into a single ICA in which the sources from multiple disparate datasets form underlying "joint sources." Joint component estimation, and hence, the estimation of individual components, can then be achieved through the performance of a single ICA on the horizontally concatenated $\mathbf{X}^{[k]}$ defined as

$$\underline{\mathbf{X}} = [\mathbf{X}^{[1]}, \mathbf{X}^{[2]}, \ldots, \mathbf{X}^{[K]}] = \mathbf{A}[\mathbf{S}^{[1]}, \mathbf{S}^{[2]}, \ldots, \mathbf{S}^{[K]}]$$
$$= \mathbf{A}\underline{\mathbf{S}}. \tag{2}$$

In this application, let $\mathbf{X}^{[1]}$ be the matrix of fMRI features for all subjects, whose dimensions are $M \times V$, where M is the number of subjects and V is number of voxels in the fMRI spatial map (60,261). Similarly, let $\mathbf{X}^{[2]}$ be the matrix of EEG features for all subjects, whose dimensions are $M \times T$, where T is number of time points in the EEG feature (451).

The columns of the estimated mixing matrix, $\hat{\mathbf{A}}$, provide the loadings of each component across subjects. The p-th column of the estimated mixing matrix, $\hat{\mathbf{a}}_p$, represents the relative weights of the p-th source estimate, $\hat{\underline{\mathbf{s}}}_p$, for each corresponding subject. Since each dataset is reduced to a feature for each subject, to look for differences in the expression of components between two groups, a two-sample t-test can be performed on the component loadings, where one group is represented by the component loadings from the first group and the second by the component loadings from the second group. However, though easy to calculate, the use of significance values to determine algorithm performance may not be ideal as discussed in [2]. Additionally, a two-sample t-test can not be performed for feature data drawn from only one group.

We should briefly describe the basic concepts behind both Infomax and EBM and how their differences translate into separate performances for the two methods. Though not initially proposed in this light, Infomax can be shown to be equivalent to using a maximum likelihood framework [7], and the use of a sigmoid noninearity implies that all sources are drawn from a distribution that is hyperbolic secant in shape [7]. Note that though this distribution is super-Gaussian, which can be a good approximation of many physical signals, such an assumption leads to very poor performance in the case when the sources vary greatly from the assumed distribution [7]. In contrast to this technique, EBM does not assume any distribution for the latent sources and instead attempts to upper bound their entropy through the use of several measuring functions [13]. These functions provide different bounds on the entropy of the latent sources, with the tightest bound being closest to the true entropy. The measuring functions used in this implementation of EBM describe a wide variety of distributions including those that are unimodal, bimodal, symmetric, and skewed [13]. This ability to describe a diverse set of distributions leads to more accurate estimation of all sources within the mixture and thus better optimization of the cost. Therefore, EBM is expected to provide improved performance over Infomax for achieving ICA, since the main objective is the maximization of independence.

2.2   Auditory Task and Feature Extraction

The fMRI and EEG task AOD data used in this analysis were obtained from 22 healthy controls and 14 patients with schizophrenia. The subjects were scanned while listening to three kinds of auditory stimuli: standard (500 Hz occurring with probability 0.80), novel (nonrepeating random digital sounds with probability 0.10), and target (1 kHz with probability 0.10, to which a button press was required). The fMRI data was preprocessed using: slice timing correction to account for the delay time of the sequential acquisition of horizontal slices of the brain, registration to correct for subject motion, and spatial normalization to allow for the straightforward comparison of brains of differing size. After fMRI preprocessing, task-related spatial activity maps are created for each subject and used as features for the fusion analysis [8]. The EEG data was preprocessed by amplifying the signals to correct for the minor observed changes. Event-related-potentials (ERPs), which are used as features for the EEG data, are formed by averaging the EEG matched to the stimulus of the AOD task for each subject [8]. To produce results for healthy-only data, features were drawn and analyzed from the healthy controls alone.

2.3   Mask Creation

The masks used in this analysis were created using WFU PickAtlas [14,15], which allows for the creation of masks by selecting different areas of the brain called Brodmann areas (BAs). Masks including areas that are expected to be activated during the AOD task were created for each component. This prior knowledge was obtained from previously published experimental results, see e.g. [4], as well as from knowledge about the specific BAs, [9,11]. The included areas for each mask are as follows:

1. Motor-temporal (MT): BAs 7, 8, 22, 29, and 39, which include areas corresponding auditory processing and logical reasoning.
2. Auditory-motor (AM): BAs 1, 2, 3, 7, 10, 11, 21, 22, and 31, which include areas related to motor control and auditory processing.
3. Frontal-temporal (FT): BAs 10, 30, 38, which include areas corresponding to memory and intention.

These masks are applied to the components in order to compare the relative performance of ICA algorithms in producing physically meaningful components.

3   Results and Discussion

3.1   Detection of Significant Components

Using significance values is a natural way to compare algorithm performance for the jICA fusion of data drawn from two groups, since a primary purpose of such an analysis it to produce components, which significantly differentiate between groups, e.g., patients from controls. The results of the experiment are shown in Fig. 1 and Table 1. Only the physically meaningful components, which resulted in uncorrected p-values below 0.05 are considered. These components correspond to temporal-sensorimotor and auditory-motor activation regions, respectively. We can see from Table 1 that EBM consistently produces components with greater significance than Infomax.

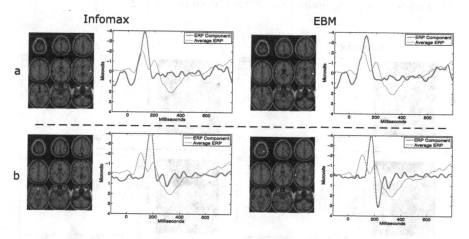

Fig. 1. Statistically significant components obtained using jICA. a. corresponds to a temporal-sensorimotor component. b. corresponds to an auditory-motor component. The spatial maps correspond to z-maps thesholded at 3.5, where red and orange represent activation while blue represents deactivation. It should be noted that Infomax found two other significant components, however neither of them was physically meaningful. EBM found no other significant components showing its ability to reject false positives.

It is also noteworthy that we can see the effect of using a dynamic nonlinearity on the ERP components in Fig. 1. The ERP components are less sparse and less unidirectional for EBM than they are for Infomax, due to the usage of a nonlinearity that does not constrain the estimated sources to be super-Gaussian.

Table 1. Statistical significance of components depicted in Fig. 1

	Infomax	EBM
Temporal-sensorimotor	0.005	0.002
Auditory-motor	0.023	0.005

3.2   Voxels in Masks

When data from two groups is not available, such as when the study is conducted on only healthy individuals, statistical significance values cannot be used to assess the algorithmic differences of jICA. Instead we use the number of voxels in physically meaningful masks, because the purpose of neurological data fusion is to produce physically meaningful components. See Fig. 2 for components that were generated using the healthy-only data. These components correspond to motor-temporal, auditory-motor, and frontal-temporal activation regions, respectively. The number of activated voxels for each component that overlap with the corresponding mask are shown in Table 2.

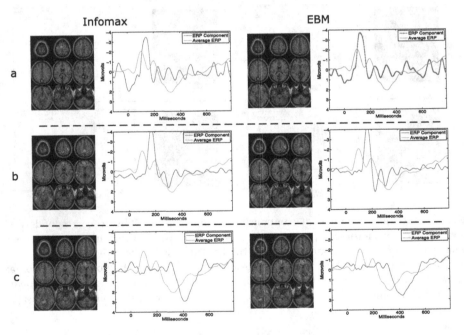

Fig. 2. Sample components obtained using jICA. a. corresponds to a motor-temporal component. b. corresponds to an auditory-motor component. c. corresponds to frontal-temporal. The spatial maps correspond to z-maps thesholded at 3.5, where red and orange represent activation while blue represents deactivation (Color figure online).

Table 2. Number of voxels in masks for the components depicted in Fig. 2

	Infomax	EBM
Motor-temporal	635	2171
Auditory-motor	1935	2391
Frontal-temporal	476	949

From Table 2, we see that EBM produces more activated voxels in spatially interpretable regions. We also note from Fig. 2 that EBM produces components with physically meaningful symmetries, such as the default mode network activation seen in c. Additionally, as observed in both Figs. 1 and 2, Infomax produces components with more conflicting signs in voxels activation, which are difficult to interpret physiologically, than those generated by EBM. Finally, as seen with Figs. 1 and 2, EBM, in general, produces components with higher activation values than Infomax. Thus, for these reasons, we conclude that EBM provides a more meaningful decomposition than Infomax.

4   Conclusions

In this paper, we discussed two methods to assess the performance of ICA algorithms for the fusion of fMRI and EEG data using jICA. For detecting differences between two groups, the use of significance values is a natural way to probe algorithmic differences, since one of the main purposes of such an analysis it to produce components that significantly differentiate between groups. However, there can be issues with this metric. Additionally, this metric cannot be used to assess the algorithmic differences of jICA for data drawn from only one group, *i.e.*, healthy-only data. Instead, we used the number of voxels in physically meaningful masks, since the main purpose of data fusion is to produce interpretable results. We applied the first metric on AOD data drawn from subjects with schizophrenia and healthy controls and the second metric on AOD data drawn from only healthy subjects in order to evaluate the performance of EBM versus Infomax. We demonstrated that EBM produces components with greater significance, *i.e.*, discriminatory power, than Infomax. We also showed that the components estimated using EBM display more physically meaningful symmetries. These components have, in general, higher activation and fewer conflicting signs in voxel activation than Infomax. These results demonstrate the effectiveness of an algorithm that employs a dynamic nonlinearity, such as EBM, for the fusion of multiple modalities using jICA.

References

1. Abney, S.: Bootstrapping. In: Proceedings of the 40th Annual Meeting of the Association for Computational Linguistics, pp. 360–367 (2002)

2. Adalı, T., Levin-Schwartz, Y., Calhoun, V.D.: Multi-modal data fusion using source separation: Application to medical imaging. In: Proceedings of the IEEE (2015, in review)
3. Bell, A., Sejnowski, T.: An information maximization approach to blind separation and blind deconvolution. Neural Comput. **7**, 1129–1159 (1995)
4. Calhoun, V.D., Adalı, T., Pearlson, G., Kiehl, K.: Neuronal chronometry of target detection: Fusion of hemodynamic and event related potential data. NeuroImage **30**, 544–553 (2006)
5. Calhoun, V.D., Adalı, T.: Feature-based fusion of medical imaging data. IEEE Trans. Inf. Technol. Biomed. **13**(5), 711–720 (2009)
6. Calhoun, V.D., Allen, E.: Extracting intrinsic functional networks with feature-based group independent component analysis. Psychometrika **78**(2), 243–259 (2013)
7. Cardoso, J.F.: Infomax and maximum likelihood for blind source separation. IEEE Signal Process. Lett. **4**(4), 112–114 (1997)
8. Correa, N., Adalı, T., Li, Y.O., Calhoun, V.D.: Canonical correlation analysis for data fusion and group inferences. IEEE Signal Process. Mag. **27**(4), 39–50 (2010)
9. Ernst, M., Nelson, E.E., McClure, E.B., Monk, C.S., Munson, S., Eshel, N., Zarahn, E., Leibenluft, E., Zametkin, A., Towbin, K., Blair, J., Charney, D., Pine, D.S.: Choice selection and reward anticipation: an fMRI study. Neuropsychologia **42**(12), 1585–1597 (2004)
10. Friston, K.J.: Modalities, modes, and models in functional neuroimaging. Science **326**(5951), 399–403 (2009)
11. Gu, X., Han, S.: Neural substrates underlying evaluation of pain in actions depicted in words. Behav. Brain Res. **181**(2), 218–223 (2007)
12. James, A.P., Dasarathy, B.V.: Medical image fusion: A survey of the state of the art. Inf. Fusion **19**, 4–19 (2014)
13. Li, X.L., Adalı, T.: Independent component analysis by entropy bound minimization. IEEE Trans. Signal Process. **58**(10), 5151–5164 (2010)
14. Maldjian, J.A., Laurienti, P.J., Burdette, J.H.: Precentral gyrus discrepancy in electronic versions of the talairach atlas. NeuroImage **21**(1), 450–455 (2004)
15. Maldjian, J.A., Laurienti, P.J., Kraft, R.A., Burdette, J.H.: An automated method for neuroanatomic and cytoarchitectonic atlas-based interrogation of fMRI data sets. NeuroImage **19**(3), 1233–1239 (2003)
16. Mangalathu-Arumana, J., Beardsley, S., Liebenthal, E.: Within-subject joint independent component analysis of simultaneous fMRI/ERP in an auditory oddball paradigm. NeuroImage **60**(4), 2247–2257 (2012)
17. Mckeown, M.J., Makeig, S., Brown, G.G., Jung, T.P., Kindermann, S.S., Bell, A.J., Sejnowski, T.J.: Analysis of fMRI Data by Blind Separation Into Independent Spatial Components. Hum. Brain Mapp. **6**, 160–188 (1998)
18. Smith, S.M., Fox, P.T., Miller, K.L., Glahn, D.C., Fox, P.M., Mackay, C.E., Filippini, N., Watkins, K.E., Toro, R., Laird, A.R.: Correspondence of the brain's functional architecture during activation and rest. Proc. Natl. Acad. Sci. U.S.A. **106**(31), 13040–13045 (2009)
19. Sui, J., Adalı, T., Yu, Q., Chen, J., Calhoun, V.D.: A review of multivariate methods for multimodal fusion of brain imaging data. J. Neurosci. Methods **204**(1), 68–81 (2012)

Blind Calibration of Mobile Sensors Using Informed Nonnegative Matrix Factorization

Clément Dorffer, Matthieu Puigt[✉], Gilles Delmaire, and Gilles Roussel

LISIC, ULCO, Université Lille Nord de France, Calais, France
{clement.dorffer,matthieu.puigt,gilles.delmaire,
gilles.roussel}@lisic.univ-littoral.fr

Abstract. In this paper, we assume several heterogeneous, geolocalized, and time-stamped sensors to observe an area over time. We also assume that most of them are uncalibrated and we propose a novel formulation of the blind calibration problem as a Nonnegative Matrix Factorization (NMF) with missing entries. Our proposed approach is generalizing our previous informed and weighted NMF method, which is shown to be accurate for the considered application and to outperform blind calibration based on matrix completion and nonnegative least squares.

Keywords: Blind calibration · Mobile sensor network · Informed nonnegative matrix factorization · Missing values

1 Introduction

Monitoring a natural or an industrial area is usually obtained from automated measurements provided by a set of sensors or from campaigns conducted by scientists. In the first case, the collected data are very accurate but the high sensor cost limits their number, hence a poor spatial sampling rate over the area. In the second case, the geographical coverage is large but—as the cost of such campaigns is high—sustaining them is difficult, hence a very low time sampling rate. Wireless Sensor Networks (WSN) were shown to solve the drawbacks of both above strategies: sensors in WSN are usually cheap and mobile, thus allowing their massive deployment for both an accurate time and spatial sampling. Moreover, even if they are individually less accurate than high cost sensors, they globally provide a similar accuracy while adding a finer spatial resolution [6]. However, their calibration is an issue—as the sensors might not be accessible—and blind sensor calibration techniques were proposed in, e.g., [1,4,9,11,12,14,16,18] for that purpose. These methods may be divided into two categories, depending if the sensor network is mobile [9,12,14,18] or not [1,4,11,16]. Blind Mobile Sensor Calibration (BMSC) techniques usually assume that sensors in the same vicinity should provide the same data while Blind Fixed Sensor Calibration (BFSC) methods need additional assumptions about the acquired signals in order to perform the calibration, namely the measurement matrix in a compressed sensing framework [4,16] or the low-rank subspace in which the observed data lie [1,11].

© Springer International Publishing Switzerland 2015
E. Vincent et al. (Eds.): LVA/ICA 2015, LNCS 9237, pp. 497–505, 2015.
DOI: 10.1007/978-3-319-22482-4_58

In this paper, we investigate the BMSC problem as an informed matrix factorization, since our formulation provides a specific structure of the matrix factors. Assuming the calibration parameters, the acquired signals, and the sensed physical phenomenon to be nonnegative, we revisit blind calibration as an informed Nonnegative Matrix Factorization (NMF) problem that we solve with an extended version of our previous work [10].

The remainder of the paper is structured as follows. We introduce the considered problem in Sect. 2, for which we propose a solution in Sect. 3. Section 4 investigates the experimental performance of our proposed method while we conclude and discuss about future work in Sect. 5.

2   Problem Statement

In this paper, we assume that a geographical area is observed by m heterogeneous, geolocalized, time-stamped, and mobile sensors along time. Such a situation arises in crowdsensing for example [8], where volunteers share some sensed information provided by their mobile device, e.g., their smartphone. The data obtained with crowdsensing are usually irregularly sampled in both the time and the space, hence the need of appropriate methods to process them. In this paper, we focus on blindly calibrating the sensors, for which we first introduce the definitions and assumptions used in this paper.

Definition 1 (*[15]*). *A* rendezvous *is a temporal and spatial vicinity between two sensors.*

Sensors in rendezvous should acquire the same phenomenon, thus providing the same data. Such rendezvous are classically used in BMSC. Most approaches—e.g., [9,12]—consider that uncalibrated data are randomly distributed around the calibrated ones, so that averaging the measurements in rendezvous performs the calibration. However, the averaging-based calibration is not always applicable—see [7, Fig. 2] for example. As an alternative, some authors consider rendezvous between both calibrated and uncalibrated sensors [14] in order to locally perform the calibration[1]. The newly calibrated sensors are then used to calibrate the still uncalibrated ones in other rendezvous and so on. Such a multi-hop calibration technique needs a dense network to be deployed, so that one can ensure each sensor to be in rendezvous with a (newly) calibrated sensor [14]. However, multi-hop calibration might suffer from propagating calibration estimation errors.

In this paper, we start from the same idea of rendezvous and the same hypothesis of dense sensor network but we propose a matrix formulation allowing to calibrate the *whole* sensor network without multiple hops. For that purpose, we first introduce the following definition.

Definition 2. *A* scene S *is a discretized area observed during a time interval* $[t, t + \Delta t)$. *A spatial pixel has a size lower than* Δd, *where* Δt *and* Δd *define the vicinity of the rendezvous.*

---

[1] Using calibrated and uncalibrated sensors has also been considered in BFSC [1,11].

A scene can thus be seen as a grid of locations where sensors go to and where they sense a physical phenomenon. When two sensors share a common position in a scene, they are in rendezvous (Fig. 1). Setting Δt and Δd in order to define a scene highly depends on the nature of the sensed phenomenon [15]. In this paper, we assume to observe one scene[2] $\mathcal{S}$ that we rearrange as a vector $\underline{y} \triangleq [y(1), \ldots, y(n)]^T$, where n is the number of space samples in $\mathcal{S}$.

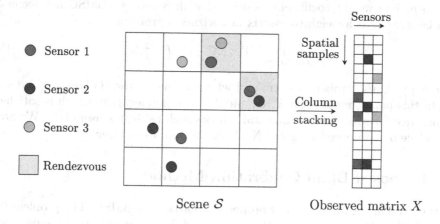

Fig. 1. From a scene $\mathcal{S}$ (with $n = 16$ spatial samples, $m = 3$ sensors and 2 rendezvous) to the data matrix X (white pixels mean no observed value).

We now assume that m heterogeneous sensors are observing $\mathcal{S}$ (see Fig. 1). Let $x(i, j)$ be the sample from Sensor j corresponding to the i-th sample in $\underline{y}$. Physically, $x(i, j)$ is a sensor-output voltage which is here assumed to be linked to the sensed phenomenon $y(i)$ according to an affine relationship, i.e.,

$$x(i, j) \simeq y(i) \cdot \alpha_j + \beta_j, \qquad (1)$$

where α_j and β_j are the unknown gain and offset associated with Sensor j, respectively. These coefficients are assumed to be constant over the scene[3].

We now define G and F, the $n \times 2$ and $2 \times m$ matrices which respectively read

$$G \triangleq \begin{bmatrix} y(1) & 1 \\ \vdots & \vdots \\ y(n) & 1 \end{bmatrix} \text{ and } F \triangleq \begin{bmatrix} \alpha_1 & \alpha_2 & \cdots & \alpha_m \\ \beta_1 & \beta_2 & \cdots & \beta_m \end{bmatrix}. \qquad (2)$$

If we assume each sensor to observe the whole scene $\mathcal{S}$, the matrix form of Eq. (1) then reads

$$X \simeq G \cdot F, \qquad (3)$$

---

[2] The case of multiple scenes—out of the scope of this paper—is discussed in Sect. 5.
[3] Some authors, e.g., [9], consider the sensor responses to drift over time.

where $X \triangleq [x(i,j)]$ is a low-rank matrix, assumed to be well-conditioned. Solving the blind calibration problem then consists of estimating F from the observed matrix X. In particular—considering the $n \times m$ weight matrix W defined as

$$W(i,j) \triangleq \begin{cases} 0 & \text{if } x(i,j) \text{ is not available,} \\ \rho_j & \text{otherwise,} \end{cases} \tag{4}$$

where ρ_j is a weight coefficient associated with Sensor j—BMSC for Scene $\mathcal{S}$ can be written as a weighted matrix factorization problem, i.e.,

$$\min_{G \geq 0, F \geq 0} \| W \circ (X - G \cdot F) \|_f^2, \tag{5}$$

where $\|.\|_f$ is the Frobenius norm and where $\circ$ denotes the Hadamard product.

In this paper, we assume X, G, and F to be nonnegative, which is satisfied in practice for several environmental sensors such as dust sensors [17]. We now introduce our proposed weighted NMF method for estimating F.

3   Proposed Blind Calibration Method

In this section, we introduce our proposed approach for BMSC. The problem (5) may yield scale ambiguities in the columns of F, as for any source separation problem. In order to solve them, we consider that one sensor—say Sensor m—is calibrated and that its calibration parameters are respectively equal to[4]

$$\alpha_m = 1 \quad \text{and} \quad \beta_m = 0. \tag{6}$$

i.e., if the value $x(i,m)$ is available, we get $x(i,m) \simeq y_i$. At this stage, it should be noticed that the factorization problem in Eq. (5) is informed. Indeed, Eqs. (2) and (6) show the last column in both G and F to be known. Taking into account such information should improve the factorization and fix the scale ambiguity inherent to blind factorization.

Recently, informed NMF methods were proposed [5,10] and considered the available information as a penalization term in the NMF optimization [5] or as a specific parameterization which sets the known parameters [10]. As a penalization term does not freeze the known entries of G—which do not depend on the observed data—the latter strategy seems better suited for the considered problem[5]. However, the parameterization in [10] only considers information on F. In this paper, we thus generalize [10] in order to apply it to the considered BMSC problem.

Using the same formalism as in [10], we define Ω^F and Ω^G, the binary matrices which inform the presence/absence of constraints in F and G, respectively.

---

[4] Actually, if no calibrated sensor is available, it is still possible to perform a *relative* calibration, thus providing some consistency in the sensor responses [1,4,11].

[5] An alternative might consist of successively (i) updating F or G with usual update rules and (ii) replacing the known entries by their actual values at each iteration. However, this strategy yielded a low performance in some preliminary tests.

We then define Φ^F and Φ^G, the sparse matrices of set entries in F and G, respectively. It should be noticed that

$$\Phi^F = \Omega^F \circ F \quad \text{and} \quad \Phi^G = \Omega^G \circ G. \tag{7}$$

Defining $\overline{\Omega}^F \triangleq 1 - \Omega^F$ (respectively $\overline{\Omega}^G \triangleq 1 - \Omega^G$) and ΔF (respectively ΔG)—the matrix of free parameters in F (respectively in G)—we extend [10] and derive

$$F = \Omega^F \circ \Phi^F + \overline{\Omega}^F \circ \Delta F \quad \text{and} \quad G = \Omega^G \circ \Phi^G + \overline{\Omega}^G \circ \Delta G. \tag{8}$$

Solving Eq. (5) is then performed using an alternating technique, where we successively aim to estimate F and G using the parameterization in Eq. (8), i.e.,

$$\min_{\Delta F \geq 0} \|W \circ (X - G \cdot \Phi^F - G \cdot \Delta F)\|_f^2 \quad \text{and} \quad \min_{\Delta G \geq 0} \|W \circ (X - \Phi^G \cdot F - \Delta G \cdot F)\|_f^2. \tag{9}$$

Using the same strategy as in [10]—the proof is omitted for space considerations but it is based on a Majoration-Minimization optimization—we derive the update rules which read, respectively:

$$F \leftarrow \Phi^F + \Delta F \circ \overline{\Omega}^F \circ \left[\frac{G^T (W \circ (X - G \cdot \Phi^F)^+)}{G^T (W \circ (G \cdot \Delta F))} \right], \tag{10}$$

$$\text{and } G \leftarrow \Phi^G + \Delta G \circ \overline{\Omega}^G \circ \left[\frac{(W \circ (X - \Phi^G \cdot F)^+) F^T}{(W \circ (\Delta G \cdot F)) F^T} \right]. \tag{11}$$

The superscript $^+$ here denotes the function defined as $(z)^+ \triangleq \max\{\epsilon, z\}$, where ϵ is a small user-defined threshold. It should be noticed that, contrary to [10], we here consider a matrix factorization problem with missing entries. Let us remind that calibration can only be done if the sensor network is dense enough, as assumed in Sect. 2.

Lastly, in order to apply informed NMF to the considered problem, we must initialize G and F, which is known to be tricky. Classical strategies consist of a random initialization while some authors propose an initialization provided by experts [10], the output of another factorization method [3], or a physical model [13]. In this paper, we take advantage of the fact that X is low rank and that some entries of X equal some of G according to Eq. (6). Our proposed strategy to initialize NMF then consists of applying a matrix completion technique [2] to X. The completed matrix is denoted $\tilde{X}$ hereafter. By construction, the last column of $\tilde{X}$ is an estimation of the first column of G. Estimating F can then be obtained from $\tilde{X}$ and G using nonnegative least-squares. In this paper, we study the enhancement provided by our proposed informed NMF with respect to the matrix-completion-based calibration, that we use for initializing NMF.

4 Experimental Validation

In this section, we aim to investigate the enhancement provided by our proposed informed NMF method for BMSC. For that purpose, we simulate a crowdsensing-like particulate matter sensing during a time interval $[t, t + \Delta t)$, which satisfies the assumptions in Sect. 2. The scene is a 10×10 discretized area (the length of $\underline{y}$ is thus equal to $n = 100$) which is observed by $m = 26$ sensors, i.e., $m - 1$ uncalibrated and mobile dust sensors [17] connected to mobile devices and one calibrated, high quality, and mobile sensor[6].

The observed concentrations in $\underline{y}$ range between 0 and $0.5\,\mathrm{mg/m^3}$, for which the sensor response is assumed to be affine [17]. For each uncalibrated sensor, each observed data point represents a nonnegative voltage linked to the corresponding ground truth point in $\underline{y}$ according to Eq. (1). In particular, following the datasheet in [17], the gain and offset coefficients α_j and β_j are randomly set according to a Gaussian distribution centered around $5\,\mathrm{V/(mg/m^3)}$ and $0.9\,\mathrm{V}$, respectively, and then projected onto their respective interval of admissible values—provided by the manufacturer [17]—i.e., $3.5 < \alpha_j < 6.5$ and $0 < \beta_j < 1.5$, $\forall j = 1, \ldots, m - 1$. We then get a 26×100 theoretical observation matrix for which we randomly keep $k+l$ samples in X only, where k (respectively, l) is the number of calibrated (respectively, uncalibrated) sensor samples—with $k \ll l$—hence providing the irregular spatial sampling over the scene. Lastly, Gaussian noise realizations may be added to the observed uncalibrated sensor data and the weight coefficients ρ_j defined in Eq. (4) are set to

$$\rho_j = 1, \forall j = 1, \ldots, m - 1, \quad \text{and} \quad \rho_m = l. \tag{12}$$

In this section, we aim to explore the influence of the number of rendezvous between calibrated and uncalibrated sensors, the number of missing entries in X and the influence of the input SNR to the BMSC performance. For each test condition—i.e., one number of rendezvous, one proportion of missing entries, or one input SNR—25 simulations are performed. In each run, we randomly set the positions of the samples in X in the three experiments and we generate different noise realizations in the last one. The number k of calibrated sensor values in the m-th column of X is set to $k = 4$ in all the tests. Except when we make these values vary, the proportion of uncalibrated sensors to have rendezvous with calibrated ones, and the proportion of missing entries in X are set to 30 % and 90 %, respectively.

Figure 2 shows the Root-Mean Square Error (RMSE) achieved by our proposed method in the above test configurations. The RMSEs are computed over the first line[7] of F, for the uncalibrated sensors only. Dark gray (respectively, light gray) areas show the RMSE envelope while the solid lines (respectively, dashed lines) represent the median RMSE obtained after 10^6 NMF iterations (respectively, at initialization). On the left plot, we first investigate the effect of the proportion of missing entries—ranging from 20 to 95 %—on BMSC in a noiseless configuration. Our proposed method is robust to the number of missing entries (with consistent RMSEs below 10^{-10} until 90 % of missing values) and outperforms the matrix-completion-based initialization (with a median RMSE

---

[6] Actually, we get k fixed, calibrated, and accurate sensors whose obtained values are modeled as those of the m-th sensor in the BMSC problem.

[7] RMSEs computed over the second line of F—not shown for space consideration—are similar to those plotted in Fig. 2.

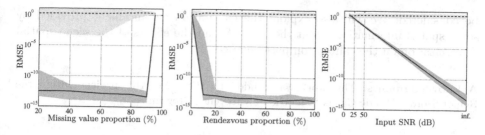

Fig. 2. Performance of the BMSC method vs *(left)* the proportion of missing values, *(center)* the proportion of rendezvous between calibrated and uncalibrated sensors, and *(right)* the input SNR.

around 3). When 5 % of the data in X are available, the dense network assumption is not satisfied anymore and the performance drastically decreases.

The central plot shows the influence of the number of uncalibrated sensors to have rendezvous with calibrated ones. The achieved performance is quite similar to the previous one, except that the upper side of the dark gray envelope is much higher than the median RMSE when the rendezvous proportion is equal to 10 %. Please note even when the calibration error is high, rows of F are correctly estimated, up to a scale coefficient which cannot be handled anymore.

The right plot shows the influence of the input SNR on the BMSC performance. In addition to the noiseless case, we make vary the SNR from 11 to 70 dB. The calibration accuracy decreases with the input SNR and is quite similar to the one obtained with matrix completion for the lowest tested input SNRs.

5   Conclusion and Discussion

In this paper, we revisited blind mobile sensor calibration as a matrix factorization problem. Assuming any of the matrices in the factorization to be non-negative, we generalized our previous informed NMF [10] for the considered application. The approach was shown to be robust to the number of missing entries and to the number of rendezvous between calibrated and uncalibrated sensors. However, some assumptions—e.g., the dense sensor network over the considered zone, or the fact that X is well-conditioned—might seem restrictive. It should be noticed that the approach proposed in this paper can be extended to the case of multiple scenes, by stacking all the observed—and sufficiently different—matrices in one unique well-conditioned matrix, so that we multiply the number of both the known sensors and the rendezvous. In that case, it is more likely that the calibration assumptions will be satisfied.

In future work, we aim to explore several directions. As mentioned above, joint-factorization will be investigated. Moreover, the NMF method proposed in this paper is an extension of the Lee and Seung multiplicative update algorithm, which is known to be slow to converge when the size of the data matrix is large. Extending recent and fast NMF methods to our informed framework will be considered. We will also explore the calibration enhancement provided by some spatial information about the scene S and the effects of the scene spatial discretization on the calibration performance.

Acknowledgments. This work was funded by the "OSCAR" project within the Région Nord – Pas de Calais "Chercheurs Citoyens" Program.

References

1. Balzano, L., Nowak, R.: Blind calibration of sensor networks. In: Proceedings of IPSN, pp. 79–88 (2007)
2. Becker, S., Candès, E., Grant, M.: Templates for convex cone problems with applications to sparse signal recovery. Math. Program. Comput. **3**(3), 165–218 (2011)
3. Benachir, D., Deville, Y., Hosseini, S., Karoui, M.S., Hameurlain, A.: Hyperspectral image unmixing by non-negative matrix factorization initialized with modified independent component analysis. In: Proceedings of WHISPERS (2013)
4. Bilen, C., Puy, G., Gribonval, R., Daudet, L.: Convex optimization approaches for blind sensor calibration using sparsity. IEEE Trans. Signal Process. **62**(18), 4847–4856 (2014)
5. Choo, J., Lee, C., Reddy, C., Park, H.: Weakly supervised nonnegative matrix factorization for user-driven clustering. Data Min. Knowl. Disc. 1–24 (2014)
6. Cochran, E., Lawrence, J., Kaiser, A., Fry, B., Chung, A., Christensen, C.: Comparison between low-cost and traditional MEMS accelerometers: a case study from the M7.1 Darfield, New Zealand, aftershock deployment. Ann. Geophys. **54**(6), 728–737 (2012)
7. D'Hondt, E., Stevens, M., Jacobs, A.: Participatory noise mapping works! An evaluation of participatory sensing as an alternative to standard techniques for environmental monitoring. Pervasive Mobile Comput. **9**(5), 681–694 (2013)
8. Ganti, R., Ye, F., Lei, H.: Mobile crowdsensing: current state and future challenges. IEEE Commun. Mag. **49**(11), 32–39 (2011)
9. Lee, B.T., Son, S.C., Kang, K.: A blind calibration scheme exploiting mutual calibration relationships for a dense mobile sensor network. IEEE Sens. J. **14**(5), 1518–1526 (2014)
10. Limem, A., Delmaire, G., Puigt, M., Roussel, G., Courcot, D.: Non-negative matrix factorization under equality constraints–a study of industrial source identification. Appl. Numer. Math. **85**, 1–15 (2014)
11. Lipor, J., Balzano, L.: Robust blind calibration via total least squares. In: Proceedings of ICASSP, pp. 4244–4248, May 2014
12. Miluzzo, E., Lane, N.D., Campbell, A.T., Olfati-Saber, R.: CaliBree: a self-calibration system for mobile sensor networks. In: Nikoletseas, S.E., Chlebus, B.S., Johnson, D.B., Krishnamachari, B. (eds.) DCOSS 2008. LNCS, vol. 5067, pp. 314–331. Springer, Heidelberg (2008)
13. Plouvin, M., Limem, A., Puigt, M., Delmaire, G., Roussel, G., Courcot, D.: Enhanced NMF initialization using a physical model for pollution source apportionment. In: Proceedings of ESANN, pp. 261–266 (2014)
14. Saukh, O., Hasenfratz, D., Thiele, L.: Reducing multi-hop calibration errors in large-scale mobile sensor networks. In: Proceedings of IPSN (2015)

15. Saukh, O., Hasenfratz, D., Walser, C., Thiele, L.: On rendezvous in mobile sensing networks. In: Langendoen, K., Hu, W., Ferrari, F., Zimmerling, M., Mottola, L. (eds.) Real-World Wireless Sensor Networks, Part I. LNEE, vol. 281, pp. 29–42. Springer, Switzerland (2014)
16. Schulke, C., Caltagirone, F., Krzakala, F., Zdeborova, L.: Blind calibration in compressed sensing using message passing algorithms. In: Proceedings of NIPS, vol. 26, pp. 566–574 (2013)
17. Sharp Corp.: GP2Y1010AU0F compact optical dust sensor (2006), datasheet
18. Wang, C., Ramanathan, P., Saluja, K.: Moments based blind calibration in mobile sensor networks. In: Proceedings of ICC 2008, pp. 896–900, May 2008

Texture Retrieval Using Scattering Coefficients and Probability Product Kernels

Alexander Sagel[✉], Dominik Meyer, and Hao Shen

Department of Electrical and Computer Engineering,
Technische Universität München, Munich, Germany
a.sagel@tum.de

Abstract. In this paper we introduce a content based image retrieval system that leverages the benefits of the scattering transform as a means of feature extraction. To measure similarity between feature vectors, we adapt a probability product kernel and derive an approximate version which can be implemented efficiently. The proposed approach achieves a retrieval performance superior to comparable filterbank transform systems.

Keywords: Statistical modeling · Hierarchical models · Wavelet-based analysis · Texture analysis · Feature extraction · Similarity measures

1 Introduction

Many pattern recognition and pattern matching problems can be solved by a general two-step approach, namely, *feature extraction* (FE) and applying a *similarity measure* (SM). In image processing, the FE step typically refers to an algorithm that produces a *feature vector* of few numerical values from an input image such that it describes the contents of the image sufficiently well in a given context of application, while the SM step assigns a real value to each pair of feature vectors corresponding to the similarity of the respective images. Often, the SM is a non-negative real value.

Without loss of generality, we assume that a higher degree of similarity corresponds to a lower similarity measure. The present work focuses on *Content-based image retrieval (CBIR)*. In CBIR, a signature database stores the features extracted from a set of images. When a search query is initiated, the system evaluates the SM and returns the images with the lowest SM values.

Many recent works about CBIR of texture images follow a common scheme introduced in [5]. For the FE, the involved images are subjected to a filterbank transform. The SM evaluates the histogram similarity of the respective transformed data, typically by applying a parametrized version of the Kullback-Leibler Divergence (KLD). This idea was further developed in several later publications [3,4,9,16], often by varying the subband representation or the statistical model. Linear filterbank transforms feed the input image to a bank of frequency-selective filters, yielding a set of band-pass signals as a representation. One difficulty in constructing FE based on such a decomposition is that the higher frequency subbands are prone to deformations in the spatial domain.

© Springer International Publishing Switzerland 2015
E. Vincent et al. (Eds.): LVA/ICA 2015, LNCS 9237, pp. 506–513, 2015.
DOI: 10.1007/978-3-319-22482-4_59

This work introduces an FE/SM framework for texture images which is descriptive, yet robust with respect to translations and deformations. Unlike the KLD based SM methods inspired by generative models, we motivate and derive an SM by geometrical intuitions. Namely, we adopt ideas from sub-band histogram estimation [5] and kernel machines [7] in order to construct low-dimensional representations of Scattering transforms [11].

2   Feature Extraction with Scattering Transforms

2.1   Notations

In this work, we consider a *signal* as an element of the Lebesgue space $L^2(\mathbb{R}^2)$. Bold-faced lowercase letters $\boldsymbol{x}$ or $\boldsymbol{x}^i$ denote vectors, while regular lowercase letters like x or x_i denote scalar values. Depending on the context, uppercase letters stand either for scalar values or for matrices. An asterisk denotes the convolution $f * g$ of two signals f and g.

2.2   The Scattering Transform

Let $\theta \in L^2(\mathbb{R}^2)$ be a rotationally symmetric window function with low-pass characteristics. Let $\boldsymbol{\eta} \in \mathbb{R}^2 \setminus \{0\}$ and $J \in \mathbb{Z}$ be fixed and $\mathcal{R} \subset SO(2)$ a finite subgroup of rotation matrices. With $\psi(\boldsymbol{x}) = \theta(\boldsymbol{x})e^{i\boldsymbol{\eta}^{\mathsf{T}}\boldsymbol{x}}$, we define the wavelet $\psi_{j,R}$ as

$$\psi_{j,R}(\boldsymbol{x}) = 4^{-j}\psi(2^{-j}R\boldsymbol{x}), \; j \in \{J, J-1, ...\}, R \in \mathcal{R}. \tag{1}$$

Furthermore, let $\phi \in L^2(\mathbb{R}^2)$ be a low-pass and rotationally symmetric *scaling function*, and define $\phi_J \in L^2(\mathbb{R}^2)$ as

$$\phi_J(\boldsymbol{x}) = 4^{-J}\phi(2^{-J}\boldsymbol{x}), \tag{2}$$

such that for the respective Fourier transforms $\hat{\psi}, \hat{\phi}$, the equality

$$|\hat{\phi}(2^J\boldsymbol{\omega})|^2 + \sum_{j=-\infty}^{J} \sum_{R\in\mathcal{R}} |\hat{\psi}(2^J R\boldsymbol{\omega})|^2 = 1 \tag{3}$$

holds, for almost all $\boldsymbol{\omega} \in \mathbb{R}^2$.

The key building block of the *Windowed Scattering transform* (WST) [11] is a dyadic wavelet decomposition $U_{\phi,\psi,J,\mathcal{R}}[f; j, R]$ of the input signal $f \in L^2(\mathbb{R}^2)$ with the complex modulus performed on the band-pass components, defined as

$$U_{\phi,\psi,J,\mathcal{R}}[f; j, R] = \begin{cases} |\psi_{j,R} * f|, & j \leq J, \\ \phi_J * f, & j > J. \end{cases} \tag{4}$$

The modulus operation $|\cdot|$ traverses some of the energy of the band-pass signals towards lower frequencies. Therefore, $U_{\phi,\psi,J,\mathcal{R}}$ can be applied to the output signals $|\psi_{j,R} * f|$ again. Basically, the idea of the WST is to apply $U_{\phi,\psi,J,\mathcal{R}}$

successively to the input signal and to keep the low-pass signals only. This yields a tree-like structure of low-pass signals. The WST along the *path* $p = ((j_1, R_1), ..., (j_m, R_m))$ of scaling factors and rotations is defined as

$$S_{\phi,\psi,J,\mathcal{R}}[f;p] = \phi_J * |\psi_{j_m,R_m} * | \cdots * |\psi_{j_1,R_1} * f| \cdots ||. \qquad (5)$$

Scattering representations are known to be robust with respect to additive noise and spatial translations and deformations [11]. Their expressiveness increases with the maximum path length M.

In order to reduce redundancy and to increase invariance to distortions, the *Normalized WST* (NWST) [1] was introduced. Let $\tilde{p}$ be the predecessor of p, i.e. $p = ((j_1, R_1), ..., (j_m, R_m))$ implies $\tilde{p} = ((j_1, R_1), ..., (j_{m-1}, R_{m-1}))$. Let φ denote a narrow-band low-pass blurring filter. For the layers $m \geq 1$, the NWST is defined as

$$\bar{S}_{\varphi,\phi,\psi,J,\mathcal{R}}[f;p] = \begin{cases} \frac{S_{\phi,\psi,J,\mathcal{R}}[f;p]}{|f|*\varphi}, & \text{if } p \text{ is in layer } m = 1, \\ \frac{S_{\phi,\psi,J,\mathcal{R}}[f;p]}{S_{\phi,\psi,J,\mathcal{R}}[f;\tilde{p}]}, & \text{otherwise.} \end{cases} \qquad (6)$$

Specifically, each subband of the WST is normalized by the respective parent subband, except for the subbands in the first layer which are normalized by the mean of the modulus of the input signal. In practice, a small constant is added to the denominator in order to avoid division by zero.

2.3   Subband Modeling

In what follows, we propose to model the gray-value distributions of the different WST subbands with parametrized probability density functions (PDFs) and describe the images in terms of their respective parameters to obtain a complete FE mechanism on top of the WST.

The most distinctive features in textures are those of higher frequencies and are thus carried by the layers $m \geq 1$. These layers contain signals of the form

$$S_{\phi,\psi,J,\mathcal{R}}[f;p] = \phi_J * |\psi_{j_m,R_m} * | \cdots * |\psi_{j_1,R_1} * f| \cdots ||, \quad p \neq p_0. \qquad (7)$$

Since the modulus of a band-pass filtered natural image looks a lot like a natural image under dim lighting condition, we can treat each subband as a low-resolution approximation of the modulus of a band-pass filtered natural image, provided the input signal of the WST is also a natural image. It is known, that the distribution of band-pass components of natural images are close to be Gaussian. For this reason, a generalized form of the normal distribution is typically used as a model [10]. For complex band-pass filters like those used for the WST, this implies a Rayleigh-like distribution of the respective moduli. In order to further account for variations in shape, we employ *Weibull Distribution* (WD) as a generalization of the Rayleigh distribution as a model for the WST subbands. As suggested in Fig. 1, the WD model is capable of describing the subband histograms fairly well. In fact, the WD was already successfully

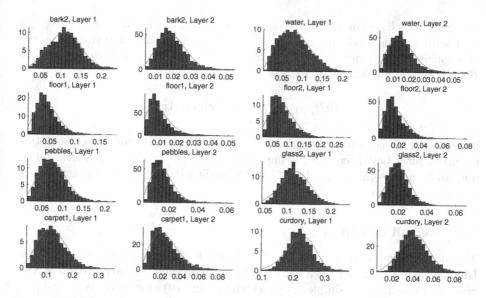

Fig. 1. Histograms (blue) and respective ML Weibull fittings (red) of WST subbands from layers $m = 1$ and $m = 2$ for different texture patches from the UIUC texture database (Color figure online)

employed in the modeling of complex wavelet coefficients [9]. The PDF of the WD is defined for $x \geq 0$ as

$$p_{\mathrm{WD}}(x|\lambda, k) = \lambda k \cdot (\lambda x)^{k-1} \mathrm{e}^{-(\lambda \cdot x)^k}, \tag{8}$$

where $\lambda \in \mathbb{R}_+$ is the scale parameter, and $k \in \mathbb{R}_+$ determines the shape of the WD. Even though it is sensible to model the WST with its Weibull coefficients, it is still questionable if this decision is justifiable for the NWST. It is certainly valid for the first layer since it only involves an overall scaling. Unfortunately, this can not be assumed for the other descendant layers. Nevertheless, experiments show, that in practice this assumption still holds. This is due to the fact that for the most important ranges of k, the multiplicative inverses of WD distributed samples exhibit histograms which again can be well modeled by the WD. Thus, the WD will dominate the subband histograms of the NWST, as well.

The respective shape and scale parameters of the (N)WST subbands of a texture image altogether constitute a feature vector. In order to extract them, a Maximum Likelihood estimation can be employed [14].

3 The Similarity Measure

3.1 Probability Product Kernels

Intuitively, perceptual dissimilarity can be interpreted geometrically as a distance of data points in a metric vector space [15]. As a consequence, it appears worthwhile to base the similarity on a positive-definite kernel. The entities we deal with in this work are probability distributions which motivates the use of Probability Product Kernels [7]. Specifically, for two PDFs p and q, The *Bhattacharyya* coefficient, defined as

$$BC(p, q) = \int_{-\infty}^{\infty} \sqrt{p(\boldsymbol{x})q(\boldsymbol{x})}\mathrm{d}\boldsymbol{x}, \tag{9}$$

is a popular choice for Kernels and as such imposes a Hilbert space structure on probability based features. It has been shown to be useful in several image and audio processing tasks [2,6,17] and will thus be the starting point for the following discussion.

3.2 Weibull Similarity

To the authors' best knowledge, no closed form expressions exist for the Bhattacharyya coefficient of a pair of WDs with different shape parameters. Our aim is to derive a simple approximation of (9) for two WDs $p_{\mathrm{WD}}(x|\lambda_1, k_1), p_{\mathrm{WD}}(x|\lambda_2, k_2)$. Let us assume that $k_1 \approx k_2$. This is a justifiable assumption since the distributions are always compared for each subband individually. Let us further define

$$k = \frac{k_1 + k_2}{2} \text{ and } \lambda = \sqrt[k]{\frac{\lambda_1^k + \lambda_2^k}{2}}. \tag{10}$$

This ultimately leads to

$$
\begin{aligned}
BC&(p_{\mathrm{WD}}(x|\lambda_1, k_1), p_{\mathrm{WD}}(x|\lambda_2, k_2)) \\
&= \sqrt{\lambda_1^{k_1}\lambda_2^{k_2}}\sqrt{k_1 k_2} \int_0^{\infty} x^{k-1} e^{-\frac{\lambda_1^{k_1}x^{k_1} + \lambda_2^{k_2}x^{k_2}}{2}} dx \\
&\approx \sqrt{\lambda_1^k \lambda_2^k}\sqrt{k_1 k_2} \int_0^{\infty} \frac{\lambda^{-k}}{k} p_{\mathrm{wbl}}(x|k, \lambda) dx = 4\frac{\sqrt{\lambda_1^k \lambda_2^k}}{\lambda_1^k + \lambda_2^k} \frac{\sqrt{k_1 k_2}}{k_1 + k_2}.
\end{aligned}
\tag{11}
$$

For the sake of convenience, let us write the arithmetical and geometrical mean of two values $y_1, y_2 \in \mathbb{R}_+$ as $\mu_a(y_1, y_2) = (y_1 + y_2)/2$ and $\mu_g(y_1, y_2) = \sqrt{y_1 y_2}$, respectively. From the last equation of (11) we define our similarity measure for pairs of Weibull PDFs $p_{\mathrm{WD}}(x|\lambda_1, k_1)$ and $p_{\mathrm{WD}}(x|\lambda_2, k_2)$ as

$$K(\lambda_1, k_1; \lambda_2, k_2) = \frac{\mu_g(\lambda_1^k, \lambda_2^k)}{\mu_a(\lambda_1^k, \lambda_2^k)} \cdot \frac{\mu_g(k_1, k_2)}{\mu_a(k_1, k_2)}. \tag{12}$$

Just like (9), the expression (12) defines a Kernel for Weibull PDFs.

For a pair of sets of N independent WDs with the parameter vectors $\boldsymbol{\lambda}^1, \boldsymbol{k}^1, \boldsymbol{\lambda}^2, \boldsymbol{k}^2 \in \mathbb{R}_+^N$, we can write

$$K(\boldsymbol{\lambda}^1, \boldsymbol{k}^1; \boldsymbol{\lambda}^2, \boldsymbol{k}^2) = \prod_{i=1}^{N} K(\lambda_{1,i}, k_{1,i}; \lambda_{2,i}, k_{2,i}). \tag{13}$$

For a pair of WST transforms with N subbands, it is therefore straightforward to derive an SM. Since by our definition a low SM indicates a high similarity we apply the logarithm and reverse the sign which finally leads us to

$$SM_{\text{Scat}}(\boldsymbol{\lambda}^1, \boldsymbol{k}^1; \boldsymbol{\lambda}^2, \boldsymbol{k}^2) = -\sum_{i=1}^{N} \ln K(\lambda_{1,i}, k_{1,i}; \lambda_{2,i}, k_{2,i}). \qquad (14)$$

4 Numerical Experiments

We evaluated our methods in an image retrieval experiment analogous to [5]. The code reproducing the key results is available online[1]. For comparison, we also implemented the FWT+GGD+KLD method according to [5]. Additionally, we implemented a method based the Dual-Tree Complex Wavelet Transform (DT-CWT), inspired by [9]. For the latter, the subband histograms were modeled by the WD, but the KLD was replaced by the proposed SM in (14) for the sake of comparability.

The database D1 is the same as used in [5]. The database D2 was generated from images of the following subset of the UIUC texture database[2]: Bark1, Bark2, Wood2, Wood3, Water, Marble, Floor1, Floor2, Pebbles, Wall, Brick1, Glass1, Glass2, Carpet1, Carpet2, Wallpaper, Fur, Knit, Curdoroy, Plaid. Two 640×480 images from each class were used to create five overlapping 256×256 patches which are then scaled down to half the edge size. Hence, we get a database of 20 different texture classes each containing ten 128×128 patches. All image patches are normalized to zero mean and unit energy, in order to avoid any bias caused by the overall lighting condition of each original texture image. The set of all patches generated from the same texture is considered a class. Its cardinality will be denoted by c in the following. Consequently, $c = 16$ for D1 and $c = 10$ for D2. For each image patch, the $c - 1$ most similar patches were retrieved. The *retrieval rate* for each patch is defined as the ratio of the number of retrieved patches from the same class to $c - 1$. The overall retrieval rate is the average of the retrieval rates for all the images in the database.

The retrieval rates are summarized in Table 1. While WST+WD+SM_{Scat} produces a similar result as DT-CWT+WD+SM_{Scat} on Database D1, NWST+WD+SM_{Scat} is able to outperform all of the competing frameworks by 4.72% for $M = 2$ and 5.12% for $M = 3$. Database D1 is widely used as a benchmark for CBIR. In order to provide a sense for the state of the art, Table 2 summarizes the results from recent publications on comparable approaches: DWT + Generalized Gamma Distribution (GΓD) [3], Wavelet Domain Hidden Markov Models (WD-HMM) [4] and Rotated Complex Wavelets [8]. Also, the original result for FWT+GGD+KLD from [5] was included, since we were not able to reproduce it. To the authors' best knowledge, the best published result for this

---

[1] http://www.gol.ei.tum.de/fileadmin/w00bhl/www/texture_retrieval_scattering_15. zip.

[2] http://www-cvr.ai.uiuc.edu/ponce_grp/data/.

Table 1. Performance of WST + WD and NWST + WD in comparison with FWT + GGD and DT-CWT + GGD on databases 1 and 2

	FWT +GGD +KLD	DT-CWT +WD $+SM_{\text{Scat}}$	WST+WD, $M=2$ $+SM_{\text{Scat}}$	WST+WD, $M=3$ $+SM_{\text{Scat}}$	NWST+WD, $M=2$ $+SM_{\text{Scat}}$	NWST+WD, $M=3$ $+SM_{\text{Scat}}$
D1	75.50 %	78.18 %	78.93 %	78.14 %	84.90 %	**85.30 %**
D2	52.39 %	59.61 %	66.50 %	63.78 %	**66.94%**	65.17 %

Table 2. Performance of state of the art methods on Database D1

FWT+GGD+KLD	FWT+G$\varGamma$D+KLD	WD-HMM	Rotated Wavelets
76.93 %	78.40 %	80.05 %	82.81 %

experiment so far was achieved by Rotated Complex Wavelets with a retrieval rate of 82.81 % which is still outperformed by 2.09 % by NWST+WD+SM_{Scat} with $M=2$ and 2.49 % with $M=3$.

Database D2 is considerably smaller than D1, but involves more variation within the classes, for instance, in terms of camera angle and deformation. Again, the WST greatly improves the retrieval performance. However, this time the regular WST does not fall behind the NWST. Also, increasing the maximum path length up to $M=3$ harms the performance. However, we can conclude that NWST+WD+SM_{Scat} performs comparably well and produces the best results in both our test settings.

5   Conclusion and Future Work

In this work, we derive an FE based on the Scattering transform of texture images and propose an SM inspired by the Bhattacharryya kernel. In the application of image texture retrieval, our method demonstrates superior performance. Since approaches for rotation invariant Scattering representations are already available [12,13], it appears feasible to extend the presented ideas towards a rotation invariant CBIR framework which would allow for more general problem settings. The Kernel property of (12) enables us to transfer the idea to other machine learning tasks, such as classification via support vector machines. Since the Scattering transforms of other data types can be expected to have similar statistics, it could be viable to apply the proposed techniques on other signal types with repetitive characteristics, for instance in audio processing.

References

1. Anden, J., Mallat, S.: Deep scattering spectrum. IEEE Trans. Signal Process. **62**(16), 4114–4128 (2014)

2. Choi, E., Lee, C.: Feature extraction based on the bhattacharyya distance. Pattern Recogn. **36**(8), 1703–1709 (2003)
3. Choy, S., Tong, C.: Statistical wavelet subband characterization based on generalized gamma density and its application in texture retrieval. IEEE Trans. Image Process. **19**(2), 281–289 (2010)
4. Do, M.N., Vetterli, M.: Rotation invariant texture characterization and retrieval using steerable wavelet-domain hidden markov models. IEEE Trans. Multimedia **4**(4), 517–527 (2002)
5. Do, M.N., Vetterli, M.: Wavelet-based texture retrieval using generalized gaussian density and kullback-leibler distance. IEEE Trans. Image Process. **11**(2), 146–158 (2002)
6. Goudail, F., Réfrégier, P., Delyon, G.: Bhattacharyya distance as a contrast parameter for statistical processing of noisy optical images. J. Opt. Soc. Am. A **21**(7), 1231–1240 (2004)
7. Jebara, T., Kondor, R., Howard, A.: Probability product kernels. J. Mach. Learn. Res. **5**, 819–844 (2004)
8. Kokare, M., Biswas, P.K., Chatterji, B.N.: Texture image retrieval using new rotated complex wavelet filters. IEEE Trans. Syst. Man Cybern. **35**(6), 1168–1178 (2005)
9. Kwitt, R., Uhl, A.: Image similarity measurement by kullback-leibler divergences between complex wavelet subband statistics for texture retrieval. In: ICIP 2008. 15th IEEE International Conference on Image Processing, pp. 933–936 (2008)
10. Mallat, S.: A theory for multiresolution signal decomposition: the wavelet representation. IEEE Trans. Pattern Anal. Mach. Intell. **11**(7), 674–693 (1989)
11. Mallat, S.: Group invariant scattering. Commun. Pure Appl. Math. **65**(10), 1331–1398 (2012)
12. Sifre, L., Mallat, S.: Combined scattering for rotation invariant texture analysis. In: ESANN 2012. 20th European Symposium on Artificial Neural Networks, pp. 127–132 (2012)
13. Sifre, L., Mallat, S.: Rotation, scaling and deformation invariant scattering for texture discrimination. In: CVPR 2013. 26th IEEE Conference on Computer Vision and Pattern Recognition, pp. 1233–1240 (2013)
14. Sornette, D.: Critical Phenomena in Natural Sciences: Chaos, Fractals, Selforganization and Disorder: Concepts and Tools. Springer Series in Synergetics, pp. 185–194. Springer, Heidelberg (2006)
15. Torgerson, W.: Theory and Methods of Scaling. Wiley, New York (1958)
16. Tzagkarakis, G., Beferull-Lozano, B., Tsakalides, P.: Rotation-invariant texture retrieval with gaussianized steerable pyramids. IEEE Trans. Image Process. **15**(9), 2702–2718 (2006)
17. You, C.H., Lee, K.A., Li, H.: An svm kernel with gmm-supervector based on the bhattacharyya distance for speaker recognition. IEEE Signal Process. Lett. **16**(1), 49–52 (2009)

Acceleration of Perfusion MRI Using Locally Low-Rank Plus Sparse Model

Marie Daňková[1,2]([⊠]), Pavel Rajmic[1], and Radovan Jiřík[3]

[1] SPLab, Brno University of Technology, Brno, Czech Republic
m.dankova@phd.feec.vutbr.cz, rajmic@feec.vutbr.cz
[2] CEITEC, Masaryk University, Brno, Czech Republic
[3] ASCR, Institute of Scientific Instruments, Brno, Czech Republic
jirik@isibrno.cz

Abstract. Perfusion magnetic resonance imaging is a technique used in diagnostics and evaluation of therapy response, where the quantification is done by analyzing the perfusion curves. Perfusion- and permeability-related tissue parameters can be obtained using advanced pharmacokinetic models, but, these models require high spatial and temporal resolution of the acquisition simultaneously. The resolution is usually increased by means of compressed sensing: the acquisition is accelerated by under-sampling. However, these techniques need to be improved to achieve higher spatial resolution and/or to allow multislice acquisition. We propose a modification of the L+S model for the reconstruction of perfusion curves from the under-sampled data. This model assumes that perfusion data can be modelled as a superposition of locally low-rank data and data that are sparse in the spectral domain. We show that our model leads to a better performance compared to the other methods.

Keywords: Perfusion · MRI · DCE-MRI · Compressed sensing · Sparsity · Locally low-rank

1 Introduction

Perfusion magnetic resonance imaging (MRI), more specifically the dynamic contrast enhanced MRI (DCE-MRI) [1–4], is nowadays a promising method for medical diagnosis and evaluation of therapy response. Using perfusion MRI, oncological and cardiovascular diseases can be diagnosed and their effective treatment can be monitored. In perfusion MRI, a suitable contrast agent is administered intravenously. Due to the cardiovascular system, the contrast agent is distributed within the organism and its temporal and spatial distribution can be observed and analyzed. The time dependency of contrast agent concentration in a region of interest is called *perfusion curve(s)*. The tissue-specific perfusion parameters, necessary for the diagnosis, are estimated from the perfusion curves by approximation using a pharmacokinetic model.

The usual pharmacokinetic models in use are the Tofts and the extended Tofts models [5], which only allow estimating of a limited number of perfusion

© Springer International Publishing Switzerland 2015
E. Vincent et al. (Eds.): LVA/ICA 2015, LNCS 9237, pp. 514–521, 2015.
DOI: 10.1007/978-3-319-22482-4_60

parameters. To allow estimating of additional highly relevant perfusion parameters such as the blood flow and vessel permeability, more complex models need to be used [5]. However, successful application of such models assumes high temporal resolution of the acquisition to capture the vascular-distribution phase of the contrast agent. These requirements substantially limit the achievable spatial resolution of the data acquired or the ability to acquire multiple slices (i.e. perfusion curves in 3D volume). Using the classic (Nyquist-rate) acquisition in MRI, it is impossible for both resolutions to be high at the same time.

Therefore, much effort is devoted to using compressed sensing (CS) in MRI. In MRI, the images are acquired in their Fourier domain (also termed k-space). CS comes into play naturally by under-sampling of the k-space (sampling below the Nyquist rate). The k-space sampling trajectories used include: cartesian sampling [6], radial sampling [7] (including the flexible golden-angle technique [8]) and spiral trajectories [9].

Good quality of reconstruction is strongly dependent on good a priori knowledge of the signal. Many CS reconstruction techniques applied to DCE-MRI use basic formulations of single priors, like the total variation (TV) in the spatial [8] or the temporal [10] domain or the wavelet transform in the spatial and/or the temporal domains [6]. The assumption that different tissues have different but consistent perfusion characteristics is usually mathematically described by the low-rank property of the so-called Casorati matrix, which is an image sequence reshaped such that each image in time forms a column vector [11,12].

More sophisticated approaches in DCE-MRI combine several priors. Article [13] assumes that the Casorati matrix is low-rank and sparse in the spectral domain (row-by-row spectra). A more efficient approach [14] regards the data as the sum of a low-rank component and a component sparse in the spectral domain (L+S model). Superior results have been achieved by regularization via low-rank penalty block-by-block, like in [15,16].

In the present article, we propose to use the sum of locally low-rank matrix and a component which is sparse in the spectral domain (local L+S model) instead of the (global) L+S model in [14]. We show that we can achieve better signal-to-noise ratio (SNR) with our local L+S model than when using global L+S model. We compare this approach with a model using locally low-rank and component sparse in the spectral domain simultaneously (locally L&S model) as presented in [16].

2 Materials and Methods

To work with perfusion data, image sequences are reformatted to the so-called Casorati matrix [11], where each column of this matrix represents a single image in one temporal phase. Such a matrix is used in all models mentioned below.

2.1 Locally Low-Rank and Sparse Matrix Model (Local L&S Model)

The locally L&S model [16] utilizes the fact that perfusion curves have similar time courses in some space regions and have a sparse row spectrum. The similar time-dependency of perfusion curves in each tissue of desired data reconstruction $\mathbf{M}$ is enforced by assuming a low rank of non-overlapping blocks of matrix $\mathbf{M}$ (with a few non-zero singular values in each block) and a sparse row spectrum with a few non-zero elements. Matrix $\mathbf{M}$ is of size $N_1 N_2 \times N_f$, where N_1, N_2 represent the size of each image/frame and N_f is their overall number. The reconstruction can be achieved by solving the following convex optimization problem:

$$\min_{\mathbf{M}} \frac{1}{2} \|E\mathbf{M} - \mathbf{d}\|_F^2 + \lambda_L \sum_{i=1}^{N_b} \|B_i \mathbf{M}\|_* + \lambda_S \|T\mathbf{M}\|_1, \tag{1}$$

where T is the operator of 1D Fourier transform applied to the matrix rows, E is the under-sampled 2D Fourier transform for each image (representing the measurement process), $\mathbf{d}$ is the under-sampled (acquired) data in k-space, and λ_L, λ_S are suitable regularization parameters. B_i is an operator choosing the block i out of the entire matrix, and N_b is the number of blocks. The quadratic term is the data fidelity term, $\|\mathbf{M}\|_*$ is the nuclear norm, which is the sum of singular values of $\mathbf{M}$ (enforces low-rank), and $\|T\mathbf{M}\|_1$ is the ℓ_1-norm, which is the sum of absolute values of entries in $T\mathbf{M}$ (enforces sparsity of perfusion curves in the Fourier domain).

2.2 Low-Rank and Sparse Matrix Decomposition (Global L+S Model)

In the context of perfusion MRI, Low-Rank and Sparse Matrix Decomposition (L+S model) was first introduced in [14]. Unlike the L&S model (either local or global), which aims at promoting low-rank and spectral-domain sparsity of the solution simultaneously, the L+S model aims to compose the desired reconstruction matrix as a sum of low-rank matrix $\mathbf{L}$ and matrix $\mathbf{S}$ with sparse row spectrum, i.e. in this model the desired reconstruction is $\mathbf{M} = \mathbf{L} + \mathbf{S}$. This decomposition can be obtained by solving the following convex problem:

$$\min_{\mathbf{L}, \mathbf{S}} \frac{1}{2} \|E(\mathbf{L} + \mathbf{S}) - \mathbf{d}\|_F^2 + \lambda_L \|\mathbf{L}\|_* + \lambda_S \|T\mathbf{S}\|_1. \tag{2}$$

2.3 Locally Low-Rank and Sparse Matrix Decomposition (Local L+S Model)

In [15,16], the locally low-rank constraint is used instead of the (global) low-rank and it can improve the quality of desired reconstruction data if applied to areas with similar perfusion curves. We propose using this locally low-rank prior in the previously introduced L+S model instead of using the global low-rank of matrix $\mathbf{L}$. The local L+S model can be formulated as

$$\min_{\mathbf{L}, \mathbf{S}} \frac{1}{2} \|E(\mathbf{L} + \mathbf{S}) - \mathbf{d}\|_F^2 + \lambda_L \sum_{i=1}^{N_b} \|B_i \mathbf{L}\|_* + \lambda_S \|T\mathbf{S}\|_1. \tag{3}$$

A crucial question is choosing the appropriate size of the blocks in order to capture areas with similar perfusion parameters and improve the reconstruction as will be shown in the following.

3 Experimental Results

3.1 Perfusion Phantom

For the purpose of simulation, we have created a perfusion phantom using Matlab. We have utilized the modified Shepp-Logan phantom [17], which simulates a brain slice, see Fig. 1. We use a perfusion phantom of 100×100 px $\times 100$ time points in size. The perfusion curves assigned to clearly separated areas of the phantom share the same behaviour in time, following the log-normal model [3]. At every time instant, the k-space values have been perturbed by additive Gaussian noise with a standard deviation of 0.05 to simulate the measurement noise. This model is simplified but it allows us to accurately compare the behaviour of methods.

To capture perfusion data, we use radial trajectories, where random slopes of halflines starting at the origin of k-space are used independently for each time frame.

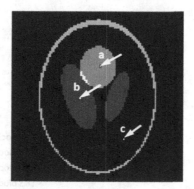

Fig. 1. Single temporal frame of perfusion phantom (arrows mark areas in which reconstruction of perfusion parameters using different methods is further compared) (Color figure online).

3.2 Under-Sampled Perfusion Data Reconstruction

For under-sampled perfusion-phantom data, three types of reconstruction were performed, based on: local L&S model (1), global L+S model (2) and local L+S model (3). The respective optimization problems were solved using the proximal gradient method [18]. Non-overlapping square blocks of different sizes were used for local methods. The parameters λ_L, λ_S were empirically chosen (dependent on the method used, size of blocks and resolution, see Figs. 2 and 3) such that the reconstruction gave the highest possible SNR. In all cases, algorithms were stopped when the relative change in the solution was less than 10^{-5}.

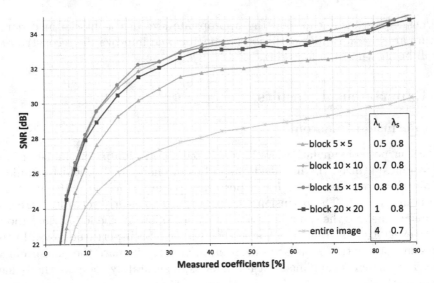

Fig. 2. Comparison of global and local L+S models — dependency of SNR on percentage of measured coefficients.

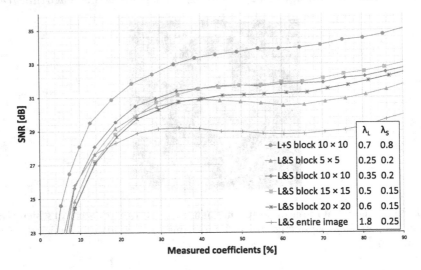

Fig. 3. Comparison of local L+S model and local L&S models — dependency of SNR on percentage of measured coefficients.

Reconstructions of perfusion curves were compared for the global and local L+S models, using different sizes of the blocks and different percentages of measured coefficients, see Fig. 2. Partitioning into blocks brought an increase in the SNR. It can be seen that block size affects the reconstruction quality. The optimal block size depends on the contents of the image used — better performance is achieved for blocks small enough to include only a tissue with similar perfusion

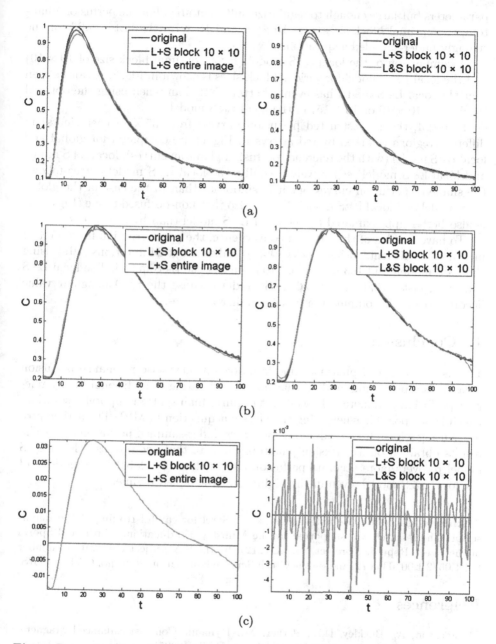

Fig. 4. Reconstruction of perfusion curves (from 35 % of the Fourier coefficients) in different regions (marked by red arrows in Fig. 1) using global and local L+S model (left) and local L+S and local L&S model (right), the last pair represents non-perfused tissue (the blue line, i.e. the original curve, is hidden under the red line, representing the local L+S model) (Color figure online).

parameters but large enough to regularize sufficient area. For the perfusion phantom, the optimal block sizes were 10×10 and 15×15, the latter valid for the strongly under-sampled acquisition.

A comparison of the local L+S model (for an optimal block size of 10×10) and the local L&S model for various block sizes is shown in Fig. 3. It can be seen that the local L+S model has even a better SNR than when using the optimal block size (10×10 or 15×15) in the local L&S model.

In Fig. 4, the reconstructed perfusion curves (from 35 % of coefficients) in different regions (marked by red arrows in Fig. 1) are compared for global and local L+S model (with the reference perfusion phantom curves), local L+S model and local L&S model respectively. Usually, the local L+S model better follows the shape of perfusion curves but it is smoother than in the case of the global L+S model and local L&S model. Notice also that non-perfused tissue (Fig. 4(c)) is also better approximated by the local L+S model than by the others.

To have an idea of the speed of convergence, the reconstruction from 35 % of measured coefficients, using blocks 10×10, takes about 24 iterations (30 s) using the L&S model, and 163 iterations (105 s) using the L+S model. The local L+S converges faster than the global L+S model because the SVD (singular value decomposition) is computed on much smaller matrices.

4   Conclusion

The use of advanced pharmacokinetic models in magnetic resonance perfusion imaging is a promising method that allows estimating of additional highly relevant perfusion parameters. However, it requires high spatio-temporal resolution, which is not possible when using the classic acquisition in MRI. The article proposes using a local L+S model to reconstruct under-sampled perfusion data. The results obtained on a perfusion phantom indicate that the proposed local L+S model recovers more accurate perfusion curves than the global L+S model and the local L&S model, even in the highly under-sampled regime.

Acknowledgement. The authors thank M. Šorel for careful reading of the manuscript. The research was supported by the Ministry of Education, Youth, and Sports of the Czech Republic (project No. LO1212), by the SIX project registration number CZ.1.05/2.1.00/03.0072, and by the Czech Science Foundation grant no. GA15-12607S.

References

1. Jackson, A., Buckley, D.L., Parker, M.: Dynamic Contrast-Enhanced Magnetic Resonance Imaging in Oncology. Springer, Berlin (2005)
2. Mézl, M., Jiřík, R., Harabiš, V.: Acquisition and data processing in ultrasound perfusion analysis. In: New trends in biomedical engineering, pp. 44–51. Brno University of Technology, Czech (2013)
3. Harabiš, V., Kolář, R., Mézl, M., Jiřík, R.: Comparison and evaluation of indicator dilution models for bolus ofultrasound contrast agents. Physiol. Meas. **34**, 151–162 (2013)

4. Bartoš, M.: Advanced signal processing methods in dynamic contrast enhanced magnetic resonance imaging. Doctoral thesis, Brno University of Technology (2014)
5. Sourbron, S.P., Buckley, D.L.: Tracer kinetic modelling in MRI: estimating perfusion and capillary permeability. Phys. Med. Biol. **57**(2), R1–R33 (2012)
6. Han, S.H., Paulsen, J.L., Zhu, G., Song, Y., Chun, S.I., Cho, G., Ackerstaff, E., Koutcher, J.A., Cho, H.J.: Temporal/spatial resolution improvement of in vivo DCE-MRI with compressed sensing-optimized FLASH. Magn. Reson. Imaging **30**(6), 741–752 (2012)
7. Block, K.T., Uecker, M., Frahm, J.: Undersampled radial MRI with multiple coils. Iterative image reconstruction using a total variation constraint. Magn. Reson. Med. **57**(6), 1086–1098 (2007)
8. Li, Z., Berman, B.P., Altbach, M.I., et al.: Highly accelerated 3D dynamic imaging with variable density golden angle stack-of-stars sampling. In: Proceedings of the International Society for Magnetic Resonance in Medicine, vol. 21, p. 3797 (2013)
9. Santos, J.M., Cunningham, C.H., Lustig, M., Hargreaves, B.A., Hu, B.S., Nishimura, D.G., Pauly, J.M.: Single breath-hold whole-heart MRA using variable-density spirals at 3T. Magn. Reson. Med. **55**(2), 371–379 (2006)
10. Sungheon, K., Feng, L., Moy, L., et al.: Highly accelerated golden-angle radial acquisition with joint compressed sensing and parallel imaging reconstruction for breast DCE-MRI. In: Proceedings of the International Society for Magnetic Resonance in Medicine, vol. 21, p. 1468 (2012)
11. Zhang, N., Song, G., Liao, W., et al.: Accelerating dynamic contrast-enhanced MRI using K-T ISD. In: Proceedings of the International Society for Magnetic Resonance in Medicine, vol. 21, p. 4221 (2012)
12. Wang, H., Miao, Y., Zhou, K., Yanming, Y., Bao, S., He, Q., Dai, Y., Xuan, S.Y., Tarabishy, B., Ye, Y., Jiani, H.: Feasibility of high temporal resolution breast DCE-MRI using compressed sensing theory. Med. Phys. **37**(9), 4971–4981 (2010)
13. Lingala, S.G., Hu, Y., DiBella, E., Jacob, M.: Accelerated dynamic MRI exploiting sparsity and low-rank structure: k-t SLR. IEEE Trans. Med. Imaging **30**(5), 1042–1054 (2011)
14. Otazo, R., Candes, E., Sodickson, D.K.: Low-rank plus sparse matrix decomposition for accelerated dynamic MRI with separation of background and dynamic components. Magn. Reson. Med. **73**(3), 1125–1136 (2015)
15. Zhang, T., Cheng, J.Y., Potnick, A.G., Barth, R.A., Alley, M.T., Uecker, M., Lustig, M., Pauly, J.M., Vasanawala, S.S.: Fast pediatric 3D free-breathing abdominal dynamic contrast enhanced MRI with high spatiotemporal resolution. J. Magn. Reson. Imaging **41**(2), 460–473 (2015)
16. Zhang, T., Alley, M., Lustig, M., Li, X., Pauly, J., Vasanawala, S.: Fast 3D DCE-MRI with sparsity and low-rank enhanced SPIRiT (SLR-SPIRiT). In: Proceedings of the 21st Annual Meeting of ISMRM, Salt Lake City, Utah, USA, p. 2624 (2013)
17. Shepp, L.A., Logan, B.F.: The Fourier reconstruction of a head section. IEEE Trans. Nucl. Sci. **21**, 21–43 (1974)
18. Combettes, P.L., Pesquet, J.C.: Proximal splitting methods in signal processing. In: Bauschke, H.H., Burachik, R.S., Combettes, P.L., Elser, V., Russell Luke, D., Wolkowicz, H. (eds.) Fixed-Point Algorithms forInverse Problems in Science and Engineering, pp. 185–212. Springer, New York (2011)

Decomposition-Based Compression
of Ultrasound Raw-Data

Yael Yankelevsky[✉], Arie Feuer, and Zvi Friedman

Department of Electrical Engineering, Technion - I.I.T, Haifa 32000, Israel
yaelyan@tx.technion.ac.il, feuer@ee.technion.ac.il,
zvi.friedman@med.ge.com

Abstract. Sonography techniques use multiple transducer elements for tissue visualization. The signals detected at each element are combined in the process of digital beamforming, requiring that large amounts of data be acquired, transferred and processed. One of the main challenges is reducing the data size while retaining the image contents. For this purpose, we propose a component based model for the raw ultrasonic signals. We show that a decomposition based approach with a suited processing scheme for each component individually, can achieve over twenty-fold reduction of needed data size.

Keywords: Biomedical ultrasound · Signal modeling · Sparse representation · Dictionary learning

1 Introduction

Medical ultrasound imaging allows visualization of internal body structures by radiating them with acoustic energy and analyzing the returned echoes. The two-dimensional image typically comprises of multiple one-dimensional scan lines, each constructed by integrating the data collected by the transducer elements following the transmission of an acoustic pulse along a narrow beam. As the transmitted pulse propagates through the body, echoes are scattered by acoustic impedance perturbations in the tissue. These back-scattered echoes are detected by the transducer elements and combined, after aligning them with the appropriate time delays, in a process referred to as beamforming, which results in Signal-to-Noise Ratio (SNR) enhancement. Each resulting beamformed signal forms a line in the image.

Taking into account the high frequency used for ultrasound imaging, the number of transducer elements and the number of lines in an image, the amount of data needed to be transferred and processed is very large, motivating methods to reduce the amount of needed data without compromising the reconstructed image quality and its diagnostic credibility. In recent years, there has been growing interest in reducing the amounts of data in general signal processing applications, and significant research efforts have been focused at the field of sparse

© Springer International Publishing Switzerland 2015
E. Vincent et al. (Eds.): LVA/ICA 2015, LNCS 9237, pp. 522–529, 2015.
DOI: 10.1007/978-3-319-22482-4_61

representations and Compressed Sensing (CS) [9,10]. Ideas arising from CS theory have been successfully implemented in diverse applications such as radar [3,5], Synthetic Aperture Radar (SAR) [17,20], and MRI [16].

Several preliminary works have recently adapted these methods to ultrasound signals [4,11–15,18,19,21].

Another set of works attempting to reduce the amount of sampled data in ultrasound signals, based on complementary ideas arising from the Finite Rate of Innovation (FRI) framework [6,25], was carried out by Tur et al. [24] and later extended by Wagner et al. [26].

The authors modeled the ultrasonic echo as a finite stream of strong pulses, which are replicas of a known-shape pulse with unknown time-delays and amplitudes:

$$x(t) = \sum_{\ell=1}^{L} a_\ell h(t - t_\ell) \tag{1}$$

Assuming that overall L pulses were reflected back to the transducer from the pulse's propagation path, the detected signal is completely defined by $2L$ degrees of freedom, corresponding to the unknown parameters $\{a_\ell, t_\ell\}_{\ell=1}^{L}$. Based on the FRI framework, these $2L$ parameters are estimated and the signal recovered from a minimal subset of $2L$ of the signal's Fourier series coefficients. The needed coefficients are recovered from low rate samples of the analog signals, as the sampling frequency is now determined by the number of pulses L, which is rather small compared with the bandwidth of the transmitted pulse, leading to a substantial sample rate reduction.

These works achieve an almost eight-fold reduction of sample rate, however the reconstructed data is partial as it only contains the macroscopic reflections while disregarding the speckle.

As the focus of our work, we aim at reducing the large amount of data needed to be stored and processed while preserving the image contents including the speckle.

Our proposed digital processing scheme is based on the separation of the received ultrasonic echoes into two components, both carrying valuable information: the strong reflectors component, that is highly important for tracking purposes in cardiac ultrasound imaging [7,23], and the speckle, also referred to in this work as the background component, that characterizes the microscopic structure of the tissue [8].

We then show that each component on its own is compressible, and derive the suitable representation bases. Thereafter, the compressed background signals are integrated in a digital beamforming process. To conclude the proposed algorithm, the strong reflectors may be reconstructed from their sparse coefficients obtained during the decomposition stage, combined with the beamformed background signals and processed to form an image.

Applying the proposed processing schemes to real cardiac ultrasound data, we successfully reconstruct both macroscopic and microscopic reflections from the scanned region, such that the image contents are highly preserved, while achieving over twenty-fold reduction of the data size.

Throughout this paper, we use the term *raw signals* to refer to the ultrasonic signals detected by each sensor immediately after their sampling.

2 Signal Decomposition

As previously stated, each received signal x is initially decomposed into a background signal x_b and strong reflectors component x_s. The decomposition algorithm is based on a greedy detection of the strong reflectors followed by their separation from the original signal.

Modeling the strong reflectors component, we mostly adopt the "stream of pulses" signal model [24, 26] according to which this component is composed of a limited number of strong pulses, that are amplified and delayed replicas of a known-shape pulse (1). This pulse $h(t)$ has the form of a sinusoid signal oscillating at the transmission frequency f_0 in a Gaussian envelope.

As an extension to this model, we suggest that the returning pulse shape is somewhat corrupted with respect to the transmitted pulse. This corruption may be manifested in either a frequency shift, resulting from the frequency dependent attenuation [2], or a phase shift formed between the carrier wave and the Gaussian envelope. In order to account for those possible corruptions, we propose to represent the strong reflectors in a time-frequency domain using the Short-Time Fourier Transform (STFT). This will allow simultaneous optimization of both the time-delay and frequency shift.

Denoting the STFT of $x(t)$ by $\mathbf{X}(t,\omega)$, the STFT decomposition is

$$\mathbf{X}(t,\omega) = \mathbf{X_b}(t,\omega) + \mathbf{X_s}(t,\omega) = \mathbf{X_b}(t,\omega) + \sum_{k=1}^{L} a_k \mathbf{H}(t - t_k, \omega - \omega_k) \qquad (2)$$

The proposed decomposition algorithm is described in Algorithm 1.

In the presented algorithm, the strong reflectors are detected as the maximal magnitude peaks of the STFT. In practice, the detection can be improved by matching the known pulse pattern in a narrow region around each peak. The maximal number of pulses L and error threshold ϵ_0 are chosen empirically.

It can be observed that the strong reflectors are naturally compressed by saving only the pulse model parameters $\{a_k, t_k, \omega_k\}_{k=1}^{L}$ along the decomposition process. This may be thought of as sparse coding over a very large dictionary whose atoms represent all the possible time and frequency shifts of the known pulse. However, due to the high sampling rate of the signals and the enormous dimensions of such dictionary, standard OMP-like techniques are not feasible and an alternative amplitude-based pulse matching was here performed.

3 Background Data Compression

Resulting from intereferences of weak ultrasonic reflections, speckle is typically characterized by a statistical model with few parameters, indicating that it could

---

Algorithm 1. STFT-based Decomposition

Task: Decompose a given signal x into the strong reflectors and background components (x_s, x_b respectively)

Inputs: The signal's STFT $\mathbf{X}(t, \omega)$, the STFT signature of the pulse model $\mathbf{H}(t, \omega)$ centered such that $\arg\max_{(t,\omega)}|\mathbf{H}(t, \omega)| = (0, 0)$, the maximal number of pulses L, and an error threshold ϵ_0.

Initialization: Set the initial residual $\mathbf{r}^0 = \mathbf{X}(t, \omega)$

Main Iteration: for $k = 1, ..., L$ perform the following:

- Locate the strongest reflection (t_k, ω_k) with magnitude a_k:

$$(t_k, \omega_k) = \arg\max_{(t,\omega)}|\mathbf{X}(t, \omega)| \quad ; \quad a_k = \mathbf{X}(t_k, \omega_k)$$

- Residual update: $\mathbf{r}^k = \mathbf{r}^{k-1} - a_k\mathbf{H}(t - t_k, \omega - \omega_k)$
- Stopping Rule: If $\|\mathbf{r}^k\|_\infty < \epsilon_0$, stop. Otherwise, apply another iteration.

Output: $x_s(t) = ISTFT\left(\sum_{j=1}^{k} a_j\mathbf{H}(t - t_j, \omega - \omega_j)\right)$, $x_b(t) = ISTFT(\mathbf{r}^k)$.

---

be easily sparsified. Having separated the two components, and considering that the strong reflectors are readily compressed, we next want to compress the background component as well.

For that purpose, each such background signal will be sparsely represented over an optimized dictionary that is trained offline from prototype examples using the K-SVD algorithm [1]. For the training set we use a small, randomly-chosen subset of the signals constituting a single frame of real cardiac imaging data, each of them divided into one-dimensional, non-overlapping patches. Although our goal is to compress raw signals, i.e. signals detected by each sensor prior to receive beamforming, the training set signals are chosen to be beamformed scan lines, since those were shown to have improved SNR [22].

It should be emphasized that the aforementioned dictionary learning process is only performed once for every imaging system settings, and do not need to be repeated for every analyzed signal or even for every imaged frame.

Returning to the online processing cycle, each separated background component is next divided into non-overlapping patches and sparsely represented over the trained dictionary.

Denote by $\boldsymbol{\varphi}_m \in \mathbb{R}^N$ the background component of the signal received by the m-th sensor. Each such component is divided into P patches $\mathbf{y}_{m,p} \in \mathbb{R}^Q$ of length Q:

$$\boldsymbol{\varphi}_m = \left[\mathbf{y}_{m,1}^T \; \mathbf{y}_{m,2}^T \cdots \mathbf{y}_{m,P}^T\right]^T \tag{3}$$

Denote the dictionary by $\mathbf{A} \in \mathbb{R}^{Q \times K}$ ($K > Q$), then using Orthogonal Matching Pursuit (OMP) we solve for each patch ($\forall 1 \leq m \leq M, 1 \leq p \leq P$)

$$\arg\min_{\mathbf{z}_{m,p}} \|\mathbf{z}_{m,p}\|_0 \quad \text{subject to} \quad \|\mathbf{A}\mathbf{z}_{m,p} - \mathbf{y}_{m,p}\|_2 \leq \epsilon \tag{4}$$

Following, each patch is reconstructed by $\hat{\mathbf{y}}_{m,p} = \mathbf{A}\mathbf{z}_{m,p}$ and the recovered patches are plugged back to reassemble the full signals $\{\varphi_m\}_{m=1}^M$. Afterwards, these signals are combined to produce the beamformed background signal, which could then be further processed to form the image. Moreover, we note that a simplified beamforming process may be carried out in the representation domain, as a weighted combination of the sparse representation coefficients. The weights are data independent and can be pre-computed, so that only the sparse coefficients should be transferred to the beamformer.

4   Simulation and Results

Our proposed method was evaluated on several sets of consecutive frames of cardiac ultrasound data provided by GE Healthcare.

The results obtained for a typical frame are illustrated in Fig. 1. The original frame is presented along with the corresponding background estimation and its 24-fold compressed version.

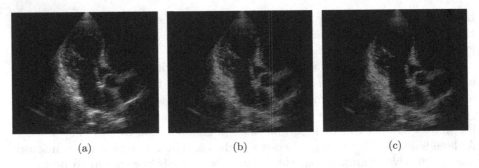

(a)   (b)   (c)

Fig. 1. Background estimation and compression results. (a) Original image. (b) STFT background image. (c) Compressed STFT background image (PSNR = 29.16[dB]).

It can be observed that the proposed decomposition successfully detects and removes the strong reflections, producing a background image with relatively homogeneous regions. Moreover, the compression scheme produces a visually good image that preserve even the subtle image features. These results were obtained despite the formerly mentioned challenges, and while achieving a compression ratio of 24.56, implying that the number of coefficients needed for reconstruction is only 4 % of the number of time samples in the received RF signal.

Similar results were obtained for other cardiac ultrasound frames and for computer simulated phantoms.

Recall that the dictionary used for sparse coding was learned from a subset of a single frame, yet our results indicate that it is suitable for representing data of other frames obtained with the same imaging settings (not necessarily from the same consecutive set as the image used for training).

Comparing our decomposition-based compression with a direct compression of the raw signals, that was performed using a similar K-SVD dictionary learning approach, we found that in order to obtain a compressed image of comparable quality in terms of the visible amount of saved features, a compression factor of only 10.99 was obtained in the direct compression.

In this regard, it should be pointed out that for analyzing the total amount of saved data, a fair comparison demands that the amount of coefficients needed for representing the strong reflectors is added to those used for representing the background signal. By doing so, a slightly reduced compression ratio of 21.4 is achieved. Nonetheless, this achieved compression ratio is still twice as high as the one achieved for the original raw data.

Moreover, assuming a known pulse shape, the strong reflectors reconstruction is straightforward and does not require beamforming or any additional processing. Therefore, in terms of the data needed to be employed in beamforming computations, the higher compression factor (that only considers the background) is still applicable.

The decomposition-based compression is thus significantly advantageous to a direct compression of the raw data in terms of the achieved compression ratio.

5 Conclusions

In this work, we extended previous models proposed in [24, 26] by integrating the speckle reflections and assembling a direct sum of two components, each of which carries valuable information and could be characterized by a limited amount of parameters. In accordance with this model, we developed a decomposition-based processing scheme for raw ultrasound signals that exploits the inherent redundancy of the data, and achieves an improved compression ratio while preserving the image information.

The proposed decomposition-based compression is equivalent to sparse coding over a two-dictionary set (union of subspaces), that is a mixture of a fixed dictionary for the strong reflectors, based on apriori knowledge of the pulse shape, and a data-driven dictionary for the background component.

The novelty of this model lies in the component-based approach, especially as it concerns the raw signals rather than the beamformed ones or the resulting image. It is clearly desirable to compress the data as early in the processing chain as possible. As far as digital compression is concerned, our approach operates on raw signals "close to the source", i.e. immediately after sampling. Though not yet attempted in the scope of our work, we believe that utilizing the proposed two-component model and learned dictionary, a low rate sampling scheme can be established, such that our algorithm may be extended to the compressed sensing framework. Doing so, our results could be compared with other ultrasound compression techniques currently employed in the analog domain. Furthermore, the potential gain of the component-based approach goes beyond compression. Our experiments indicate that by appropriate alterations to the proposed processing scheme, the resulting image quality may be enhanced, for example by suppressing side-lobe artifacts. Further elaboration on this matter is beyond the scope of this paper.

Finally, we note that the component-based modeling may open more possibilities for analyzing ultrasonic signals. While we identified two main components, other decomposition ideas may be investigated, such as separating the first- and second- harmonic echoes, or detecting more than two components related to various artifacts which require special processing.

References

1. Aharon, M., Elad, M., Bruckstein, A.: K-SVD: an algorithm for designing overcomplete dictionaries for sparse representation. IEEE Trans. Signal Proc. **54**(11), 4311–4322 (2006)
2. Azhari, H.: Basics of Biomedical Ultrasound for Engineers. Wiley, Hoboken (2010)
3. Baraniuk, R.: Compressive radar imaging. In: Proceedings of the IEEE Radar Conference, pp. 128–133 (2007)
4. Basarab, A., Liebgott, H., Bernard, O., Friboulet, D., Kouame, D.: Medical ultrasound image reconstruction using distributed compressive sampling. In: IEEE 10th International Symposium on Biomedical Imaging (ISBI), pp. 628–631, April 2013
5. Bhattacharya, S., Blumensath, T., Mulgrew, B., Davies, M.: Fast encoding of synthetic aperture radar raw data using compressed sensing. In: IEEE Workshop on Statistical Signal Processing (2007)
6. Blu, T., Dragotti, P.L., Vetterli, M., Marziliano, P., Coulot, L.: Sparse sampling of signal innovations: theory, algorithms and performance bounds. IEEE Signal Process. Mag. **25**(2), 31–40 (2008)
7. Bohs, L.N., Trahey, G.E.: A novel method for angle independent ultrasonic imaging of blood flow and tissue motion. IEEE Trans. Biomed. Eng. **38**(3), 280–286 (1991)
8. Burckhardt, C.B.: Speckle in ultrasound B-mode scans. IEEE Trans. Sonics Ultrason. **25**(1), 1–6 (1978)
9. Candes, E.J., Wakin, M.B.: An introduction to compressive sampling. IEEE Signal Process. Mag. **25**(2), 21–30 (2008)
10. Donoho, D.L.: Compressed sensing. IEEE Trans. Inf. Theory **52**(4), 1289–1306 (2006)
11. Friboulet, D., Liebgott, H., Prost, R.: Compressive sensing for raw RF signals reconstruction in ultrasound. In: IEEE Ultrasonics Symposium (IUS), pp. 367–370 (2010)
12. Li, Y.F., Li, P.C.: Ultrasound beamforming using compressed data. IEEE Trans. Inf. Technol. Biomed. **16**(3), 308–313 (2012)
13. Liebgott, H., Basarab, A., Kouame, D., Bernard, O., Friboulet, D.: Compressive sensing in medical ultrasound. In: IEEE International Ultrasonics Symposium (IUS), pp. 1–6 (2012)
14. Liebgott, H., Prost, R., Friboulet, D.: Pre-beamformed RF signal reconstruction in medical ultrasound using compressive sensing. Ultrasonics **53**(2), 525–533 (2013)
15. Lorintiu, O., Liebgott, H., Bernard, O., Friboulet, D.: Compressive sensing ultrasound imaging using overcomplete dictionaries. In: IEEE International Ultrasonics Symposium, pp. 45–48 (2013)
16. Lustig, M., Donoho, D.L., Santos, J.M., Pauly, J.M.: Compressed sensing MRI. In: IEEE Signal Processing Magazine (2007)

17. Patel, V.M., Easley, G.R., Healy, D.M., Chellappa, R.: Compressed synthetic aperture radar. IEEE J. Sel. Top. Signal Proces. **4**(2), 244–254 (2010)
18. Quinsac, C., Basarab, A., Girault, J.M., Kouame, D.: Compressed sensing of ultrasound images: sampling of spatial and frequency domains. In: IEEE Workshop on Signal Processing Systems (SiPS), Oct 2010
19. Quinsac, C., Basarab, A., Kouame, D.: Frequency domain compressive sampling for ultrasound imaging. Adv. Acoust. Vibr. **12**, 1–16 (2012)
20. Rilling, G., Davies, M., Mulgrew, B.: Compressed sensing based compression of SAR raw data. In: SPARS 2009 - Signal Processing with Adaptive Sparse Structured Representations, Saint Malo, France (2009)
21. Schiffner, M.F., Schmitz, G.: Fast pulse-echo ultrasound imaging employing compressive sensing. In: IEEE International Ultrasonics Symposium (IUS), pp. 688–691 (2011)
22. Szabo, T.L.: Diagnostic Ultrasound Imaging: Inside Out. Academic Press Series in Biomedical Engineering. Elsevier Academic Press, Burlington (2004)
23. Trahey, G.E., Allison, J.W., Von Ramm, O.T.: Angle independent ultrasonic detection of blood flow. IEEE Trans. Biomed. Eng. **BME–34**(12), 965–967 (1987)
24. Tur, R., Eldar, Y.C., Friedman, Z.: Innovation rate sampling of pulse streams with application to ultrasound imaging. IEEE Trans. Signal Proc. **59**(4), 1827–1842 (2011)
25. Vetterli, M., Marziliano, P., Blu, T.: Sampling signals with finite rate of innovation. IEEE Trans. Signal Proc. **50**(6), 1417–1428 (2002)
26. Wagner, N., Eldar, Y.C., Feuer, A., Friedman, Z.: Compressed beamforming applied to B-mode ultrasound imaging. In: Proceedings of the 9th IEEE International Symposium on Biomedical Imaging (ISBI) (2012)

Author Index

Printed in the United States
by Booksmasters

Printed in the United States
By Bookmasters